Estrogen Action, Selective Estrogen Receptor Modulators and Women's Health

Progress and Promise

Estrogen Action, Selective Estrogen Receptor Modulators and Women's Health

Progress and Promise

Editor

V Craig Jordan

Georgetown University, USA

ICP

Imperial College Press

Published by

Imperial College Press
57 Shelton Street
Covent Garden
London WC2H 9HE

Distributed by

World Scientific Publishing Co. Pte. Ltd.
5 Toh Tuck Link, Singapore 596224
USA office: 27 Warren Street, Suite 401-402, Hackensack, NJ 07601
UK office: 57 Shelton Street, Covent Garden, London WC2H 9HE

Library of Congress Cataloging-in-Publication Data
Estrogen action, selective estrogen receptor modulators, and women's health : progress and promise / edited by V. Craig Jordan.
p. ; cm.
Includes bibliographical references and index.
ISBN 978-1-84816-957-9 (hardcover : alk. paper)
I. Jordan, V. Craig (Virgil Craig)
[DNLM: 1. Estrogens--pharmacology. 2. Estrogen Receptor Modulators--pharmacology. 3. Receptors, Estrogen--physiology. 4. Women's Health. WP 522]

618.1'75061--dc23

2012046673

British Library Cataloguing-in-Publication Data
A catalogue record for this book is available from the British Library.

Typeset by Stallion Press
Email: enquiries@stallionpress.com

Printed by FuIsland Offset Printing (S) Pte Ltd Singapore

For my daughters Helen & Alexandra

— they lived the development of tamoxifen
and SERMs every day

CONTENTS

ABOUT THE AUTHOR AND EDITOR

Professor V. Craig Jordan was educated at Leeds University, U.K. (B.Sc., 1969; Ph.D., 1973; D.Sc., 1985; Hon. D.M., 2001) and is referred to by the media and by his professional colleagues as the "father of tamoxifen." He defined the scientific strategy for reinventing a failed contraceptive, ICI 46,474, to become the gold standard for the long-term, adjuvant treatment for breast cancer and a pioneering medicine for the prevention of breast cancer. He also was the first to recognize the new drug group (of which tamoxifen is a member) Selective Estrogen Receptor Modulators (SERMs), which switch on and switch off estrogen target sites throughout a woman's body. This knowledge advanced the clinical development of raloxifene for the prevention of both osteoporosis and breast cancer. His recent research on estrogen-induced apoptosis in breast cancer is the result of a 20-year study of the vulnerability of breast cancer during the evolution of acquired resistance to antihormone therapy.

As a result of his contributions to medical science and therapeutics, Professor Jordan has received numerous international awards, including the Bristol-Myers Squibb Award and Medal for Distinguished Achievement in Cancer Research (2001), the American Cancer Society Medal of Honor (Basic Research Award) (2002), the Charles F. Kettering Prize of the General Motors Cancer Research Foundation (2003), the David A. Karnofsky Award from the American Society for Clinical

Oncology (2008) and the St. Gallen International Clinical Breast Cancer Prize (2011). In 2002, Her Majesty Queen Elizabeth II appointed Professor Jordan as Officer of the Most Excellent Order of the British Empire for his contributions to international breast cancer research.

He has served on the Board of Directors of the American Association of Cancer Research (2007–2010), the Board of Trustees for the Worcester Foundation for Biomedical Research, University of Massachusetts (1995–2005), and is currently the President of the Royal Society of Medicine Foundation, North America.

Professor Jordan is an elected Honorary Fellow of the Royal Society of Medicine, U.K. (2008), a Fellow of the Academy of Medical Sciences, U.K. (2009) and a Member of the National Academy of Sciences, U.S.A. (2009). He has held three inaugural named chairs: the Diana, Princess of Wales Professorship of Cancer Research (R.H. Lurie Comprehensive Cancer Center, Northwestern University, Chicago), the Alfred G. Knudson Chair for Basic Cancer Research (Fox Chase Cancer Center, Philadelphia) and the Vincent T. Lombardi Chair for Translational Cancer Research (Georgetown University, Washington, D.C.).

Professor Jordan is Scientific Director of the Lombardi Comprehensive Cancer Center, Georgetown University, Professor of Oncology and Pharmacology and Vice Chair of the Department of Oncology at Georgetown University Medical Center, Washington, D.C.

CONTRIBUTORS

Hiltrud Brauch, Deputy Head of Dr. Margarete Fischer-Bosch Institute of Clinical Pharmacology and Professor of Molecular Pathology, completed her Ph.D. thesis in Immunology at the University of Heidelberg in Germany in 1985. She received further education as a Visiting Fellow at the Laboratory of Immunobiology at the National Institutes of Health, National Cancer Institute — Frederick Cancer Research and Development Center in Frederick, MD. Her current research focuses on the investigation of biomarkers for breast cancer drug treatment with the goal of enhancing personalized medicine. Her group concentrates on the association between CYP2D6 polymorphisms and outcomes among women with early stage postmenopausal breast cancer treated with adjuvant tamoxifen. She is the principal investigator of a European Union/Marie Curie Ph.D. training program, "Fighting Drug Failure." She collaborates with Dr. Jordan on the pharmacology of CYP2D6 and tamoxifen metabolism, and Dr. Jordan is the scientific advisor to the "Fighting Drug Failure" EU training grant.

Angela M. Hartley Brodie, Professor of Pharmacology and Experimental Therapeutics in the School of Medicine and the University of Maryland Greenbaum Cancer Center, received her Ph.D. from the University of Manchester, UK in 1961. She was awarded an NIH postdoctoral fellowship to join the Steroid Biochemistry Training Program at Clark University and at the Worcester Foundation for Experimental Biology in Massachusetts. She remained at the Worcester Foundation as Staff and Senior Scientist until 1979, when she moved to the University of Maryland School of Medicine in Baltimore, MD. Dr. Brodie is considered to be the pioneer for the targeted application of aromatase inhibitors and took the concept from the laboratory to clinical trials. She has received a number of awards, including the Brinker Award from the Susan G. Komen Foundation in 2000, the General Motors Kettering Prize for Translational Research in 2005 and the American Association for Cancer Research Dorothy P. Landon Prize for Translational Research. Drs. Brodie and Jordan jointly received the Gregory Pincus Award and Medal from the Worcester Foundation in 2007. Dr. Brodie has published 190 peer-reviewed research articles, book chapters, and reviews and has six patents.

V. Daniel Castracane, Director of Research and Professor at Texas Tech University HSC in Odessa, TX, obtained his Ph.D. in Reproductive Endocrinology from Rutgers University in New Brunswick, NJ. He completed a postdoctorate in Reproductive Physiology from Case Western Reserve University School of Medicine in Cleveland, OH and in Steroid Biochemistry from the Worchester Foundation for Experimental Biology in Shrewsbury, MA. It was at the Worcester Foundation that Dr. Castracane and Dr. Jordan became scientific colleagues, publishing a number of papers together. Dr. Castracane has published more than 150 papers on reproductive endocrinology.

Christopher Chien, a Postdoctoral Fellow at the National Cancer Institute, Pediatric Oncology Branch in Bethesda, MD, obtained his Ph.D. in Tumor Biology from Georgetown University with Dr. Anna Riegel. His research project was centered on the splice variant amplified in breast cancer 1-Δ4 (AIG 1-Δ4, also known as AIB 1-Δ3 and SRC-3Δ4) and characterizing the potent nuclear coactivation function of the protein. He has already contributed more than a dozen publications to the literature.

Saranya Chumsri, an Assistant Professor in Medicine at the University of Maryland School of Medicine, obtained her M.D. at Chulalongkorn University in Bangkok, Thailand. She completed her internal medicine residency training at the Albert Einstein Medical Center in Philadelphia, PA, and hematology/oncology/fellowship training at the University of Maryland Medical Center in Baltimore, MD. She is particularly interested in laboratory research in breast and prostate cancer, and has published more than 20 scientific papers.

Robert Clarke, Professor with the Department of Oncology at Georgetown University and Dean for Research at the Georgetown University Medical Center, obtained his Ph.D. in Biochemistry at Queen's University of Belfast in Northern Ireland in 1986 and a prestigious D.Sc. degree in 1999. He completed a postdoctoral fellowship at the Medicine Branch of the National Cancer Institute prior to moving to Georgetown University. He is Co-Director of the Breast Cancer Program (with Dr. Claudine Isaacs) at the Lombardi Comprehensive Cancer Center. He is internationally recognized for his seminal research on antihormone resistance in breast cancer. His current research focus is on the application of *in silico* methods for network-based prediction of antiestrogen resistance, and the use of informatics-based analysis for the integration of ER-related signaling in breast cancer. Dr. Clarke has contributed more than 225 articles to the medical literature. He was elected Fellow of the Royal Society of Medicine (U.K.) in 2001 and is on the Board of the Royal Society of Medicine Foundation of North America as Vice President (Dr. Jordan is the elected President).

Christopher C. Coss is a Research Scientist at GTx, Inc. in Memphis, TN. He received Bachelor of Science degrees in Molecular Genetics and Computer Science from The Ohio State University in Columbus, OH in 2003 and a Ph.D. in Pharmaceutics in 2008 under the mentorship of Dr. James T. Dalton. In 2006 he joined GTx, where his research continues on selective nuclear hormone receptor modulators, receptor tyrosine kinase inhibitors and novel anticancer agents.

James T. Dalton, Chief Scientific Officer, GTx, Inc. in Memphis, TN, obtained his Ph.D. in Pharmaceutics and Pharmaceutical Chemistry at The Ohio State University in Columbus, OH. He completed a postdoctoral fellowship in the Division of Pharmaceutics and Pharmaceutical Chemistry at The Ohio State University. Following faculty appointments at The Ohio State University and the University of Tennessee, he became Kimberly Professor and Chair at The Ohio State University, College of Pharmacy, Division of Pharmaceutics and Pharmaceutical Chemistry before moving to GTx in 2005.

Jennifer Eng-Wong is a medical oncologist who specializes in breast cancer prevention clinical research. She was an Assistant Professor at the Lombardi Comprehensive Cancer Center, Georgetown University (2007–2011). She has evaluated endocrine and lifestyle interventions for breast cancer prevention. She also served as Medical Director of the Capital Breast Care Center, Washington, D.C. Dr. Eng-Wong attended medical school at SUNY Stony Brook, New York and completed her internal medicine residency at UC Davis, Sacramento, CA. She completed her medical oncology and cancer prevention fellowships at the National Cancer Institute, NIH, Bethesda, MD. She is currently Associate Medical Director, Genentech, Inc.

Ping Fan, Research Assistant Professor at the Lombardi Comprehensive Cancer Center, received her M.D. and Ph.D. from Nanjing Medical University in China. She completed postdoctoral training in Dr. Richard Santen's laboratory at the University of Virginia with an investigation of the mechanisms underlying tamoxifen resistance in breast cancer. This was followed by further training in Dr. Jordan's laboratory at the Fox Chase Cancer Center. Her work focuses on the mechanisms of SERM action and the events involved in estrogen-induced apoptosis.

Geoffrey L. Greene, Virginia and D.K. Ludwig Professor and Vice Chair, The Ben May Department for Cancer Research, received his Ph.D. in Organic Chemistry from Northwestern University in 1974. He is internationally recognized for his work on the function of female hormones and SERMs in breast cancer, and for the development of estrogen and progesterone receptor antibodies, which have had major diagnostic applications in breast cancer throughout the world. Dr. Greene has received several prestigious awards for his research accomplishments, including the Ernst Oppenheimer Award from the Endocrine Society, and the John Brewer Distinguished Alumni Lectureship at Northwestern University. He recently received the NAMS/Wyeth Pharmaceutical SERMs award from the North American Menopausal Society and the Brinker International Breast Cancer Award from Susan G. Komen for the Cure.

Anthony Howell is Professor of Medical Oncology and the Director of the Breakthrough Breast Cancer Research Unit, University of

Manchester. Formerly he was Chairman of the UKCCCR, the British Breast Group and the Manchester Breast Centre, as well as Research & Development Director of The Christie NHS Foundation Trust and the Cancer Research Network. His interests are the endocrine therapy and biology of the breast and breast cancer with a particular interest in prevention. He has published over 500 papers, mainly in these areas.

Claudine Isaacs is Professor of Medicine and Oncology and Director of the Clinical Breast Cancer Program at the Georgetown University Medical Center in Washington, D.C. She earned her medical degree from McGill University in Montreal, Quebec, Canada. She completed an internal medicine residency as well as a hematology and medical oncology residency/fellowship at McGill University and a fellowship in breast cancer at the Georgetown University Medical Center. Dr. Isaacs is a Fellow of the Royal College of Physicians and Surgeons of Canada and Co-Director of the Fisher Center for Familiar Cancer Research at the Georgetown University Medical Center.

Armina Alcasabas Kazi, Assistant Professor at Loyola University in Maryland and Baltimore, MD, obtained her Ph.D. in Physiology-Reproductive Endocrinology from the University of Maryland School of Medicine in Baltimore, MD. She was also a postdoctoral fellow in Physiology-Reproductive Endocrinology and Breast Cancer at the University of Maryland School of Medicine. She has particular interest in the potential of nonhypoxic HIF-1, miRNAs, and cancer stem cells as potential diagnostic and therapeutic targets for drug-resistant breast cancer. Currently, she is working on elucidating the underlying mechanisms of drug-resistant breast cancer and has a dozen publications.

Joan Lewis–Wambi, Assistant Professor in the Department of Cancer Biology, Kansas University Medical Center in Kansas City, received her Ph.D. in Cancer Biology and Biochemistry from Rutgers University, NJ. A postdoctoral fellowship followed in 2004 at Northwestern University, at the Robert H. Lurie Comprehensive Cancer Center in Chicago, IL with Dr. Jordan, and a second postdoctoral fellowship at the Fox Chase Cancer Center. Dr. Lewis-Wambi's current area of research focuses on antihormone drug resistance and the mechanism(s) of estrogen-induced apoptosis in breast cancer with a particular emphasis on resistance to aromatase inhibitors and selective estrogen receptor modulators such as tamoxifen and raloxifene. She has over 20 peer-reviewed articles and is a past Avon Scholar, and has a career development K01 grant from NIH and a grant from the American Cancer Society.

Russell E. McDaniel received his B.S. in Biochemistry in 2007 from Temple University in Philadelphia, PA. He served as a Peace Corps Volunteer for two years, teaching high school chemistry in southern Mozambique, before joining Professor Jordan's laboratory in 2009. At present, he is laboratory manager and is also pursuing a Master of Science degree in Biotechnology at Georgetown.

Ifeyinwa E. Obiorah received her M.B.B.S. in July 2000 from Nnamdi Azikiwe University in Nnewi, Nigeria. In November 2006 she earned a diploma and an M.Sc. in Thoracic Medicine/NHLI from Imperial College in London, England.Then, in May 2009, she received her M.Sc. in Tumor Biology from Georgetown University in Washington, D.C. Currently she is studying for a Ph.D. in Tumor Biology at Georgetown University with Dr. Jordan. Her thesis topic is the role of the shape of the estrogen — estrogen receptor complex for the modulation of apoptosis in estrogen-deprived breast cancer cells.

Trevor J. Powles is Emeritus Professor of Breast Oncology at the Institute of Cancer Research, London. Until 2003 he was Head of Breast Cancer at the Royal Marsden Hospital. He was also Divisional Medical Director of Common Cancers and previously Chairman of the Clinical Research Committee at the Royal Marsden. He is now Consultant Breast Oncologist and Medical Director at Cancer Centre London at Parkside. Professor Powles holds medical specialist accreditations in endocrinology and medical oncology and a Ph.D. in Molecular Pathology of Breast Cancer. His clinical and research work has focused on the prevention, risk assessment, diagnosis and early treatment of breast cancer. He has published and lectured widely on the diagnosis, prevention and treatment of breast cancer. He has won awards, including the Komen Brinker award for outstanding contribution to breast cancer research and the All Party Parliamentary Group award for his lifetime contribution to breast cancer care. In 2002 he was awarded a C.B.E. in recognition of his services to breast cancer.

Anna Riegel is a Professor of Oncology and Pharmacology, and Associate Director for Cancer Research Education at the Lombardi Comprehensive Cancer Center, Georgetown University. She obtained an M.Sc., with distinction, in Steroid Endocrinology from the University of Leeds (U.K.) and a Ph.D. in Oncology from the University of Wisconsin, McArdle Laboratory. Dr. Jordan was her thesis advisor. She was a Fulbright Scholar at the University of Wisonsin, where she won the James Price Award for Cancer Research. She is currently the Cecilia Fisher Rudman Endowed Professor in Breast Cancer Research at the Lombardi Comprehensive Cancer Center. Throughout her career she has focused on hormone action in breast cancer and she is currently involved with coactivator signal transduction pathways. She has contributed nearly 100 publications to the scientific literature.

Eliot M. Rosen, Charlotte Gragnani Chair in Oncology and Radiation Biology, Lombardi Comprehensive Cancer Center, obtained his M.D. and Ph.D. at the University of Pennsylvania, Philadelphia in 1979 and 1980, respectively. He was a resident in Radiation Therapy, Joint Center at Harvard Medical School in Boston, MA, followed by faculty positions at Harvard Medical School and Yale University School of Medicine. He moved to Georgetown University as Professor of Oncology, Biochemistry, and Radiation Medicine in 2003. His innovative research focuses on the role of BRCA-1 in progesterone and estrogen receptor signaling, and has resulted in more than 150 publications in the medical/scientific literature.

Gauri J. Sabnis, Assistant Professor in the Department of Pharmacology and Experimental Therapeutics at the University of Maryland, Baltimore, obtained her Ph.D. in Pharmacology from that university. There, she completed her postdoctoral training in pharmacology. She was awarded first place in Clinical/Translational Research at the Breast Cancer Program Retreat at The Sidney Kimmel Comprehensive

Cancer Center at Johns Hopkins in Baltimore, MD. She has more than a dozen refereed publications.

Surojeet Sengupta, Research Assistant Professor in the Department of Oncology at Georgetown University, Washington, D.C., did his M.Sc. in Biotechnology at the Indian Institute of Technology (IIT), Roorkee, India and his Ph.D. at the Central Drug Research Institute (CDRI), Lucknow, India. This was followed by postdoctoral research in Professor Benita S. Katzenellenbogen's laboratory at the University of Illinois at Urbana-Champaign, IL, and in Dr. Jordan's laboratory at the Fox Chase Cancer Center, Philadelphia, PA. He was awarded a fellowship by the department of biotechnology, India during his Masters study and a five-year prestigious research fellowship by the Council of Scientific and Industrial Research (CSIR), India during his Ph.D. study. He received the Edward David Lustbader Award for oral presentation at the 13th Annual Postdoctoral & Graduate Student Research Conference, 2008, at the Fox Chase Cancer Center.

Anton Wellstein, Professor of Pharmacology and Medicine and Associate Director for Basic Research at the Lombardi Cancer Center at Georgetown University, obtained his M.D. in Medicine from J. Gutenberg University in Mainz, Germany in 1973 and his Ph.D. in Pharmacology from J.W. Goethe University in Frankfurt, Germany in 1985. He has a special interest in fibroblast growth factors and in coactivator signal transduction, with a contribution of more than 153 peer-reviewed publications in the literature. His work has been recognized internationally with the International Paul Martin Award in Clinical Pharmacology, and the International Boehringer Ingelheim Research Award.

ACKNOWLEDGEMENTS

I am honored to have Ambassador Nancy Brinker, C.E.O. of Susan G. Komen for the Cure Foundation, and Professor Bert O'Malley, M.D., Chairman of the Department of Molecular Biology at the Baylor School of Medicine, each offer a Foreword from a different perspective. Both are pioneers of international standing. Ambassador Brinker created a global organization from nothing to raise awareness internationally of the need to invest in breast cancer research to cure this deadly disease. She did something about it by raising resources and mobilizing the community. Professor O'Malley, early in his career, rapidly advanced the science of sex steroid receptor action to become the "father of molecular endocrinology." Both have received recognition for their international contributions to women's health with the award of the Medal of Freedom and the Medal of Science, respectively, from the President of the United States of America. Thank you both for your support of this book.

The invitation to create this book came from a visit to Imperial College London in 2005, as part of a two-week lecture tour of chemistry departments in Great Britain, at Oxford, Cardiff, Manchester, York, Dundee and Imperial College. It was the occasion of my work being recognized by the Third George and Christine Sosnovsky Award in Cancer Therapy from the Royal Society of Chemistry. The prize was a fine medal, a certificate and the opportunity to reach out to young chemists in the next generation through lectures and numerous informal dinners. An excellent award strategy. Sincere thanks to the Royal Society of Chemistry, the Sosnovsky Family in Milwaukee and my publisher for their patience, especially Joy Quek. A lot happened to delay the process, but I believe the resulting book is better.

I thank my secretaries, Julia Jessup Tijerina and Amy Botello, and my laboratory manager, Helen Kim, for their indispensable assistance in creating order out of chaos during this process of assembling the manuscripts for this book. I thank Russell McDaniel for artwork. Julia Jauch, M.D., deserves a special thank you for photographing the butterfly displayed on the cover. This maintains my tradition of having a butterfly on the cover of each of my books on tamoxifen. The reason for this is that the advertising slogan for tamoxifen in the early 1970's in the United Kingdom was "its gentle as a butterfly" with the picture of a butterfly.

The staff and I at the Georgetown-Lombardi Comprehensive Cancer Center were supported by my Department of Defense, Center of Excellence Grant W81XWH-06-1-0590, Susan G. Komen for the Cure Grant SAC100009, the AACR/SU2C Grant DT0409 and the CCSG Grant P30CA051008 (to Louis Weiner, M.D.).

This book could not have been written without the members of my "Tamoxifen Teams" and my academic colleagues. We made this journey together over the past 40 years. Our book, therefore, is not only a record of progress in women's health, but also a tribute to their individual contributions. I am deeply grateful for their support and friendship over the years and their commitment to this project. I especially thank Nance Guilmartin of the Landon Foundation for suggesting to me that I should recognize the contribution of my graduate students as an encouragement for future generations (Appendix 3). For reasons of space, I could only focus on my graduate students as they transitioned from their doctoral degrees to careers of their own making. However, this book would not have been possible but for the scores of visiting faculty from overseas, postdoctoral fellows, surgical residents, medical oncology fellows and superb technical staff, whose ideas, practical skill and innovation in the laboratory ultimately saved women's lives. Thank you all!

V. Craig Jordan

FOREWORDS

Nancy Brinker
Founding Chairwoman
Susan G. Komen Breast Cancer Foundation
Dallas, TX

Tamoxifen changed the treatment of breast cancer and brought hope to millions of women with the disease. It was the first effective therapy targeted to the tumor estrogen receptor (ER). In 1973, tamoxifen was approved in the United Kingdom for the treatment of advanced breast cancer in postmenopausal women, and in 1978 it was finally introduced in the United States. Though originally indicated for the treatment of advanced or late-stage disease, it has since been approved by the Food and Drug Administration (FDA) for the treatment of all stages of breast cancer. In 1998, the FDA also approved it for reducing the incidence of breast cancer in healthy women at high risk for developing the disease.

Tamoxifen's success was the result of a close collaboration between AstraZeneca Pharmaceuticals — the agent's discoverer and manufacturer — and the academic and clinical communities. However, it was the investigators themselves who believed in tamoxifen's strength and took the time to probe it to its full potential. It was their steadfastness and their will to fight the battle against cancer that enabled tamoxifen to reign, which in turn saved the lives of millions of women worldwide. However, this book is more than being about tamoxifen.

To use the words of Winston Churchill, "This is not the end, not even the beginning of the end, but it is perhaps the end of the beginning" (1943). In the history of hormones and women's health, tamoxifen was the first effective weapon in our fight against breast cancer. But, as in any other fight, if we are to win, we must use our weapons wisely, constantly seek new opportunities for success and discover new weapons to prevent this disease and improve women's health. This book is about the evolution of innovation and the advances in new approaches and investigations to cure breast cancer. Without the success of tamoxifen, we would not have selective estrogen receptor modulators (SERMs), and without SERMs and aromatase inhibitors, we would not have the new science of estrogen-induced apoptosis. This is a remarkable story of medicine that goes from strength to strength in the service of women worldwide.

We at Susan G. Komen for the Cure Foundation are committed to the fight against breast cancer. We have now passed our 20th anniversary and our commitment has raised funds to support thousands of grants to advance research. This research is to discover key vulnerabilities in the relentless growth of cancer that can be targeted to destroy the malignancy and lead to cures. We have made a pledge to raise the resources to support "the best and the brightest" young researchers studying at the world's top universities as they build a career of excellence in cancer research. We also are committed to partnering the most accomplished medical scientists and clinicians to guide the future of breast cancer research. We recognize that sustained success in cancer research is a mark of excellence that we choose to support. These medical scientists are the role models who continue to have a positive influence on the next generation.

Achievements in research that improves women's health are the milestones by which we can measure our progress. We believe in recognizing great achievements, and in 1992, to mark the 10th anniversary of our Foundation, we established the Brinker International Breast Cancer Award for completed research. Drs. Craig Jordan and

Bert O'Malley are both recipients of our highest award in basic research and both are Susan G. Komen for the Cure Scholars.

Susan G. Komen for the Cure is pleased to have supported translational research and has invested in young, productive medical scientists in Dr. Jordan's laboratory (Appendix 3). In 2010, we again recognized Dr. Jordan for his contribution of the new group of medicines called SERMs which improve women's health. This event took place at the Kennedy Center for the Performing Arts in Washington, D.C. His sustained proactive contributions to women's health made Dr. Jordan the inaugural inductee to the Susan G. Komen Investigator Hall of Fame.

I am honored to contribute this foreword on behalf of Susan G. Komen for the Cure Foundation. This book represents the progress we continue to make in the fight against breast cancer and in global women's health. The chapters in this book demonstrate that much has been done, but there is much still to do. You have our support and our commitment to the Cure.

Bert W. O'Malley, M.D.
Thompson Distinguished Professor and Chair
Department of Molecular and Cellular Biology
Baylor College of Medicine
Houston, TX

The existence of a receptor for estrogen was first recognized in the 1960s by the pioneering work of Jensen and Gorski in the rat uterus. In the early 1970s, my laboratory elucidated the general mechanism of steroid hormone action in studies on the function of estrogen and progesterone in the chicken oviduct; this work culminated in an accepted model of hormone action — steroid hormones bind receptors, translocate to target genes, and induce mRNA and protein synthesis. It meant that receptors were transcription factors and spawned the development of a new field termed "molecular endocrinology." The explosive nature of these early pioneering events in hormone action eventually seduced over 100,000 worldwide workers into the field and resulted in an explosion of publications numbering now in excess of 500,000. The first nuclear hormone receptor (glucocorticoid) was cloned in 1985 by the Evans lab group; this was soon followed by the cloning of many other steroid receptors, leading to the establishment of a structurally related nuclear receptor superfamily of transcription factors that now comprises 48 members in humans, and that includes steroid and thyroid receptors, vitamin A and D receptors, and many metabolic and orphan receptors. The next major advance in the mechanism appeared in 1995, when the first steroid receptor coactivator (SRC-1) was cloned in our lab. We soon realized that all nuclear receptor functions were mediated by large numbers of coregulators (coactivators and corepressors) that were recruited to the target gene by the receptors to amplify or repress transcription. The work on coregulators would later be critical in deciphering the mechanism of action of mixed antagonist/agonist ligands for receptors.

The family of nuclear receptor coregulators also experienced an explosive growth in number over the ensuing 15 years; the number of coactivators and corepressors has grown to approximately 500 published molecules, and may be over 11,000 when the entire nuclear transcriptome is considered. The molecular biology of coactivators has informed us of a cadre of diverse and interesting mechanisms of transcriptional action, including chromatin modification and remodeling, initiation of transcription, elongation, alternative RNA splicing and, finally, protein degradation. Of the currently discovered coregulators, approximately 200 have already been demonstrated to result in human pathologies, and heritable dysfunction has been demonstrated to be causal in numerous instances of embryonic lethality, growth retardation and maturation, mental retardation, metabolic and endocrine disorders, inflammatory disorders, reproductive and cardiovascular abnormalities, and, especially, in oncogenic disorders.

It was in the "middle ages" of these five decades of explosive molecular endocrinology that, through the work of Jordan on tamoxifen and his subsequent description of the potential of SERMs, we first realized the extent of our power to develop drugs to modify receptor functions for therapies of cancers and other human diseases. The combined knowledge of structural changes in receptors and coregulator recruitment induced by ligands opened up a panoply of new ideas for functional extensions of both old and new drugs. It is precisely this "period of enlightenment" that is the particular focus of this book. Within the book, you will find a treasured history of accomplishments by major contributors to our field that includes descriptions of the deductive power of ligand-induced structural modifications of receptor/coregulator for cell physiologies, the discovery and evolution of selective receptor modulator agents, the role of environmental agents in human health, the evolution of adjuvant drug therapies for human breast and prostate cancers, hormone synthesis inhibitors, and future perspectives for the field.

"To know nothing of what happened before you were born is to remain forever a child"

Cicero

46 B.C.

"The past is never dead. It is not even the past"

William Faulkner

Nobel Prize for Literature 1949

INTRODUCTION: THE FOUNDATION OF A CAREER

V. Craig Jordan
Lombardi Comprehensive Cancer Center
Georgetown University
Washington, D.C.

When I was a teenager, a love of chemistry and what chemistry could do to cure disease drew me toward pharmacology, with a specific interest in cancer therapeutics. Unfortunately, at the time in the late 1960s, this was not a popular choice for a career and, in my case, would require some effort to become involved with the science behind developing drugs to treat cancer. There was little or no interest, either in academia or in the pharmaceutical industry. However, in 1969, armed with a First Class Honors degree from the Department of Pharmacology at the University of Leeds and supported by a research scholarship from the Medical Research Council, I decided to study the estrogen receptor (ER) with Dr. Edward Clark in the Department of Pharmacology at the university. Dr. Jack Gorski had published[1,2] an exciting series of reports showing that the ER could easily be extracted from the rat uterus and isolated by sucrose density gradient analysis. My project was going to be simple: I was to establish the new technique of sucrose density gradient analysis, isolate the receptor, and crystallize the protein with an estrogen and an antiestrogen. Through x-ray crystallography in the Astbury Department of Biophysics at the University of Leeds, we would then establish the three-dimensional shape of the complexes to explain antiestrogenic action. The goal was to solve a fundamental problem in pharmacology: What is the molecular mechanism of action for a drug?

Progress was slow in establishing the receptor purification technique of sucrose gradient analysis, and I switched from my thesis topic to the study of the structure–activity relationships of antiestrogens. As it turned out, this was a good strategic decision, as it took the best efforts of the research community nearly 30 years to achieve partial success. The problem of the structure of the ligand-binding domain of the ER complex was solved by scientists at York University, England, in 1997.[3] However, the whole ligand–ER complex has yet to be crystallized and resolved.

In 1972, however, there was little academic interest in the pharmacology of antiestrogens. Enthusiasm in the pharmaceutical industry had chilled because these drugs had not fulfilled their promise as contraceptive "morning after" pills. It was clear that no one was recommending antiestrogen research as a sound career choice; it was perceived as a dead end. To make matters worse, the University of Leeds encountered difficulty in securing a qualified examiner for my thesis. Sir Charles Dodds, the discoverer of the synthetic estrogen, diethylstilbestrol (DES),[4] declined with regret because he had not kept up with the literature during the 20 years since his retirement! This led the chairman of the Department of Pharmacology, Professor Barrett, to select Dr. Arthur Walpole, the Head of Fertility Control Research from ICI Pharmaceutical Division. Professor Barrett had worked at ICI Pharmaceuticals with Dr. Walpole. The research facility was located in the area of northwest Cheshire where I grew up and I had met Dr. Walpole previously in 1967 when I was a summer student at ICI's Alderley Park, Cheshire. Thus, indirectly, the door was opened for the future development of ICI 46,474 into tamoxifen.

I was recruited to a tenure track faculty position at the University of Leeds in 1972 when I was still a graduate student, but I was first required to obtain additional research experience elsewhere. Professor Barrett solved this problem by arranging for me to work with Dr. Michael Harper at the Worcester Foundation for Experimental Biology in Massachusetts. Dr. Harper is a reproductive biologist, and had some years earlier worked with Dr. Arthur Walpole at ICI Pharmaceutical Division. He is also the co-patent-holder for ICI 46,474.[5-7] However, at that time, Dr. Harper was heading a team at the Foundation working on the potential of prostaglandins to be used as a "once-a-month" contraceptive pill. I remember the transatlantic telephone call when he offered me a two-year Visiting Scientist job: "... can you come in September, will $12,000 a year be OK, and will you work on prostaglandins...?" "YES, YES, YES," I said and headed for the library to find out what prostaglandins were!

When I arrived at the Worcester Foundation in September 1972 — incidentally, not knowing anything much about prostaglandins — I discovered that Dr. Harper had accepted a job with the World Health Organization in Geneva. I was told I could do anything I liked as long as some of my work included prostaglandins. I immediately found myself to be an independent investigator and planned my work on prostaglandins. However, my new circumstances would also allow me to explore my passion — to develop a drug for breast cancer.

By 1971, ICI 46,474 had shown modest activity in a preliminary clinical study for the treatment of advanced breast cancer.[8] It was still clear, however, that much remained to be done to develop antiestrogens as acceptable therapeutic agents. ICI did not have a breast cancer research program, but Dr. Walpole agreed to help with my request for assistance to conduct a systematic laboratory study of the antitumor properties of ICI 46,474 at the Worcester Foundation. What I did not know at the time was that in April 1972 ICI scientists had reviewed all the data on ICI 46,474 and by the end of the year the Research Director had decided to abandon clinical development. Dr. Walpole said he would take early retirement, but stay on if resources could be used to discover an appropriate application for the drug in breast cancer. He had met a young Ph.D. candidate, Craig Jordan, who was keen to help, so why not give him funds? However, I had no experience with laboratory models of breast cancer. Fortunately, the Worcester Foundation was a marvelous place to learn, as well as to find help in new areas of scientific endeavor.

Dr. Elwood Jensen, then Director of the Ben May Laboratory for Cancer Research in Chicago, was a member of the scientific advisory board for the Foundation. He had been appointed to enhance opportunities in endocrinology and cancer following the passing of the National Cancer Act the year before, in 1971. Quite generously, he agreed to help me prepare for my study of ICI 46,474.The short time I spent in his laboratory, with the help of Dr. Gene DeSombre, was exceptional and provided me with the skills to evaluate the antitumor actions of ICI 46, 474 back in Massachusetts.

To facilitate progress, Dr. Walpole suggested I contact Ms. Lois Trench, the clinical drug monitor for ICI 46,474, at one of ICI's companies in Wilmington, Delaware (Stuart Pharmaceuticals). She admitted, however, that no one knew much about the new drug in the United States clinical department, and insisted that I become a consultant to encourage clinical trials with ICI 46,474.

Full of energy and drive, Lois arranged for me to study the interaction of ICI 46,474 with the human tumor ER. In 1974, she also arranged for me to explain the antitumor effects of ICI 46,474, which was now tamoxifen, at meetings of the Eastern Cooperative Oncology Group (ECOG).[9,10] Thereafter, she invited me to introduce tamoxifen to the National Surgical Adjuvant Breast and Bowel Project (NSABP) at an international symposium she organized in 1976 in Key Biscayne, Florida.[11]

The work at the Worcester Foundation went exceptionally well. The controversy of whether tamoxifen inhibited the binding of estrogens to the human ER was resolved.[12] The first laboratory studies of tamoxifen as a breast cancer preventive were also successfully completed in 1974, and Lois sponsored me to present the work at the International Congress of Steroidal Hormones meeting in Mexico City.[13] There was still little interest in the concept of prevention at that point (it was 10 years too early!) but at least the first, translational scientific research had commenced.

Some of that work was included in the handbook given to clinicians in America to support their clinical studies in 1974. The work was also used to support licensing of tamoxifen in Japan and Germany. But I felt no urgency to publish the results, since there was still no particular clinical interest in antiestrogens as breast cancer drugs. Cytotoxic combination chemotherapy was king!

Contributing to the clinical development of tamoxifen was my interest and goal. However, my strategic error was ultimately pointed out by Dr. Eliahu Caspi, a rather scary senior scientist at the Worcester Foundation. He taught me his simple adage: "Tell them the story so far." He explained that each paper should take no more than two weeks to complete after one has the data. Heeding his advice, I have not stopped writing since that day in 1974. Twenty-eight years after my interview and mentoring session with Dr. Caspi, I was honored to be invited to present the Inaugural Dr. Eliahu Caspi Memorial Lecturer at the Worcester Foundation for Biomedical Research in 2002. It was then that I discovered from his family that Dr. Caspi had survived Russian prison camps when he was captured as a young man in Poland at the start of the Second World War. He was there for the birth of Israel and joined the Hagana (local Israeli defense force) before he emigrated to America. He subsequently obtained a Ph.D. at Clark University and started his distinguished career in steroid endocrinology at the Foundation.

A NEW STRATEGY: LONG-TERM ADJUVANT TAMOXIFEN

Upon my return to the Department of Pharmacology at Leeds University in September 1974, my principal interest was to understand the pharmacology of tamoxifen and to devise a strategy for the best clinical application for an antiestrogenic treatment. The drug had been available for the treatment of advanced breast cancer in Britain since 1973, but similar approval would not occur in the United States until December 1977. Throughout this period, ICI Pharmaceutical Division and the Yorkshire Cancer Research Campaign supported my laboratory at the University of Leeds in England. There, the "tamoxifen team" devised the strategy of long-term (greater than one year)[14,15] adjuvant tamoxifen therapy that was taken into clinical trials on both sides of the Atlantic. I also continued my studies on the chemoprevention of breast cancer with tamoxifen. My first paper, submitted in 1974, was rejected by the three referees invited to review my work. However, one of the three provided me with sound scientific advice and I persevered. The work was published in 1976 by the *European Journal of Cancer*.[16] My group also discovered that a metabolite of tamoxifen, 4-hydroxytamoxifen,[17] had a high affinity for the ER, but still retained antiestrogenic activity. These data would provide the rationale for the clinical development of new antiestrogens for the next 25 years as the SERMs, following their recognition and development for the treatment and prevention of osteoporosis and breast cancer. This occurred during the 1980s and 1990s, when my laboratory was at the Wisconsin Comprehensive Cancer Center, University of Wisconsin (Madison).

This book is primarily intended for young, medical scientists in training at the Ph.D. or postdoctoral level who want to discover new biological knowledge of medical significance. Discoveries will make a difference in our society. The story of antihormone therapy will never be repeated in this form, as it is "outside the box" and was dependent on chance meetings. But that is what discovery is — change and the skill to recognize the discovery, but the principles will remain the same. It is people who make discoveries, people who advocate change and people who change healthcare. This book will also be of value to the physician whose primary discipline is not medical oncology but whose practice includes women at risk for breast cancer and/or women who are taking either tamoxifen or aromatase inhibitors to treat breast cancer, or women who are taking raloxifene to prevent osteoporosis. It is also recommended for nurses in breast cancer centers who are responsible for the day-to-day care of women.

Additionally, this book has been designed to provide a foundation of the most current information for the surgeon or the physician-in-training, who, by the time they come to practice, will be called upon for advice from their patients about the risks and benefits of a diverse menu of selective estrogen receptor modulators for a number of clinical applications.

The story of the endocrine treatment and prevention of breast cancer, over the past 60 years, is a well-documented success story. Nevertheless, the changes in healthcare may seem to have been "glacial" to many without the huge fanfare of today's targeted therapies, but the impact on women's health has been, and continues to be, profound and far exceeds the clinical significance of other new targeted drugs. The lives extended by antihormone breast cancer therapies are perhaps now in the millions around the world.

This book documents a personal journey in medical science, with many students and colleagues coauthoring the chapters herein. This is a landmark scientific success story, with many twists and turns never anticipated or the outcomes predicted before 1970. Only teams of people make discoveries and progress. The successive tamoxifen teams built on the scientific contributions of others who came before and the result was to advance health around the world by controlling breast cancer and increasing breast cancer survivorship for millions of women.

REFERENCES

1. Toft D, Gorski J. (1966) A receptor molecule for estrogens: isolation from the rat uterus and preliminary characterization. *Proc Natl Acad Sci USA* **55**(6): 1574–1581.
2. Toft D, Shyamala G, Gorski J. (1967) A receptor molecule for estrogens: studies using a cell-free system. *Proc Natl Acad Sci USA* **57**(6):1740–1743.
3. Brzozowski AM *et al.* (1997) Molecular basis of agonism and antagonism in the oestrogen receptor. *Nature* **389**(6652):753–758.
4. Dodds EC, Goldberg L, Lawson W, Robinson R. (1938) Estrogenic activity of certain synthetic compounds. *Nature* **141**:247–248.
5. Harper MJ, Walpole AL. (1966) Contrasting endocrine activities of cis and trans isomers in a series of substituted triphenylethylenes. *Nature* **212**(5057):87.
6. Harper MJ, Walpole AL. (1967) A new derivative of triphenylethylene: effect on implantation and mode of action in rats. *J Reprod Fertil* **13**(1):101–119.
7. Harper MJ, Walpole AL. (1967) Mode of action of I.C.I. 46,474 in preventing implantation in rats. *J Endocrinol* **37**(1):83–92.
8. Cole MP, Jones CT, Todd ID. (1971) A new anti-oestrogenic agent in late breast cancer. An early clinical appraisal of ICI46474. *Br J Cancer* **25**(2):270–275.
9. Jordan VC. (1974) The antiestrogen tamoxifen (ICI 46,474) as an antitumor agent. In *Proc. Eastern Cooperative Oncology Group Meeting*, Miami, FL; February 11–12.
10. Jordan VC. (1974) Tamoxifen mechanism of antitumour activity in animals and man. *Proc. Eastern Cooperative Oncology Group Meeting*, Jasper, Alberta, Canada; June 22–25.
11. Jordan VC. (1976) Antiestrogenic and antitumor properties of tamoxifen in laboratory animals. *Cancer Treat Rep* **60**(10):1409–1419.

12. Jordan VC, Koerner S. (1975) Tamoxifen (ICI 46,474) and the human carcinoma 8S oestrogen receptor. *Eur J Cancer* **11**(3):205–206 (in Eng.).
13. Jordan VC. (1974) Antitumour activity of the anti-oestrogen ICI 46,474 (tamoxifen) in the dimethylbenzanthracene (DMBA)-induced rat mammary carcinoma model. *J Steroid Biochem* **5**:354.
14. Jordan V, Dix CJ, Allen KE. (1979) The effectiveness of long term tamoxifen treatment in a laboratory model for adjuvant hormone therapy of breast cancer. *Adjuvant Ther Cancer* **2**:19–26.
15. Jordan VC, Allen KE. (1980) Evaluation of the antitumour activity of the non-steroidal antioestrogen monohydroxytamoxifen in the DMBA-induced rat mammary carcinoma model. *Eur J Cancer* **16**(2):239–251.
16. Jordan VC. (1976) Effect of tamoxifen (ICI 46,474) on initiation and growth of DMBA-induced rat mammary carcinomata. *Eur J Cancer* **12**(6):419–424.
17. Jordan VC, Collins MM, Rowsby L, Prestwich G. (1977) A monohydroxylated metabolite of tamoxifen with potent antioestrogenic activity. *J Endocrinol* **75**(2):305–316.

1

A CENTURY OF DECIPHERING THE CONTROL MECHANISMS OF ESTROGEN ACTION IN BREAST CANCER: THE ORIGINS OF TARGETED THERAPY AND CHEMOPREVENTION

V. Craig Jordan

Abstract

The story of deciphering the mechanisms that control the growth of sex-hormone-dependent cancers started more than a hundred years ago. Clinical observations of the apparently random responsiveness of breast cancer to endocrine ablation (hormonal withdrawal) provoked scientific enquiry in the laboratory that resulted in the development of effective strategies for targeting therapy to the estrogen receptor (ER) (or the androgen receptor in the case of prostate cancer), the development of antihormonal treatments that dramatically enhanced patient survival, and the first successful testing of agents to reduce the risk of developing any cancer. Most importantly, elucidating the receptor-mediated mechanisms of sex-steroid-dependent growth and the clinical success of antihormones has had broad implications for medicinal chemistry with the synthesis of new selective hormone receptor modulators for numerous clinical applications. Indeed, the successful translational research on the ER was the catalyst for the current strategy for developing targeted therapies for the tumor and the start of "individualized medicine." Over the past 50 years, ideas about the value of antihormones have translated effectively from the laboratory to improvement of clinical care, increase in national survival rates and significant reduction in the burden of cancer.

Keywords: Breast Cancer; Estrogen Receptor; Tamoxifen; Antiestrogen; Selective Estrogen Receptor Modulators; Antiandrogens; Prostate Cancer.

BEGINNINGS AT THE DAWN OF THE 20TH CENTURY

Schinzinger[1] is credited with suggesting that oophorectomy could be used to treat breast cancer; however, this suggestion does not appear to have been adopted. In contrast, the report by Beatson[2] that oophorectomy could initiate a regression of metastatic breast cancer in two premenopausal women was a landmark achievement. Although it is often stated that Beaston's work was empirical clinical research, the rationale for conducting an oophorectomy was, in fact, an example of early translational research. Beaston was aware of the essential role of removing the ovary in maximizing milk production in cows. He reasoned that there was potentially some factor that traveled in the blood supply to the breast as there was no known connection through the nerves. Interestingly enough, he also conducted laboratory experiments on rabbits before his clinical experiment, so the work was bench-to-bedside.[2] By 1900, Boyd[3] had assembled the results of all the available clinical cases of oophorectomy to treat breast cancer in Great Britain in perhaps the first "clinical trial." He concluded that only a third of metastatic breast tumors responded to oophorectomy. This clinical result and overall response rate has remained the same to this day.

Unfortunately, responses were of limited duration and enthusiasm waned as to whether this approach was the answer for cancer treatment. Also, the approach of endocrine ablation was relevant only to breast cancer (and subsequently prostate cancer[4]); thus, the approach was effective only in a small subset of cases of all cancer types. At the dawn of the 20th century, there was no understanding of the endocrine system or hormones. Nevertheless, laboratory studies started to decipher the biological control mechanisms responsible for the clinical observations.

LINKS BETWEEN SEX STEROIDS AND CANCER

The fashion in breast cancer research in the early years of the 20th century was to use inbred strains of mice to study the growth and incidence of spontaneous mammary cancer. Lathrop and Loeb[5] found that before three months of age was the optimal time for oophorectomy to *prevent* the development of mammary cancer, but obviously this knowledge could not be translated to the clinical setting: who would one treat? The mechanism was also unknown until Allen and Doisy,[6] using an ovariectomized mouse vaginal cornification assay, demonstrated that a principle, which they called estrogen (identified as estrone, the principal steroid), was present in ovarian follicular fluid. Their major advance set the scene for the subsequent breakthroughs in molecular endocrinology and therapeutics in the latter half of the 20th century (Fig. 1).

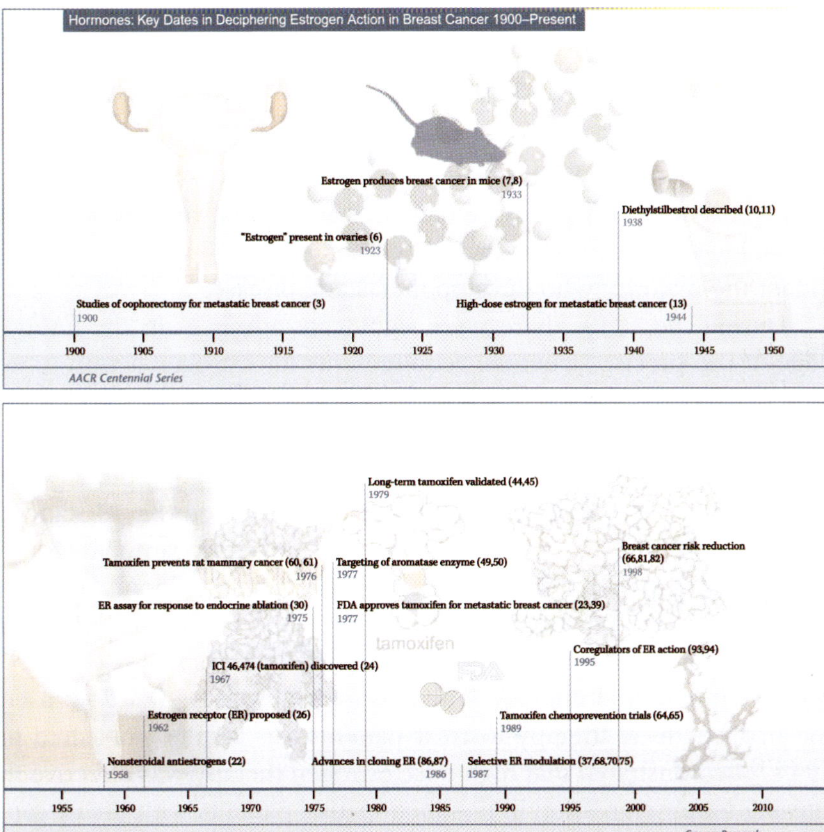

Figure 1. Timeline of the major landmarks in estrogen action and its application for the treatment and prevention of breast cancer.

The idea that breast cancer might be a preventable disease was extended by Professor Antoine Lacassagne, who first demonstrated that estrogen could induce mammary tumors in mice.[7,8] Lacassagne[9] hypothesized: "If one accepts the consideration of adenocarcinoma of the breast as the consequence of a special hereditary sensibility to the proliferative action of oestrone, one is led to imagine a therapeutic preventive for subjects predisposed by their heredity to this cancer. It would consist — perhaps in the very near future when the knowledge and use of hormones will be better understood — in the suitable use of a hormone, antagonistic or excretory, to prevent the stagnation of oestrone in the ducts of the breasts." However, when Lacassagne stated his vision at the annual meeting of the American Association for Cancer Research (Boston) in 1936, there were no lead compounds that antagonized estrogen action but the Allen Doisy mouse assay could be used to study structure-activity relationships so as to find synthetic estrogens. Within a decade, a landmark discovery was to occur in "chemical therapy," which was to expand the treatment of metastatic breast cancer to include postmenopausal women, who are in fact the majority who develop metastatic disease.

During the 1930s there were significant advances in the knowledge of the precise structural requirements for estrogen action in its target tissue, the vagina. Synthetic compounds based on stilbene[10,11] and triphenylethylene[12] were screened using the Allen Doisy ovariectomized mouse vaginal cornification assay to define compounds with optimal structures and duration of estrogen action. Sir Alexander Haddow found that carcinogenic polycyclic hydrocarbons would cause tumor regression in animals. However, they could not be used to treat humans. The nonsteroidal triphenylethylene-based estrogens had structures similar to those of polycyclic hydrocarbons and also caused tumor regression in animals. With this clue, Alexander Haddow used the first chemical therapy to treat patients. His results, published in 1944,[13] demonstrated that high dose estrogen therapy was effective in causing tumor regression in postmenopausal patients with breast cancer and men with prostate cancer. There was, however, no understanding of a mechanism. Indeed, he stated in 1970: "In spite of the extremely limited practicability of such a measure [high dose estrogen], the extraordinary extent of tumor regression observed in perhaps 1% of post-menopausal cases has always been regarded as of major theoretical importance, and it is a matter for some disappointment that so much of the underlying mechanisms continues to elude us."[14] These experimental data were also a paradox, as endocrine ablation to *remove* estrogens and their precursors was the dogma of the time.[15]

In the past 50 years, the progress in deciphering the control mechanisms of estrogen action in breast cancer (and androgen action in prostate cancer) has accelerated with advances in technology and an understanding of cell biology. However, progress in research does not occur in straight lines but chance observations can create a major breakthrough. This has happened repeatedly in the story of the treatment and prevention of breast cancer.

CONCEPTUAL PROGRESS THROUGH SCIENTIFIC SERENDIPITY

It is perhaps relevant to illustrate a few astute observations by scientists that immensely accelerated progress in deciphering the complexities of hormone action and the control of breast cancer growth.

Sir Charles Dodds is credited with the synthesis of the potent synthetic estrogen diethylstilbestrol[11] (Fig. 2), which was subsequently used for the treatment of both prostate cancer and breast cancer and, regrettably, was also applied to prevent recurrent abortions[16] which caused a rise in clear cell carcinoma of the vagina in the children.[17] During the race to describe the minimal molecular structure that would trigger vaginal cornification in the ovariectomized mouse vagina, controversy erupted in the 1930s over the reproducibility of results concerning the compound anethole. The authors were minimalistic in reporting the synthetic methodology, so replication proved impossible to create the correct biology. Rather, the product was correct, but the method used by the *original* authors was not reported accurately and actually caused dimerization of anethole to an impurity, dianethole an estrogen. This active impurity was structurally similar with the parallel research endeavors that concluded with the

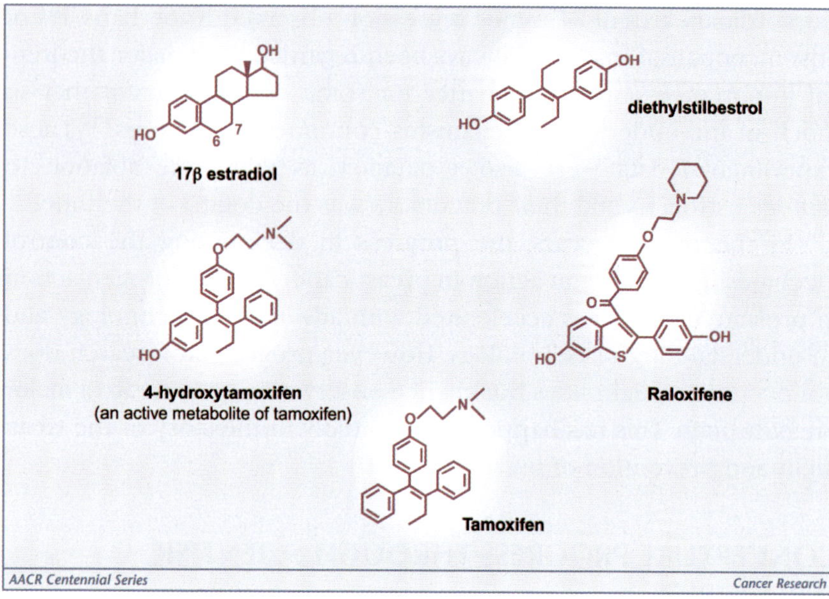

Figure 2. The structures of estrogens, antiestrogens and SERMs mentioned in the text. The positions 6 and 7 on the estradiol molecule indicate where tritium atoms were inserted to first describe estrogen binding to target tissue.[26] The metabolite 4-hydroxytamoxifen[112] is an active metabolite of tamoxifen that has been the standard laboratory antiestrogen and crystallized with the ligand binding domain of the ER.[96]

synthesis of the potent estrogen diethylstilbestrol. Thus, the purity of chemicals for testing was critical for successful science.

A similar story was also immensely important in allowing scientists to understand the direct actions of estrogen on the breast cancer cell *in vitro*. The MCF-7 ER-positive breast cancer cell line[18] has been the workhorse for the study of estrogen-stimulated growth. However, early examination of MCF-7 cells in the 1970s could not uniformly demonstrate estrogen-stimulated growth. Antiestrogens inhibited the apparently constitutive growth of MCF-7 cells but estradiol did reverse the inhibitory actions of antiestrogens on growth.[19] The mystery deepened when studies *in vitro* could not demonstrate estrogen-stimulated growth but MCF-7 cells inoculated into athymic mice would grow into tumors only with estrogen treatment. There was clearly a second factor required for estrogen-stimulated tumor growth *in vivo*![20]

The astute observations of John and Benita Katzenellenbogen solved the mystery of why estrogen did not stimulate MCF-7 breast cancer cell growth *in vitro*. It appears that all cells had been grown for more than a decade in standard media containing large concentrations of a pH indicator called phenol red. The Katzenellenbogens realized that the structure of phenol red was similar to that of nonsteroidal estrogens, and that removal of the indicator from cell culture media caused the cell growth rate to fall and only then would exogenous estrogen cause growth.[21] In other words, the cells were already growing maximally in phenol-red-containing media. Subsequent studies revealed that the culprit was, in fact, a partially dimerized chemical contaminant of phenol red. This critical technical advance permitted all of the subsequent understanding of the molecular biology of direct estrogen action.

Leonard Lerner was a young research endocrinologist employed by Merrell Dowe to study nonsteroidal estrogen pharmacology. He noticed that the structure of one of the compounds being tested for the control of coronary artery disease was a triphenylethanol similar to the estrogenic triphenylethylenes and he asked to test this chemical as an estrogen. To his surprise the compound, subsequently renamed MER25 or ethamoxy-triphetol, was antiestrogenic in all species tested and had no estrogen-like actions in any animal tests.[22] Lerner had discovered the first nonsteroidal antiestrogen.[22] Although the compound was too toxic and not potent enough for clinical use, he went on to be involved in the discovery of the first triphenylethylene antiestrogen, called chloramiphene (MRL41) and later to be known as clomiphene.[23] Originally, the nonsteroidal antiestrogens were predicted, based on animal studies, to be potent postcoital contraceptives, which in the early 1960s had a huge potential market as "morning-after pills." However, clomiphene did exactly the opposite: it induced ovulation in women.[23] Enthusiasm waned and there was general disinterest in this area of research until ICI 46,464, another nonsteroidal antiestrogen discovered in the fertility program of ICI Pharmaceutical Ltd. (now AstraZeneca),[24] was reinvented as the first targeted therapy for breast cancer and the first chemopreventive for any cancer.[25]

A TARGET FOR TREATMENT AND PREVENTION

The early theory for estrogen action in its target tissues, e.g. the uterus and the vagina, was that there was chemical transformation between estrone and the less abundant 17β estradiol (Fig. 2) to control the redox potential of the tissue environment. In the late 1950s, Elwood Jensen (Fig. 3) and Herb Jacobsen[26] chose another approach at the Ben May Laboratories of the University of Chicago. They synthesized[6,7] $[^{3}\mathrm{H}]$ estradiol (Fig. 2) with very high specific activity and, following injection into the immature female rats, the unchanged steroid bound to and was retained by the estrogen target tissues, i.e. the uterus, vagina and pituitary gland. In contrast, $[^{3}\mathrm{H}]$ estradiol bound to but was not retained by nontarget tissues, e.g. the muscle, lung and heart. There was clearly a receptor mechanism at play that could be blocked[27] by the coadministration of the first nonsteroidal antiestrogen, MER-25.[22]

The mystery of why only about a third of advanced breast cancers responded to either endocrine ablation[3] or high dose estrogen

Figure 3. Professor Charles Huggins (*left*) and Elwood Jensen were to receive the Nobel Prize in Physiology or Medicine (1966) and the Lasker Award for their work on androgen action in cancer and the role of ER in physiology and cancer, respectively.

therapy[15] was solved by the application of basic endocrinology to the practical issue of excluding women with metastatic breast cancer who *would not* significantly benefit from unnecessary endocrine ablative surgery (oophorectomy, adrenalectomy or hypophysectomy). The ER was found to be an extractable protein from the rat uterus that would bind [^{3}H] estradiol in the extraction cytosol.[28,29] During the late 1960s, numerous methods were described to identify and quantitative ER levels in tumor biopsies[30] and these data were subsequently correlated with clinical outcomes in metastatic breast cancer.[31] Breast tumors without the ER were unlikely to respond to endocrine ablation and therefore should not be treated with this modality. The ER assay was introduced as the standard of care in the mid-1970s to predict endocrine responsiveness to endocrine ablation. It should be stressed that tamoxifen was not available in medical practice until the FDA approved this "hormone therapy" in December 1977 for the treatment of metastatic breast cancer in postmenopausal women.[23] Indeed, research with the value of the ER assay to predict responsiveness to antiestrogens was unconvincing[23] and the value of adding another "hormone therapy" to the treatment armamentarium was uncertain. The 1970s were a time when all hopes in medical oncology were focused on discovering the correct combination of high dose cytotoxic therapies to cure breast cancer much in the same way as both childhood leukemias and Hodgkin's disease had been cured. This was not to be but translational research took another route, using the ER as a drug target and not as a predictive test for endocrine ablation.[32]

AN UNLIKELY THERAPEUTIC SOLUTION

Professor Paul Ehrlich (1854–1915) established a model for the development of chemical therapies (chemotherapy) to treat infectious disease. A range of chemical therapies would be synthesized to study structure-function relationships in appropriate laboratory models that replicated human disease.[33] A clinical study would then be performed on the most promising candidate. Ehrlich's pioneering work to develop Salvarsan for the successful treatment of syphilis is a landmark achievement.[33] He was, however, unsuccessful in applying the same principles to cancer chemotherapy. Indeed, even as recently as 1970, Sir Alexander Haddow[14] stated that there was unlikely to be a "*chemotherapia specifica*" in the sense that Ehrlich envisioned because cancer was so similar to the tissue of origin. There was also no target or effective tests or models to predict efficacy in cancer treatment prior to administration to the patient. The key to the successful development of tamoxifen, a failed contraceptive,[23] was the application of Ehrlich's principles of developing an effective treatment strategy by employing disease-specific laboratory models and the utilization of the tumor ER as a target for drug action.[25]

Available laboratory models for the study of the antitumor actions of antiestrogenic drugs were strains of mice with a high incidence of spontaneous mammary tumors[5] or the carcinogen-induced rat mammary carcinoma.[34] The mouse models had fallen out of fashion with the discovery of the "Bittner milk factor," a virus that transmits mammary carcinogenesis to subsequent generations through the mother's milk.[35] The research community also began to realize that breast cancer was not a viral disease. Nevertheless, the knowledge of mouse mammary carcinogenesis proved to be pivotal for developing precise and targeted promoters to initiate mammary cancer with oncogenes using transgenic mice.[36] Another problem with tumor testing of tamoxifen in mice was the unusual observation that tamoxifen — or ICI 46,474, as it was then called — was an estrogen in the mouse.[24,37] This pharmacological peculiarity was later to become important with the recognition of selective ER modulation.[38] Most importantly, work did not advance quickly in the 1960s and early 1970s, as there was no enthusiasm about introducing a new "hormonal therapy" into clinical practice.[25] All early compounds had failed to advance past early clinical studies and only tamoxifen was marketed[23] for the induction of ovulation or the general treatment of late stage breast cancer in postmenopausal women.[39-41]

In the late 1960s, the 7,12-dimethybenz[a]anthracene-induced (DMBA) rat mammary carcinoma model[34] was extremely fashionable for research on the endocrinology of rat mammary carcinogenesis.[42,43] However, the parallels with breast cancer are few, as the tumors do not metastasize and are regulated primarily by prolactin secreted by the pituitary gland in direct response to estrogen action.[44] Be that as it may, there was no alternative and so the DMBA rat mammary carcinoma model was adapted to determine the appropriate strategy for the use of antihormonal therapy as an adjuvant. At that time in the mid-1970s, the early adjuvant trials with tamoxifen did not target patients with ER-positive breast cancer and used only short term (one-year) tamoxifen treatment to avoid premature drug resistance. This duration of tamoxifen was selected as the antiestrogen only controlled the growth of metastatic breast cancer for about a year.[40] The value of short and long term (one or six months' treatment, equivalent to one or six years of adjuvant treatment in patients) antihormone administration was determined starting treatment one month after DMBA administration to 60-day-old Sprague Dawley rats. Long term therapy was remarkably effective in controlling the appearance of mammary tumors and was far superior to short term treatment.[45,46] The concepts of targeting the ER and using long term adjuvant therapy effectively translated through clinical trials to improvement of the national survival rates for breast cancer.[47,48]

TARGETING TREATMENT FOR BREAST CANCER

The early clinical work of Santen[49] established the practical feasibility of employing aminogluthemide, an agent that blocks both adrenal steroidogenesis and the CYP19 aromatase enzyme to stop conversion of testosterone and androstenedione to estradiol and estrone, respectively. Unfortunately, aminogluthemide must be given with a natural glucocorticoid and therefore long term therapy is not a practical possibility. Brodie and coworkers[50,51] advanced knowledge of the specific targeting of the CYP19 aromatase enzyme with the identification and subsequent development of 4-hydroxyandrostenedione[52] as the first practical suicide inhibitor of the aromatase enzyme (Fig. 4). Incidentally, the pivotal work with both tamoxifen and 4-hydroxyandrostenedione (Figs. 2 and 4) was initiated at the Worcester Foundation for Experimental Biology in Massachusetts in the early 1970s.[53] Brodie's contribution eventually became the catalyst for creating a whole range of agents [e.g. anastrozole (Fig. 3)] targeted to the aromatase enzyme for the treatment of breast cancer in postmenopausal

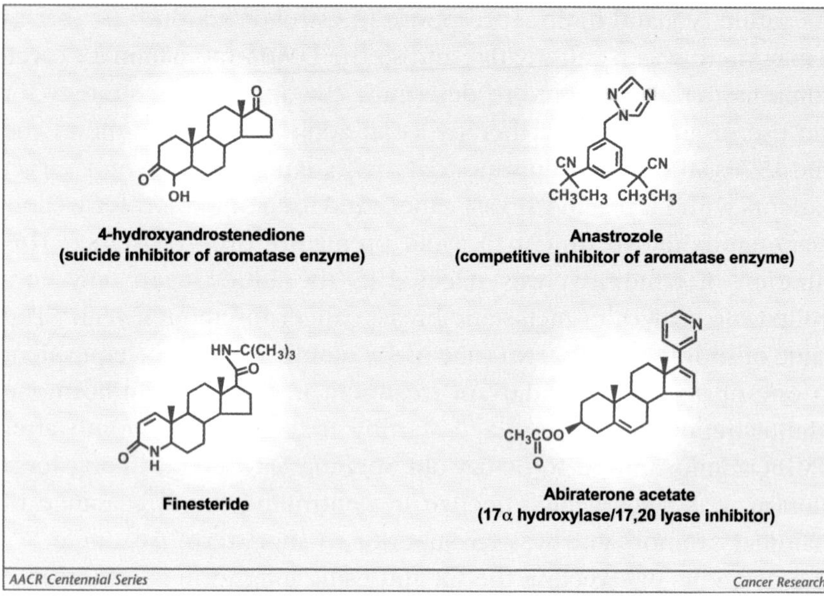

Figure 4. Structures of inhibitors of estrogen and androgen biosynthesis.

women.[54] The clinical application of aromatase inhibitors has reduced the side effects noted with tamoxifen in postmenopausal women, such as blood clots and endometrial cancer, and there has been a small but significant improvement in disease control for the postmenopausal patient when results are compared with tamoxifen.[55,56]

However, recent research into the pharmacogenetics of tamoxifen has suggested that the CYP2D6 enzyme product is important for metabolism to the active antiestrogen endoxifene (4-hydroxy-N-desmethyltamoxifen)[57] and the use of certain selective serotonin reuptake inhibitors (SSRIs) to reduce hot flashes appears to be contraindicated because of drug interaction at the CYP2D6 enzyme.[58,59] Current research is also exploring the hypothesis that a mutated and ineffective CYP2D6 gene product undermines the therapeutic activity of tamoxifen.[58,59] It may be that patients could eventually be selected for optimal effective tamoxifen treatment in cases of ER-positive breast cancer. This would be worthwhile for the chemoprevention of breast cancer. Clearly, the identification of patients for optimal long term use of tamoxifen should exclude those high risk women with a mutant CYP2D6 gene who choose to use chemoprevention, as tamoxifen treatment may possibly be suboptimal.

CHEMOPREVENTION OF BREAST CANCER

In the middle of the 1970s, Sporn[60] advanced the concept of the chemoprevention of cancer and strongly advocated this approach as the optimal and clearly most rational way to reduce the burden of cancer. Practical chemoprevention articulated by Lacassagne[9] has its foundations with the finding that tamoxifen prevents DMBA-induced rat mammary carcinogenesis.[61,62] These laboratory findings[46,61,62] and the subsequent clinical finding that adjuvant tamoxifen treatment reduces the incidence of contralateral breast cancer[63] prompted Powles[64,65] to initiate the first exploratory trial to test the worth of tamoxifen in reducing the incidence of breast cancer in high risk women. Although numbers were small, the Powles study did ultimately demonstrate the ability of tamoxifen to reduce breast cancer incidence many years after the treatment had stopped.[66] In contrast, the large study by Fisher[67,68] definitively demonstrated the efficacy of tamoxifen in reducing the incidence of ER-positive breast cancer initially and continuing to do so after therapy stops in both pre- and postmenopausal women at high risk. Tamoxifen became the first medicine approved by the Food and Drug Administration for risk reduction of any cancer. However, concerns — based on laboratory findings[69] — about the potential of tamoxifen to increase the risk of endometrial cancer in postmenopausal women and the carcinogenic potential of tamoxifen as a hepatocarcinogen[70] demanded that there be a better way to reduce the risk of breast cancer as a public health initiative.

The recognition of SERM action by nonsteroidal antiestrogens that stimulate some estrogen target tissues but block estrogen-stimulated tumor growth in others[71] introduced a new dimension into therapeutics and advanced chemoprevention. Raloxifene has its origins as a nonsteroidal antiestrogen for the treatment of breast cancer[72,73] — LY156758 or keoxifene. The drug failed in that indication,[74] and further development was abandoned.[75] The discovery that both tamoxifen and keoxifene would maintain bone density in ovariectomized rats[76] and block rat mammary carcinogenesis,[77] but that keoxifene was less estrogen-like than tamoxifen in the rodent uterus[72] and was less effective in stimulating the growth of endometrial cancer,[78] suggested a new therapeutic strategy.[79] Simply stated[80]: "We have obtained valuable clinical information about this group of drugs that can be applied in other disease states. Research does not travel in straight lines and observations in one field of science often become major discoveries in another. Important clues have been garnered about the effects of tamoxifen on bone and lipids so it is possible that derivatives could find targeted applications to retard osteoporosis or atherosclerosis. The ubiquitous application of novel compounds to prevent diseases associated with the progressive changes after menopause may, as a side effect, significantly retard the development of breast cancer. The target population would be postmenopausal women in general, thereby avoiding the requirement to select a high-risk group to prevent breast cancer."

Several years later, keoxifene was renamed raloxifene (Fig. 2) and was shown to maintain bone density in osteoporotic or osteopenic women,[81] and simultaneously reduce the incidence of invasive breast cancer without causing an increase in the incidence of endometrial cancer.[82] Raloxifene went on to be tested against tamoxifen in the Study of Tamoxifen and Raloxifene (STAR) trial[83] and was FDA-approved both for the treatment and prevention of osteoporosis in postmenopausal women and for the reduction of invasive breast cancer incidence in postmenopausal women at elevated risk. The clinical advances with SERMS to modulate estrogen target tissues have provided exceptional opportunities to treat and prevent multiple diseases. However, for the future it is the study of the molecular events of estrogen action that holds the promise of further breakthroughs in patient care.

MOLECULAR MECHANISMS OF ESTROGEN AND SERM ACTION

It is not possible to provide a review of the explosion of interest in receptor-mediated molecular mechanisms of action of estrogen, so the reader is referred to significant reviews to appreciate the evolution of the topic.[84,85] What will be presented is an evolving guide to current thinking. There are two ERs, referred to as α and β (Figs. 5 and 6). The receptor ERα is the traditional ER[26,28] but it should be stressed that the development of monoclonal antibodies to ER[86] was the essential step for ERα cloning[87,88] that provided the clues to discovering ERβ.[89] The receptor proteins encode on different chromosomes and have homology as members of the steroid receptor superfamily but there are distinct patterns of distribution and distinct and subtle differences in structure and ligand binding affinity. An additional dimension that may be significant for tissue modulation is the ratio of ERα and ERβ at a target site. A high ERα:ERβ ratio correlates well with very high levels of cellular proliferation, whereas the

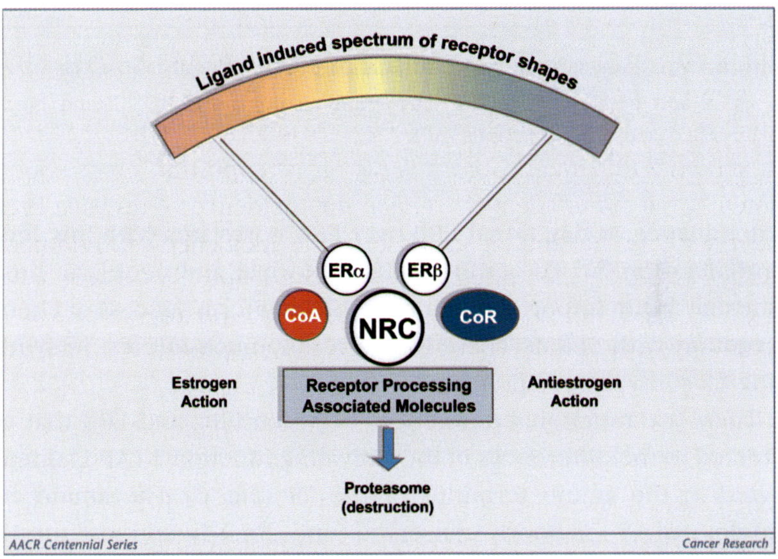

Figure 5. Molecular mechanisms of estrogens, antiestrogens and SERMs in estrogen target tissues. The nuclear receptor complex (NRC) that results from ligand binding to either ERα or ERβ can interact with either coactivators (CoA) or corepressors (CoR) to initiate estrogenic or antiestrogenic responses, respectively. The activation of transcription at a promoter site of an estrogen-responsive gene or curves through the binding of the complex that is cyclically destroyed through the proteosome and then a new complex is reassembled.

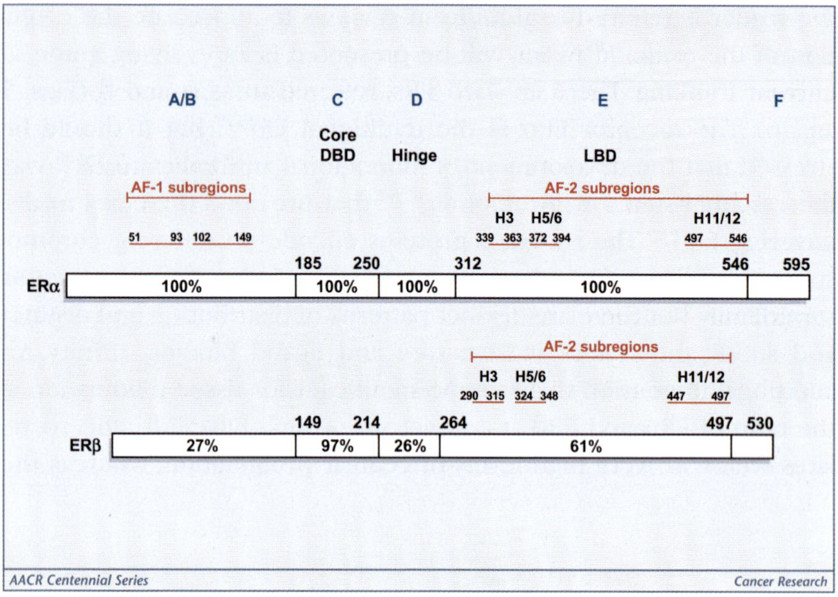

Figure 6. Comparison of the percent homology of the domains of estrogen receptors alpha and beta. Abbreviations: DNA binding domain (DBD), ligand binding domain (LBD), activating functions (AF).

predominance of functional ERβ over ERα correlates with low levels of proliferation.[90,91] The ratio of ERs in normal and neoplasic breast tissue may be an important factor for the long term success of chemoprevention with SERMs. There is, as a result, much interest in synthesizing ER-subtype-specific ligands.

There are functional differences between ERα and ERβ that can be traced to the differences in the activating function 1 (AF-1) domain located in the amino terminus of the ER (Fig. 6). The amino acid homology of AF-1 is poorly conserved (only 20%). In contrast, the AF-2 region, located at the C terminus of the ligand binding domain, differs only by one amino acid: D545 in ERα and N496 in ERβ. Since the AF-1 and AF-2 regions are critical for the interaction with other coregulatory proteins and gene transcription, the structural differences between AF-1 provide a clue about the potential functional differences between ERα and ERβ. Studies using chimeras of ERα and ERβ by switching the AF-1 regions demonstrate that this region contributes to the cell- and promoter-specific differences in transcriptional activity. In general, SERMs can partially activate engineered genes regulated by an estrogen response element through ERα but not ERβ.[92,93] In contrast, 4-hydroxytamoxifen and raloxifene can stimulate activating protein 1 (AP-1)–regulated reporter genes with both ERα and ERβ in a cell-dependent fashion.

The simple model for estrogen action, with either ERα or ERβ controlling estrogen-regulated events, has now evolved into a fascinating mix of protein partners that have the potential to modulate gene transcription (Fig. 5). It is more than a decade since the first steroid receptor coactivator (SRC-1) was first described.[94] Now, dozens of coactivator molecules are known and also corepressor molecules exist to prevent the gene transcription by unliganded receptors.[95]

It is reasonable to ask: How does the ligand program the receptor complex to interact with other proteins? X-ray crystallography of the ligand binding domains of the ER liganded with either estrogens or antiestrogens demonstrates the potential of ligands to promote coactivator binding or prevent coactivator binding based on the shape of the estrogen or antiestrogen receptor complex.[96,97] Evidence has accumulated that the broad spectrum of ligands that bind to the ER can create a broad range of ER complexes that are either fully estrogenic or antioestrogenic at a particular target site.[98] Thus, a mechanistic model of estrogen action and antiestrogen action (Fig. 5) has emerged based on the shape of the ligand that programs the complex to adopt a particular shape that ultimately interacts with coactivators or corepressors in target cells to determine the estrogenic or antiestrogenic response, respectively. But how does transcription become initiated?

Not surprisingly, the coactivator model of steroid hormone action has now become enhanced into multiple layers of complexity, thereby amplifying the molecular mechanisms of modulation.[99] Associated molecules in the complex at the estrogen-regulated promoter site cause the complex to be built and then destroyed in a dynamic cycle of DNA remodeling that initiates transcription. The cyclical stimulation of activated receptor complexes appears to be necessary for orchestrating a consistent estrogenic response at a target gene.

DRUG RESISTANCE TO SERMS

The acceptance of the concept of long term antihormonal therapy to target, treat and prevent breast cancer[25] raised the specter of drug resistance to SERMs. However, the early models of SERM resistance did not reflect the majority of clinical experience. The natural laboratory models of antihormone resistance caused stimulation of tumor growth during a year of therapy[100] and therefore reflected drug resistance in patients with metastatic breast cancer who are only treated successfully for a year. However, the earlier laboratory models of drug resistance did not replicate clinical experience with adjuvant therapy for five years. Remarkably, drug resistance evolves (Fig. 7) and the survival signaling pathway in tamoxifen-resistant tumors becomes reorganized so that instead of estrogen being a survival signal, physiologic estrogen now inhibits tumor growth.[101] This discovery provides an invaluable insight into the evolution of drug resistance to SERMS and prompted the reclassification of the process through

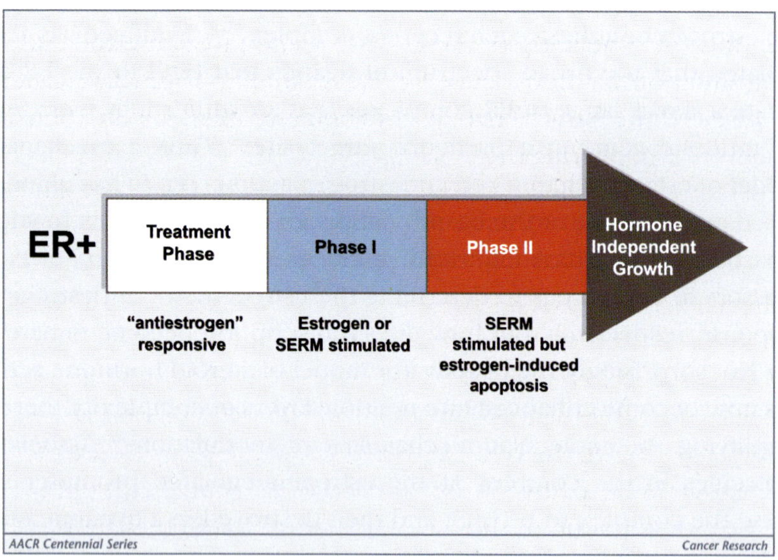

Figure 7. The evolution of resistance to selective estrogen receptor modulators (SERMS: tamoxifen or raloxifene) long term therapy.

Phase I (SERM/estrogen-stimulated) to Phase II (SERMs-stimulated/ estrogen)-inhibited growth.[102]

This model would also explain the earlier observations[13] as to why high dose estrogen therapy was effective only as a treatment for breast cancer in women many years after menopause. Natural estrogen deprivation had occurred. The process is accelerated and enhanced, however, in patients treated long term with SERMs or aromatase inhibitors so that only low doses of estrogen are necessary for causing experimental tumors to regress. The new knowledge of the apoptotic action of estrogen (or androgen — see next section) could potentially lead to the discovery of a precise apoptotic trigger initiated naturally by steroid hormone receptors.[102] Discovery of this apoptotic trigger might result in applications to target critical survival signals with new drugs.

PARALLEL PATH OF THE PROSTATE

Charles Huggins (Fig. 3) resurrected the use of endocrine ablation for the treatment-hormone-dependent breast cancers.[103] His focus, however, was the regulation of the growth of the prostate gland and the application of that knowledge for the treatment of prostate cancer.[4] He received the Nobel Prize in Physiology or Medicine in 1966. The process of deciphering the molecular mechanisms of androgen action in its target tissues and prostate cancer has tended to lag behind the pathfinder estrogen. Nevertheless, the basic model for the regulation of nuclear hormone receptor action is consistent but the details of androgen action are distinctly different than those of estrogen action, which in turn created novel therapeutic opportunities to stop the biosynthesis of each active steroidal agent. The similarities and differences in the molecular actions of estrogen and androgen action are illustrated in Fig. 8. The two significant differences (yet similarities) in the biosynthetic pathways between estrogens and androgens are: (1) the aromatization of the A ring of testosterone to create the high affinity ER binding ligand 17β-estradiol in women — this bioactivation led to the development of aromatase inhibitors to block estrogen synthesis[51]; and (2) the reduction of testosterone to the high afffinity

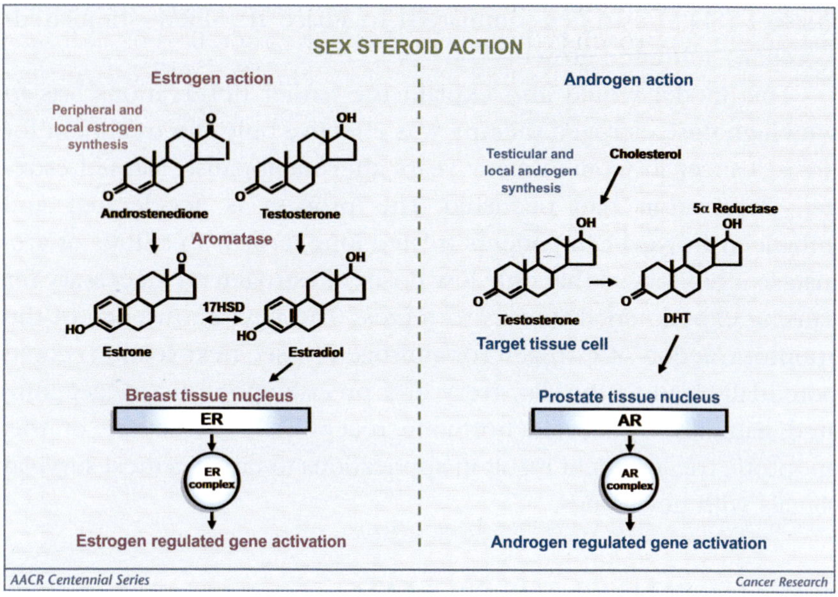

Figure 8. Comparison of the molecular mechanisms of estrogens and androgens in their respective target tissues. The transformations of the respective steroids are necessary for high binding affinity for their receptors, but the activation from prohormones occurs in different tissue sites relative to their target.

AR binding ligand dihydotestosterone in men — this knowledge led to the development of the 5α reductase inhibitor finesteride (Fig. 4), which was tested successfully for risk reduction for prostate cancer in men.[104] Unfortunately, as yet, finesteride has failed to advance for use as a chemopreventive for prostate cancer because of overstated concerns about the accelerated development of potentially more aggressive prostate cancers in those men who did not have tumorigenesis prevented. In contrast, aromatase inhibitors have advanced to test their worth as chemopreventive agents.[54]

A range of antiandrogenic drugs that competitively block the AR are available in clinical practice.[105] Drug resistance to antiandrogen therapy parallels antiestrogen drug resistance[106] and, following long term antihormonal therapy with antiandrogens, androgen induces apoptosis in antiandrogen-resistant prostate cancer cells.[107] Recent research has identified high local levels of androgen prediction as a major form of antihormonal drug resistance.[108] As a result, a new therapeutic approach is the development of an inhibitor of androgen biosynthesis from cholesterol (Fig. 8) by blocking 17-hydroxylase/17,20-lyase (CYP17). A promising compound, abiraterone acetate (Fig. 4), is currently being evaluated in clinical trials.[109] However, there is also a need to coadminister glucocorticoids, so long term therapy must be monitored carefully.

THE SUCCESSFUL EVOLUTION OF TARGETED ANTIHORMONAL THERAPY IN THE 20TH CENTURY AND BEYOND

The identification of the ER and subsequently the AR as the conduit for hormone-mediated development and growth in breast and prostate cancer, respectively, has had a profound effect on the approach to the treatment and prevention of cancers. These hormone-mediating molecules have proven to be the pathfinders for the development of targeted therapies that transformed the approach to cancer treatment away from the nonspecific cytotoxic chemotherapy approach during the 1950s to 1990s. As a result, there is today enthusiasm for the promise of individualized medicine and tumor-specific therapeutics.[25,110]

The impact of antihormonal therapy on breast cancer has been profound, with improvements in patient survival, a menu of medicines now available to suit individual patient needs and a decrease in national mortality rates in numerous countries.[48] Additionally, there are now two SERMS (tamoxifen and raloxifene) available to reduce the incidence of breast cancer.[68,83] But progress in our understanding and application of SERMs is more than chemoprevention. The SERM concept[71] has spread to the development of tissue-selective drugs for all members of the hormone receptor superfamily.[25,111] An enormous interest in developing selective glucocorticoid receptor modulators, selective progesterone receptor modulators, selective androgen receptor modulators, and even agents to treat rheumatoid arthritis is an ongoing therapeutic outcome of translational research for the chemoprevention of breast cancer.

ACKNOWLEDGEMENTS

Dr. Jordan is supported by the Department of Defense Breast Program under award No. W81XWH-06-1-0590 Center of Excellence [views and opinions of and endorsements by the author(s) do not reflect those of the U.S. Army or the Department of Defense], FCCC Core Grant NIH P30 CA006927, the Avon Foundation, Genuardi's Foundation and the Weg Fund of the Fox Chase Cancer Center. This article is republished from Jordan VC (2009) *Cancer Res* **69**:1248–1253 with permission from the American Association for Cancer Research.

REFERENCES

1. Schinzinger A. (1889) Ueber carcinoma mammae [abstract]. Beilage zum Centralblatt fur Chirurgie. *18th Congress of the German Society for Surgery* **16**:55–56.
2. Beatson GT. (1896) On treatment of inoperable cases of carcinoma of the mamma: suggestions for a new method of treatment with illustrative cases. *Lancet* **2**:162–165.
3. Boyd S. (1900) On oophorectomy in cancer of the breast. *Br Med J* **ii**:1161–1167.
4. Huggins C, Hodges CV. (1941) Studies on prostatic cancer. I. The effect of castration, of estrogen and of androgen injection on serum phosphatases in metastatic carcinoma of the prostate. *Cancer Res* **1**:293–297.
5. Lathrop AE, Loeb L. (1916) Further investigations on the origins of tumors in mice III on the part played by internal secretions in the spontaneous development of tumors. *J Cancer Res* **1**:1–19.
6. Allen E, Doisy EA. (1923) An ovarian hormone: preliminary reports on its localization, extraction and partial purification and action in test animals. *JAMA* **81**:810–821.
7. Lacassagne A. (1933) Influence d'un facteur familial dans la production par la folliculine, de cancers mammaires chez la souris male. *Comptes rendus des seances de la Societe de biologie et de ses filiales* **114**:427–429.
8. Lacassagne A. (1936) A comparative study of the carcinogenic action of certain oestrogenic hormones. *Am J Cancer* **28**:735–740.
9. Lacassagne A. (1936) Hormonal pathogenesis of adenocarcinoma of the breast. *Am J Cancer* **27**:217–225.
10. Dodds EC, Goldberg L, Lawson W, Robinson R. (1938) Oestrogenic activity of certain synthetic compounds. *Nature* **141**:247–248.
11. Dodds EC, Lawson W, Noble RL. (1938) Biological effects of the synthetic oestrogenic substance 4:4′dihydroxy-a:b-diethylstilbene. *Lancet* **1**:1389–1391.

12. Robson JM, Schonberg A. (1937) Oestrous reactions including mating produced by triphenylethylene. *Nature* **140**:196.
13. Haddow A, Watkinson JM, Paterson E. (1944) Influence of synthetic oestrogens upon advanced malignant disease. *Br Med J* **2**:393–398.
14. Haddow A, David A. (1970) Karnofsky memorial lecture: thoughts on chemical therapy. *Cancer* **26**:737–754.
15. Huggins C. (1956) Control of cancers of man by endocrinologic methods: a review. *Cancer Res* **16**:825–830.
16. Smith OW. (1948) Diethylstilbestrol in the prevention and treatment of complications of pregnancy. *Am J Obstet Gynec* **56**:821.
17. Herbst AL, Ulfelder H, Poskanzer DC. (1971) Adenocarcinoma of the vagina. *N Engl J Med* **284**:878–881.
18. Levenson AS, Jordan VC. (1997) MCF-7: the first hormone-responsive breast can cer cell line. *Cancer Res* **57**:3071–3078.
19. Lippman M, Bolan G, Huff K. (1976) The effects of estrogens and antiestrogens on hormone-responsive human breast cancer in long-term tissue culture. *Cancer Res* **36**:4595–4601.
20. Shafie SM. (1980) Estrogen and the growth of breast cancer: new evidence suggests indirect action. *Science* **209**:701–702.
21. Berthois Y, Katzenellenbogen JA, Katzenellenbogen BS. (1986) Phenol red in tissue culture media is a weak estrogen: implications concerning the study of estrogen-responsive cells in culture. *Proc Natl Acad Sci USA* **83**:2496–2500.
22. Lerner LJ, Holthaus JF, Thompson CR. (1958) A non-steroidal estrogen antagonist 1-(p-2-diethylaminoethoxyphenyl)-1-phenyl-2-p-methoxyphenylethanol. *Endocrinology* **63**:295–318.
23. Jordan VC. (2003) Tamoxifen: a most unlikely pioneering medicine. *Nat Rev Drug Discov* **2**:205–213.
24. Harper MJ, Walpole AL. (1967) A new derivative of triphenylethylene: effect on implantation and mode of action in rats. *J Reprod Fertil* **13**:101–119.
25. Jordan VC. (2008) Tamoxifen: catalyst for the change to targeted therapy. *Eur J Cancer* **44**:30–38.
26. Jensen EV, Jacobson HI. (1962) Basic guides to the mechanism of estrogen action. *Recent Prog Horm Res* **18**:387–414.
27. Jensen EV. (1962) Antiestrogens. *Recent Prog Horm Res* **18**:418.
28. Toft D, Gorski J. (1966) A receptor molecule for estrogens: isolation from the rat uterus and preliminary characterization. *Proc Natl Acad Sci USA* **55**:1574–1581.
29. Toft D, Shyamala G, Gorski J. (1967) A receptor molecule for estrogens: studies using a cell-free system. *Proc Natl Acad Sci USA* **57**:1740–1743.
30. McGuire WL, Carbone PP, Volmer EP (eds.). (1975) In: *Estrogen Receptor in Human Breast Cancer*. New York: Raven.

31. McGuire WL, Carbone PP, Sears ME, Escher GC. (1975) Estrogen receptors in human breast cancer: an overview. In: *Estrogen Receptor in Human Breast Cancer*, eds. McGuire WL, Carbone PP, Volmer EP. New York: Raven, pp. 1-7.
32. Jensen EV, Jordan VC. (2003) The estrogen receptor: a model for molecular medicine. The Dorothy P. Landon AACR Prize for Translational Research. *Clin Cancer Res* **9**:1980-1989.
33. Baumler E. (1984) *Paul Ehrlich, Scientist for Life*. New York: Holmes & Meier.
34. Huggins C, Grand LC, Brillantes, FP. (1961) Mammary cancer induced by a single feeding of polynuclear hydrocarbons and their suppression. *Nature* 189-204.
35. Bittner JJ. (1936) Some possible effects of nursing on the mammary gland tumor incidence in mice. *Science* **84**:162.
36. Stewart TA, Pattengale PK, Leder P. (1984) Spontaneous mammary adenocarcinomas in transgenic mice that carry and express MTV/myc fusion genes. *Cell* **38**:627-637.
37. Harper MJ, Walpole AL. (1966) Contrasting endocrine activities of cis and trans isomers in a series of substituted triphenylethylenes. *Nature* **212**:87.
38. Jordan VC, Robinson SP. (1987) Species-specific pharmacology of antiestrogens: role of metabolism. *Fed Proc* **46**:1870-1874.
39. Klopper A, Hall M. (1971) New synthetic agent for the induction of ovulation. Preliminary trial in women. *Br Med J* **1**:152-154.
40. Cole MP, Jones CT, Todd ID. (1971) A new anti-oestrogenic agent in late breast cancer. An early clinical appraisal of ICI 46474. *Br J Cancer* **25**:270-275.
41. Ward HW. (1973) Anti-oestrogen therapy for breast cancer: a trial of tamoxifen at two dose levels. *Br Med J* **1**:13-14.
42. Welsch CW. (1985) Host factors affecting the growth of carcinogen-induced rat mammary carcinomas: a review and tribute to Charles Brenton Huggins. *Cancer Res* **45**:3415-3443.
43. Russo J, Tay LK, Russo IH. (1982) Differentiation of the mammary gland and susceptibility to carcinogenesis. *Breast Cancer Res Treat* **2**:5-73.
44. Lieberman ME, Maurer RA, Gorski J. (1978) Estrogen control of prolactin synthesis *in vitro*. *Proc Natl Acad Sci USA* **75**:5946-5949.
45. Jordan VC, Dix CJ, Allen KE. (1979) The effectiveness of long term tamoxifen treatment in a laboratory model for adjuvant hormone therapy of breast cancer. *Adjuv Ther Cancer* **2**:19-26.
46. Jordan VC, Allen KE. (1980) Evaluation of the antitumour activity of the non-steroidal antioestrogen monohydroxytamoxifen in the DMBA-induced rat mammary carcinoma model. *Eur J Cancer* **16**:239-251.
47. EBCTCG. (1998) Tamoxifen for early breast cancer: an overview of the randomised trials. *Lancet* **354**:1451-1467.
48. EBCTCG. (2005) Effects of chemotherapy and hormonal therapy for early breast cancer on recurrence and 15-year survival: an overview of the randomised trials. *Lancet* **365**:1687-1717.

49. Santen RJ, Lipton A. (1974) Successful medical adrenalectomy with aminoglutethimide. Role of altered drug metabolism. *JAMA* **230**:1661–1665.
50. Schwarzel WC, Kruggel WG, Brodie HJ. (1973) Studies on the mechanism of estrogen biosynthesis. 8. The development of inhibitors of the enzyme system in human placenta. *Endocrinology* **92**:866–880.
51. Brodie AM, Schwarzel WC, Shaikh AA, Brodie HJ. (1977) The effect of an aromatase inhibitor, 4-hydroxy-4-androstene-3,17-dione, on estrogen-dependent processes in reproduction and breast cancer. *Endocrinology* **100**:1684–1695.
52. Coombes CR, Goss P, Dowsett M, Gazet JC, Brodie A. (1984) 4-hydroxyandostenedione in treatment of postmenopausal patients with advanced breast cancer. *Lancet* **i**:1237–1239.
53. Jordan VC, Brodie AM. (2007) Development and evolution of therapies targeted to the estrogen receptor for the treatment and prevention of breast cancer. *Steroids* **72**:7–25.
54. Goss P, Strasser K. (2001) Aromatase inhibitors in the treatment and prevention of breast cancer. *J Clin Oncol* **19**:881–894.
55. Howell A, Cuzick J, Baum M, *et al.* (2005) Results of the ATAC (Arimidex, Tamoxifen, Alone or in Combination) trial after completion of 5 years adjuvant treatment for breast cancer. *Lancet* **365**:60–62.
56. Thurlimann B, Keshaviah A, Coates AS, *et al.* (2005) A comparison of letrozole and tamoxifen in postmenopausal women with early breast cancer. *N Engl J Med* **353**:2747–27457.
57. Lien EA, Solheim E, Kvinnsland S, Ueland PM. (1988) Identification of 4-hydroxy-N-desmethyltamoxifen as a metabolite of tamoxifen in human bile. *Cancer Res* **48**:2304–2308.
58. Goetz MP, Rae JM, Suman VJ, *et al.* (2005) Pharmacogenetics of tamoxifen biotransformation is associated with clinical outcomes of efficacy and hot flashes. *J Clin Oncol* **23**:9312–9318.
59. Jin Y, Desta Z, Stearns V, Ward B, Ho H, Lee KH, Skaar T, Storniolo AM, Li L, Araba A, Blanchard R, Nguyen A, Ullmer L, Hayden J, Lemler S, Weinshilboum RM, Rae JM, Hayes DF, Flockhart DA. (2005) CYP2D6 genotype, antidepressant use, and tamoxifen metabolism during adjuvant breast cancer treatment. *J Natl Cancer Inst* **97**:30–39.
60. Sporn MB. (1976) Approaches to prevention of epithelial cancer during the preneoplastic period. *Cancer Res* **36**:2699–2702.
61. Jordan VC. (1974) Antitumour activity of the antioestrogen ICI 46,474 (tamoxifen) in the dimethylbenzanthracene (DMBA)-induced rat mammary carcinoma model. *J Steroid Biochem* **5**:354.
62. Jordan VC. (1976) Effect of tamoxifen (ICI 46,474) on initiation and growth of DMBA- induced rat mammary carcinoma. *Eur J Cancer* **12**:419–424.
63. Cuzick J, Baum M. (1985) Tamoxifen and contralateral breast cancer [letter]. *Lancet* **2**:282.

94. Onate SA, Tsai SY, Tsai MJ, O'Malley BW. (1995) Sequence and characterization of a coactivator for the steroid hormone receptor superfamily. *Science* **270**:1354–1357.
95. Smith CL, O'Malley BW. (2004) Coregulator function: a key to understanding tissue specificity of selective receptor modulators. *Endocr Rev* **25**:45–71.
96. Shiau AK, Barstad D, Loria PM, Cheng L, Kushner PJ, Agard DA, Greene GL. (1998) The structural basis of estrogen receptor/co-activator recognition and the antagonism of this interaction by tamoxifen. *Cell* **95**:927–937.
97. Brzozowski AM, Pike AC, Dauter Z, *et al.* (1997) Molecular basis of agonism and antagonism in the oestrogen receptor. *Nature* **389**:753–758.
98. Wijayaratne AL, Nagel SC, Paige LA, *et al.* (1999) Comparative analyses of mechanistic differences among antiestrogens. *Endocrinology* **140**:5828–5840.
99. O'Malley BW. (2006) Little molecules with big goals. *Science* **313**:1749–1750.
100. Gottardis MM, Jordan VC. (1988) Development of tamoxifen-stimulated growth of MCF-7 tumors in athymic mice after long-term antiestrogen administration. *Cancer Res* **48**:5183–8187.
101. Yao K, Lee ES, Bentrem DJ, *et al.* (2000) Antitumor action of physiological estradiol on tamoxifen-stimulated breast tumors grown in athymic mice. *Clin Cancer Res* **6**:2028–2036.
102. Jordan VC. (2008) The 38th David A. Karnofsky Lecture. The paradoxical actions of estrogen in breast cancer: survival or death? *J Clin Oncol* **26**:3073–3082.
103. Huggins C, Dao TL. (1953) Adrenalectomy and oophorectomy in treatment of advanced carcinoma of the breast. *J Am Med Assoc* **151**:1388–1394.
104. Thompson IM, Goodman PJ, Tangen CM, *et al.* (2003) The influence of finasteride on the development of prostate cancer. *N Engl J Med* **349**:215–224.
105. Kolvenbag GJCM, Furr BJA. (2002) Nonsteroidal antiandrogens. In: *Hormone Therapy in Breast and Prostate Cancer* eds. Jordan VC and Furr BJA. Totowa, NJ: Humana Press, pp. 347–368.
106. Scher HI, Sawyers CL. (2005) Biology of progressive, castration-resistant prostate cancer: directed therapies targeting the androgen-receptor signaling axis. *J Clin Oncol* **23**:8253–8261.
107. Umekita Y, Hiipakka RA, Kokontis JM, Liao S. (1996) Human prostate tumor growth in athymic mice: inhibition by androgens and stimulation by finasteride. *Proc Natl Acad Sci USA* **93**:11802–11807.
108. Locke JA, Guns ES, Lubik AA, *et al.* (2008) Androgen levels increase by intratumoral *de novo* steroidogenesis during progression of castration-resistant prostate cancer. *Cancer Res* **68**:6407–6415.
109. Attard G, Reid AH, Yap TA, *et al.* (2008) Phase I clinical trial of a selective inhibitor of CYP17, abiraterone acetate, confirms that castration-resistant prostate cancer commonly remains hormone-driven. *J Clin Oncol* **28**: http://jco/ascopubs.org/cgi/doi/10.1200/jco.2007.15.9749.
110. Strebhardt K, Ullrich A. (2008) Paul Ehrlich's magic bullet concept: 100 years of progress. *Nat Rev Cancer* **8**:473–480.

111. Jordan VC, O'Malley BW. (2007) Selective estrogen-receptor modulators and antihormonal resistance in breast cancer. *J Clin Oncol* **25**:5815–5824.
112. Jordan VC, Collins MM, Rowsby L, Prestwich G. (1977) A monohydroxylated metabolite of tamoxifen with potent antioestrogenic activity. *J Endocrinol* **75**:305–316.

2

DIETHYLSTILBESTROL: A TRAGEDY IN REPRODUCTIVE ENDOCRINOLOGY BUT A PIONEERING CANCER TREATMENT

Joan Lewis-Wambi and V. Craig Jordan

Abstract

Diethylstilbestrol (DES) is a synthetic nonsteroidal estrogen that was used by millions of pregnant women in the United States to prevent miscarriages and other pregnancy-related complications between 1938 and 1971. In 1971, the U.S. Food and Drug Administration issued a warning against the use of DES during pregnancy after a relationship was found between exposure to DES during the first trimester of pregnancy and the development of clear cell adenocarcinoma of the vagina and cervix in young women whose mothers had taken DES during pregnancy. Additional side effects associated with DES included an increased risk for breast cancer in women who took DES during pregnancy and structural reproductive tract anomalies, increased infertility, and poor pregnancy outcomes for women who were exposed to DES *in utero*. Compilation of the various sites of DES toxicity in humans and experimental animals indicates that lesions appear predominantly in estrogen-responsive target tissues suggesting that the presence of the estrogen receptor (ER) in such target tissues may help to govern the toxic effects of DES. Interestingly, it should be noted that DES, despite its adverse effects in reproductive endocrinology, was the first successful "chemical therapy" for cancer. Specifically, DES was available for the treatment of advanced prostate and breast cancer from the 1940s up until the 1990s. This review discusses the long-term health implications of DES for mothers, DES daughters and DES sons, and the possible side effects of DES on the third generation. We also propose possible mechanism(s) of actions of DES as a carcinogen and its paradoxical action as the first effective hormonal therapy for prostate and breast cancer.

Keywords: Breast Cancer; Cancer Therapy; Carcinogenesis; Diethylstilbestrol; Estradiol; Estrogen Receptor; Prostate Cancer; Clear Cell Adenocarcinoma.

INTRODUCTION

It is an unfortunate fact that many medications and treatments that are first thought to be safe and effective are later found to be harmful. Although instances of this are rare, they do happen. The story of the synthetic estrogen diethylstilbestrol (DES) is one of those unfortunate stories. From 1945 to 1971, doctors prescribed DES to an estimated five million pregnant women in the United States. DES is now recognized as a potent endocrine disruptor capable of distorting fetal development, but doctors originally encouraged pregnant women to take it to prevent miscarriages. In the early-1970s, doctors began to notice reproductive abnormalities in the daughters and sons of DES mothers. Specifically, DES daughters developed clear cell adenocarcinoma of the vagina and cervix (CCAC) before the age of 30 and reproductive tract abnormalities such as malformed reproductive organs and reduced fertility. DES sons suffered from small and undescended testicles, hypospadias, and abnormal semen.[1] DES is no longer used clinically to prevent miscarriage, but a major concern remains that, as DES-exposed women age and reach the time at which the incidence of reproductive organ cancers normally increases, they will show a much higher incidence of cancer than unexposed individuals. Furthermore, the possibility of second- and third-generation effects of DES has to be considered. This review will discuss the history of DES in terms of its adverse effects on human health, its mechanism of action, and its pioneering role as the first effective chemical therapy against advanced prostate and breast cancer.

THE HISTORY OF DES

DES is a nonsteroidal synthetic estrogen that was discovered by Sir Charles Dodds and colleagues[2-4] during the late 1930s. When it was first released on the market in 1941, it was approved by the Food and Drug Administration (FDA) for four indications: gonorrheal vaginitis, atrophic vaginitis, menopausal symptoms, and postpartum lactation suppression to prevent breast engorgement.[1] The gonorrheal vaginitis indication was dropped when the antibiotic penicillin became available. From the 1940s until the late 1980s, DES was FDA-approved as estrogen replacement therapy for estrogen deficiency states such as ovarian dysgenesis, premature ovarian failure, and postoophorectomy. In 1947, the FDA approved DES for the prevention of miscarriages and pregnancy complications.[1,5,6] Its usage as an anti-abortive agent was based on the assumption that the hormonal balance in problematic pregnancies is disturbed. Specifically, studies showed that late pregnancy toxemia, premature delivery, and miscarriages were preceded by low urinary estrogen and progesterone levels. Since the biological properties of DES are quite similar to those of estrogens, it was thought that the administration of this compound would restore the hormonal balance by compensation for low levels of natural steroid hormones. Hence, miscarriages would be prevented, or at least the severity of the complications would be reduced.[6] Unfortunately, the results of the first randomized controlled clinical trials on the effectiveness of DES in preventing miscarriage and premature birth, published in 1953, showed no protective effect of DES.[7] This outcome was supported by the result of another study, in which increasing doses of DES were shown to have no therapeutic advantage on pre-eclampsia, prematurity, fetal weight and survival, or the size of the placenta.[8] Despite these findings, however, DES usage in pregnant women continued in the United States through 1971, when the first negative effects of DES were reported. A case control study showed a highly significant association between *in utero* exposure to DES and subsequent development of vaginal clear cell adenocarcinoma (CCAC) in females in their late teens or early 20s.[9-11] Before the use of DES, this cancer was rarely seen in women younger than age 50. Some additional adverse effects associated with *in utero* DES exposure in women include infertility,[12] tissue and structural anomalies of the reproductive tract[13] (Table 1), pregnancy loss, premature delivery,[14,15] a possible increased risk of breast cancer in women over 40,[16] and squamous neoplasia of the cervix.[12,16] Although

Table 1. Structural abnormalities in women with *in utero* DES exposure.

Cervix	Vagina
Hypoplastic cervix	Clear cell adenocarcinoma
Cockscomb cervix	Adenosis
Cervical collar	Uterus
Pseudopolyp	T-shaped uterus

no increased risk of cancer has been observed in DES-exposed males, several teratogenic effects have been reported in the reproductive tract, including testicular hypoplasia, cryptorchidism, and epididymal cysts.[14,17,18]

It is estimated that 5–10 million individuals (including women who were prescribed DES while pregnant and the female and male children born of these pregnancies) were exposed to DES over the 30 years (1941–1971) that it was available in America. Numerous studies have confirmed that DES is a teratogen, an agent that can cause malformations of an embryo or fetus. Unfortunately, many healthcare providers are unaware of the long-term implications of DES exposure for their patients. Furthermore, females who may have been exposed to DES *in utero* are now of childbearing age and researchers are continuing to look at its potential for health problems on a third generation (i.e. grandchildren of women who took DES during pregnancy).

In 1971, physicians in the U.S. were advised by the FDA to stop prescribing DES to pregnant women because it was linked to the development of CCAC in women exposed prenatally to the drug. On November 10, 1971, the FDA ordered that prevention of miscarriage be removed from indications and that pregnancy be added to contraindications in the physician prescribing information for DES. In the 1990s, the only approved indications for DES were treatment of advanced prostate cancer and treatment of advanced breast cancer in postmenopausal women. In 1997, Eli Lilly stopped manufacturing DES, and the product is no longer available in the U.S.

humans, data on these species are not nearly as abundant as those on the mouse. Indeed, the mouse has been the most extensively studied species, and the size of the data set in mice is superior to those for all other species combined. The perinatal DES-treated mouse correlates remarkably well with the adverse effects observed in both male and female humans exposed *in utero* to DES.[39]

A variety of receptor-mediated and non-receptor-mediated mechanisms have been put forward to explain DES-induced toxicity. Indeed, the generation of mice lacking ERα (αERKO) or ERβ (βERKO) has helped to provide important insight into the contribution of each ER-dependent pathway in mediating the effects of neonatal DES exposure in the female and male reproductive tract tissues of the mouse. Specifically, studies[40] using the neonatal mouse model have shown that αERKO female mice exhibit complete resistance to the chronic effects of neonatal DES exposure compared to exposed wild-type females. These chronic effects of DES include epithelial squamous metaplasia in the uterus, proliferative lesions of the oviduct, and persistent cornification of the vaginal epithelium. Similar experiments were performed in males to determine whether DES-mediated effects in the male reproductive tract (i.e. seminal vesicles and prostate) were ER-mediated. It was found that DES treatment of wild-type males resulted in a significant decrease in seminal vesicle size at all ages, whereas no differences in the seminal vesicle sizes of DES-treated αERKO males were observed.[41] With respect to the prostate gland, histological analyses revealed that neonatal DES exposure in WT and βERKO mice resulted in changes in the aged ventral and dorsolateral prostates, including increased stromal mass, epithelial hyperplasia and dysplasia, and interstitial lymphocyte infiltration, all of which are characteristic of the phenotype of an estrogenized prostate gland.[42] None of these effects were observed in the prostates of αERKO males treated with DES, thus indicating that ERα is required for DES-mediated effects in the prostate. Collectively, these data suggest that DES acts through an ERα-mediated mechanism in the female and the male reproductive tract. Indeed, these findings have been supported by additional studies[40,41,43] performed in transgenic mouse models that overexpress ERα (MT-mER) or express variant forms of ERα. Results from these studies show that mice overexpressing ERα are at a higher risk of developing abnormalities including uterine carcinoma in response to neonatal DES compared with DES-treated wild-type mice. Furthermore, these lesions occurred at an earlier age compared to wild-type DES mice.

There is evidence that developmental exposure to DES interferes with normal differentiation of the Müllerian duct and regression of the Wolffian duct.[44] In particular, the ERKO mouse models have been used to determine whether DES acts through an ER-mediated mechanism to disrupt the expression of genes that are required for normal differentiation and organization of the reproductive tract.[40] The *Hox* and *Wnt* families of genes represent potential targets for DES-mediated effects, because their expression is critical for the development and organization of the female reproductive tract and the reproductive tract phenotypes of null mouse models of the *Hox* and *Wnt* gene families are similar to those found in DES-exposed mice. Furthermore, neonatal DES exposure reduces the expression of *Hoxa9*, *Hoxa10*, *Hoxa11*, and *Wnt7a* in the murine female reproductive tract.[45,46] Based on these observations, experiments were performed on neonatal αERKO female mice to determine the role of ERα in mediating DES-induced down-regulation of *Hox* and *Wnt* gene expression. The results of this study[40] revealed that, in contrast to wild-type mice, DES treatment had no effect on *Hoxa10*, *Hoxa11*, and *Wnt7a* expression in the αERKO mice, thus indicating that the DES-mediated reduction in the expression of these genes is regulated through ERα. Initially, it was thought that one potential mechanism by which DES decreases *Hox* and *Wnt* gene expression might involve epigenetic modifications such as DNA methylation; however, DES treatment does not affect the methylation state of the *Hoxa10* and *Hoxa11* proximal promoters in the mouse uterus.[47] But neonatal exposure to DES does appear to cause demethylation of the estrogen-responsive gene lactoferrin (LF) in the mouse uterus.[47] LF is a protein that is regulated by estrogen in the female mouse reproductive tract. In the uterus of the adult mouse, LF transcripts are stimulated approximately 300-fold by estradiol (E2) or DES.[48] The expression of LF is induced by DES in the uterus prenatally, neonatally, and in adult stages, making it a particularly attractive biomarker of DES action in the mouse.

While the estrogenic activity of DES is important, there is evidence that oxidative metabolism also plays a role in its toxicity. DES is oxidatively metabolized to catecholestrogens including diethylstilbestrol-4′,4″-quinone (DQ), an unstable and reactive intermediate compound[49] that binds to the ER with a relative binding affinity (RBA) of 3.6 compared to E2 and DES, which have RBAs of 100 and 286, respectively.[50] DQ binds irreversibly and specifically to ER, suggesting that it may be biologically active despite its rapid metabolism and lower binding affinity compared to the parent compound, DES. It is believed that DQ and other catecholestrogens, unless they are detoxified, can undergo a process called redox cycling, which leads to the formation of reactive oxygen species that can lead to DNA damage and lipid peroxidation, as well as to the formation of reactive intermediates that can directly adduct to DNA, potentially resulting in mutagenesis. There is evidence that this happens in tissue culture and in animal models of estrogen (E2)-induced cancer, as well as a limited amount of evidence from studies with human breast tissues.[51] Studies using the DES mouse model showed that catecholestrogens were very effective at inducing uterine tumors, with 4-hydroxyestradiol (4-OHE2) being the most carcinogenic, with a nine fold higher tumor incidence compared with the parent hormone, E2.[52] For prostate cancer, there is evidence from studies with the NBL rat model that estrogens can have genotoxic activity in the rat prostate and that this might be related to induction of cancer by E2 plus testosterone.[52]

ESTROGEN (DES) THERAPY AND PROSTATE CANCER

Until the introduction of luteinizing hormone-releasing hormone (LHRH) agonist therapy in the 1980s, estrogen therapy in the form of DES was used effectively to treat advanced prostate cancer (PCa). Charles Huggins pioneered the use of DES in the treatment of advanced PCa in the early 1940s.[53] The action of DES was thought to be mediated via a blockade of the pituitary-testicular axis, which effectively lowered circulating levels of androgen and caused tumor regression. However, recent investigations have demonstrated that DES exerts direct growth-inhibitory effects on prostatic cancer cells via induction of mitotic arrest or apoptosis through a mechanism not mediated by ERs.[54] In particular, it was found that the intensity of the growth-inhibitory effect of DES was dependent on the androgen sensitivity of the prostate cancer cells, with androgen-insensitive prostate cancer cells being more susceptible to DES-induced apoptosis than androgen-sensitive prostate cancer cells. Despite its benefits, however, there were serious adverse effects associated with DES (i.e. feminization, exacerbation of heart failure, vascular complications, gynecomastia, and impotence). Hence, it slowly lost its attractiveness as a mainstay treatment for advanced PCa. The clinical use of tamoxifen (TAM), a nonsteroidal estrogen mixed agonist/antagonist, was introduced in the 1980s as an alternative to DES in the treatment of PCa; however, it produced low response rates compared to DES despite being better-tolerated.[55] With the discovery of ERβ as a new estrogen signaling pathway and the availability of the pure antiestrogen fulvestrant (ICI 182,780) (Fig. 1),[56] there has been renewed interest in determining whether pure antiestrogens could be considered in the treatment of prostate cancer. Indeed, a previous study by Lau and coworkers[57] showed that ICI 182,780 was more effective than 4-hydroxytamoxifen at inhibiting the growth of ERβ-expressing DU145 prostate cancer cells and that the antiproliferative effects of ICI 182,780 could be reversed by cotreatment of these cells with an ERβ antisense oligonucleotide, which validates, to some extent, the potential benefits of pure antiestrogens in the treatment of advanced PCa.

It should be noted that low-dose DES treatment is still used today as first-line hormonal therapy, albeit not in the United States, and can provide benefit as second-line hormonal therapy.[58]

ESTROGEN (DES) THERAPY AND BREAST CANCER

The link between estrogen and the development and proliferation of breast cancer is well documented, but paradoxically estrogen is also capable of causing tumor regression. In 1944, Sir Alexander Haddow[59] published the results of his clinical trial with the synthetic estrogens triphenylchlorethylene, triphenylmethylethylene, and DES administered at high doses. He found that 10 out of 22 postmenopausal patients with advanced mammary carcinomas, who were treated with triphenylchlorethylene, had significant regression of tumor growth. Five patients out of 14 who were treated with high-dose DES produced similar responses. Interestingly, the duration of the postmenopausal period was found to be a critical factor affecting the success of this therapy. For example, when DES was administered at 150 mg per day, women who had experienced the onset of menopause less than one year prior to therapy did not respond to DES women who had experienced the onset of menopause within five years of menopause experienced a 7.9% objective response rate and women who reached menopause more than 10 years earlier experienced a 22% response rate.[59] By the 1960s, based on data from clinical trials, it was recommended that postmenopausal patients with advanced breast cancer be treated with high-dose DES. There were, however, significant systemic side effects associated with high-dose DES therapy.[59]

In 1971, Cole and coworkers[60] reported the first clinical trial of the antiestrogen tamoxifen in women with late or recurrent breast cancer, and compared their findings with historical data from women receiving DES. They concluded that the level of response was similar for DES and tamoxifen; however, tamoxifen had a lower incidence of side effects. Ingle and coworkers[61] compared tamoxifen with DES directly and noted that response rates were similar but tamoxifen had fewer side effects. Based on these data, the use of high-dose estrogen (DES) for treatment of advanced breast cancer fell out of favor, and tamoxifen became the standard first-line endocrine therapy. The Ingle study[61] which compared DES-treated and tamoxifen-treated patients was followed up by Peethambaram and coworkers,[62] and surprisingly showed a survival advantage for DES-treated patients. Another small trial conducted by Lønning *et al.*[63] on postmenopausal patients with advanced breast cancer exposed to multiple endocrine therapies revealed a 31% objective response rate with DES (15 mg daily) therapy. More recently, Ellis *et al.*[64] reported that a daily dose of 6 mg E2 could stop the growth of tumors or even cause them to shrink in about 25% of women with metastatic breast cancer that had developed resistance to standard antihormonal therapy. These clinical observations clearly show that estrogen can induce tumor regression after several years of antihormonal therapy, and they suggest that the adaptation of cancer cells to low levels of estrogen might sensitize them to the apoptotic effect of estrogen.

In addition to the clinical data, the ability of estrogen to inhibit tumor growth and induce apoptosis has been demonstrated previously in preclinical models.[65-72] Specifically, our laboratory has demonstrated that when ER-positive breast cancer cells are grown and maintained in an estrogen-deprived environment for prolonged periods, due to antihormonal therapy (i.e. tamoxifen, raloxifene, or aromatase inhibitors), they can ultimately develop enhanced responsiveness to greatly diminished levels of estrogen. In particular, preclinical animal model studies show that, initially, ER expressing tumors are stimulated by estrogen and respond appropriately to tamoxifen or raloxifene with tumor regression; however, with continued exposure to either tamoxifen or raloxifene, the tumors become resistant and regrow. The subsequent treatment of these resistant tumors with postmenopausal levels of E2 inhibits tumor growth as well as causes regression of established tumors.[65,66,68,69] This phenomenon in which E2 switches from being a potent agonist to a killing agent has also been successfully demonstrated *in vitro* and *in vivo* using long-term estrogen-deprived (LTED) breast cancer cells that are resistant to estrogen withdrawal (i.e. aromatase-inhibitor-resistant),[67,70,71] thus establishing a universal principle. Interestingly, similar experiments performed with DES and ethinylestradiol (EE2) have shown that these two compounds, which are potent agonists in hormone-dependent MCF-7 breast cancer cells, are also capable of inducing apoptosis in LTED resistant breast cancer cells with a potency that is either equivalent to (in the case of DES) or greater than (in the case of EE2) that of E2 (Fig. 2). It should be noted that EE2 is an orally bioactive synthetic steroidal estrogen that is currently used in almost all modern formulations of combined oral contraceptive pills (see Chapter 3).

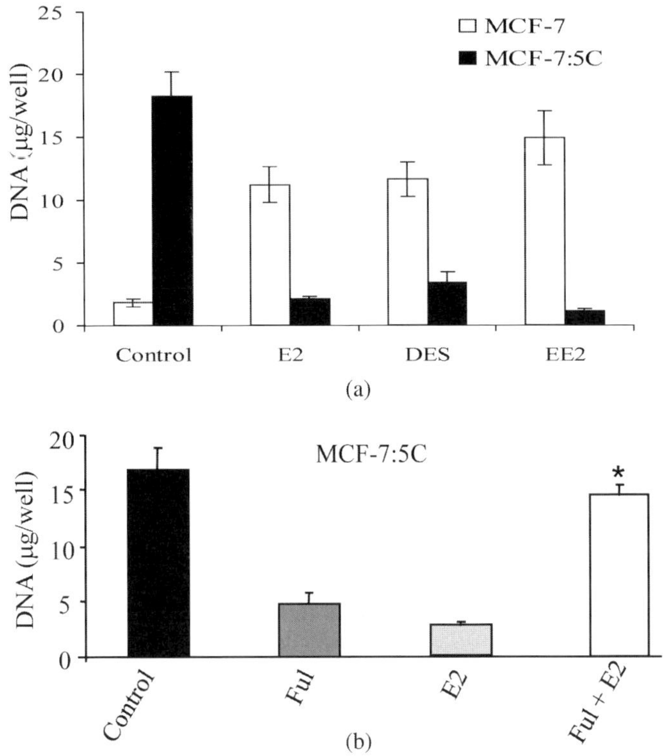

Figure 2. Effects of natural and synthetic estrogens on the growth of wild-type MCF-7 breast cancer cells and long-term estrogen-deprived (LTED) antihormone-resistant MCF-7:5C cells. (**a**) MCF-7 cells were grown in phenol-red-free RPMI media containing 10% charcoal stripped fetal bovine serum (i.e. SFS) for three days prior to the start of the growth assay. On the day of the experiment, MCF-7 and MCF-7:5C cells were seeded in phenol-red-free RPMI media containing 10% SFS 24-well plates (30,000/well) and after 24 hr were treated with <0.1% ethanol vehicle (control), 10^{-9} M E2, 10^{-9} M DES, or 10^{-9} M EE2 for seven days with retreatment on days 3 and 5. At the conclusion of the experiment, cells were harvested and total DNA (µg/well) was determined using a fluorescent DNA quantitation kit (Bio-Rad). (**b**) The effect of the pure antiestrogen fulvestrant on the growth of MCF-7:5C breast cancer cells in the absence or presence of estradiol. Cells were seeded in phenol-red-free media containing 10% SFS and treated with 10^{-9} M E2, 10^{-8} M fulvestrant (Ful), or 10^{-9} M E2 + 10^{-8} M Ful for seven days with retreatment on days 3 and 5. At the conclusion of the experiment, cells were harvested and total DNA (µg/well) was determined using a fluorescent DNA quantitation kit (Bio-Rad). The data represent the mean of three independent experiments; bars, ±SE. * $P < 0.001$ compared to E2-treated or Ful-treated cells.

MECHANISM OF ESTROGEN-INDUCED APOPTOSIS IN BREAST CANCER

Sir Alexander Haddow, during the presentation of the inaugural Karnofsky lecture,[73] illustrated his talk with reference to the value of DES as the first "chemical therapy" for any cancer. He did, however, make the following statement: "In spite of the extremely limited practicability of such a measure [high dose estrogen], the extraordinary extent of tumor regression observed in perhaps 1% of postmenopausal cases has always been regarded as of major theoretical importance, and it is a matter for some disappointment that so much of the underlying mechanisms continues to elude us."[73]

Since that time much progress has been made in describing the events that herald estradiol-induced apoptosis.[74] Regrettably, the precise molecular mechanism by which estrogen exerts its proapoptotic/antitumor effect in breast cancer remains unknown. Nevertheless, a growing body of evidence suggests the involvement of the extrinsic (death receptor) and intrinsic (mitochondrial) pathways in this process. Apoptosis is a form of programmed cell death that is executed by a family of proteases called caspases, which can be activated either by cell-surface death receptors (i.e. the extrinsic pathway) or by perturbation of the mitochondrial membrane (i.e. the intrinsic pathway)[75] (Fig. 3). Components of the extrinsic pathway include the death receptors FasR/FasL, DR4/DR5, and tumor necrosis factor (TNFR), whereas the intrinsic pathway centers on the mitochondria, which contain key apoptogenic factors such as cytochrome *c* and apoptosis-inducing factor (AIF)[76] (Fig. 3). In the intrinsic pathway, the integrity of mitochondrial membranes is controlled primarily by a balance between the antagonistic actions of the proapoptotic and anti-apoptotic members of the Bcl-2 family. The proteins belonging to the Bcl-2 family comprise three principal subfamilies: (1) anti-apoptotic members, including Bcl-2/Bcl-x_L, which possess the Bcl-2 homology (BH) domains BH1, BH2, BH3, and BH4; (2) proapoptotic members, such as Bax, Bak, and Bok, which have the BH1, BH2, and BH3 domains; and (3) proapoptotic members such as Bid, Bim, Bad, Bik, and Puma, which generally possess only the BH3 domain.[76] These proteins

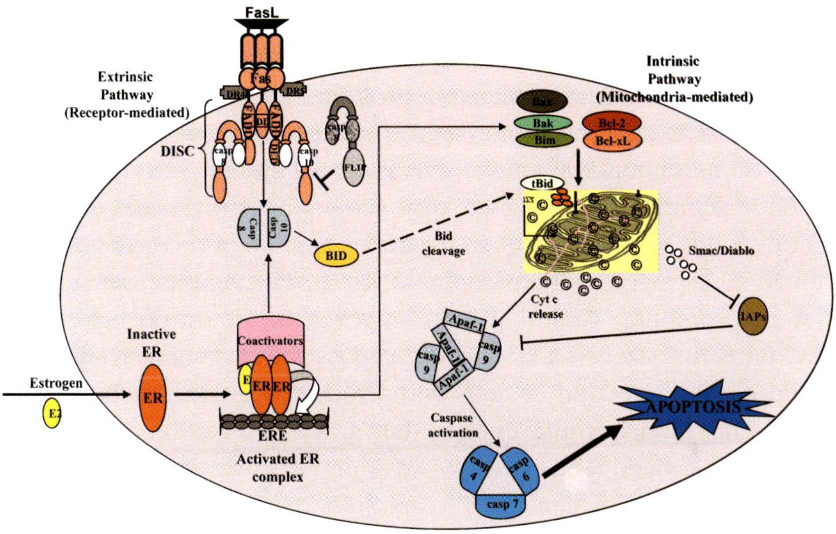

Figure 3. The two main pathways involved in estrogen-induced apoptosis in breast cancer cells that are resistant to long term estrogen deprivation. (**1**) The extrinsic pathway begins outside the cell through the activation of receptors on the cell surface by specific molecules known as proapoptotic ligands, including CD95L/FasL (receptor CD95/Fas). Once they are activated, the death domains of these receptors bind to the adapter protein FADD, resulting in the assembly of DISC, and recruitment and assembly of initiator caspases 8 and 10. The two caspases are stimulated and processed, releasing active enzyme molecules into the cytosol, where they activate caspases 3, 6, and 7, thereby converging on the intrinsic pathway. (**2**) The intrinsic (mitochondrial) pathway is initiated in response to cellular signals resulting from DNA damage, a defective cell cycle, detachment from the extracellular matrix, hypoxia, loss of cell survival factors, or other types of severe cell stress. This triggers activation of specific members of the proapoptotic Bcl-2 protein family involved in the promotion of apoptosis. In LTED resistant breast cancer cells, E2 acts through the ER to upregulate proapoptotic proteins Bax, Bim, Bak, along with p53, and downregulate Bcl-2 and Bcl-xl, which causes mitochondrial membrane disruption and cytochrome c release, followed by activation of caspases (i.e. caspases 7 and 9), and utimately apoptosis (cell death).

regulate apoptosis by altering mitochondrial membrane permeabilization and controlling the release of cytochrome *c*.

Based on various laboratory models that are currently available, it is clear that antihormone-resistant breast cancer tumors and cells have reconfigured the ER signal transduction pathway; so, despite the fact that the ER still regulates the appropriate estrogen-regulated genes (including pS_2 and *myc*)[77] there is a profound effect of estrogen to activate the Fas (death) receptor system[67,69] or to alternatively have a direct effect on mitochondrial function via the bcl-2 system[69,70] (Fig. 3). Thus, an understanding of the paradoxical actions of estrogen has emerged that depends on the state of estrogen deprivation of the breast cancer cell. In an estrogen rich environment, the E2-ER complex is a survival system promoting tumor growth, however, in an estrogen-deprived environment, estrogen action is replaced by internal survival signaling based on the selection of cells with enhanced growth factor receptors.[78] This model would explain the earlier observations as to why high-dose estrogen (DES) therapy was only effective as a treatment for breast cancer in women many years after the menopause.[59] Natural estrogen deprivation had occurred. In patients treated long-term with SERMs (tamoxifen and raloxifene) or aromatase inhibitors, however, the process is accelerated and enhanced so that only low doses of estrogen are necessary for causing experimental tumor regression.

CONCLUSION

The story of DES is indeed a medical tragedy. Initially thought to be safe, with the ability to prevent miscarriages and other pregnancy complications, DES was prescribed to an estimated 4–5 million pregnant women by their physicians between 1938 and 1971. Unfortunately, despite initial evidence disproving its effectiveness in preventing miscarriages and other pregnancy-related disorders, the prescription of DES continued for decades, resulting in major health consequences for the women who took DES while pregnant, their offspring who were exposed to DES during the first trimester of pregnancy, and possibly the third generation (i.e. the offspring of DES sons and daughters). The fact that DES was administered to pregnant women is particularly troubling, especially since DES is able to cross the placenta just like other estrogens,[79] and is therefore capable of interfering with the development of the fetus. Indeed, sufficient evidence has been accumulated through the years in experimental animals and humans to show that the developing fetus and neonate are uniquely sensitive to exogenous estrogen exposure. If exposure occurs during critical periods of differentiation, permanent adverse effects are well documented to occur. Some of these effects, such as reproductive tract abnormalities and uterine tumors, may not be observed until much later in life, long after exposure occurs. Hence, great precaution should be taken in administration of drugs during a pregnancy, especially because side effects in future generations might be revealed only years after exposure. In particular, drugs with biological properties similar to those of DES are a point of concern and caution, as is the case for fertility drugs that are used to help women become pregnant.

ACKNOWLEDGEMENTS

This work was supported by National Institutes of Health Career Development Grant K01CA120051-01A2 (J.L.W), American Cancer Society Grant IRG-92-027-14 (J.L.W), the Hollenbach Family Fund (J.L.W), and Department of Defense Grant W81XWH-06-1-0590 Center of Excellence [V.C.J.; views and opinions of and endorsements by the author(s) do not reflect those of the U.S. Army or the Department of Defense], the Weg and Genuardis Funds (V.C.J), and NIH Grants P30CA006927 (Fox Chase Cancer Center) and NIH P30 CA051008 (Lombardi Comprehensive Cancer Center).

In addition, this work (V.C.J) was supported by the Department of Defense Breast Program under Award No. W81XWH-06-1-0590 Center of Excellence; subcontract under the SU2C (AACR) Grant No. SU2C-AACR-DT0409; the Susan G. Komen for the Cure Foundation under Award No. SAC100009; GHUCCTS CTSA (Grant #UL1RR031975) and the Lombardi Comprehensive Cancer Center Support Grant (CCSG) Core Grant NIH P30 CA051008. The views and opinions of the author(s) do not reflect those of the U.S. Army or the Department of Defense.

REFERENCES

1. Giusti RM, Iwamoto K, Hatch EE. (1995) Diethylstilbestrol revisited: a review of the long-term health effects. *Ann Intern Med* **122**: 778–788.

2. Dodds EC, Goldberg L, Lawson W, Robinson R. (1938) Oestrogenic activity of certain synthetic compounds. *Nature* **141**:247–248.
3. Dodds EC, Lawson W, Noble RL. (1938) Biological effects of the synthetic oestrogenic substance 4:4'dihydroxy-a: b-diethylstilbene. *Lancet* **1**:1389–1391.
4. Dodds EC. (1949) Synthetic oestrogens. *J Pharm Pharmacol* **1**:137–147.
5. Smith OW, Smith GV. (1949) The influence of diethylstilbestrol on the progress and outcome of pregnancy as based on a comparison of treated with untreated primigravidas. *Am J Obstet Gynecol* **58**:994–1009.
6. Smith OW, Smith GV. (1949) Use of diethylstilbestrol to prevent fetal loss from complications of late pregnancy. *N Engl J Med* **241**:562–568.
7. Dieckmann WJ, Davis ME, Rynkiewicz LM, Pottinger RE. (1953) Does the administration of diethylstilbestrol during pregnancy have therapeutic value? *Am J Obstet Gynecol* **66**:1062–1081.
8. Ferguson JH. (1953) Effect of stilbestrol on pregnancy compared to the effect of a placebo. *Am J Obstet Gynecol* **65**:592–601.
9. Herbst AL, Ulfelder H, Poskanzer DC. (1971) Adenocarcinoma of the vagina. Association of maternal stilbestrol therapy with tumor appearance in young women. *N Engl J Med* **284**:878–881.
10. Herbst AL. (1972) Stilbestrol and vaginal cancer in young women. *CA Cancer J Clin* **22**:292–295.
11. Herbst AL, Kurman RJ, Scully RE, Poskanzer DC. (1972) Clear-cell adenocarcinoma of the genital tract in young females. Registry report. *N Engl J Med* **287**: 1259–1264.
12. Hatch EE, Herbst AL, Hoover RN, *et al.* (2001) Incidence of squamous neoplasia of the cervix and vagina in women exposed prenatally to diethylstilbestrol (United States). *Cancer Causes Control* **12**:837–845.
13. Herbst AL, Kurman RJ, Scully RE. (1972) Vaginal and cervical abnormalities after exposure to stilbestrol *in utero*. *Obstet Gynecol* **40**:287–298.
14. Bibbo M, Haenszel WM, Wied GL, Hubby M, Herbst AL. (1978) A twenty-five-year follow-up study of women exposed to diethylstilbestrol during pregnancy. *N Engl J Med* **298**:763–767.
15. Kaufman RH, Adam E, Hatch EE, *et al.* (2000) Continued follow-up of pregnancy outcomes in diethylstilbestrol-exposed offspring. *Obstet Gynecol* **96**:483–489.
16. Titus-Ernstoff L, Hatch EE, Hoover RN, *et al.* (2001) Long-term cancer risk in women given diethylstilbestrol (DES) during pregnancy. *Br J Cancer* **84**: 126–133.
17. Gill WB, Schumacher GF, Bibbo M, Straus FH, 2nd, Schoenberg HW. (1979) Association of diethylstilbestrol exposure *in utero* with cryptorchidism, testicular hypoplasia and semen abnormalities. *J Urol* **122**: 6–39.
18. Wilcox AJ, Baird DD, Weinberg CR, Hornsby PP, Herbst AL. (1995) Fertility in men exposed prenatally to diethylstilbestrol. *N Engl J Med* **332**:1411–1416.

19. Beral V, Colwell L. (1980) Randomised trial of high doses of stilboestrol and ethisterone in pregnancy: long-term follow-up of mothers. *Br Med J* **281**: 1098–1101.
20. Colton T, Greenberg ER, Noller K, *et al.* (1993) Breast cancer in mothers prescribed diethylstilbestrol in pregnancy. Further follow-up. *JAMA* **269**: 2096–2100.
21. Rossouw JE, Anderson GL, Prentice RL, *et al.* (2002) Risks and benefits of estrogen plus progestin in healthy postmenopausal women: principal results from the Women's Health Initiative randomized controlled trial. *JAMA* **288**:321–333.
22. Pharoah PD, Day NE, Duffy S, Easton DF, Ponder BA. (1997) Family history and the risk of breast cancer: a systematic review and meta-analysis. *Int J Cancer* **71**: 800–809.
23. Melnick S, Cole P, Anderson D, Herbst A. (1987) Rates and risks of diethylstilbestrol-related clear-cell adenocarcinoma of the vagina and cervix. An update. *N Engl J Med* **316**:514–516.
24. Henderson BE, Benton B, Cosgrove M, *et al.* (1976) Urogenital tract abnormalities in sons of women treated with diethylstilbestrol. *Pediatrics* **58**:505–507.
25. Docimo SG, Silver RI, Cromie W. (2000) The undescended testicle: diagnosis and management. *Am Fam Physician* **62**:2037–2044, 2047–2048.
26. Driscoll SG, Taylor SH. (1980) Effects of prenatal maternal estrogen on the male urogenital system. *Obstet Gynecol* **56**:537–542.
27. Newbold RR, Padilla-Banks E, Jefferson WN. (2006) Adverse effects of the model environmental estrogen diethylstilbestrol are transmitted to subsequent generations. *Endocrinology* **147**:S11–S17.
28. Newbold RR, Hanson RB, Jefferson WN, Bullock BC, Haseman J, McLachlan JA. (1998) Increased tumors but uncompromised fertility in the female descendants of mice exposed developmentally to diethylstilbestrol. *Carcinogenesis* **19**: 1655–1663.
29. Newbold RR, Hanson RB, Jefferson WN, Bullock BC, Haseman J, McLachlan JA. (2000) Proliferative lesions and reproductive tract tumors in male descendants of mice exposed developmentally to diethylstilbestrol. *Carcinogenesis* **21**: 1355–1363.
30. Kuiper GG, Carlsson B, Grandien K, *et al.* (1997) Comparison of the ligand binding specificity and transcript tissue distribution of estrogen receptors alpha and beta. *Endocrinology* **138**:863–870.
31. Cordera F, Jordan VC. (2006) Steroid receptors and their role in the biology and control of breast cancer growth. *Semin Oncol* **33**:631–641.
32. Nilsson S, Makela S, Treuter E, *et al.* (2001) Mechanisms of estrogen action. *Physiol Rev* **81**:1535–1565.
33. Kushner PJ, Agard DA, Greene GL, *et al.* (2000) Estrogen receptor pathways to AP-1. *J Steroid Biochem Mol Biol* **74**:311–317.

64. Ellis MJ, Gao F, Dehdashti F, *et al.* (2009) Lower-dose vs. high-dose oral estradiol therapy of hormone receptor-positive, aromatase inhibitor-resistant advanced breast cancer: a phase 2 randomized study. *JAMA* **302**:774–780.
65. Wolf DM, Jordan VC. (1993) A laboratory model to explain the survival advantage observed in patients taking adjuvant tamoxifen therapy. *Recent Results Cancer Res* **127**:23–33.
66. Yao K, Lee ES, Bentrem DJ, *et al.* (2000) Antitumor action of physiological estradiol on tamoxifen-stimulated breast tumors grown in athymic mice. *Clin Cancer Res* **6**:2028–2036.
67. Song RX, Mor G, Naftolin F, *et al.* (2001) Effect of long-term estrogen deprivation on apoptotic responses of breast cancer cells to 17beta-estradiol. *J Natl Cancer Inst* **93**:1714–1723.
68. Liu H, Lee ES, Gajdos C, *et al.* (2003) Apoptotic action of 17beta-estradiol in raloxifene-resistant MCF-7 cells *in vitro* and *in vivo*. *J Natl Cancer Inst* **95**:1586–1597.
69. Osipo C, Gajdos C, Liu H, Chen B, Jordan VC. (2003) Paradoxical action of fulvestrant in estradiol-induced regression of tamoxifen-stimulated breast cancer. *J Natl Cancer Inst* **95**:1597–1608.
70. Lewis JS, Meeke K, Osipo C, *et al.* (2005) Intrinsic mechanism of estradiol-induced apoptosis in breast cancer cells resistant to estrogen deprivation. *J Natl Cancer Inst* **97**:1746–1759.
71. Lewis JS, Osipo C, Meeke K, Jordan VC. (2005) Estrogen-induced apoptosis in a breast cancer model resistant to long-term estrogen withdrawal. *J Steroid Biochem Mol Biol* **94**:131–141.
72. Balaburski GM, Dardes RC, Johnson M, *et al.* (2010) Raloxifene-stimulated experimental breast cancer with the paradoxical actions of estrogen to promote or prevent tumor growth: a unifying concept in anti-hormone resistance. *Int J Oncol* **37**:387–398.
73. Haddow A. (1970) David A. Karnofsky memorial lecture. Thoughts on chemical therapy. *Cancer* **26**:737–754.
74. Jordan VC. (2008) The 38th David A. Karnofsky lecture: the paradoxical actions of estrogen in breast cancer — survival or death? *J Clin Oncol* **26**:3073–3082.
75. Ziegler DS, Kung AL. (2008) Therapeutic targeting of apoptosis pathways in cancer. *Curr Opin Oncol* **20**:97–103.
76. Peter ME, Krammer PH. (2003) The CD95(APO-1/Fas) DISC and beyond. *Cell Death Differ* **10**:26–35.
77. Osipo C, Meeke K, Cheng D, *et al.* (2007) Role for HER2/neu and HER3 in fulvestrant-resistant breast cancer. *Int J Oncol* **30**:509–520.
78. Ali S, Coombes RC. (2002) Endocrine-responsive breast cancer and strategies for combating resistance. *Nat Rev Cancer* **2**:101–112.
79. Shah HC, McLachlan JA. (1976) The fate of diethylstilbestrol in the pregnant mouse. *J Pharmacol Exp Ther* **197**:687–696.

3

ORAL CONTRACEPTIVES AND WOMEN'S HEALTH

V. Daniel Castracane and V. Craig Jordan

"*Birth control, Ho Chi Minh, Richard Nixon back again. . . .*"
We Didn't Start the Fire — Billy Joel

"No egg, no baby" — Gregory Pincus

Abstract

The initiation of studies starting the train of development of the hormonal contraceptives occurred over 100 years ago with animal studies which demonstrated that ovarian transplants and ovarian extracts were able to inhibit ovulation. Subsequent studies defined the endocrinology of the ovarian cycle and factors involved in ovulation. This was followed by a dramatic era of pharmaceutical chemistry during which the orally active estrogens and progestins were discovered. Early studies demonstrated that a combination of the steroids could produce an effective inhibition of ovulation and would serve as an acceptable contraceptive for women. Over the years dramatic changes have been made to these products such that the levels of both the estrogen and progestin components have been greatly lowered and a variety of new steroidal compounds (progestins) have been produced along the way to reduce adverse effects, probably imparting some pharmaceutical patent protection for the manufacturers. Most recently, the utilization of these same products formulated for nonoral administration has turned out to be a fine-tuning of the use of estrogen-progestin contraception for better utility and compliance. Hormonal contraceptives have been an important sociological discovery having a great impact on the sexual liberation of women, allowing reasonable family planning and initiating controversies along the way in politics, religion and clinical medicine. They have been the most thoroughly studied pharmaceutical agents in history and have brought tremendous benefits to mankind in their effective and, by all standards, relatively safe usage in many different clinical situations. They are currently used by millions of women around the world as the major form of contraception. Perhaps the most surprising aspect of such pharmaceutical agents to inhibit ovulation is that we now recognize a broad spectrum of noncontraceptive benefits which have extended the utility of these agents into many areas of women's health. This chapter briefly discusses the development, utility and relationship of estrogen-progestin contraception to women's health.

Keywords: Oral Contraceptive; Worcester Foundation; Progestin.

INTRODUCTION AND HISTORY

Early animal studies at the turn of the 20th century demonstrated that ovarian heterotransplants would result in cessation of ovarian cyclicity and ovulation. Ovarian extracts were found to have the same effect. The observation that pregnancy was associated with a cessation of ovulatory function was also noted. Professor Ludwig Haberlandt, in the 1920s, was the first to suggest that ovarian and placental hormones could be used for fertility regulation.[1] Many of these studies took place before we knew what estradiol or progesterone was and, more importantly, which ovarian tissue was responsible for their production. These seminal studies have been described in more detail in several reviews.[2-6]

From the 1920s through the 1940s, the identification of steroids was an active research area and the structure of estradiol and of progesterone were discovered, although much more had to be learned in order to understand their role in reproductive endocrinology. The synthesis of steroids in the laboratory required long and laborious techniques to produce small quantities of steroids and cost was astronomical. In the 1940s, after devising new methods for synthesis of steroids from a diosgenin precursor, Russell Marker went to Mexico to find adequate supplies of this precursor in the Mexican yam and started a small company, later to become Syntex. He was able to synthesize progesterone — at the time a very expensive steroid — in four steps and he presented two large vials of progesterone to Searle for a fraction of the going price. Syntex eventually became the major contributor of steroid precursors to large pharmaceutical companies that produced the synthetic hormones in the development and ultimately production of the estrogen-progestin birth control pills. The fascinating tale of Russell Marker and the synthesis of steroid hormones has been told many times.[5,7]

In the late 1930s ethinyl estradiol and mestranol were discovered as orally active estrogens (Fig. 1). An ethinyl group at the 17 position of the steroid molecule rendered these compounds orally active. This side group is also present in all of the early progestational agents to impart oral activity. Ethinyl estradiol is simply estradiol-17B, the most important natural estrogen, with an ethinyl side group at the 17a

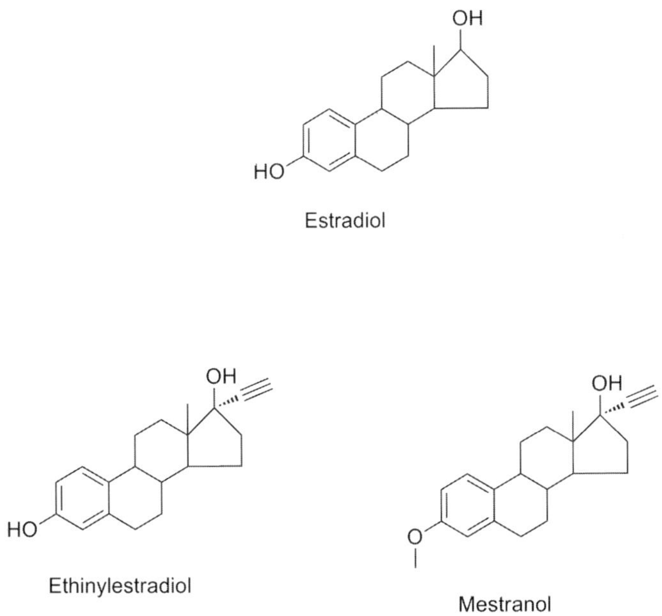

Figure 1. Steroid structure of estradiol, the most potent naturally occurring ovarian estrogenic steroid, and the two synthetic orally active estrogens, ethinyl estradiol and mestranol, which are the only two synthetic estrogens that have been used in combination oral contraceptives over the last 50 years.

position. Mestranol is ethinyl estradiol methylated at the 3 phenolic position. Mestranol, after a first pass through the liver, has the methyl ether group removed and is converted to ethinyl estradiol. It is not an active estrogen until the methyl group is removed and, consequently, ethinyl estradiol is the active estrogen in all birth control pills. The pharmacokinetics of both compounds are similar but not identical. At the present time, mestranol is rarely used as the estrogen component, but ethinyl estradiol has continued to be the major estrogen used in the estrogen–progestin combined pills for over 50 years.[8]

Another major impetus to the development of a pharmaceutical product to control ovulation was the development of orally active progestins. The steroid synthesis group at Syntex, under the direction of Carl Djerassi, was awarded the first patent for an orally active progestin, norethindrone, in 1951 (Fig. 2). Two years later the second

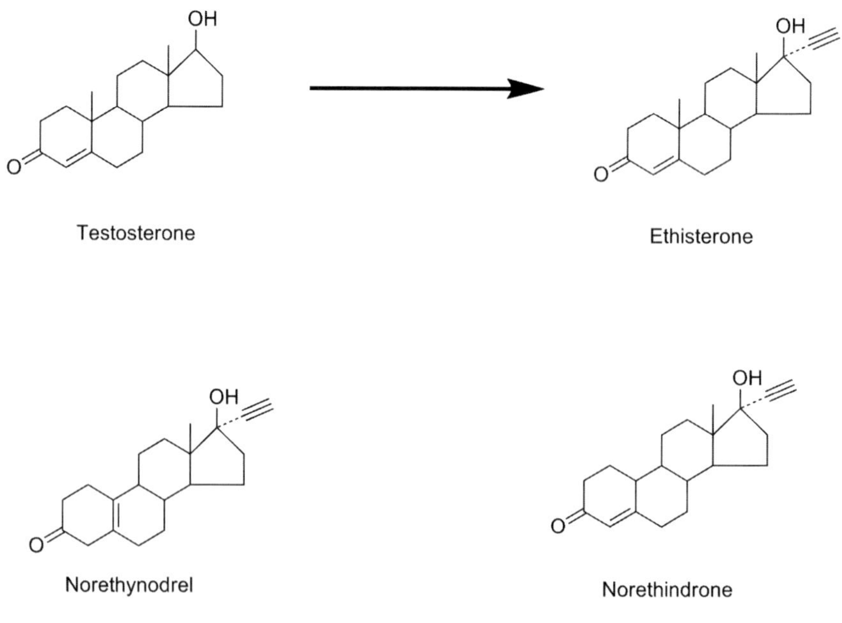

Figure 2. The simple conversion of testosterone to ethisterone, which serves as the precursor for the synthesis of the first two synthetic progestins, norethindrone and norethynodrel.

Figure 3. Gregory (Goode) Pincus, Ph.D.

patent for a synthetic progestin, developed under the direction of Fred Colson, was awarded to the Searle Pharmaceutical Company for norethynodrel.

Gregory (Goode) Pincus (Fig. 3), working for many years at the Worcester Foundation for Experimental Biology in Shrewsbury, MA (Fig. 4), starting with animal studies with his colleague M.-C. Chang, had demonstrated that progesterone administration could block ovulation in several animal species. He continued to work on this approach as a contraceptive means for humans. He received support and encouragement from Margaret Sanger, who started the Planned Parenthood Foundation, and generous financial support for several decades from Katherine McCormack. The availability of orally active progestational agents now became *an important discovery to allow contraceptive research to proceed.* Clinical trials had used these

Figure 4. The Worcester Foundation for Experimental Biology (WFEB), in Shrewsbury, MA.

orally active progestational agents and they were clearly effective in limited trials in suppressing ovulation.

Dr. John Rock, a physician at Harvard University, conducted a small clinical trial with his own patients and then initiated major clinical trials in Puerto Rico and Haiti with Dr. Edris Rice-Wray from Puerto Rico. Dr. Rock was a Catholic physician who had to fight battles with the Catholic Church as well as the laws of the state of Massachusetts, where the Comstock Law prohibited the use or development of contraceptive methodologies. Without his dedication to the development of these preparations, the trio of Pincus, Chang and Rock would not have been successful in bringing the first birth control pill (Enovid) to market in 1960 as a clinically useful contraceptive modality. The decades of work by Dr. Pincus and his colleagues has been reviewed in numerous publications.[2,4–6,9]

Dr. Pincus's work was supported to a large extent by the Searle Company and it is not surprising that the first combination pill contained norethynodrel. To this day Dr. Djerassi constantly questions why "his steroid" was not the one selected by Gregory Pincus for the first combination pill, Enovid.[10] Following clinical trials in Puerto Rico and Haiti, Enovid was released in 1960, and it contained 150 μg of mestranol and 9.85 mg of norethynodrel.

While the discovery of the natural steroids and the synthesis of new steroid analogs, leading to the development of the combination estrogen–progestin birth control pill, were going on over several decades, it was thought that natural steroids, such as estradiol and progesterone, were not orally active. Only years later was it demonstrated that micronized estradiol and progesterone could be orally active. The lesser potency of estradiol and progesterone than the new synthetic analogs would probably not have been effective in oral contraceptive preparations. Surprisingly, the FDA has just approved the use of estradiol valerate, which allows the constant release of estradiol, in a new estrogen–progestin contraceptive.

In early studies it was determined that initial progestin products had a small contamination with mestranol, about 1%, which was a significant amount of estrogen (50–500 μg). When the progestin was purified to remove the estrogen contamination, this resulted in breakthrough bleeding and it became necessary to restore some mestranol for cycle control, and ethinyl estradiol still fulfills that purpose today. Early clinical trials on the estrogen–progestin formulation were done by J. W. Goldzieher in San Antonio and E. T. Tyler in Los Angeles.[5]

Over the years many studies have been done to understand the action of combination contraceptives which prevents ovulation. It is now clear that there are multiple mechanisms but the predominant function is to prevent ovulation. The progestational component prevents the LH surge required for ovulation and it also serves to thicken cervical mucus, decrease tubal motility and thin the endometrium. These changes suppress sperm and oocyte transport, and result in a uterus less prepared for implantation as ancillary mechanisms in this process. The estrogen component of the pill also improves cycle control. Estrogen inhibits the release of FSH, preventing development of the dominant follicle, but also synergizes with the progestational inhibition of the LH surge. An important function of estrogen is to improve menstrual cycle control by stabilizing the endometrium, minimizing irregular shedding, generally considered by women as "breakthrough bleeding," and representing one of the most common patient complaints during treatment cycles.[11]

The question of compliance becomes particularly important in the use of oral contraceptives where pills should be taken each day, preferably at the same time. When we refer to the usage of birth control pills we often refer to perfect use, where pills are taken on the schedule exactly as prescribed. In real life, there are exceptions to this perfect use and women may forget to take a pill for a day or two or use the pills inconsistently, which may lead to variable results. When looking at the effectiveness of estrogen-progestin pills in preventing pregnancy, if we carefully ascertain that perfect compliance is maintained, then the failure rate is remarkably low, much less than 1%. Any failure (pregnancy) that occurs with perfect use will be considered a method failure, and not a result of lack of compliance. Unfortunately, in actual usage, the failure rate can be several percent, depending on the populations and a variety of factors, but these failures are generally considered to be user failures, due to the lack of daily compliance. There are now a number of recent approaches which circumvent this compliance problem by formulations that do not require daily pill-taking and these would include the vaginal ring and transdermal patches.[12]

Over the years since the introduction of combination oral contraceptives in 1960, there have been many changes. The principal change has been the decrease in estrogen content, originally at 150 μg of mestranol or ethinyl estradiol, until now, when 35 μg or as low as 20 μg represents the standard estrogen content of pills available today. These estrogen changes were done to decrease the risk of vascular thromboembolic disease (VTE), a relatively rare but serious clinical condition. The original estrogen dose (150 μg) caused a sixfold increase in VTE risk and now, with the current lower dose pills, there is only a twofold increase. Perhaps it is most important to remember that pulmonary embolism is the leading cause of pregnancy-related maternal mortality.[13] Similar changes in the progestin content have also followed this trend, but the variety of different progestins over this time has been more important. The major rationale for the decrease in steroid content of these pills was to decrease the risk of VTE.

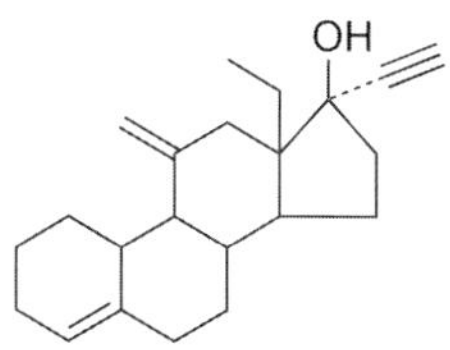

Desogestrel

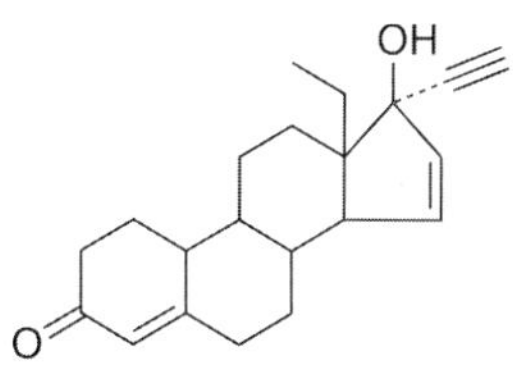

Gestodene

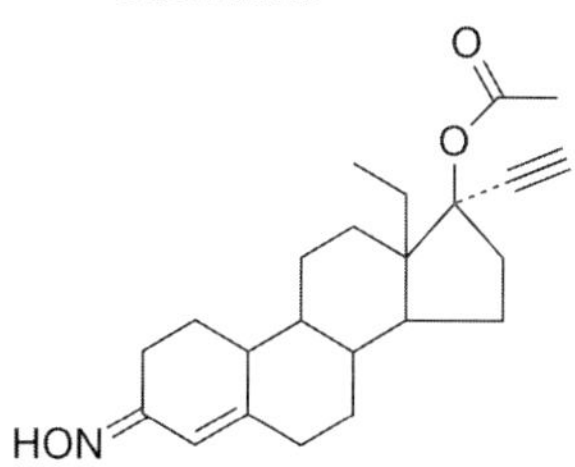

Norgestimate

Figure 5. Desogestrel and gestodene are third generation progestins. Norgestimate is a recent progestin, but is metabolized to levonorgestrel and metabolites, and therefore is still considered a second generation progestin.

The progestin component has changed to decrease the androgenic action of the progestin component. The pharmaceutical industry undoubtedly also developed these new progestins with the prospect of new patent protection for its contraceptive products and to offset the competition of generic preparations.

First generation pills are now generally considered to be those containing 50 μg or more of ethinyl estradiol, which no longer remain on the market. Second generation contraceptives are those products containing levonorgestrel, norgestimate and other members of the norethindrone family (19-nortestosterone progestins) and 30 or 35 μg of ethinyl estradiol. With the development of new progestins, a third generation group of contraceptives appeared on the market and these were desogestrel or gestodene, with 20 or 30 μg of ethinyl estradiol.[5] At the same time norgestimate was introduced, but it is not a third generation contraceptive since it is metabolized to levonorgestrel as the active progestin and therefore is considered a second generation contraceptive (Fig. 5).

The latest progestin on this list is drospirenone (Fig. 6), which is derived from spironolactone and not through the usual sequence from 19 norsteroid precursors as for the earlier progestins (norethindrone, norethynodrel or levonorgestrel). Some might consider this a fourth generation progestin, and contraceptives with drospirenone are now the most frequently prescribed combination pills in the United States.

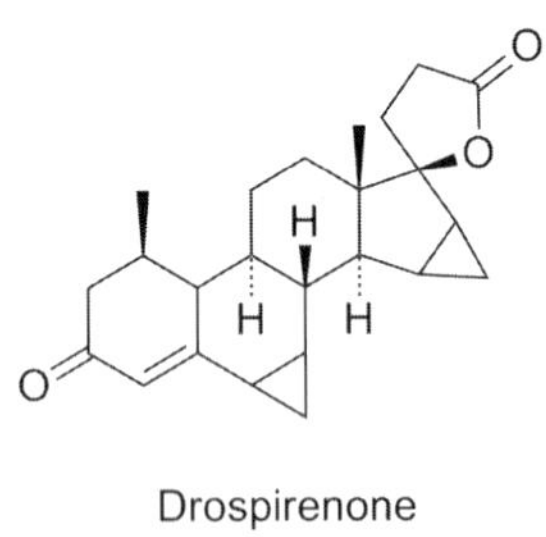

Figure 6. Drospirenone is the most popular progestin in combination oral contraceptives in the United States, and is an analog of spironolactone. In addition to its progestational activity, it binds to the mineralocorticoid receptor and has a mineralocorticoid effect.

Because of its relationship to spironolactone, drospirenone would be of benefit in those women who have a problem with fluid retention or weight gain. Since drospirenone is spironolactone-like with antiandrogenic and antimineralocorticoid activity, caution is recommended in regard to serum potassium levels and avoiding its use in women with renal, adrenal or hepatic dysfunction. It has been suggested that oral contraceptives with drospirenone are effective in treating PMS/premenstrual dysphoric disorder (PMDD).[5]

NONCONTRACEPTIVE BENEFITS

Combination oral contraceptives have served very well their original intended purpose, and over the years a number of clinical situations have been identified in which combination pills provide a noncontraceptive benefit. This has now become quite a long list of benefits, so that in some cases combination pills may be used with no relation to any need for contraception. Perhaps the most significant finding, documented in several reports, is that oral contraceptives will protect against the risk of *ovarian and endometrial cancer* and indeed for many years after the cessation of contraceptive use. The longer the duration of contraceptive use, the more prolonged the protection. Some studies have demonstrated this protection for at least 15–20 years after the cessation of contraceptive use. A number of individuals have suggested that this prophylactic usage may be a reason for women to take several years of combination pills, purely for protection against either of these two reproductive cancers.[14–17]

In addition to ovarian cancer, an emerging benefit of combination contraceptives seems to be protection against benign ovarian tumors, and one month or more of ever use of oral contraceptives has been reported to decrease the risk of serous cystadenomas, endometriomas and mucinous cystadenomas by as much as 20%. The reduction in risk was greatest in those women with 24 or more months of contraceptive use. Twelve studies were reviewed in a meta-analysis, suggesting that women on oral contraceptives were 18% less likely to be diagnosed with cancer of the colon or rectum and this protection was most likely in those with recent usage.

There are several clinical conditions associated with the normal menstrual cycle which represent difficult situations that can be alleviated with combination contraceptives. Dysmenorrhea is a common menstrual disorder frequently reported in greater than 50% of young women. This painful condition results from the release of prostaglandins which leads to increased myometrial activity. Combination oral contraceptives are reported to reduce menstrual prostaglandin release and thereby decrease uterine contractions and alleviate dysmenorrhea. Oral contraceptives have been reported to provide effective pain relief for 70–80% of women with primary dysmenorrhea and this effect is seen even with low dose contraceptives. Approximately 10% of fertile women suffer from menorraghia, which has been defined as a menstrual blood loss of greater than 80 ml. The prevalence of this condition rises with increasing age. This blood loss may become so severe that it may result in iron deficiency anemia and ultimately may require hysterectomy. This severe blood loss has been reported to be decreased by as much as 50% in users of both high dose and low dose contraceptives. In these studies the duration of menstruation was also reduced.

Relieving the pain of dysmenorrhea, decreasing heavy menstrual periods and the possibility of anemia, and providing a more regular menstrual cycle through the effective use of oral contraceptives are important noncontraceptive benefits of combination oral contraceptives.[18]

The use of oral contraceptives, results in a substantial reduction — sometimes estimated to be as much as 90% — in the risk of ectopic pregnancy. The reduction in risk of combination contraceptives is even greater than the reduction seen with other reversible contraceptive methods or tubal sterilization. This is an important benefit of oral contraceptives.[19]

Pelvic inflammatory disease frequently requires hospitalization of affected women, but oral contraceptives can be effective in reducing this risk and are even effective when using those contraceptives containing less than 35 µg of estrogen. This protective effect may be related to the progestational changes in cervical mucus, endometrium and reduced blood loss resulting from combination contraceptives.

In the earliest studies when oral contraceptives contained greater than 50 μg of estrogen, a decrease in the development of ovarian cysts was noted. We now know that this effect is related to the higher estrogen dose, and with today's low estrogen formulation this benefit is almost lost.

The use of combination oral contraceptives has also been noted to reduce the risk of benign breast disease. Oral contraceptive use prior to the first full term pregnancy decreased this risk by 40% with 1–4 years of contraceptive use, 60% or more with 4–8 years of use and 70% with use beyond 8 years. Surprisingly, oral contraceptive use beyond the first full term pregnancy was without effect on this risk.[20]

Because of the role of estrogen in *breast cancer*, it did not take long before concern regarding the relationship of combination oral contraceptives to breast cancer and other gynecological cancers became a focus of interest. Since the release of the first combination oral contraceptive in 1960, many studies have investigated the effect of combination oral contraceptives on many aspects of general women's health, but always including breast and gynecological cancers. In the earliest forms the dose of estrogen was quite high, and the significant change has been the decrease in estrogen concentration, from 150 μg to as low as 20 μg today. Numerous studies over the last few decades have indicated no risk or minimal risk for the development of breast cancer. Perhaps one of the largest studies is the Royal College of General Practitioners oral contraceptive study in the U.K., which was initiated in 1968. A recent report from this database included 744,000 woman-years for ever users of oral contraception, compared to 339,000 woman-years of observation for never users. With this extensive database it was demonstrated that oral contraception was not associated with an overall increased incidence of cancer and indeed, because of the beneficial effects on decreasing mortality, the oral contraceptives may produce an overall net health gain for women. The main dataset for ever users of oral contraceptives had a 12% reduction in the risk of any cancer when compared against never users. Statistically significant reductions were found in rates of cancer of the large bowel or rectum, uterine, or ovaries, as well as "other" sites. In this study no material difference was found in the risk for breast cancer and there was a 29% combined reduced risk for the main gynecological cancers.[21] There are now numerous studies demonstrating that there is a protective effect of combination oral contraceptives on ovarian and endometrial cancers and the duration of this protection continues for as long as 20 years. The longer the duration of oral contraceptive use, the longer the duration of protection.[16,22] It is difficult in these large databases to distinguish which birth control pill or what dose of estrogen was used, since these studies go on for a long time and pill usage by manufacturers and patients' selections change over the years, so that these remain as general conclusions. It is thought that the reduction in estrogen in the combination pills represents a beneficial change for the decreased risk of breast and gynecological cancers.

In a study conducted in the United States,[23] subjects between the ages of 35 and 64 with breast cancer ($n = 4{,}575$) were compared with a similar group of women who did not have breast cancer ($n = 4{,}682$) and were analyzed for the use of oral contraceptives in earlier life. Current or former use of oral contraceptives was not associated with an increased risk of breast cancer. There were no differences in breast cancer risk for family history of breast cancer, ethnicity or the age at initiation of oral contraceptive use.

Similar results were obtained by the Oxford Family Planning Association, another large database following breast cancer cases, and were very reassuring in that oral contraceptive use was not associated with any nonreproductive cancer or breast cancer. *Endometrial and ovarian cancers* showed a distinct protection of more than 20 years after cessation of oral contraceptive use.[24] *Cervical cancer*, however, is an exception, as no protection was noted. It is well known that human papilloma virus (HPV), age at first coitus and number of sexual partners are strong predictors of cervical cancer risk. It may be that other aspects of the sexual history of women are more important in cervical cancer risk than is the use of oral contraceptives.

Bernstein[25] reported in a meta-analysis that oral contraceptives do indeed show the prolonged effectiveness in the prevention of both ovarian and endometrial cancer, but there is a slight risk for breast cancer to be increased on oral contraceptive use. Only current or recent users were at risk and this risk disappeared rapidly with the cessation of oral contraceptive use.

Acne has often been studied in oral contraceptive subjects and is generally found to be improved. Many years ago the androgenic nature of some 19-nor-derived progestational agents was a matter for concern and the reduced androgenicity of third generation progestins would have been expected to have a more effective outcome for subjects with acne. The FDA in the United States has approved only two oral contraceptive formulas, both triphasic preparations — most likely related to the high cost of clinical trials. More realistically, probably all oral contraceptives decreased serum levels of androgens related to the suppression of gonadotrophins, which results in a decrease in ovarian steroidogenesis, including androgens and androgenic precursors. The estrogen component increases sex-hormone-binding globulin (SHBG) and would result in a decrease in the levels of free androgen in the circulation.[26,27]

ADVERSE EFFECTS

As for noncontraceptive benefits, some adverse side effects have also been associated with oral contraceptive use. Perhaps the most important is *vascular thromboembolic disease*, which was noted early in the history of oral contraceptive usage and was the great impetus for initiating the decrease in steroid hormone content, so that today the increased risk of oral contraceptives is markedly less than for the original preparation introduced in 1960. This is a serious adverse event, but it should be kept in mind that the risk of VTE is relatively rare, although serious, and the risk of VTE when on oral contraceptives is far less than the risk during pregnancy. Women with a prior history of VTE, or those who are obese or smokers, are at enhanced risk for VTE and may not be appropriate candidates for estrogen–progestin contraceptives. The incidence of VTE in women not using hormonal contraception is 5–10 per 100,000 woman-years. With low dose oral contraceptives the incidence increases to 12–20/100,000 woman-years, and it is further increased with high dose contraceptives (EE2 equal to or greater than 50 μg) to

24–50/100,000 woman-years. Perhaps the most important statistic is that in pregnancy the incidence is dramatically increased to 60–70 per 100,000 women per year, emphasizing the importance of oral contraceptives in preventing the risk of VTE, which increases with pregnancy.[11]

Four papers were published in 1995–1996 on the incidence of VTE in women on the new third generation oral contraceptives. They reported results before it was appropriate, without adequate peer review, and the information appeared in the lay press and caused what is now referred to as the "*pill scare*" *of 1996*. This resulted in many women stopping the use of these third generation oral contraceptives, and some of them became pregnant. Abortion statistics suggest that there was an increase in abortions related to the cessation of these third generation contraceptives.[28] Countless papers have been published on this topic and, while some still support the original contentions, the overwhelming majority now conclude that these were hastily done reports and that these articles should not have been published without better review and the British government should not have released warnings advising caution with these third generation contraceptives. Now numerous reports agree that the risk of VTE was not appreciably increased over second generation pills and that this was just a bump in the roadmap of contraceptive history.[29-31]

In the last few years concern has focused on newer oral contraceptives and their role in deep vein thrombosis and thromboembolic disease. Since the estrogen (ethinyl estradiol) remains unchanged, emphasis has been placed on the newer progestins in the contraceptive formulation. Gronick *et al.*[32] using computerized healthcare records in Israel, reported that the use of drospirenone was associated with an increased, albeit small, risk of VTE in third generation pills. Lidegaard *et al.*[33] found that for the same dose of estrogen and the same length of treatment, oral contraceptives containing desogestrel, gestodene and drospirenone were associated with a higher incidence of VTE than oral contraceptives containing levonorgestrel. Similarly, Vlieg *et al.*[34] reported that the risk of vascular thrombosis was associated with oral contraceptives by differing progestins.

Third generation progestins (gestodene, desogestrel, norgestimate and drospirenone) demonstrate an increased risk of deep vein thrombosis or pulmonary edema over oral contraceptives containing levonorgestrel and low dose ethinyl estradiol.

In contrast to the above reports, other studies suggest little, if any, increased VTE risk for third generation progestins or drospirenone-containing oral contraceptives. Several literature searches have questioned the increased risk of these new progestins for VTE. Brown and Vartan[35] reported that some studies showed a weak increased risk, but that several confounders had not been considered and that the majority of studies did not support the conclusion that drospirenone-containing oral contraceptives had any increased risk of VTE when compared with other oral contraceptives. Seeger *et al.*[36] and Dinger *et al.*[37] reported two large studies which did not support any increased risk of drospirenone oral contraceptives for VTE or TE.

Sehovic and Smith[38] conducted a literature search and found five studies with the use of drospirenone-containing oral contraceptives which were compared with other oral contraceptives with progestins, such as levonorgestrel. They also made the point that most studies failed to consider known confounders for VTE, such as prolonged immobility, obesity, smoking history or family history of VTE, and it was not always clear if equivalent doses of estrogens were used in these comparisons.

The question of any role for third generation progestins or drospirenone in newer oral contraceptives is certainly not resolved. The FDA has issued a warning (www.fda.gov/Drugs/DrugSafety/ucm257164.htm) to be more attentive until it completes its investigation of multiple studies. Clearly, better-controlled studies are needed to answer such questions. At the present time, too much confusion exists to answer this question but caution is advised.

One of the well-known effects of estrogens is to induce *nausea* and sometimes vomiting, and approximately 50% of women on combination oral contraceptives report nausea, though rarely vomiting. This effect is most pronounced in the first few months of use and diminishes after several months of use. The incidence of nausea has also been decreased over time with the lowering of the estrogen dosage, and today it represents a minor adverse condition of oral contraceptive use.

Estrogens are also known to stimulate breast tissue and can result in fluid accumulation and discomfort. As for other side effects, the lowering of estrogen doses has decreased this side effect over the years. It is important to remember that this *breast discomfort* is in no way related to development of breast cancer and patients should always be reminded that these are not related.

Headache is a common complaint among oral contraceptive users and the incidence increases in the pill-free interval. In the case of migraine headaches, combination pills should be used with caution. There is some concern that in some women the incidence of stroke may be increased. If migraine headaches present without aura in the absence of other risk factors, women can be given a trial of oral contraceptives. If headaches occur more frequently in the hormone-free interval, taking extended use contraceptives without the hormone-free interval may be effective in preventing migraines and still maintain effective contraception.

Many women on oral contraceptives will experience *irregular bleeding* and *midcycle spotting*, and these typically occur in the first few months after initiation of treatment and generally decrease over time, often in 3-6 months. If bleeding episodes are not diminished, the use of a higher estrogen dose pill (50 μg) may be necessary.

NEW FORMULATIONS OF THE ESTROGEN–PROGESTIN CONTRACEPTIVES

The effectiveness of the combination pills has kept this approach in the forefront of contraceptive methodology. Recent changes have not developed new approaches to hormonal contraception; however, more recent modifications of the estrogen–progestin contraceptives have focused on different methods of administration. These now include an *intravaginal ring*, which allows the sustained release of low doses of estrogen and progestin. The ring is inserted in the vagina and releases hormones for 21 days and is then removed for 7 days without a vaginal ring, allowing the uterus to slough off, and normal menses occur. The hormonal action of the vaginal ring functions in the same fashion as the original birth control pills: 21 active days and 7 days of no hormone. The ring facilitates hormonal contraception over the daily pill regimen and assures compliance and prevents the missed pill or two, which is so common with daily pill-taking. The use of this delivery system avoids the first pass effect through the liver and maintains constant low dose serum steroid levels.[39,40] This same advantage is also available with a *transdermal patch*, which releases a constant tonic dose of estrogen and progestin. A patch is applied to the skin for 7 days, and then a second and a third patch at 7-day intervals, resulting in 21 days of active estrogen–progestin administration. No patch is applied for the next 7 days, allowing menses to occur.[41,42]

CONTINUOUS ORAL CONTRACEPTIVES

The combination pills have been very successful over the years but some issues needed to be resolved. Women would prefer less frequent menstrual cycles per year, and this has been a recent development in available combination contraceptives. Gynecologists have known for many years that increasing the number of active pills, before the week of no pills or placebo pills (pill-free interval), could prolong the time before menses. These approaches required mixing pills from several packets and the instructions were sometimes difficult to follow. It was also known that hormone-related symptoms, such as pelvic pain, bleeding, headaches, nausea or vomiting, breast tenderness, and the use of analgesics were increased during the pill-free interval.[43] Changes in the pill-free interval were developed to decrease these symptoms, and they would be either a shorter pill-free interval (4 days; 24:4) or the use of a very low estrogen dose (10 μg EE2) in lieu of a placebo, which results in less adverse events in this interval.[44]

Long term or continuous estrogen–progestin contraception has now become available and involves prolonged use of active pills. These extended protocols can have active pills (estrogen and progestin) from day 1 to day 84, followed by a week of HFI, or reduced estrogen during that week, resulting in about four menses a year. Other combinations can use a continuous daily regimen of active pills for 12 months in an attempt to prevent menses for one year. The incidence of intramenstrual bleeding or spotting decreases over time. While this approach is acceptable to women, discussion about whether the elimination of all menstrual bleeding for a year remains controversial. The unpredictable nuisance of breakthrough bleeding/spotting may interfere with compliance and result in discontinuation. It has been suggested that it may be appropriate to take a three-day break from the active pills should breakthrough bleeding become troublesome.[45-47]

CONTRACEPTION AT MORE THAN 40 YEARS OF AGE

During the perimenopausal transition (ages 40–49), most women continue to have ovulatory cycles but do not generally wish to become pregnant. This is a segment of the population that is increasing with the larger baby boomer population. Unwanted pregnancies in this age group result in a higher voluntary abortion rate in the over-40 group (26 per 100 known pregnancies) compared to the 30–34 group (6 per 100 pregnancies), which can be prevented with the use of oral contraceptives. The use of low dose contraceptives in this transition phase before menopause is a convenient method for preventing pregnancy. This would maintain regular menstrual cycles and generally provide most of the noncontraceptive benefits as for younger women. Data on the noncontraceptive benefits are lacking and some age-related risks must be considered prior to advising oral contraceptive use for this group of women.[48,49]

CLOSING THOUGHTS

In the most recent report from the Royal College of General Practitioners study, the use of oral contraceptives was associated with a significantly lower rate of all-cause mortality, and also had significantly lower death rates from all cancers. It presented the caveat that the balance of risks and benefits around the world would be expected to vary from the results of this healthy U.K. population and might depend on patterns of contraceptive use and the background risk of disease.[50]

If compliance is maintained, the combination oral contraceptive is remarkably effective and has remained the standard which has not been surpassed. In the past 50 years, waves of new progestational steroids have been synthesized, and similar steroidal developments would be expected to continue, especially in the progestational component. The pharmaceutical industry may synthesize new progestins, which is now a predictable process, and may even be directed by computer modeling of possible steroid structures that retain progestational activity and de-emphasize androgenic activity. The use of estradiol and progesterone as the "natural sex steroids" always seems more desirable than for synthetic compounds, which can be associated with unexpected adverse actions. Recent studies using estradiol valerate to maintain tonic serum levels of estradiol may be a first step in utilizing these natural steroids.

Attempts to develop a hormonal male contraceptive which would produce azoospermia have been in development for decades and even in clinical trials. These approaches use either GnRH agonists to suppress LH and FSH or long-acting testosterone esters (e.g. testosterone enanthate) for the same purpose. While these studies indicate some success, the possibility of absolute azoospermia in a large population seems just out of reach and precludes FDA approval with such a built-in failure rate.

The earlier unfortunate history of intrauterine contraceptive devices (IUDs), particularly the Dalkon shield, still affects the acceptability of newer and highly effective IUDs without serious adverse events. These IUDs, either with or without a progestin content, may become more acceptable over time and fill a much-felt void — specifically, an effective and reversible contraceptive modality not dependent on patient compliance.

Permanent sterilization with either tubal ligation in women or vasectomy in men represents a large proportion of the adults who have completed their family size. Newer techniques which block the tubes may provide the opportunity for a reversible means which might be another new alternative in the future. These approaches would be nonhormonal and avoid the side effects — menstrual bleeding and libido.

There are currently many contraceptive approaches available to the clinical community and it is becoming increasingly difficult and expensive for pharmaceutical discovery and development, clinical trials and marketing to introduce new contraceptive approaches, with the bleak prospect of a small piece of the available market for new contraceptives. Very few pharmaceutical companies maintain an active division for contraceptive development, and new contraceptive approaches would not be in prospect.

Malcolm Potts was the President of the International Planned Parenthood Federation for many years and has gained a broad range of experience with contraceptives during his long tenure in this area. Years ago, Dr. Potts said: "The invention of the Pill was the greatest revolution that has ever occurred, or will ever occur, in family planning. It disassociated contraception from sexual intercourse; it was virtually 100% effective and totally reversible." "The Pill is going to be with us for a long time. We can expect to see incremental improvements in dose and formulation. A wider variety of progestational agents are likely to be used. In short, there is every reason to believe that more women will use oral contraception in the future than have ever used them in the past 30 years."[51]

Table 1. Noncontraceptive benefits of combination oral contraceptives.

Decreased risk of ectopic pregnancies
Decreased risk of pelvic inflammatory disease
Decreased risk of endometrial cancer, ovarian cancer and colorectal cancer
Decreased menstrual flow and dysmenorrhea
Decreased benign breast disease
Improved bone mineral density
Improved acne or hirsutism
Improvement in rheumatoid arthritis
Prevention of menstrual migraines
Treatment of pelvic pain due to endometriosis
Treatment of bleeding due to fibroids
Maintaining menstrual regularity
Treatment of premenstrual syndrome
Inducing amenorrhea for lifestyle considerations

REFERENCES

1. Haberlandt E. (2009) Ludwig Haberlandt — a pioneer in hormonal contraception. *Wien Klin Wochenschr* **121**:746–749.
2. Diczfalusy E. (1979) Gregory Pincus and steroidal contraception: a new departure in the history of medicine. *J Steroid Biochem* **11**:3–11.
3. Perone N. (1994) The progestins. In: *Pharmacology of the Contraceptive Steroids*, eds. Goldzieher JW, Fotherby K. New York: Raven, pp. 5–19.
4. Goldzieher JW, Rudel HW. (1974) How the oral contraceptives came to be developed. *J Amer Med Assoc* **230**:421–425.
5. Speroff L, Fritz MA. (2005) In: *Clinical Gynecologic Endocrinology and Infertility*, 7th edn., eds. Speroff L, Fritz MA. Lippincott Williams & Wilkins, Chap. 22, pp. 861–942.
6. Speroff L. (2009) *A Good Man, Gregory Goodwin Pincus: The Man, His Story, the Birth Control Pill.* Portland, Oregon: Arnica.
7. Maisel AQ. (1965) *The Hormone Quest*. Random House.
8. Goldzieher JW. (1994) The estrogens. In: *Pharmacology of the Contraceptive Steroids*, eds. Goldzieher JW, Fotherby K. New York: Raven, pp. 21–25.
9. Halberstam D. (1993) *The Fifties*. New York: Villard Books, pp. 282–294, 599–606.
10. Djerassi C. (1995) The mother of the pill. *Rec Prog Horm Res* **50**:1–17.
11. Kiley JA, Hammond C. (2007) Combined oral contraceptives: a comprehensive review. *Clin Obstet Gynecol* **50**:868–877.
12. Rosenberg MJ, Burnhill MS, Waugh MS, Grimes DA, Hillard PJ. (1995) Compliance and oral contraceptives: a review. *Contraception* **52**:137–141.
13. James AH, Jamison MG, Brancazio LR, Myers ER. (2006) Venous thromboembolism during pregnancy and the postpartum period: incidence, risk factors and mortality. *Am J Obstet Gynecol* **194**:1311–1315.
14. Kaufman DW. Shapiro S, Slone D, Rosenberg L, Miettinen OS, *et al.* (1980) Decreased risk of endometrial cancer among oral contraceptive users. *N Engl J Med* **303**:1045–1047.
15. Jick SS, Walker AM, Jick H. (1993) Oral contraceptives and endometrial cancer. *Obstet Gynecol* **82**:931–935.
16. Collaborative Group on Epidemiological Studies of Ovarian Cancer. (2008) Ovarian cancer and oral contraceptives: collaborative reanalysis of data from 45 epidemiological studies including 23,257 women with ovarian cancer and 87,303 controls. *Lancet* **371**:303–314.
17. Ness RB, Grisso JA, Klapper J, Schlesselman JJ, *et al.* (2000) Risk of ovarian cancer in relation to estrogen and progestin dose and use characteristics of oral contraceptives. *Am J Epidemiol* **152**:233–241.
18. ESHRE Capri Workshop Group. (2005) Noncontraceptive health benefits of combined oral contraceptives. *Human Reprod Update* **5**:513–525.

19. Mol BW, Ankum WM, Bossuyt PM, *et al.* (1995) Contraception and the risk of ectopic pregnancy: a meta-analysis. *Contraception* **52**:337–341.
20. Westhoff C, Britton JA, Gammon MD, Wright T, Kelsey JL. (2000). Oral contraceptive and benign ovarian tumors. *Am J Epidemiol* **152**:242–246.
21. Hannaford PC, Selvaraj S, Elliot AM, *et al.* (2007) Cancer risk among users of oral contraceptives: cohort data from the Royal College of General Practitioner's oral contraception study. *Br Med J* **335**:651–658.
22. Grimbizis GF, Tarlatzis BC. (2010) The use of hormonal contraception and its protective role against endometrial and ovarian cancer. *Best Pract Res Clin Obstet Gynecol* **24**:29–38.
23. Marchbanks PA, McDonald JA, Wilson HG, Folger SG, *et al.* (2002) Oral contraceptives and the risk of breast cancer. *New Engl J Med* **346**:2025–2032.
24. Vessey M, Painter R. (2006). Oral contraceptive use and cancer: findings in a large cohort study, 1968–2004. *Br J Cancer* **95**:385–389.
25. Bernstein L. (2006) The risk of breast, endometrial and ovarian cancer in users of hormonal preparations. *Basic Clin Pharmacol Toxicol* **98**:288–296.
26. Thorneycroft IH. (1999) Update on androgenicity. *Am J Obstet Gynecol* **180**:S288–S294.
27. O'Connell K, Westhoff C. (2008) Pharmacology of hormonal contraceptives and acne. *Cuytis* **81**(Suppl. 1):8–12.
28. Ramsay S. (1996) UK "pill scare" led to abortion increase. *Lancet* **347**:1109.
29. Cohen J, Edwards RG. (1999) Conclusions: the relative safety of modern oral contraceptives. *Hum Reprod Update* **5**:756–771.
30. Spitzer WO. (1999) The aftermath of a pill scare: regression to reassurance. *Hum Reprod Update* **5**:736–745.
31. Spitzer WO. (1997) The 1995 pill scare revisited: anatomy of a non-epidemic. *Hum Reprod* **12**:2347–2357.
32. Gronich MD, Lavi I, Rennert G. (2011) Higher risk of venous thrombosis associated with drospirenone-containing oral contraceptives: a population-based study. *Can Med Assoc J* **183**:E1319–E1325.
33. Lidegaard OL, Lokkgaard E, Svendsen AL., Agger C. (2009) Hormonal contraception and risk of venous thromboembolism: national follow-up study. *Br Med J* **339**:b2890.
34. van Hylckama Vlieg A, Helmerhorst FM, Vandenbroucke JP, Doggen CJM, Rosendaal FR. (2009) The venous thrombotic risk of oral contraceptives, effects of oestrogen dose and progestogen type: results of the MEGA case-control study. *Br Med J* **339**: 2921.
35. Brown DA, Vartan CM. (2011) Risk of venous thromboembolism with drospirenone-containing oral contraceptives. *Am J Health-Syst Pharm* **68**:1003–1010.

36. Seeger JD, Loughlin J, Eng PM, Clifford CR, Cutone J, Walker AM. (2007) Risk of thromboembolism in women taking ethinylestradiol/drospirenone and other oral contraceptives. *Obstet Gynecol* **110**:587–593.
37. Dinger JC, Heinemann LAJ, Kuhl-Habich D. (2007) The safety of a drospirenone-containing oral contraceptive: final results from the European Active Surveillance study on oral contraceptives based on 142,475 women years of observation. *Contraception* **75**:344–354.
38. Sehovic N, Smith KP. (2010) Risk of venous thromboembolism with drospirenone in combined oral contraceptive products. *Ann Pharmacother* **44**:898–903.
39. Madden T, Blumenthal P. (2007) Contraceptive vaginal ring. *Clin Obstet Gynecol* **50**:878–885.
40. Wagner MS, Arias RD, Nucatola DL. (2007) The combined etonogestrel/ethinyl estradiol vaginal ring. *Expert Opin Pharmacotherap* **8**:1769–1777.
41. Audet MC, Moreau M, Koltun WD, *et al.* (2001) Evaluation of contraceptive efficacy and cycle control of a transdermal contraceptive patch vs. an oral contraceptive: a randomized control trial. *JAMA* **285**:2347–2354.
42. Smallwood GH, Meador ML, Lenihan JP, *et al.* (2001) Efficacy and safety of a transdermal contraceptive system. *Obstet Gynecol* **98**:799–805.
43. Sulak PJ, Scow RD, Preece C, Riggs MW, Kuehl TJ. (2000) Hormone withdrawal symptoms in oral contraceptive users. *Obstet Gynecol* **95**:261–266.
44. Kaunitz AM, Burkman RT, Fisher AC, LaGuardia KD. (2009) Cycle control with a 21-day compared with a 24-day oral contraceptive pill. *A randomized controlled trial. Obstet Gynecol* **114**:1205–1212.
45. Sulak PJ. (2008) Continuous oral contraception: changing times. *Best Pract Res Clin Obstet Gynaecol* **22**:355–374.
46. Shrader SP, Dickerson LM. (2008) Extended- and continuous-cycle oral contraceptives. *Pharmacotherap* **28**:1033–1040.
47. Kroll R, Reape KZ, Margolis M. (2010) The efficacy and safety of a low-dose, 91-day extended-regimen oral contraceptive with continuous ethinyl estradiol. *Contraception* **81**:41–48.
48. ESHRE Capri Workshop Group. (2009) Female contraception over 40. *Hum Reprod Update* **15**:599–612.
49. Kaunitz AM. (2008) Hormonal contraception in women of older reproductive age. *New Engl J Med* **358**:1262–1270.
50. Hannaford PC, Iversen L, Macfarlane TV, Elliott AM, Angus V, Lee AJ. (2010) Mortality among contraceptive pill users: cohort evidence from Royal College of General Practitioners' Oral Contraceptive Study. *Brit Med J* **340**:927–935.
51. Potts M. (1993) Introduction. In: *Pharmacology of the Contraceptive Steroids*, eds. Goldzieher JW, Fotherby K. New York: Raven, pp. 1–4.

4

HORMONE REPLACEMENT THERAPY AND WOMEN'S HEALTH

Jennifer Eng-Wong,
Ifeyinwa Obiorah and V. Craig Jordan

Abstract

Hormone replacement therapy (HRT) is the most efficient treatment for the relief of menopausal symptoms. Prophylactic administration of HRT has been used in clinical practice in postmenopausal women with increased risk for osteoporosis, cardiovascular disease or Alzheimer's disease. However, potential risks of cancer, thromboembolism and stroke have created some insecurity in the prolonged use of therapy. In this chapter, an extensive review of clinical trials will be made to assess the potential benefits and risks of HRT.

Keywords: Hormone Replacement Therapy; Osteoporosis; Coronary Heart Disease; Women's Health Initiative; Breast Cancer.

STRATEGIES IN DISEASE PREVENTION

Hormone replacement therapy (HRT), a combination of oral estrogens and progestins, evolved with a goal of preventing two prevalent diseases in postmenopausal women: heart disease and osteoporosis. Heart disease is the leading cause of morbidity and mortality for postmenopausal women in the United States, with more than 8.1 million women with coronary heart disease (CHD) and 37% of all deaths in women attributed to CHD.[1] Similarly, osteoporosis is a pervasive problem, with one in two women over age 50 expected to have an osteoporotic fracture in their lifetime.[2] Fractures are associated with significant morbidity and mortality: 20% cent of patients admitted with an acute hip fracture die in the following six months and another 30% become permanently disabled.[3] Thus, a drug to reduce the incidence and related morbidity and mortality of these common diseases is appealing, and yet the path of scientific clinical evaluation led to unexpected results. Clinical trials of HRT did not confirm projected results in CHD; however, they did raise the possibility of new approaches in unanticipated areas.

USAGE OF ESTROGEN REPLACEMENT THERAPY, APPROVALS IN THE UNITED STATES

Commercial production of estrogens as drugs to combat menopause began in the early 20th century in the U.S. However, these preparations were expensive, which translated into a costly therapy for patients. In order to obtain a less expensive estrogen preparation, scientists at Ayerst (now Pfizer Pharmaceuticals) in Canada extracted conjugated estrogens from pregnant horses' urine.[4] These estrogens were more potent than unconjugated estrogens already on the market and side effects were minimal. Thus, in 1941, Ayerst introduced Premarin (*Pre*gnant, *mar*es' ur*in*e or PMU) in Canada and received FDA approval in the U.S. in 1942 for the treatment of menopausal symptoms and related conditions.[5] The initial indication for Premarin was for the short term treatment of menopause. The use of long term estrogen replacement therapy (ERT) did not become popular until the 1960s, when there were numerous publications on the beneficial effects of these estrogens, especially on postmenopausal osteoporosis. A most influential publication was *Feminine Forever*, a book written by a prominent gynecologist, Dr. Robert Wilson, in 1966. On the book he classified menopause as a pathological disease and said that a wonder pill was available to keep women young forever.[4] This message was spread with the help of the media and Ayerst drug company, leading to a dramatic rise in the annual prescription of estrogens from 1966 to 1975.[4]

Despite the worldwide acceptance of ERT, there were increasing concerns about the medical risks of long term ERT. This was substantiated by a series of publications from *The New England Journal of Medicine* in 1975 which showed increased risks of developing endometrial cancer in women on ERT.[6-7] Hence, in 1976, the FDA mandated that all estrogen products should contain a package insert warning of the risks of cancer and thrombosis.[8] The association of estrogens with endometrial cancer led to a decline in the prescription of these drugs to postmenopausal women. In the early 1980s, there was another generation of interest in the use of estrogen in the treatment of postmenopausal osteoporosis.[4] However, a second scientific development was needed to resolve the issue of its carcinogenic effect in the endometrium. Several reports showed that women who took progestins along with estrogens had a lower incidence of endometrial cancer.[9-10] Histologic studies of endometrial biopsies on women on estrogen alone and on estrogen and progestins revealed that estrogen caused proliferation of the endometrial lining while progestins prevented this effect.[10] Therefore, the new recommendation was that women with an intact uterus were to take estrogens in combination with progestins or HRT. In 1986, based on the reports on the beneficial effects of estrogens in the treatment and prevention of osteoporosis, the FDA deemed estrogens effective therapy for osteoporosis.[8]

PREPARATION AND CONTENT OF ERT

Premarin is made up of conjugated estrogens extracted from the urine of pregnant mares. The urine is collected by attaching containers to the horses in order to obtain stable water concentrate of conjugated estrogens. The FDA[5] background information on conjugated estrogen revealed that in 1942, Premarin was known to contain two estrogens, estrone and equilin, as well as additional estrogens in smaller amounts. In 1970, the United States Pharmacopeia (USP) published the first comprehensive standards for conjugated estrogens. The drug was shown to contain sodium estrone sulfate and sodium equilin sulfate but little data was available on the detailed composition of Premarin or the pharmacologic activity of its components.

In 1972, based on the effectiveness of Premarin in the treatment of menopausal symptoms, the FDA approved abbreviated new drug applications (ANDAs) for cost-effective generic versions of Premarin.[8] Modern compositional analysis shows that Premarin is a cocktail of a significant number of compounds with potential pharmacologic activity.[5] However, the steroidal component of Premarin is still not yet completely defined. Furthermore, the FDA withdrew approval of all ANDAs based on an updated amendment in 1992 to the 1970 USP monograph.[8] The updated version showed that in addition to sodium estrone sulfate and sodium equilin sulfate, conjugated estrogens contain additional estrogenic compounds in their sulfate ester form. Analysis of two batches of Premarin 0.625 mg tablets by the FDA's Center for Drug Evaluation and Research (CDER) revealed that a tablet of Premarin consists of at least 10 estrogens which include estrone (59.2%), equilin (26.9%), 17α-dihydroequilin (16.3%), 17α-estradiol (4.32%), 17β-dihydroequilin (1.76%), 17α-dihydroequilenin (1.76%), 17β-dihydroequilenin (3.36%), equilenin (2.4%), 17β-estradiol (0.8%) and $\Delta^{8,9}$-dehydroestrone (4.16%)[11] (see Fig. 1). This led to the announcement by the CDER in 1997 that it would not approve generic synthetic versions of Premarin because they did not contain the active components of Premarin. Presently, the Federal Food, Drug and Cosmetic (FD&C) Act requires that all generic versions contain the same active ingredients, bioequivalence, safety and effectiveness as the reference listed drug.[5]

RESULTS OF HRT RANDOMIZED TRIALS

Observational studies provided strong support that HRT had a protective role against cardiovascular disease. Pooled estimates from cohort and case control studies for relative risk of CHD in HRT users ranged from 0.50 to 0.65.[12] Preclinical data also supported the beneficial effects of estrogen therapy in retarding the development of heart disease. In ovariectomized animal models, estrogen replacement reduced atherosclerosis and the effects were mediated via lipoproteins as well as direct vascular effects.[13,14] Additionally, randomized trials evaluating biomarkers of CHD provided further evidence of the potential

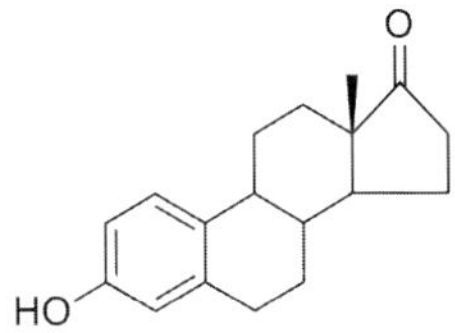

Oestrone

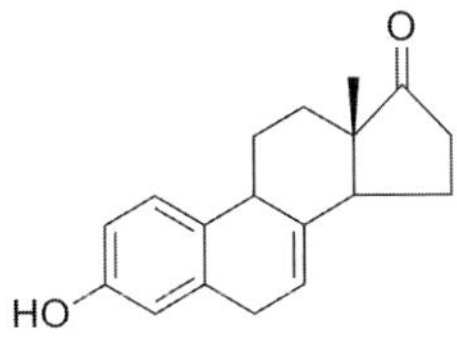

Equilin

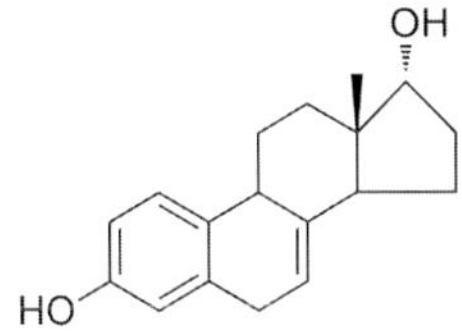

17α-Dihydroequilin

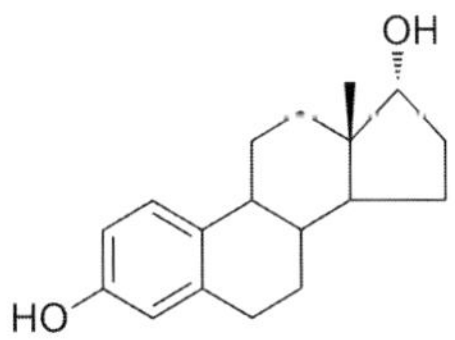

17α-Oestradiol

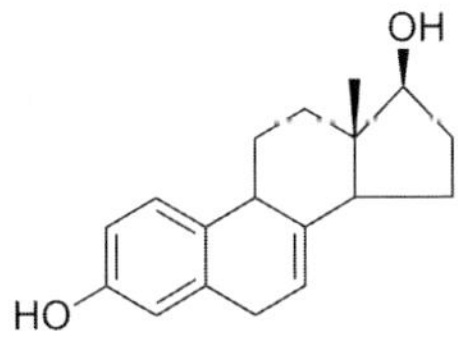

17β-Dihydroequilin

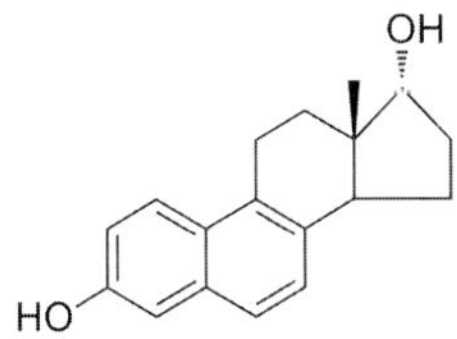

17α-Dihydroequilenin

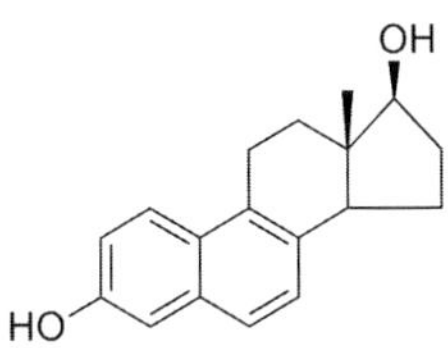

17β-Dihydroequilenin

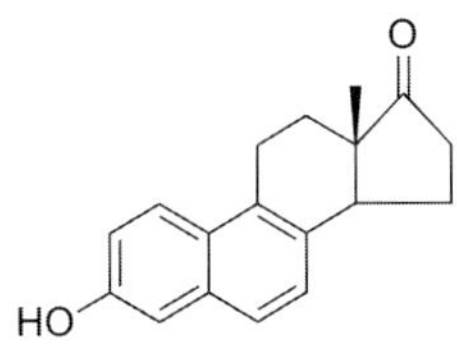

Equilenin

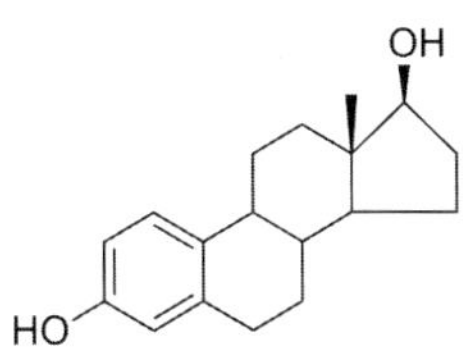

17β-Oestradiol

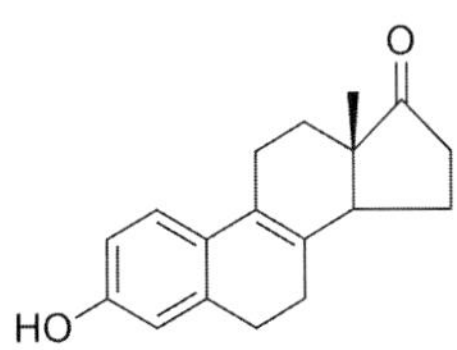

$\Delta^{8,9}$-Dehydroestrone

Figure 1. Chemical structures of the estrogenic components of Premarin.

benefits of HRT. The Postmenopausal Estrogen/Progestin Interventions Trial (PEPI), published in 1995, compared placebo, estrogen alone and three different estrogen/progestin regimens in 875 healthy postmenopausal women who were treated for three years. The primary endpoint of the trial was heart disease risk factors: high density lipoprotein (HDL), systolic blood pressure, serum insulin and fibrinogen. The investigators found that any combination of HRT improved lipoproteins and fibrinogen levels, however, the trial did not evaluate the incidence of coronary events.

Starting in the late 1980s, a number of phase III placebo-controlled randomized trials of HRT were undertaken. The primary endpoint in many of these studies was the incidence of CHD events. Secondary endpoints included the incidence of stroke, thromboembolic events (TEEs), cancer and overall mortality, among others. In 2009, the Cochrane group published a meta-analysis of 19 randomized, double-blind, placebo-controlled trials with 41,904 women and concluded that the use of combined HRT (estrogen and progesterone) increased the risk of both CHD and TEE after one year's use, and stroke after three years use, as well as breast cancer and gall-bladder disease. Single agent estrogen replacement also increased the risk of TEE, stroke and gallbladder disease but did not increase the risk of breast cancer. The only benefits found were a decreased risk of fractures for either combination or single agent HRT and a decrease in colon cancer with combination HRT. The effects of HRT on cognitive health were also analyzed. In women over age 65, combined HRT use resulted in a statistically significant increased risk of dementia.[15] We will examine the design and outcomes in several of the larger trials to more clearly define the role of HRT, if any, in preventing chronic disease or menopausal symptoms and propose opportunities for further study.

HRT and Vascular Disease

The first randomized trial to show that HRT did not have the expected cardiac protective effects was the Heart and Estrogen/Progestin Replacement Study (HERS). This trial randomized 2763 postmenopausal women with coronary disease to combined HRT or placebo.[16] The average age of participants was 67 years and the study was designed to identify a 24% reduction in coronary events. At 4.1 years' median follow-up there was no benefit to HRT on the primary endpoints of myocardial infarction or CHD death (HR 0.99; 95% CI 0.80, 1.22) despite favorable effects on serum low density lipoprotein and HDL. Additionally, there was an increased risk of TEE (HR 2.89, 95% CI 1.50, 5.59) and gallbladder disease (HR 1.38, 95% CI 1.00, 1.92). Since this trial enrolled women with a history of CAD, it did not address the issue of primary prevention in a healthy population. The Women's Health Initiative (WHI) Trials were designed to assess health effects of HRT in healthy women.

The largest of the randomized trials is the Women's Health Initiative (WHI), which conducted two parallel studies enrolling more than 27,000 subjects. Women with an intact uterus were randomized to conjugated equine estrogen (CEE) 0.625 mg daily and medroxyprogesterone acetate (MPA) 2.5 mg daily or placebo, while women who had undergone a hysterectomy were randomized to CEE or placebo in a second trial. The primary outcome for both trials was CHD, which included both nonfatal myocardial infarction and CHD death. Secondary endpoints included stroke, pulmonary embolus, breast cancer, colorectal cancer, hip fracture, total death and a global health index designed to reflect the overall risk/benefit of HRT. Previous use of HRT was allowed.

In the combined HRT trial, 16,608 women aged 50–79 years were enrolled with a planned 8.5 years of treatment. It was designed to identify a 21% reduction in CHD events at an average of 9 years of follow-up.[17] The mean age of the subjects was 63 years, with 33% between ages 50 and 59. In May 2002, after 5.2 years of follow-up, the trial was stopped due to the crossing of a predefined safety boundary for an increased risk of invasive breast cancer (see below). Contrary to expectations, there was a small increase in CHD, HR 1.29 (95% CI 1.02, 1.63), as well as stroke, HR 1.41 (95% CI 1.07, 1.85), and pulmonary embolus (PE), HR 2.13 (95% CI 1.39, 3.25). The increased risk in CHD was primarily seen in the first year of therapy, while the risk of stroke and PE remained elevated throughout the 5 years of treatment. The global health index was elevated (HR 1.15, 95% CI 1.03, 1.28),

Table 1. Hazards ratio from selected phase III randomized trials of HRT.

	WHI (estrogen and progesterone)[18,53]	WHI (estrogen alone)[19,20]	WISDOM (estrogen and progesterone)[25]	HERS (estrogen and progesterone)[16,89]
Median follow-up (years)	5.2	6.8	1	4.1, 6.8
N	16,608	10,739	4385	2763
CHD	1.29*	0.91	X*	0.99
Stroke	1.41*	1.39*	0.73	NR
PE/VTE	2.13*	1.34*	7.36*	2.89*
Fracture risk	0.76*	0.70*	0.69	0.95
Cancer Incidence				
Breast	1.26*	0.77		1.30
Endometrial	0.83	-		0.49
Ovarian	1.58	NR	0.88 (any cancer)	NR
Colorectal	0.63*	1.08		0.69
Lung	1.23	NR		1.28
Lung cancer mortality	1.71*	NR		NR
Death (all causes)	0.98	1.04	1.60	1.08

*Statistically significant.
NR: Not reported.
X: No HR reported; p value 0.016.

suggesting more harm than benefit. It was concluded that given the increased odds of CHD events, PE and breast cancer, the risks outweighed any benefits seen in fracture reduction and colorectal cancer reduction.[18] (See Table 1.) Additionally, 42% of women on HRT and 38% on placebo stopped the drug during the trial, and thus the reported risk may underestimate the actual negative outcomes.

In the CEE-alone trial 10,739 women 50–79 years old were enrolled. The average age of the subjects was 64 years, with 31% between ages 50 and 59. In February 2004, the trial was stopped early by the NIH, due to lack of benefit. With an average follow-up of 6.8 years there was no decline in CHD (HR 0.91, 95% CI 0.75, 1.12) and stroke incidence was increased (HR 1.39, 95% CI 1.1.0,1.77); additionally, there was no difference in PE. The only positive finding was a decline in hip fracture, HR 0.61 (95% CI 0.41, 0.91).[19] The results were perplexing given the hypothesis developed from multiple preclinical and epidemiology studies.

An additional 3.6 years' postintervention follow-up after the trial was stopped revealed that the risk of stroke was no longer elevated and was 0.36% in women on CEE compared with 0.41% in the placebo group (HR 0.89; 95% Cl, 0.64–1.24).[20] Additionally, the risk of deep vein thrombosis was decreased at 0.17% and 0.27%, respectively (HR 0.63; 95% Cl, 0.41–0.98). However, the postintervention risk and HR for CHD were similar to those observed during the intervention period.

Subgroup analyses as well as ancillary studies were conducted to try to tease out any group in the WHI study with a potential benefit. In the combined HRT trial, only women with an elevated low density lipoprotein at baseline had an increased risk of CHD,[21] while multiple other CHD biomarkers had no association with risk. In the CEE trial, women with elevated C-reactive protein had an increased risk of heart disease with CEE use.[22] Although these analyses were planned *a priori,* the number of cases is small and results should be interpreted with caution. An ancillary study evaluated coronary artery calcium on heart computed tomography (CT) scan as a biomarker for CV events in a younger cohort of women ($N = 1064$; 50–59 years old). Calcified plaque burden is a marker of total coronary artery plaque burden and is associated with future CHD events independent of clinical cardiac risk factors.[23] It was lower in the women on CEE ($p = 0.02$) at an average of 8.7 years of follow-up, suggesting that there may be a cardioprotective effect in this cohort; however, no older group was evaluated for comparison.[24] Additionally, the ancillary study was not powered to look at clinical events. No evaluation of coronary artery calcium was done in the combined HRT WHI. Thus, the conclusion from these large well-conducted trials is that HRT has no overall benefit in preventing CAD.

The Women's International Study of Long Duration of Estrogen after Menopause (WISDOM) trial, another randomized, placebo-controlled phase III trial of HRT conducted in the UK, Australia and New Zealand, opened to accrual in 1999 and subsequently closed early after release of the WHI findings. The intervention included both CEE and combined estrogen and progesterone (E + P). Even with only 11.9 months of follow-up on 5,692 subjects (average age 62.8 years), the investigators reported a statistically significant increase in cardiovascular events and TEE, and no difference in the number of cancers, stroke, fracture or death.[25]

HRT and Cancer Risk

Breast Cancer

In the combined HRT WHI trial, the HR for breast cancer was 1.26 (1.00, 1.59) and exceeded the predefined safety limits, resulting in early trial closure. Although baseline breast cancer risk factors were similar between the two arms (Gail risk model,[26] family history, race/ethnicity, obesity, alcohol use), breast cancers diagnosed on HRT were more often at an advanced stage (regional/metastatic 25.4% vs. 16.0%) and slightly larger.[27] There were no differences in tumor histology, grade or hormone receptor status. Other evaluations have reported differences by histologic subtype. A number of cohort studies reported an increase in lobular cancers over ductal cancers in women on HRT.[28-30] In evaluating the time trends of breast cancer diagnosis on HRT, investigators found that after the first two years on HRT the number of diagnoses increased and then rapidly fell off after stopping the drug; mammographic screening rates did not differ between intervention and placebo arm during treatment or following it.[31] Survival outcomes have not yet been reported.

This risk estimate of 1.26 for the WHI was somewhat lower than that predicted from a number of observational studies. In the Million

Women Study, a nonrandomized, prospective cohort study of HRT conducted in the U.K., breast cancer risk for E + P was markedly greater (HR 2.00, 95% CI 1.88–2.12),[32] and in a pooled analysis of 51 observational studies across 21 countries the risk was also higher (RR 1.53), though it should be noted that combination HRT was uncommon in these studies and was used by only 12% of the women.[33] The time from menopause to the start of HRT ("gap time") has been proposed to explain the differences in risk between the clinical and observational studies. When evaluating women who started HRT soon after menopause (<5 years vs. ≥ 5 years), the HR for invasive breast cancer was 1.77 and 0.99, respectively, with a p value of 0.02 for HR and interaction with gap time. Duration of use was also important, with longer use increasing the risk of breast cancer. Women who took HRT with <5 years' gap time had a HR of 1.85 at 2–5 years, which increased to an HR of 2.75 with more than 5 years of use.[34] In addition to the growth stimulant properties of E + P on breast cancer cells, additional explanations for these observations include that HRT increased mammographic density (MD), which decreases the sensitivity of mammography. Increased MD is also an independent risk factor for breast cancer. Women on placebo had a small decline in MD, which is expected with age and fatty replacement of breast tissue.[35] Abnormal mammograms requiring short interval follow-up or suspicious finding were comparable at baseline but increased in the HRT group after 1 year (9.4% vs. 5.4% $p < 0.001$) and continued to be higher with longer duration.[27] This again points to increased breast tissue activity mediated by HRT leading to increased diagnostic procedures and ultimately carcinogenesis.

Fournier *et al.*[36] evaluated the risk of breast cancer with timing of HRT related to end of menopause and duration of HRT in 53,000 postmenopausal women in the prospective cohort French E3N study, with a mean follow-up of 8.1 years. For women who had used combination HRT for less than 2 years and started it within 3 years of menopause, there was an increased risk of breast cancer (HR 1.54, 95% CI 1.28, 1.86), while for women who had used combination HRT for less than 2 years and started it after 3 years of menopause, there was no increased risk of breast cancer (HR 1.00, 95% CI 0.68, 1.47). With longer duration of use (>2 years) in women within 3 years of menopause, there a nonsignificant trend toward increasing risk of breast cancer. These findings are in keeping with the WHI results,[34] suggesting that the recent hormonal milieu of the breast plays a critical role in affecting the action of exogenous hormones.

Results of the final analysis of the WHI CEE trial demonstrated that CEE use was associated with a persistent decrease in the risk of breast cancer and was 0.27% compared with 0.35% in the placebo group (HR 0.77, 95% CI 0.62–0.95[20] (Fig. 2). There were four breast cancer deaths on CEE versus eight on placebo during the intervention period.[19] The effect of E alone on breast cancer risk in epidemiology studies has been inconsistent. In the Million Women Study, a nonrandomized, prospective cohort study of HRT conducted in the UK., E alone was

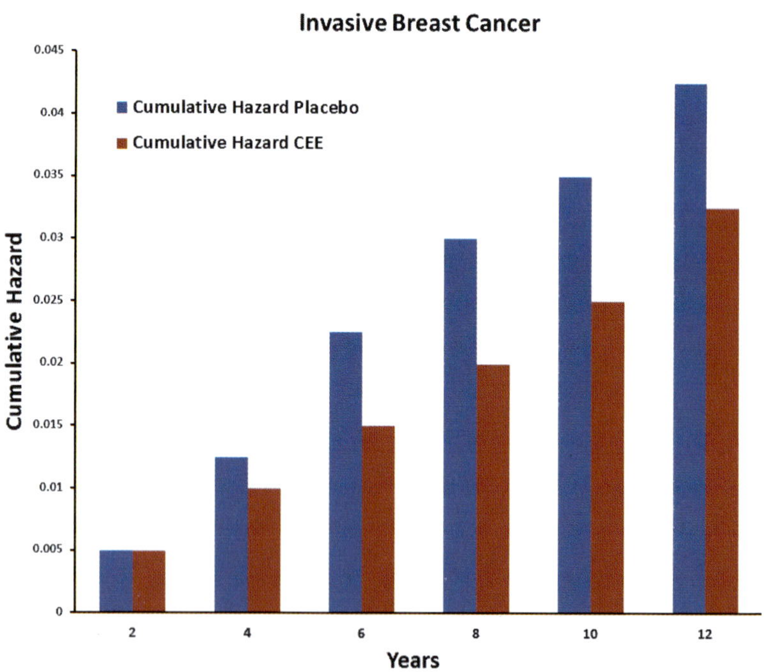

Figure 2. Cumulative hazard for invasive breast cancer by treatment group.

associated with an increased risk of breast cancer (HR 1.3, 95% CI 1.21, 1.40)[32] and longer duration of use correlated with increased breast cancer risk, though the risk was smaller than with E + P. In a case control study of HRT, Li *et al.*[37] reported no increased risk of breast cancer with E alone regardless of duration of use (range 6 months to > 25 years), while E + P consistently increases risk. Thus, the effect of E alone on breast cancer risk is minimal and there is a suggestion from the WHI data that it could be protective. In keeping with this finding, CEE has not been shown to increase MD.[38] Taken together the data suggest that progesterone with estrogen is more capable of inducing breast tumor growth than estrogen alone.

Gynecologic Cancers

In the PEPI trial, CEE alone increased the risk of endometrial hyperplasia by 31% ($p < 0.001$), and this finding mandated the design of the WHI such that E + P was required for women with a uterus and CEE was used only in hysterectomized women. As predicted by earlier studies, combined HRT in the WHI trial did not increase the risk of endometrial cancer (HR 0.83, 95% CI 0.47, 1.47). However, HRT use was associated with more endometrial biopsies (33% vs. 6%; $p < 0.001$), which limits the acceptability of this regimen.[39] Epidemiology studies assessing the risk of ovarian cancer with HRT have been mixed.[40,41] A recent meta-analysis found a small increased risk HR 1.24, 95% CI 1.15, 1.34 for cohort studies and HR 1.19, 95% CI 1.02, 1.40 for case control studies.[42] Longer use of HRT has been identified with greater risk of ovarian cancer. [41,43,44] Invasive ovarian cancer incidence was slightly higher on combined HRT in the WHI trial (HR 1.58, 95% CI 0.77–3.24) but not statistically significant.[39] No other phase III trials have yet reported on ovarian cancer incidence. Few cases of cervical cancer were reported and there was no difference by treatment arm.

Colon Cancer

Observational studies have reported a decreased risk of colon cancer with HRT. In a meta-analysis of 18 studies, women who had ever used

HRT had a 20% decreased risk of colon cancer compared with never users.[45] The majority of the studies involved CEE alone, although a number of studies (but not all) which included E + P also reported a decreased risk of colon cancer.[46–49]. In the WHI, use of combined hormones was associated with a decrease in colorectal cancer (HR 0.56, 95% CI 0.38, 0.81), with 43 vs. 72 cases at an average follow-up of 5.6 years (Fig. 3).[50] Colon cancer risk factors and screening did not differ between the two arms with the exception of a higher prevalence of family history of colon cancer in the placebo arm (an analysis which adjusted for this factor yielded similar results). Although histologic type and grade were similar between arms, women who took HRT had more advanced colon cancer at diagnosis, i.e. the decrease in cases was due largely to a smaller number of local cancers on the HRT arm. No difference in mortality rates has been observed to date.

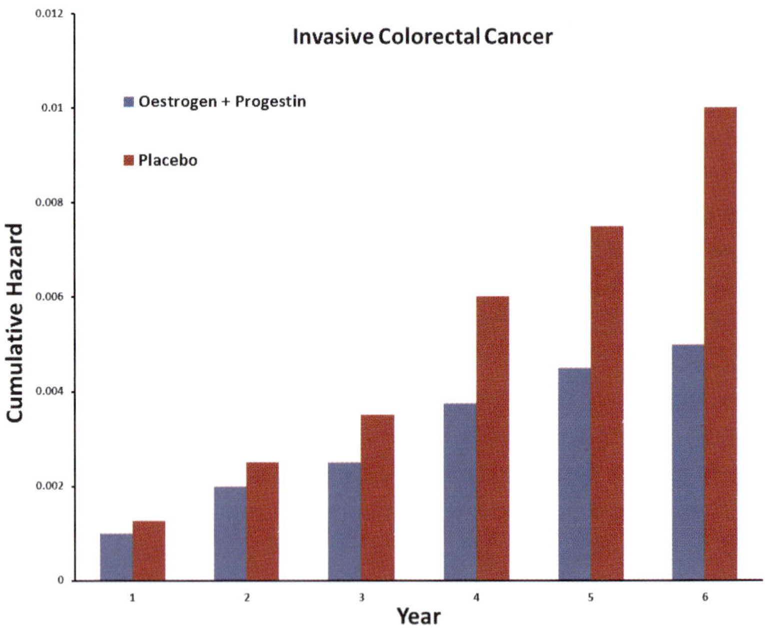

Figure 3. Incidence of invasive colorectal cancer in postmenopausal women treated with either hormone replacement therapy or placebo.

The underlying mechanism for this protective effect is unknown. The insulin/glucose pathway has been implicated in colon cancer[51] and was evaluated in a subset of WHI participants (8.6%). Interestingly, fasting glucose and insulin significantly decreased at 1 year in the HRT group. There was no interaction by HRT and body mass index or waist circumference and risk. Conversely, in the WHI CEE-alone trial, no reduction in colorectal cancer was observed (HR 1.08, 95% CI 0.75, 1.55). This lack of association between the use of CEE and colorectal cancer persisted up to 3.6 years postintervention.[20] The biology underlying this observation is not well-understood. In a laboratory evaluation of nontransformed colonocytes, investigators found that when these cells were treated with estradiol the incidence of aberrant crypt foci was reduced and apoptosis increased. The effects appear to be mediated via ERβ and illustrate a protective effect in very early stages of disease development.[52] Further work in this area is needed.

Lung Cancer

In the 2.4 years of follow-up posttreatment in WHI, an increased risk of death was noted in the combined HRT arm, and this could not be explained by excess deaths within a predefined endpoint. Thus, the cause of mortality was further examined and death from lung cancer was 43% higher in HRT users (HR 1.71, 95% CI 1.16, 2.52; $p = 0.01$). There were equal numbers of current and former smokers on each arm and the incidence of lung cancer was slightly higher in the HRT arm, with 109 cases in the HRT group versus 85 in the placebo group (HR 1.23, 95% CI 0.92, 1.63; $p = 0.16$). In regard to subtypes of lung cancer, there was a trend toward more non-small-cell lung cancer (NSCLC) on HRT (HR 1.28, 0.94–1.73, $p = 0.12$) (Fig. 4).[53] Mortality from small cell lung cancer was the same in both arms, and thus the increased mortality was attributable to more NSCLC deaths. A large cohort trial ($N = 36{,}588$), the Vitamins and Lifestyle Study (VITAL),[54] also found an increased risk of lung cancer with combined HRT (HR 1.47, 95% CI 1.06, 2.04), but not with estrogen replacement alone (HR 1.04, 95% CI 0.73, 1.48), at a mean follow-up of 5.9 years. Increasing risk was correlated with longer duration of HRT use (p for trend = 0.03). In the

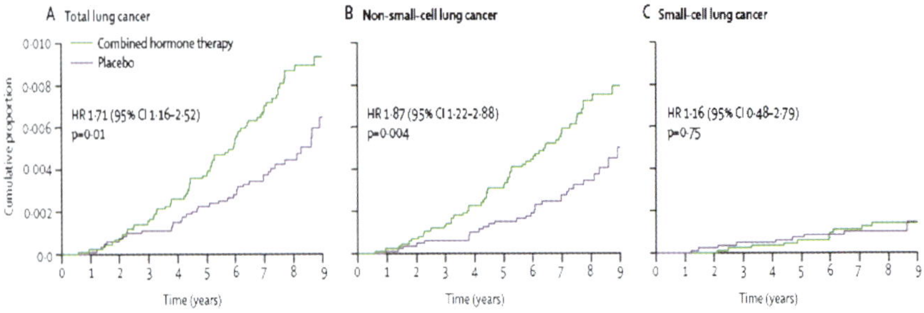

Figure 4. Deaths from lung cancer: Kaplan-Meier cumulative hazard curves for death from lung cancer by category (all lung cancer, non-small-cell lung cancer, small-cell lung cancer), study group, and time in the trial. [Reprinted from *The Lancet* **374**: 9699; Chlebowski RT, Schwartz AG, Wakelee H, *et al.* Oestrogen plus progestin and lung cancer in postmenopausal women (Women's Health Initiative trial): a post-hoc analysis of a randomised controlled trial; pp. 1248, 2011, with permission from Elsevier.

VITAL study, HRT did not preferentially increase the risk of one lung cancer subtype (NSCLC or SCLC) over the other. The HERS trial, which was smaller and had a shorter follow-up than WHI, reported a non-statistically-significant higher incidence of lung cancer (HR 1.39, 95% CI 0.84, 2.28). Other cohort trials of HRT use have reported the full spectrum of effects on lung cancer, from lower to higher risk.[55,56] Estrogen receptors are expressed in the normal lung and lung cancer and estradiol can increase proliferation of lung cancer lines. Recent analyses of NSCLC suggest that ERβ is preferentially expressed and that ERα is typically absent or found in the cytoplasm or nucleoli. In one study of 132 specimens, ERβ was found in 51%, while ERα was found in the cytoplasm of 73% of cancers.[57] In another analysis of 301 resected NSCLC cases, 45.8% expressed ERβ and none expressed ERα. Additionally, ERβ was correlated with female gender and better prognosis.[58] Abe *et al.* reported that ERβ is expressed in 84% of NSCLC and has a high concordance with aromatase expression.[59] Targeting of the estrogen pathway is being explored as lung cancer treatment.[60] There is little epidemiology evidence but two reports[61,62] have shown that increased serum estradiol is associated with increased lung cancer mortality in both men and women.

HRT and Bone Health

Combined HRT and CEE both have bone protective effects. Bone fracture incidence declined in both WHI trials, as well as hip and vertebral (osteoporosis-related) fractures. In the combined HRT WHI trial,[18] the total fracture rate was reduced by 24% (HR 0.76, 95% CI 0.69, 0.85), with a significant reduction in hip and vertebral fractures as well. In CEE alone, WHI total fractures were significantly reduced (HR 0.70, 95% CI 0.63, 0.79), as well as hip and vertebral fractures. This protective effect was lost after stopping CEE therapy and was 0.36% in the CEE group vs. 0.28% in the placebo group (HR 1.27, 95% CI, 0.88–1.82).[20] However, given the expanding pharmacopeia of drugs with bone-protective effects, the confirmation of the bone-sparing effects of estrogen does not significantly support the clinical use of hormonal agents for fracture prevention.

HRT and Menopausal Symptoms

The beneficial effect of HRT on hot flashes has been shown in multiple studies,[63-65] both for estrogen alone and in combination with progesterone. A meta-analysis of 24 placebo-controlled randomized trials[66] evaluating the effect of HRT on hot flashes and night sweats concluded that both hot flash frequency and severity declined compared to placebo. Hot flashes decreased by 75% compared to placebo (95% CI 64.3, 82.3). Interestingly, women on placebo also reported a 58% decline in hot flash frequency at end of study compared with baseline. Study durations varied from three months to three years. Note that the recruitment criteria for the WHI were written to exclude women with prevalent vasomotor symptoms due to concerns about adherence with a placebo.[17] Quality of life (QOL) was assessed in a subgroup of 1511 women, including assessment on functional status, depression, sleep, sexual and cognitive functioning as well as menopausal symptoms after one and three years on treatment. Investigators found that E + P did not have any clinical benefit for QOL.[67] A subset analysis of women aged[50-54] with moderate-to-severe vasomotor symptoms at baseline showed a benefit from E + P in regard to sleep, but the overall QOL scores were similar to those for older women. In women without vasomotor symptoms, there was no appreciable QOL benefit of HRT in other phase III trials.[64,68]

HRT and Cognitive Function

Elderly women have a higher incidence of Alzheimer's disease than elderly men and one possible explanation is a decline in the sex hormone estrogen. Despite epidemiology studies suggesting a benefit of HRT to cognition,[69] clinical trials have proven disappointing. In the Women's Health Initiative Memory Study (WHIMS), a substudy of the WHI, 2,947 women aged[65–79] participated. Neither CEE alone nor combined HRT reduced the risk of dementia or mild cognitive impairment. The pooled analysis for both types of hormone replacement showed an increased risk of probable dementia, HR 1.76, 95% CI 1.19, 2.60, $p = 0.005$. In regard to type of dementia, 52% were classified as Alzheimer's disease (AD), 9% vascular dementia and 16% mixed type; there was no significant difference by type of HRT.[70] HRT adversely affected global cognitive function.[71] Note that participants in WHIMS were older and benefit in a younger cohort has not been prospectively evaluated. The Cochrane overview included 7 trials of 351 women with AD and found that women on CEE had worse cognitive function.[72] Thus, there is no evidence to date that HRT delays or improves dementia.

HRT and Gallbladder Disease

The Nurses' Health study,[73] a prospective cohort study, reported that women who used hormones were twice as likely to undergo cholecystectomy as those who did not (RR 2.1, 95% CI 1.9, 2.4). Risk increased with longer duration of use and higher dose. While the WHI has not reported on gallbladder disease, the HERS trial in women with CAD treatment found that E + P use had a 38% increase in the relative risk for biliary tract surgery ($p = 0.05$).[74]

SUMMARY

The expectations regarding the benefits of HRT to chronic disease were high, and physicians during the 1990s were routinely recommending use of HRT to prevent common illnesses. When the clinical trial results showed lack of benefit, scientists, physicians and patients were let down as well as confused. Possible explanations for the differences included the "healthy user" effect, in which healthier women were more likely to take HRT and thus be less likely to suffer from chronic diseases. Other possible reasons included incomplete data in observational studies as well as difference in the study population at baseline or types of hormone preparations used, although the Cochrane meta-analysis argues against this last point since multiple types of hormones were studied and included in the analysis. In order to reconcile the observational data with the null or negative results of the phase III trials, investigators conducted a number of subgroup analyses focusing on the age of subjects and the time since menopause. This approach has been called the "timing hypothesis." Rossouw *et al.*[75] evaluated CHD events in both WHI trials by time from menopause.[75] When combining the two arms that received HRT (combination or single agent) in women less than 10 years from menopause, the CHD event rate was reduced by 24% (HR 0.76, 95% CI 0.50, 1.16). Conversely, for women more than 20 years from menopause, the HR was 1.28 (95% CI 1.03, 1.58). The p value for trend was significant at 0.02.[75] Although the lower event rate in the women closer to menopause is more in keeping with the expected benefits of HRT to CHD, the confidence intervals are wide, reflecting the smaller proportion of women analyzed. For stroke there was no benefit to being closer to menopause; rather, for women <10 years from menopause the stroke risk increased by 77%. These investigators also evaluated events by age group (50–59, 60–69, and 70–79) and no significant trends for CHD, stroke, total mortality or global index score were seen in the combined analysis. When evaluating the CEE-alone trial, younger age was associated with significantly better global index scores, but this was not seen in the combined HRT trial. In the Cochrane meta-analysis, HRT effects were also evaluated by age group. For the 50–59-year-old women with combined HT and E alone, the only increased risk was for TE disease, and the absolute risk was low at <1 in 500. Thus, the duration of time from menopause is important and this has been demonstrated in animal studies as well.[76] In regard to CAD, the mechanism may be related to prevention of plaque accumulation with lower baseline levels close to the time of menopause, and hence a greater benefit. The timing hypothesis for breast cancer appears to have the opposite effect, i.e. using HRT closer to menopause increases the risk of breast cancer. The preponderance of data shows a negative risk: benefit analysis suggests that, on the whole, HRT should be avoided. The results of these trials dramatically changed the use of HRT in the United States. In 2003, prescriptions for E + P decreased by 66% and for E by 33%.[77] Moreover, this change in use has been linked to an 8.6% decline in the age-adjusted incidence of breast cancer from 2001 to 2004.[78]

Currently, the use of HRT in clinical practice is largely limited to those who have recently entered menopause and is used to combat hot flushes. Professional organizations[79] recommend against the use of HRT for the general population. If employed HRT should be used for as brief a time as possible, e.g. less than 2 years, since adverse events are related to duration of use.[80] Short term use (< 2 years) of HRT does not increase the risk of breast cancer; however, CHD events, stroke and TEE still are more than expected even within the first 2 years of use.[81] In a 60-year-old woman the serious adverse event rate is 2/1000 with 1-year follow-up; for a 50-year-old woman it is 1/1000,[82] and thus short term use may be reasonable in women with menopausal symptoms not well controlled on other therapies. HRT does have positive effects on bone health, yet there are a number of other approved agents which are as efficacious and do not have the associated risks. HRT should not be used for the treatment of osteopenia/porosis and should be avoided in women at increased risk for thromboembolic events, breast cancer or lung cancer.

Challenges of the Future

Since the benefits of HRT may be tissue-specific, a number of methods have been tried to obtain target-organ-specific effects and potentially avoid unwanted systemic effects. Topical creams and intravaginal preparations have been developed specifically for vaginal symptoms. However, these formulations also are systemically absorbed and increase circulating estrogen levels.[83] There is no delivery method to date that ensures a limited tissue-specific dose of HRT. Another approach is to develop drugs that have selective estrogen agonist and antagonist, and this has been achieved with the selective estrogen receptor modulators (SERM), which include tamoxifen and raloxifene and, more recently, lasofoxifene[84] and arzoxifene.[85,86] Tamoxifen and raloxifene have been proven to reduce breast cancer incidence and improve bone health, but still have risk of stroke and TEE.[87,88] Emerging evidence suggests that a SERM-type agent may potentially have a role in prevention of colon and lung cancer. Certainly, more needs to be understood about estrogen signaling in colon and lung carcinogenesis. The search for the perfect SERM continues and review of the data suggests that perfect SERM may differ depending on a woman's baseline risk for disease, age and time since menopause.

ACKNOWLEDGEMENTS

This work (V.C.J.) was supported by the Department of Defense Breast Program under award No. W81XWH-06-1-0590 Center of Excellence; subcontract under SU2C (AACR) grant No. SU2C-AACR-DT0409; the Susan G. Komen for the Cure Foundation under award No. SAC100009; GHUCCTS CTSA (Grant #UL1RR031975) and the Lombardi Comprehensive Cancer Center Support Grant (CCSG) Core Grant NIH P30 CA051008. The views and opinions of the author(s) do not reflect those of the U.S. Army or the Department of Defense.

REFERENCES

1. *Women and Cardiovascular Diseases — Statistics.* (2009) American Heart Association.
2. *Osteoporosis Overview.* (2009) Bethesda: National Institutes of Health Osteoporosis and Related Bone Diseases.

3. Keen R. (2007) Osteoporosis: strategies for prevention and management. *Best Pract Res Clin Rheumatol* **21**: 109–122.
4. Watkins ES. (2007) *The Estrogen Elixir: A History of Hormone Replacement Therapy in America*. Baltimore, MD: The Johns Hopkins University Press, pp. 10–132.
5. *FDA Backgrounder on Conjugated Estrogens*. (2005) U.S. Food and Drug Administration Center for Drug Evaluation and Research.
6. Smith DC, Prentice R, Thompson DJ, Herrmann WL. (1975) Association of exogenous estrogen and endometrial carcinoma. *N Engl J Med* **293**:1164–1167.
7. Ziel HK, Finkle WD. (1975) Increased risk of endometrial carcinoma among users of conjugated estrogens. *N Engl J Med* **293**:1167–1170.
8. Stefanick ML. (2005) Estrogens and progestins: background and history, trends in use, and guidelines and regimens approved by the U.S. Food and Drug Administration. *Am J Med* **118**(12):64–73.
9. Nachtigall LE, Nachtigall RH, Nachtigall RB, *et al.* (1976) Estrogens and endometrial carcinoma, correspondence. *N Engl J Med* **294**:848.
10. Sturdee DW, Wade-Evans T, Paterson ME, *et al.* (1978) Relation between bleeding pattern, endometrial histology, and oestrogen treatment in menopausal women. BMJ **1**:1575–1577.
11. FDA. (1997) Conjugated Estrogens — Letter from Dr. Janet Woodcock. Memorandum, U.S. Food and Drug Administration Center for Drug Evaluation and Research.
12. Grady D, Rubin SM, Petitti DB, *et al.* (1992) Hormone therapy to prevent disease and prolong life in postmenopausal women. *Ann Intern Med* **117**:1016–1037.
13. Adams MR, Kaplan JR, Manuck SB, *et al.* (1990) Inhibition of coronary artery atherosclerosis by 17-beta estradiol in ovariectomized monkeys. Lack of an effect of added progesterone. *Arteriosclerosis* **10**:1051–1057.
14. Karas RH. (2002) Animal models of the cardiovascular effects of exogenous hormones. *Am J Cardiol* **90**:22F–25F.
15. Farquhar C, Marjoribanks J, Lethaby A, Suckling JA, Lamberts Q. (2009) Long term hormone therapy for perimenopausal and postmenopausal women. *Cochrane Database Syst Rev* CD004143.
16. Hulley S, Grady D, Bush T, *et al.* (1998) Randomized trial of estrogen plus progestin for secondary prevention of coronary heart disease in postmenopausal women. Heart and Estrogen/Progestin Replacement Study (HERS) Research Group. *JAMA* **280**:605–613.
17. Design of the Women's Health Initiative clinical trial and observational study. (1998) The Women's Health Initiative Study Group. *Control Clin Trials* **19**:61–109.
18. Rossouw JE, Anderson GL, Prentice RL, *et al.* (2002) Risks and benefits of estrogen plus progestin in healthy postmenopausal women: principal results from the Women's Health Initiative randomized controlled trial. *JAMA* **288**:321–333.

19. Anderson GL, Limacher M, Assaf AR, *et al.* (2004) Effects of conjugated equine estrogen in postmenopausal women with hysterectomy: the Women's Health Initiative randomized controlled trial. *JAMA* **291**:1701–1712.
20. LaCroix AZ, Chlebowski RT, Manson JE, Aragaki AK, Johnson KC, Martin L, Margolis KL, Stefanick ML, Brzyski R, Curb JD, Howard BV, Lewis CE, Wactawski-Wende J, for the WHI Investigators. (2011) Health outcomes after stopping conjugated equine estrogens among postmenopausal women with prior hysterectomy: a randomized controlled trial *JAMA* **305**(13):1305–1314.
21. Manson JE, Hsia J, Johnson KC, *et al.* (2003) Estrogen plus progestin and the risk of coronary heart disease. *N Engl J Med* **349**:523–534.
22. Hsia J., Langer R. D, Manson, J. E, *et al.* (2006) Conjugated equine estrogens and coronary heart disease: the Women's Health Initiative. *Arch Intern Med* **166**:357–365.
23. Greenland P, LaBree L, Azen SP, Doherty TM, Detrano RC. (2004) Coronary artery calcium score combined with Framingham score for risk prediction in asymptomatic individuals. *JAMA* **291**:210–215.
24. Manson JE, Allison MA, Rossouw JE, *et al.* (2007) Estrogen therapy and coronary-artery calcification. *N Engl J Med* **356**:2591–2602.
25. Vickers M, MacLennan AH, Lawton B, *et al.* (2007) Main morbidities recorded in the women's international study of long duration oestrogen after menopause (WISDOM): a randomised controlled trial of hormone replacement therapy in postmenopausal women. *BMJ* **335**:239.
26. Gail MH, Brinton LA, Byar DP, *et al.* (1989) Projecting individualized probabilities of developing breast cancer for white females who are being examined annually. *J Natl Cancer Inst* **81**:1879–1886.
27. Chlebowski T, Hendrix SL, Langer RD, *et al.* (2003) Influence of estrogen plus progestin on breast cancer and mammography in healthy postmenopausal women: the Women's Health Initiative randomized trial. *JAMA* **289**: 3243–3253.
28. Calle EE, Feigelson HS, Hildebrand JS, *et al.* (2009) Postmenopausal hormone use and breast cancer associations differ by hormone regimen and histologic subtype. *Cancer* **115**:936–945.
29. Fournier A, Fabre A, Mesrine S, *et al.* (2008) Use of different postmenopausal hormone therapies and risk of histology- and hormone receptor-defined invasive breast cancer. *J Clin Oncol,* **26**:1260–1268.
30. Reeves GK, Beral V, Green J, Gathani T, Bull D. (2006) Hormonal therapy for menopause and breast-cancer risk by histological type: a cohort study and meta-analysis. *Lancet Oncol* **7**:910–918.
31. Chlebowski RT, Kuller LH, Prentice RL, *et al.* (2009) Breast cancer after use of estrogen plus progestin in postmenopausal women. *N Engl J Med* **360**:573–587.
32. Beral V. (2003) Breast cancer and hormone-replacement therapy in the Million Women Study. *Lancet* **362**:419–427.

33. Collaborative Group on Hormonal Factors in Breast Cancer. (1997) Breast cancer and hormone replacement therapy: collaborative reanalysis of data from 51 epidemiological studies of 52,705 women with breast cancer and 108,411 women without breast cancer. *Lancet* **350**:1047–1059.
34. Prentice RL, Chlebowski RT, Stefanick ML, *et al.* (2008) Estrogen plus progestin therapy and breast cancer in recently postmenopausal women. *Am J Epidemiol* **167**:1207–1216.
35. McTiernan A, Martin CF, Peck JD, *et al.* (2005) Estrogen-plus-progestin use and mammographic density in postmenopausal women: Women's Health Initiative Randomized Trial. **97**:1366–1376.
36. Fournier A, Mesrine S, Boutron-Ruaul MC, Clavel-Chapelon F. (2009) Estrogen–progestagen menopausal hormone therapy and breast cancer: does delay from menopause onset to treatment initiation influence risks? *J Clin Oncol* **27**:5138–5143.
37. Li CI, Malone KE, Porter PL, *et al.* (2003) Relationship between long durations and different regimens of hormone therapy and risk of breast cancer. *JAMA* **289**:3254–3263.
38. Greendale GA, Reboussin BA, Slone S, *et al.* (2003) Postmenopausal hormone therapy and change in mammographic density. *J Natl Cancer Inst* **95**:30–37.
39. Anderson GL, Judd HL, Kaunitz AM, *et al.* (2003) Effects of estrogen plus progestin on gynecologic cancers and associated diagnostic procedures: the Women's Health Initiative randomized trial. *JAMA* **290**:1739–1748.
40. Coughlin SS, Giustozzi A, Smith SJ, Lee NC. (2000) A meta-analysis of estrogen replacement therapy and risk of epithelial ovarian cancer. *J Clin Epidemiol* **53**:367–375.
41. Garg PP, Kerlikowske K, Subak L, Grady D. (1998) Hormone replacement therapy and the risk of epithelial ovarian carcinoma: a meta-analysis. *Obstet Gynecol* **92**:472–479.
42. Zhou B, Sun Q, Cong R, *et al.* (2008) Hormone replacement therapy and ovarian cancer risk: a meta-analysis. *Gynecol Oncol* **108**:641-651.
43. Riman T, Dickman PW, Nilsson S, *et al.* (2002) Hormone replacement therapy and the risk of invasive epithelial ovarian cancer in Swedish women. *J Natl Cancer Inst* **94**:497–504.
44. Lacey JV, Jr, Mink PJ, Lubin JH, *et al.* (2002) Menopausal hormone replacement therapy and risk of ovarian cancer. *JAMA* **288**:334–341.
45. Grodstein F, Newcomb PA, Stampfer MJ. (1999) Postmenopausal hormone therapy and the risk of colorectal cancer: a review and meta-analysis. *Am J Med* **106**:574–582.
46. Newcomb PA, Storer BE. (1995) Postmenopausal hormone use and risk of large-bowel cancer. *J Natl Cancer Inst* **87**:1067–1071.
47. Persson I, Yuen J, Bergkvist L, Schairer C. (1996) Cancer incidence and mortality in women receiving estrogen and estrogen–progestin replacement therapy — long-term follow-up of a Swedish cohort. *Int J Cancer* **67**:327–332.

48. Troisi R, Schairer C, Chow WH, *et al.* (1997) A prospective study of menopausal hormones and risk of colorectal cancer (United States). *Cancer Causes Control* **8**:130–138.
49. Rennert G, Rennert HS, Pinchev M, Lavie O, Gruber SB. (2009) Use of hormone replacement therapy and the risk of colorectal cancer. *J Clin Oncol* **27**:4542–4547.
50. Chlebowski RT, Wactawski-Wende J, Ritenbaugh C, *et al.* (2004) Estrogen plus progestin and colorectal cancer in postmenopausal women. *N Engl J Med* **350**:991–1004.
51. Wolpin BM, Meyerhardt JA, Chan AT, *et al.* (2009) Insulin, the insulin-like growth factor axis, and mortality in patients with nonmetastatic colorectal cancer. *J Clin Oncol* **27**:176–185.
52. Weige CC, Allred KF, Allred CD. (2009) Estradiol alters cell growth in nonmalignant colonocytes and reduces the formation of preneoplastic lesions in the colon. *Cancer Res* **69**:9118–9124.
53. Chlebowski RT, Schwartz AG, Wakelee H, *et al.* (2009) Oestrogen plus progestin and lung cancer in postmenopausal women (Women's Health Initiative trial): a post-hoc analysis of a randomised controlled trial. *Lancet* **374**:1243–1251.
54. Slatore CG, Chien JW, Au DH, Satia JA, White E. (2010) Lung cancer and hormone replacement therapy: association in the vitamins and lifestyle study. *J Clin Oncol* **28**:1540–1546.
55. Rodriguez C, Spencer Feigelson H, Deka A, *et al.* (2008) Postmenopausal hormone therapy and lung cancer risk in the cancer prevention study II nutrition cohort. *Cancer Epidemiol Biomarkers Prev* **17**:655–660.
56. Taioli E, Wynder EL. (1994) Re: Endocrine factors and adenocarcinoma of the lung in women. *J Natl Cancer Inst* **86**:869–870.
57. Kawai H, Ishii A, Washiya K, *et al.* (2005) Estrogen receptor alpha and beta are prognostic factors in non-small cell lung cancer. *Clin Cancer Res* **11**: 5084–5089.
58. Wu CT, Chang YL, Shih JY, Lee YC. (2005) The significance of estrogen receptor beta in 301 surgically treated non-small cell lung cancers. *J Thorac Cardiovasc Surg* **130**:979–986.
59. Abe K, Miki Y, Ono K, *et al.* (2009) Highly concordant coexpression of aromatase and estrogen receptor beta in non-small cell lung cancer. *Hum Pathol* **69**(11):4598–4604.
60. McLaughlin BT, Land S, Tarhini AA, Siegfried J, Argiris A. (2011) A phase II randomized trial of anastrozole (A) and fulvestrant (F) as consolidation therapy in postmenopausal women with advanced non-small cell lung cancer who have received first-line platinum-based chemotherapy with or without bevacizumab. *J Clin Oncol* 29:(suppl; abstr TPS212).
61. Albain KS, Unger J, Gotay CC, Davies AM, Edelman M, Herbst RS, Kelly K, Williamson S, Wozniak AJ, Gandara DR, for the Southwest Oncology Group Lung Committee. (2007) Toxicity and survival by sex in patients with advanced non-small cell lung carcinoma on modern Southwest Oncology Group (SWOG) trials. *ASCO Annual Meeting* Abstr. 7549. Chicago, IL.

62. Ross H, Oldham FB, Bandstr B, Sandalic L, Bianco J, Bonomi, P, Singer JW. (2007) Serum-free estradiol (E2) levels are prognostic in men with chemotherapy-naive advanced non-small cell lung cancer (NSCLC) and performance status (PS) 2. *ASCO Annual Meeting*. Abstr. #7683. Chicago, IL, 2007.
63. Baerug U, Winge T, Nordland G, *et al.* (1998) Do combinations of 1 mg estradiol and low doses of NETA effectively control menopausal symptoms? *Climacteric* **1**:219–228.
64. Greendale GA, Reboussin BA, Hogan P, *et al.* (1998) Symptom relief and side effects of postmenopausal hormones: results from the Postmenopausal Estrogen/Progestin Interventions Trial. *Obstet Gynecol* **92**:982–988.
65. Symons J, Kempfert N, Speroff L. (2000) Vaginal bleeding in postmenopausal women taking low-dose norethindrone acetate and ethinyl estradiol combinations. The FemHRT Study Investigators. *Obstet Gynecol* **96**:366–372.
66. Maclennan AH, Broadbent JL, Lester S, Moore V. (2004) Oral oestrogen and combined oestrogen/progestogen therapy versus placebo for hot flushes. *Cochrane Database Syst Rev:* CD002978.
67. Hays J, Ockene JK, Brunner RL, *et al.* (2003) Effects of estrogen plus progestin on health-related quality of life. *N Engl J Med* **348**:1839–1854.
68. Hlatky MA, Boothroyd D, Vittinghoff E, Sharp P, Whooley MA. (2002) Quality-of-life and depressive symptoms in postmenopausal women after receiving hormone therapy: results from the Heart and Estrogen/Progestin Replacement Study (HERS) trial. *JAMA* **287**:591–597.
69. Nelson HD, Humphrey LL, Nygren P, Teutsch SM, Allan JD. (2002) Postmenopausal hormone replacement therapy: scientific review. *JAMA* **288**:872–881.
70. Shumaker SA, Legault C, Kuller L, *et al.* (2004) Conjugated equine estrogens and incidence of probable dementia and mild cognitive impairment in postmenopausal women: Women's Health Initiative Memory Study. *JAMA* **291**:2947–2958.
71. Espeland MA, Rapp SR, Shumaker SA, *et al.* (2004) Conjugated equine estrogens and global cognitive function in postmenopausal women: Women's Health Initiative Memory Study. *JAMA* **291**:2959–2968.
72. Hogervorst E, Yaffe K, Richards M, and Huppert FA. (2009) Hormone replacement therapy to maintain cognitive function in women with dementia. *Cochrane Database Syst Rev:* CD003799.
73. Grodstein F, Colditz GA, Stampfer MJ. (1994) Postmenopausal hormone use and cholecystectomy in a large prospective study. *Obstet Gynecol* **83**:5–11.
74. Simon JA, Hunninghake DB, Agarwal SK, *et al.* (2001) Effect of estrogen plus progestin on risk for biliary tract surgery in postmenopausal women with coronary artery disease. The Heart and Estrogen/Progestin Replacement Study. *Ann Intern Med* **135**:493–501.
75. Rossouw JE, Prentice RL, Manson JE, *et al.* (2007) Postmenopausal hormone therapy and risk of cardiovascular disease by age and years since menopause. *JAMA* **297**:1465–1477.

76. Mikkola TS, Clarkson TB. (2002) Estrogen replacement therapy, atherosclerosis, and vascular function. *Cardiovasc Res* **53**:605–619.
77. Hersh AL, Stefanick ML, Stafford RS. (2004) National use of postmenopausal hormone therapy: annual trends and response to recent evidence. *JAMA* **291**:47–53.
78. Ravdin PM, Cronin KA, Howlader N, *et al.* (2007) The decrease in breast-cancer incidence in 2003 in the United States. *N Engl J Med* **356**:1670–1674.
79. North American Menopause Society. (2010) Estrogen and progestogen use in postmenopausal women: 2010 position statement of The North American Menopause Society. *Menopause* **17**:242–255.
80. Force, U.S.P.S.T. (2005) Hormone Replacement Therapy for the Prevention of Chronic Conditions in Postmenopausal Women Recommendation Statement AHRQ.
81. Solomon CG, Dluhy RG. (2003) Rethinking postmenopausal hormone therapy. *N Engl J Med* **348**:579–580.
82. Grady D. (2003) Postmenopausal hormones — therapy for symptoms only. *N Engl J Med* **348**:1835–1837.
83. Johnston A. (1996) Estrogens — pharmacokinetics and pharmacodynamics with special reference to vaginal administration and the new estradiol formulation — Estring. *Acta Obstet Gynecol Scand Suppl* **163**:16–25.
84. Lewiecki EM. (2009) Lasofoxifene for the prevention and treatment of postmenopausal osteoporosis. *Ther Clin Risk Manag* **5**:817–827.
85. Bolognese M, Krege JH, Utian WH, *et al.* (2009) Effects of arzoxifene on bone mineral density and endometrium in postmenopausal women with normal or low bone mass. *J Clin Endocrinol Metab* **94**:2284–2289.
86. Sporn MB. (2004) Arzoxifene: a promising new selective estrogen receptor modulator for clinical chemoprevention of breast cancer — Commentary re C.J. Fabian *et al.*, Breast cancer chemoprevention Phase I evaluation of biomarker modulation by arzoxifene, a third generation selective estrogen receptor modulator. *Clin Cancer Res* 10:5403–5417. *Clin Cancer Res* **10**:5313–5315.
87. Fisher B, Costantino JP, Wickerham, DL, *et al.* (1998) Tamoxifen for prevention of breast cancer: report of the National Surgical Adjuvant Breast and Bowel Project P-1 Study. *J Natl Cancer Inst* **90**:1371–1388.
88. Vogel VG, Costantino JP, Wickerham DL, *et al.* (2006) Effects of tamoxifen vs. raloxifene on the risk of developing invasive breast cancer and other disease outcomes — The NSABP study of tamoxifen and raloxifene (STAR) P-2 trial. *JAMA* **295**:2727–2741.
89. Hulley S, Furberg C, Barrett-Connor E, *et al.* (2002) Noncardiovascular disease outcomes during 6.8 years of hormone therapy: Heart and Estrogen/Progestin Replacement Study follow-up (HERS II). *JAMA* **288**:58–66.
90. Anderson G, Chlebowski RT, Aragaki AK, *et al.* (2012) Conjugated equine oestrogen and breast cancer incidence and mortality in postmenopausal women with hysterectomy: extended follow-up of the Women's Health Initiative randomised placebo-controlled trial. *Lancet Oncol* **13**(5):476–486.

5

INDUCTION OF OVULATION BY NONSTEROIDAL ANTIESTROGENS

Ifeyinwa Obiorah and V. Craig Jordan

Abstract

The clinical finding that the rodent antifertility agent nonsteroidal antiestrogen clomiphene (a mixture of geometric isomers) was actually a profertility agent in subfertile women created the first practical method for enhancing fertility in women. A related compound, ICI 46,474, the pure trans isomer of a substituted triphenylethylene, was also a product of an industry fertility control program in the 1960s and it too was tested and then marketed in the United Kingdom as an inducer of ovulation in subfertile women, at the same time as an orphan drug for the treatment of metastatic breast cancer in postmenopausal women. Clomiphene has, however, remained the clinical gold standard for the induction of ovulation worldwide for 40 years. However, the fact that tamoxifen is a potent inducer of ovulation in premenopausal women remains an important consideration in breast cancer patients made infertile by combination cytotoxic chemotherapy. This chapter will trace the genesis of agents for the induction of ovulation and the current potential applications of tamoxifen for women with breast cancer who choose to preserve fertility.

Keywords: Ovarian Stimulation; Tamoxifen; Clomiphene; Pregnancy; Ethamoxytriphetol; MER-25; Polycystic Ovarian Syndrome.

INTRODUCTION

Ovulation abnormality is one of the commonest causes of subfertility as well as infertility. There are many causes of anovulation and, whenever possible, correction of the underlying cause should be the first line of treatment.[1] Etiologic diagnosis of endocrinopathies, strenuous exercise, pituitary tumors, extreme weight loss, hyperprolactinemia, polycystic ovarian disease and obesity may be determined, but most often the immediate cause of anovulation cannot be successfully defined. Before the 1960s, efficient treatment of anovulation was virtually nonexistent.[2] It was not until 1958 that ethamoxytriphetol (MER-25), an inhibitor of estrogen action, provided a new tool for laboratory and clinical investigation.[3] The discovery that MER-25 had antifertility properties in rats[4] strengthened interest in the contraceptive potential of antiestrogens. In spite of successful preliminary clinical trials with MER-25,[5] these were discontinued due to toxicity and low potency.[6] This led to the search for improved compounds for clinical application.

Clomiphene citrate (MRL.41) is a nonsteroidal triphenylethylene, with a structure related to its predecessor, MER.25 (Fig. 1). Clomiphene was initially thought to be a promising oral contraceptive, because it was found to inhibit the release of pituitary gonadotrophins in rats.[7] However, in women with long-standing anovulatory amenorrhea, Greenblatt *et al.*[8] found that it induced ovulation and allowed conception to occur. Clomiphene has both estrogen agonist and antagonist properties.[9] However, its estrogen agonist properties are manifested only when endogenous estrogen levels are extremely low.[1] Therefore, the main action of clomiphene is through its antiestrogenic effects. However, examination of clomiphene's structure (Fig. 1) suggests that there is potential for geometric isomerization. Indeed, the clinically used medicine clomiphene (Clomid) is a mixture of two distinct geometric isomers, enclomiphene and zuclomiphene (Fig. 1). It is worthy of note that when the medicinal chemistry of clomiphene and its isomers were first reported, they were misclassified. The antiestrogen isomer was identified as the cis geometric isomer and the estrogenic isomer was identified as the trans isomer. Confusions arose with the publication of the differing pharmacology of the trans ICI 46,474 (antiestrogenic properties) and its cis isomer

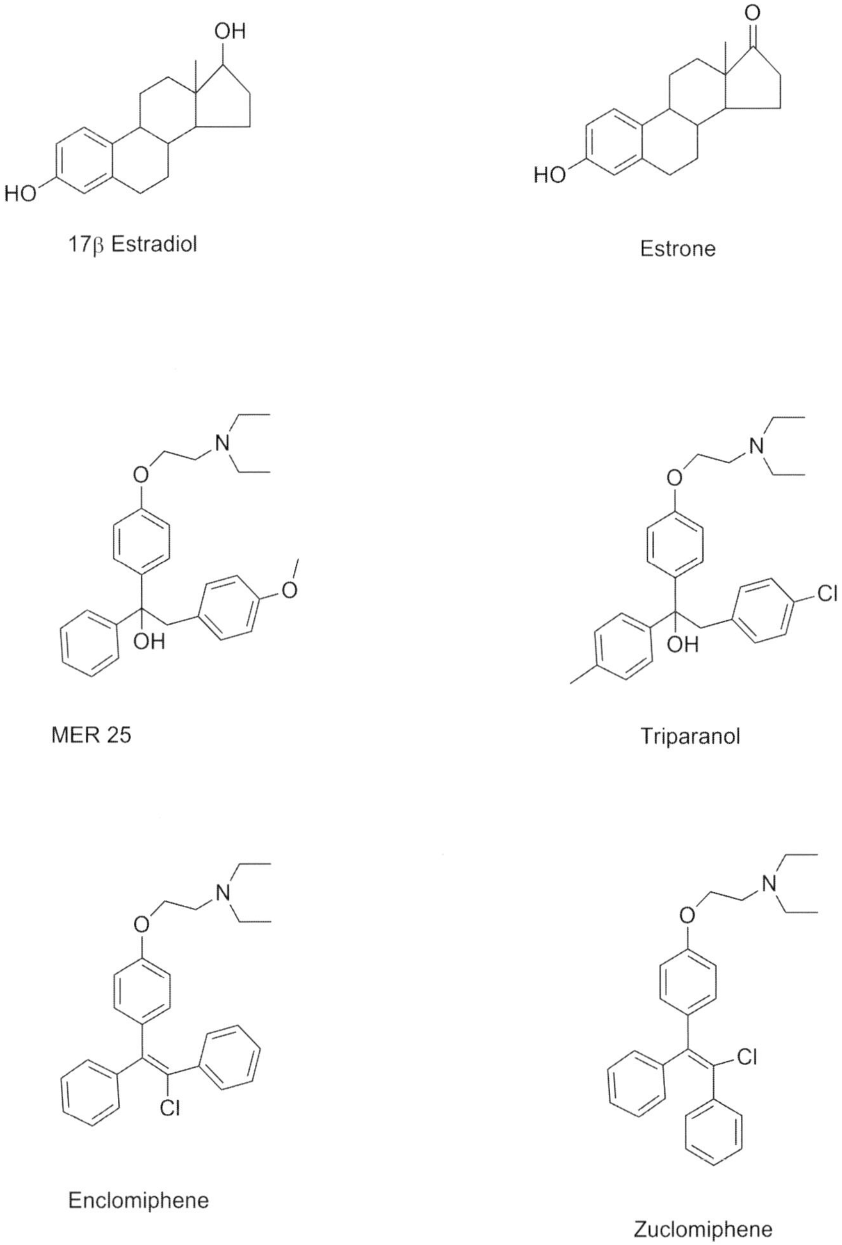

Figure 1. The chemical structure of the SERM discussed in the text. Estradiol is included for comparison.

ICI 47,699 (estrogenic properties). The situation was clarified in the early 1970s with the renaming of clomiphene isomers as enclomiphene, the antiestrogenic *trans* isomer and zuclomiphene, the estrogenic *cis* isomer. Zuclomiphene accounts for 38% of the drug. On the other hand, enclomiphene (trans isomer) is an antiestrogenic partial agonist, and it is the more potent of the two and primarily possesses the ovulation-inducing properties of clomiphene.[9,10] The mechanism of action of clomiphene in inducing ovulation has been proposed by the practice committee of the American Society for Reproductive Medicine (ARSM).[1] Clomiphene is thought to displace endogenous estrogen from hypothalamic estrogen receptor sites (see Fig. 2). This

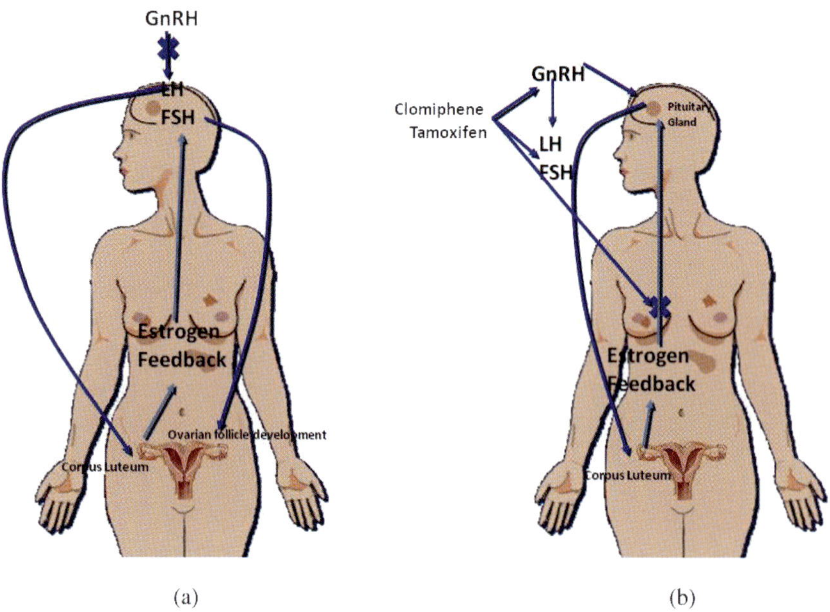

Figure 2. Hormonal regulation of the menstrual cycle. (**a**) During the menstrual cycle, GnRH stimulates the release of FSH and LH from the anterior pituitary. FSH initiates ovarian follicle development and secretion of estrogen from the follicles, while LH facilitates the maturation of the egg and subsequently induces ovulation and formation of the corpus luteum, which in turn produces estrogen and progesterone. High levels of estrogen cause a negative feedback that inhibits secretion of LH and FSH. (**b**) The antiestrogens clomiphene and tamoxifen block the estrogen negative feedback by displacing estrogen from the hypothalamic ER sites. As a result, circulating estrogens rise in the premenopausal patient.

blocks the estrogen negative feedback, resulting in a change in the release of pulsatile hypothalamic Gonadotropin Releasing Hormone (GnRH) as well as an increase in the stimulation of Follicle Stimulating Hormone (FSH) and Luteinising Hormone (LH) secretion from the pituitary gland, which subsequently drives ovarian follicle development. In women with normal ovulation, administration of clomiphene increases the frequency rather than the amplitude of GnRH pulses.[11] However, in anovulatory women with polycystic ovarian syndrome (PCOS) who have an abnormally high GnRH pulse frequency, an increase in the pulse amplitude is reported.[12] Administration of clomiphene therapy to normal women causes serum LH and FSH levels to rise and fall following a typical five-day course of treatment.[13] Therefore, it is thought that those successful treatment cycles result in one or more dominant follicles which lead to an increase in estrogen that stimulates the midcycle LH surge and, subsequently, ovulation. Since clomiphene is an antiestrogen, another proposed clinical application of the drug was in the treatment of breast cancer due to the known involvement of hormones in breast cancer development.[14,15] However, this path was discontinued by the Merrell Company as the accumulation of the cholesterol precursor, desmosterol, was noted with clomiphene administration in clinical studies. Merrell had previously marketed a drug, triparanol, a compound structurally related to MER-25 (see Fig. 1), for the reduction of circulating cholesterol. Regrettably, triparanol caused cataracts in young women and it was believed that the drug blocked cholesterol biosynthesis but desmosterol levels increased.[16] The company subsequently chose not to develop any drug that increased demosterol levels.

The early development of tamoxifen (ICI 46,474), a triphenylethylene derivative, is similar to that of clomiphene. This nonsteroidal antiestrogenic drug was developed and introduced shortly after clomiphene was established as an ovulation-inducing drug. However, the main focus of the clinical development of tamoxifen subsequently became breast cancer; but its ability to stimulate ovulation has been recognized for many years.[17] Tamoxifen is the pure Z (trans) isomer of p-β-dimethyaminoethoxy-1,2-diphenylbut-1-ene and is a partial estrogen agonist with antagonistic properties. Even though tamoxifen is structurally related to clomiphene, it should be noted that early laboratory studies showed that tamoxifen did not increase desmosterol levels in rats.[17] Many uncontrolled studies of ovulation induction have been done over the years, suggesting its safety and efficacy.[18] However, the reason why all clinical trials were initially stopped in 1972 was not toxicity but rather the economic concerns of insignificant projected profits that would occur from marketing a drug for induction of ovulation and a few thousand patients with advanced breast cancer, treated for a year.[16] Once the path for the development of tamoxifen (or ICI 46,474) as a breast cancer drug was determined outside the company in the early 1970s, this became the main focus for clinical development and marketing worldwide. Although tamoxifen is less commonly used than clomiphene, it is available in a number of countries for induction of ovulation in women with anovulatory infertility.

EARLY TRIALS WITH CLOMIPHENE CITRATE

Clomiphene citrate represented the first realization of the use of chemical therapy for induction of ovulation. A chemically synthesized compound rather than a naturally occurring hormone could now be used used to initiate ovulation in humans.[19] Clomiphene was synthesized by Dr. Frank P. Papoli,[20] a synthetic chemist with the William S. Merrell Company, and was introduced into clinical trials in 1960. The clinical studies were conducted by a group of investigators, including Dr. Robert Greenblatt, Dr. Robert Kistner and Dr. Charles Lloyd, which ultimately led to the successful introduction of clomiphene into medical practice for the induction of ovulation. Clomiphene was initially shown to induce ovulation in 28 of 36 women with chronic anovulation.[8] In normal cycling women, administration of the drug led to a delay of menses and the extension of the 14-day postovulatory thermogenic shift to 20–25 days. Subsequent confirmatory studies quickly followed. An investigational new drug application was filed with the Food and Drug Administration in 1962 and the approval for its widespread clinical use was given in 1967.[19]

The probability of successful induction of ovulation by clomiphene seemed to depend on the cause of ovulation failure. Bishop *et al.*[21] treated 102 patients with various types of amenorrhea. Their results showed that clomiphene was unlikely to be effective in primary amenorrhea, where there was no response in 7 out of 9 cases. However, 50% of patients with long-standing amenorrhea achieved ovulation. Clomiphene proved to be a more promising form of therapy in amenorrhca following pregnancy (10 out of 12 cases), in oligomenorrhoea (24 out of 26 cases) and in metropathia (12 out of 15 cases). Ten pregnancies occurred among 68 patients who achieved ovulation, giving rise to a pregnancy rate of 15%. In 35 patients whose urinary estrogen levels were estimated before and after a course of clomiphene, 20 patients responded with a rise in estrogen, 15 patients did not show clinical response and 4 of these patients did not show a rise in estrogen output. Another early study done on the use of clomiphene in 55 women with amenorrhea and varying degrees of oligomenorrhea was reported by Ferriman, Purdie and Corns.[22,23] The aim was to determine the effectiveness of the drug in different types of patients. A satisfactory response was found to be more likely in patients with a period of amenorrhea shorter than six months or if the menstrual cycle did not last longer than six months over a period of time. A better prognosis was expected if some ovarian activity was present which was evidenced by a basal estrogen excretion of more than 21 μg per 24 hr or an increase greater than 40 μg per 24 hr following administration of clomiphene. Thirty-eight out of the 55 patients achieved ovulation in response to clomiphene and 9 out of 41 women complaining of infertility got pregnant.

In another report, Kistner *et al.*[10,24] selected patients for clomiphene therapy. This included those who had some evidence of pituitary function; for instance, low normal range of total urinary gonadotrophins and clinical evidence of estrogen activity. Patients that were excluded from the study presented with primary ovulation failure, pituitary tumors, thyroid and adrenal abnormalities, as well as deficient nutritional status. Patients were given either 50 mg of clomiphene 3–5 days, 100 mg 3–5 days or 200 mg 2 days. The first course of therapy was initiated after the induction of withdrawal bleeding by progesterone or progestational agent. This was done to preclude giving clomiphene to a pregnant patient and also suppressed the hypothalamic LHRH, which theoretically reduces endogenous LH and the possibility of ovarian cyst formation.[10] If no clinical response occurred, as evidenced by the basal body temperature chart or onset of menses, a second course of therapy was initiated after exclusion of pregnancy. Results showed that ovulation was induced in 77% of 1731 patients following the first course of therapy, while 191 ovulated after repeated attempts. Most patients who responded to clomiphene continued to do so in a regular fashion; however, there was the occasional patient who responded erratically. A review of the literature.[25] revealed that the Merrell Company published a brochure about clomiphene in September 1970 that had reports on 5413 patients in the literature or cases reported to them. The overall percentage of ovulation achieved was 72.4%, while the pregnancy rate was 30.6%. Patients with PCOS — having enlarged ovaries and possibly cystic follicles, — were found to be extremely sensitive to clomiphene. Southam and Janovsky.[26] in 1962 induced excessive stimulation of both ovaries when 100 mg of clomiphene was given to one patient daily for 14 days; however, the ovaries were larger than the normal size, round and very tense before therapy was started. Jones'[10] review of clomiphene in 1968 revealed that an investigator's report in the physicians' drug monograph on clomiphene showed that patients with PCOS responded remarkably to the drug. Three hundred and five out of 391 patients achieved ovulation. At this stage it was clear that clomiphene was effective only in patients with an intact hypothalamic-pituitary system with some ovarian function. Generally, poor results are achieved in patients with primary amenorrhea. Similarly, patients with severe pituitary dysfunction were relatively unresponsive to therapy. However, a report of 35 patients with severe pituitary deficiencies revealed that 80% ovulated while 35% became pregnant.[27] Therefore, it was concluded that no patient should be totally excluded from clomiphene treatment on this basis.

CLOMIPHENE RESISTANCE

Despite the relative success of clomiphene, about 20%–25% of anovulatory women with normal FSH levels will not respond to clomiphene therapy and are deemed to be clomiphene-resistant.[2,28] Determination of patients who will respond to clomiphene is rather complex. A review by Polson *et al.*[29] showed that an absent ovarian response to an appropriate rise in serum FSH was the most common reason for anovulation. Therefore, many phenotypic factors have been implicated in the clinical response to clomiphene therapy. Imani *et al.*[28] reported that failure of ovulation induction with clomiphene is more likely in hyperandrogenic obese women then in those who do respond. Following extensive data analysis, the prospective study determined a high free androgen index and BMI was the best predictor of nonresponse to FSH stimulation induced by clomiphene. More recent reviews[30,31] indicate that approximately 70%–90% of patients achieve ovulation on clomiphene, however, a lower pregnancy rate of 30%–40% is realized in those who ovulated.[32] It seems quite discouraging that clomiphene therapy does not achieve a much higher pregnancy rate following a high ovulation induction rate. An extensive review by Homburg[30] concluded that this discrepancy between ovulation and pregnancy rates, where less than 50% of those who ovulate will conceive, may be partly explained by the antiestrogenic effects of clomiphene at the level of the endometrium and cervical mucus or by elevated levels of LH.

Although the reduction of the cervical mucus, occurring in about 15% of patients, may be surmounted by performing other procedures such as intrauterine insemination, the issue of suppression of endometrial proliferation may be harder to overcome. This apparently idiosyncratic reaction indicated a poor prognosis for pregnancy if the ultrasound scanning revealed an endometrial thickness of less than 8 mm at ovulation. The prevalence of endometrial suppression in this report was one in every 6–7 patients and, if this occurred in the first cycle of clomiphene therapy, it would almost definitely be seen in repeated cycles in the same patient.[30] Thus, there seemed to be no point in continuing therapy after even one cycle, and moving on to other forms of ovulation restoration was recommended.

CURRENT CLOMIPHENE THERAPY AND TREATMENT MONITORING

Clomiphene therapy is the recommended first-line treatment for normally estrogenized women with anovulation due to hypothalamic–pituitary dysfunction (WHO group II), whereby a vast majority of the group have PCOS. Therefore, ovulation dysfunction associated with

PCOS remains the most frequent clinical indication for clomiphene therapy.[33] Further recommended clinical use of clomiphene is to induce ovulation in women with unexplained fertility problems.[33] Since the success of clomiphene therapy in these women seems to be limited, therapy is generally combined with other forms of assisted reproductive measures such as intrauterine insemination in order to increase the chances of conception.[30] In order to improve the outcome of therapy with clomiphene, several adjuvants have been proposed. Anovulatory patients with PCOS who are unresponsive to clomiphene and have a body mass index greater than 25 are offered metformin combined with clomiphene, because this reduces hyperinsulinemia associated with PCOS and may subsequently increase ovulation and pregnancy rates.[33,34] Other adjuvant therapies combined with clomiphene include HCG and dexamethasone. However, implementing these drugs in an ovulation-inducing regimen that includes clomiphene therapy is yet to be widely accepted, due to lack of beneficial evidence.

Clomiphene is administered orally, typically in a dose of 50–150 mg for five days, between day 2 and day 5 of spontaneous or progestin-induced menses.[31] Typically, treatment should be started with 50 mg and the dose can be gradually increased until ovulation is achieved. Generally, treatment with clomiphene should be limited to a course of six ovulatory cycles, due to increased possible risk of ovarian cancer, which has been reported after use for 12 months.[32] While more studies are required, it is best to limit a patient's lifetime exposure to clomiphene to 12 treatment cycles, as additional cycles may at least place the patient at increased risk of borderline ovarian tumors.[31,33] It is recommended that second-line treatment such as FSH or laparoscopic ovarian drilling should be considered after 6 months of therapy.[32] Objective evidence of ovulation is necessary in order to establish successful therapy with clomiphene.[2] Ovulatory response can be determined by biphasic pattern on basal body temperature recording, a serum progesterone level in the expected luteal phase of >10 nmol/L when tested 6–8 days before the onset of menses, urinary LH surge and secretory changes in the endometrium via endometrial biopsy also provide evidence of ovulation.[35] Additional proof of ovulation can be obtained by serial transvaginal ultrasound, which can also be used to monitor the ovarian follicles and perform endometrial assessment.

GESTATIONAL OUTCOME WITH CLOMIPHENE AND SIDE EFFECTS

As previously discussed, clomiphene prevents the estrogen-mediated negative feedback mechanism that would reduce the stimulation of FSH secretion. This results in multiple ovarian follicle production, which in turn increases the risk of multiple pregnancies and is estimated at 8%–13%.[30,32] The majority of multiple births occurring as a result of clomiphene treatment are twin pregnancies; triplet and higher-order pregnancy are rare but can occur. Although there are conflicting reports, a comprehensive review by Kousta *et al.*[32] reported the prevalence of spontaneous miscarriages following clomiphene treatment to be between 13% and 25%, which is slightly higher than the sporadic abortion rate of 15%. The report revealed that the serum LH levels post-clomiphene-therapy were significantly higher than in the patients who had a successful pregnancy [15.5 (4.4) versus 12.0 (7.9); $p = 0.04$]. Although there was significant overlap in LH values between the two groups, just 37% of the ongoing pregnancy group had LH >10 IU/l compared with 75% of those who had a miscarriage. The suggested implication of a high serum concentration of LH in these women was that it adversely affects the timing of the maturation of the oocyte, therefore leading to the release of an "aged" oocyte that is likely to result in abortion in the event of fertilization.[36] More recently, the ASRM committee reported in their review that the prevalence of spontaneous abortions as well as congenital abnormalities following clomiphene therapy is thought to be no different from those observed in spontaneously conceived pregnancies.

Clomiphene is generally a well-tolerated drug and unpleasant side effects are few; however, some women will complain of vasomotor hot flushes and unusual visual symptoms.[35] Although mild ovarian enlargement is relatively common, severe ovarian hyperstimulation syndrome (OHSS) (massive ovarian enlargement, progressive weight gain, severe abdominal pain, nausea and vomiting, hypovolemia, ascites and oliguria) is rarely seen.[2]

TAMOXIFEN

Following the moderate success of clomiphene in inducing ovulation, tamoxifen (ICI 46,474) was expected to have ovulation-inducing properties. As noted previously, clomiphene is a mixture of the antiestrogenic (enclomiphene) and estrogenic (zuclomiphene) geometric isomers. In contrast, ICI 46, 474 is the pure trans isomer that is antiestrogenic in laboratory rats.[18] Klopper and Hall[37] established that tamoxifen could induce ovulation in women. Twenty women in whom failure to ovulate was the only known cause of infertility were selected. Based on the fact that different causes of anovulation respond differently to treatment, the patients were classified. This included the patients with cystic glandular hyperplasia; the disordered ovarian steroidogenesis group, where patients had features of polycystic ovarian disease; and the hypothalamic amenorrhea group. The initial five treatments, which consisted of 5 mg twice daily for four days, were considered to be inadequate and, thereafter, all patients received 10 mg daily for four days. Six patients had previous treatments with clomiphene; three of them ovulated on at least one occasion following clomiphene therapy. All three also ovulated after tamoxifen, and two of the three patients with negative responses to clomiphene ovulated when treated with tamoxifen. The success rate of tamoxifen therapy varied according to the category of the patients. In the three patients with polycystic-disease-like features, ovulation was induced three times and one patient became pregnant. The hypothalamic amenorrhea group received 23 treatments and ovulation occurred 10 times, while two achieved conception. Thirteen courses of treatment were given to the group with cystic glandular hyperplasia, resulting in three ovulations, one presumed anovular cycle and one presumed follicular luteinization.

Data on pregnancy rates with tamoxifen therapy are more difficult to assess, since it has been poorly documented in several studies. However, several clinical studies showed moderate success.

Williamson and Ellis[38] assessed tamoxifen's value in the treatment of anovulatory infertility in 32 patients. A total of 26 patients ovulated, 16 patients got pregnant during a treatment cycle, and 2 patients conceived during the first spontaneous cycle thereafter.

In another early series,[39] consisting of 33 patients, the induction rate of ovulation recorded 100% in the patients with sporadic anovulatory cycle, 83.3% in those with persistent anovulatory cycle, 70% in those with first-grade amenorrhea type I and 66.7% in those with first-grade amenorrhea type II. However, tamoxifen was absolutely ineffective in patients with second-grade amenorrhea.

The patients with sporadic or persistent anovulatory cycle responded to 40-mg-daily tamoxifen therapy, resulting in 100% of the ovulatory induction rate. The patients with the first-grade amenorrhea also responded to 60-mg-daily treatment, recording 70% of the ovulatory induction rate. The rate of pregnancy established marked 15.4%. Out of 4 patients who did not respond to the previous clomiphene treatment, 3 achieved ovulatory induction with tamoxifen. Hence, a total ovulatory induction rate of 73.3% was consistent with results reported by many investigators.[40–43] Presently, tamoxifen is given at a daily dose of 20–40 mg for 4 or 5 days, starting either shortly after menstruation or any time in women with amenorrhea once pregnancy has been excluded.

TAMOXIFEN AND CLOMIPHENE INDUCTION OF OVULATION

The efficacy of tamoxifen has been compared to that of clomiphene in achieving induction of ovulation and pregnancy. At first the evaluation of these two drugs was difficult, due to limited and poorly documented data. However, some studies reported that tamoxifen treatment resulted in successful induction of ovulation at a rate of 41%–66% of the cases that failed to respond to clomiphene.[37,40–43] Steiner *et al.*[44] performed a meta-analysis of four prospective clinical trials which compared the effectiveness of tamoxifen to that of clomiphene for ovulation induction and achievement of pregnancy in infertile couples with isolated anovulatory infertility. A total of

273 women were included in the analysis and a total of 921 cycles were analyzed. Induction of ovulation and pregnancy per cycle were the reported endpoints. In all four trials, no significant difference was reported between the tamoxifen and the clomiphene groups in the odds of ovulation induction (OR 0.755, 95% CI 0.513–1.111).[45–48] There was no statistically significant heterogeneity detected between the trials ($p = 0.141$). Only three studies evaluated pregnancy as an outcome for a total of 743 cycles, of which 504 achieved ovulation.[46–48] Similarly, there was no significant difference between the tamoxifen and control groups in the odds of pregnancy per cycle (OR 1.056, 95% CI 0.583–1.912) or per ovulatory cycle (OR 1.162, 95% CI 0.632–2.134), and there was no statistical heterogeneity detected ($p = 0.815$). Only one out of the four trials[48] reported all pregnancy outcomes, including miscarriages and live births. The report revealed no significant difference between the two groups in the odds of live birth per cycle (OR 0.261, 95% CI 0.005–2.711). This study also showed a higher miscarriage rate and a lower live birth rate among those women on tamoxifen. Similar reports[42,49] found that miscarriage rates among those women on tamoxifen were ~35% compared to miscarriage rates between 11% and 14% in the clomiphene group.[44] However, this has not been documented by the other investigators.

The meta-analysis failed to demonstrate any significant difference in ovulation rates between clomiphene and tamoxifen. This is supported by Brown *et al.*'s [50] report on antiestrogen and ovulation induction, where after reviewing extensive literature including 15 randomized controlled trials, no difference was shown in the efficacy between tamoxifen and clomiphene. However, there was limited evidence on gestational outcomes such as miscarriage and live birth. Therefore, the recommendations in the United Kingdom is that either clomiphene or tamoxifen should be offered to women with World Health Organization Group II ovulation disorders (hypothalamic–pituitary dysfunction) such as polycystic ovary syndrome as the first line of treatment for up to 12 months.[33] However, clomiphene is a more likely favorite with the clinicians, due to the fact that they have been familiar with the therapeutic use of the drug for many years. Although tamoxifen may be less expensive and some clinical studies have shown that it is better tolerated with fewer side effects,[43,49] patients may have concerns about taking a drug commonly used as a chemotherapeutic agent for breast cancer. Lastly, none of the trials was powered or designed to detect differences in side effects,[44,49] and therefore at this stage one can conclude that either clomiphene or tamoxifen can be used to induce ovulation in women with WHO group II ovulation disorder. However, the choice between the two will be dependent on both the clinician and the patient involved.

THE USE OF TAMOXIFEN IN OVARIAN STIMULATION IN BREAST CANCER PATIENTS

Breast cancer is the most common cause of malignancy in women of reproductive age. In the United States, the vast majority of premenopausal patients will receive combination cytotoxic chemotherapy either following surgery in early invasive breast cancer or when a diagnosis is metastatic breast cancer. Although cytotoxic chemotherapy decreases recurrence and death from various types of cancer, it is also a known cause of significant loss of ovarian function.[51] Alkylating agents, especially cyclophosphamide, pose the greatest risk for ovarian failure, when compared to other cancer chemotherapeutic agents.[51] A meta-analysis by Bines *et al.*[52] revealed that 68% of patients who received the combination of cyclophosphamide, methotrexate and fluorouracil (CMF) developed premature ovarian failure. Similarly, a mean total of ~24% of women receiving cyclophosphamide for systemic lupus erythematosis developed ovarian failure, and major risk factors were age at the start of treatment and cumulative cyclophosphamide dose.[50] Therefore, patients should be made aware of the consequences of treatment-related infertility and should be offered options of fertility preservation. A review by Schattman and Navarro[53] proposed several strategies that included ovarian hyperstimulation with embryo or oocyte cryopreservation or ovarian tissue cryopreservation. Embryo cryopreservation is a well-established form of assisted reproduction that is routinely used in clinical practice for storing spare embryos realized following ovarian stimulation and *in vitro* fertilization. On the other hand, ovarian tissue cryopreservation does not require ovarian stimulation; rather, ovaries are removed and frozen until needed, when they are retransplanted into the ovarian fossa.[53] The caveat to the use of ovarian cryopreservation as an option for fertility preservation is that there is minimal success in achieving conception in humans.[54] Therefore, it is recommended that ovarian cryopreservation should only be considered when embryo cryopreservation cannot be performed. The aim of ovarian stimulation in the process of embryo cryopreservation is to increase the number of embryos created and subsequently increase the chances of pregnancy in these patients. This is further supported by a report by Sonmezer and Oktay[55] on fertility preservation in young breast cancer patients. Here, it was pointed out that conventional ovarian stimulation drugs are usually avoided in breast cancer patients because they cause an increase in estrogen levels, which may further promote breast cancer cell proliferation.

Tamoxifen, in addition to being an ovulation-stimulating drug, evolved into the gold standard for the hormonal treatment of breast cancer. Millions of breast cancer patients, both pre- and postmenopausal, have been treated successfully and safely with tamoxifen. Therefore, tamoxifen would seem to be the appropriate drug as well as a safe way of increasing the number of embryos for the breast cancer patients; unfortunately, data are limited.[55] Oktay *et al.*[55,56] performed a pilot study on tamoxifen as an ovarian stimulating agent for IVF, embryo cryopreservation and embryo transfer in breast cancer patients in order to increase embryo yield. Twelve women with breast cancer received 40–60 mg of tamoxifen, starting on day 2–3 of the menstrual cycle for an average duration of 6.9 days, resulting in a total of 15 cycles. Here, patients underwent IVF and embryo transfer (Tam IVF) with either fresh cycles (six cycles) or cryopreserved (nine cycles) embryos. These were compared with a retrospective group of five patients who had undergone natural cycles of IVF (NCIVF). The total number of mature oocytes was significantly higher in the tamoxifen group ($p = 0.03$), which resulted in a higher number of embryos (1.6 ± 0.3 versus 0.6 ± 0.2; $p = 0.02$) when compared with the natural cycle group. Therefore, Tam IVF resulted in the production of an embryo in all 12 patients; 8 underwent embryo cryopreservation prior to chemotherapy and 4 attempted pregnancy with fresh embryos after being cured of breast cancer. In the natural cycle group, only 3 of 5 patients had an embryo. For the fresh transfer cycles, 2/6 in the Tam IVF group and 2/5 in the NCIVF group got pregnant. Of the 2 patients who conceived in the tamoxifen group, 1 delivered a healthy set of twins. Pregnancy rates could not be compared between the two groups, because most patients receiving tamoxifen cryopreserved their embryos and had not yet undergone embryo transfer. However, this was the first report of an IVF pregnancy following tamoxifen-induced ovarian stimulation. On the other hand, most patients in the NCIVF group had a fresh embryo transfer, and no deliveries were reported by the end of the study. One of the major concerns of the study was that the peak estradiol levels were higher in the TamIVF group than in the NCIVF group and whether this might have any detrimental effects on breast cancer. Although tamoxifen causes an increase in estrogen levels, it also prevents the effects of supraphysiological levels of estrogen on breast tissue and inhibits the proliferation of breast tumors by competitive antagonism of estrogen at its receptor site.[57] Indeed, it has been demonstrated that premenopausal women receiving tamoxifen therapy alone for the treatment[58] or chemoprevention[59] have either effective increased survivorship or reduction in the incidence of breast cancer, respectively. Furthermore, mean estradiol levels are chronically elevated in breast cancer patients on long-term tamoxifen treatment and can be higher than the levels seen in the patients undergoing ovulation induction with tamoxifen.[57] Low-dose FSH has been combined with either tamoxifen or letrozole in breast cancer patients wishing to preserve their fertility via embryo cryopreservation before chemotherapy.[60] The combination of low-dose FSH with tamoxifen or letrozole resulted in higher embryo yield compared to Tam IVF. Recurrence rates were not increased in both groups but the letrozole protocol seemed to be preferred because it results in lower peak estradiol levels.

Ovarian stimulation with tamoxifen appears to be a feasible safe method of IVF and fertility preservation in breast cancer patients.[56] However, it must be stressed that the pilot study was limited by small study size and time constraints. Therefore, larger studies as well as randomized controlled trials with longer follow-up periods are needed to determine the pregnancy rates with cryopreserved embryo after Tam IVF.

CONCLUSION

The introduction of clomiphene into clinical practice was a therapeutic landmark in the 1960s and this medicine, despite the fact that it is an impure mixture of geometric isomers with opposing pharmacological properties, has stood the test of time. Although the market was modest in the 1960s because most women were developing their families in their early 20s, this is not the case today. Families are deferred for educational, career and economic reasons. Since fertility rates are reduced with age, a woman deciding to start a family in her late 30s or early 40s may, as often as not, need to turn to assisted methods of reproduction. Techniques for induction of ovulation and IVF are now a multimillion-dollar business. Additionally, the fact that breast cancer therapy is successful in young women who choose preservation of reproductive capacity routinely means the dual function of the SERM tamoxifen: the induction of ovulation and life-saving breast cancer therapy have now merged and could potentially be enhanced in the future. However, let us end on a note of caution: because tamoxifen can induce ovulation and there is a risk of pregnancy during the treatment of premenopausal patients, barrier contraception must be practiced in order to reduce the potential risks of teratogenesis.

ACKNOWLEDGEMENTS

This work V.C.J. was supported by the Department of Defense Breast Program under award No. W81XWH-06-1-0590 Center of Excellence; subcontract under SU2C (AACR) grant No. SU2C-AACR-DT0409; the Susan G. Komen for the Cure Foundation under award No. SAC100009; GHUCCTS CTSA (Grant #UL1RR031975) and the Lombardi Comprehensive Cancer Center Support Grant (CCSG) Core Grant NIH P30 CA051008. The views and opinions of the author(s) do not reflect those of the U.S. Army or the Department of Defense.

REFERENCES

1. ASRM Practice Committee. (2006) Use of clomiphene citrate in women. *Fertil Steril* **86**(4):S187–S193.
2. Homburg R, Insler V. (2002) Ovulation induction in perspective. *Hum Reprod Update* **8**:449–462.
3. Lerner LJ, Holthaus JF, Thompson CR. (1958) A non-steroidal estrogen antagonist (1-(p-2-diethylaminoethoxyphenyl)-1-phenyl-2-p-methoxophenyl-ethanol. *Endocrinol* **63**:295–318.
4. Segal JS, Nelson WO. (1958) An orally active compound with antifertility effects in rats. *Proc Soc Exp Biol Med* **98**:431–436.
5. Smith OW, Kistner RW. (1963) Action of MER-25 and clomiphene on the human ovary. *J Am Med Assoc* **184**:122–130.
6. Furr BJ, Jordan VC. (1984) The pharmacology and clinical uses of tamoxifen. *Pharmac Ther* **25**:127–205.
7. Holtkamp DE, Greslin JG, Root CA, Lerner LJ. (1960) Gonadotrophin inhibiting and antifecundity effects of cloramiphene. *Proc Soc Exp Biol Med* **105**:197–201.
8. Greenblatt RB, Barfield WE, Jungck EC, Ray AW. (1961) Induction of ovulation with MRL-41. *J Am Med Assoc* **178**:101–104.
9. Clark JH, Markaverich BM. (1982) The agonistic-antagonistic properties of clomiphene: a review. *Pharmacol Ther* **15**:467–519.
10. Jones GS. (1968) Induction of ovulation. *Annu Rev Med* **19**:351–372.
11. Kerin JF, Liu JH, Phillipou G, Yen SS. (1985) Evidence for a hypothalamic site of action of clomiphene citrate in women. *J Clin Endocrinol Metab* **61**:265–268.
12. Kettel LM, Roseff SJ, Berga SL, Mortola JF, Yen SS. (1993) Hypothalamic pituitary ovarian response to clomiphene citrate in women with polycystic ovary syndrome. *Fertil Steril* **59**:532–538.
13. Rebar R, Judd HL, Yen SS, Rakoff J, VandenBerg G, Naftolin F. (1976) Characterization of the inappropriate gonadotropin secretion in polycystic ovary syndrome. *J Clin Invest* **57**:1320–1329.
14. Herbst AL, Griffiths CT, Kistner RW. (1964) Clomiphene citrate (NSC-35770) in disseminated mammary carcinoma. *Can Chemother Rep* **43**:39–41.
15. Hecker E, Vegh I, Levy CM, Magin CA, Martinez JC, Loureiro J, Garola RE. (1974) Clinical trial of clomiphene in advanced breast cancer. *Eur J Cancer* **10**:747–749.
16. Jordan VC. (2003) Tamoxifen: a most unlikely pioneering medicine. *Nat Rev Drug Discov* **2**:205–213.
17. Harper MJ, Wapole AL. (1967) A new derivative of triphenyethylene: effect on implantation and mode of action in rats. *J Reprod Fert* **13**:101.
18. Laven JS, Fauser BC. (2006) What role of estrogens in ovarian stimulation, *Maturitas* **54**:356–362.

19. Adashi EY. (1993) Clomiphene citrate: the case for a monomeric preparation. *Bailliere's Clin Obstet Gynaecol* 7(2):331–347.
20. Palopoli FP, Feil VJ, Allen RE, *et al.* (1967) Subtituted aminoalkoxytriaryl-hadoethylenes. *J Med Chem* **10:**84–86
21. Bishop PM. (1965) Clinical effects of clomiphene. *Proc R Soc Med* **58**(11 Pt 1):905–906.
22. [No authors listed]. (1968) Clomiphene. *Can Med Assoc J* (1968) **98**(13):653–654.
23. Ferrumanm D, Purdiem AW, Cornsm M. (1965) *Brit Med J* **4**:444.
24. Kistner RW. (1965) Induction of ovulation with clomiphene citrate (clomid). *Obstet Gynaecol Survey* **20**:873–900.
25. Bishop PM. (1972) Clomiphene therapy. *Postgrad Med J* **48**:24–26.
26. Southam AL, Janovski NA. (1962) Massive ovarian hyperstimulation with clomiphene citrate. *J Am Med Assoc* **181**:443.
27. Spellacy W N, Cohen WD. (1967) Clomiphene treatment of prolonged secondary amenorrhea associated with pituitary gonadotropin deficiency. *Am J Obstet Gynecol* **97**(7):943–948.
28. Imani B, Eijkemans MJ, Te Velde ER, Habbema JD, Fauser BC. (1998) Predictors of patients remaining anovulatory during clomiphene citrate induction of ovulation in normogonadotropic oligoamenorrheic infertility. *J Clin Endocrinol Metab* **83**:2361–2365.
29. Polson DW, Kiddy DS, Mason HD, Franks S. (1989) Induction of ovulation with clomiphene citrate in women with polycystic ovary syndrome: the difference between responders and nonresponders. *Fertil Steril* **51**:30–34.
30. Homburg R. (2005) Clomiphene citrate — end of an era? A mini-review. *Hum Reprod* **20**:2043–2051.
31. Messinis IE. (2005). Ovulation induction: A mini review. *Hum Reprod* **20**:2688–2697.
32. Kousta E, White DM, Frank S. (1997) Modern use of clomiphene citrate in induction of ovulation. *Hum Reprod Update* **3**:359–365.
33. NICE Guidelines. (2004) Fertility: assessment and treatment of people with fertility problems. National Collaborating Centre for Women's and Children's Health. *NICE.* London. RCOG Press.
34. Vandermolen DT, Ratts VS, Evans WS, Stovall DW, Kauma SW, Nestler JE. (2001) Metformin increases the ovulatory rate and pregnancy rate with clomiphene citrate in patients with polycystic ovary syndrome who are resistant to clomiphene citrate alone. *Fertil Steril* **75**:310–315.
35. SOGC Clinical Practice Guideline. (2010) Ovulation induction in polycystic ovary syndrome. *JOGC Mai* **242**:496–500.
36. Homburg R, Armar NA, Eshel A, Adams J, Jacobs HS. (1988) Influence of serum luteinising hormone concentrations on ovulation, conception, and early pregnancy loss in polycystic ovary syndrome. *Br Med J* **297**:1024–1026.
37. Klopper A, Hall M. (1971) New synthetic agent for the induction of ovulation: preliminary trials in women. *Br Med J* **1**:152–154.

38. Williamson JG, Ellis JD. (1973) The induction of ovulation by tamoxifen. *J Obstet Gynecol Brit Cwlth* **80**:844–847.
39. Tsuiki A, Uchara S, Kyono K, Satio A, Hoshi K, Hoshiai H, Hirano M, Suzuki, M. (1984) Induction of ovulation with an estrogen antagonist, tamoxifen. *J Exp Med* **144**:21–31.
40. Tajima C, Tamaki Y, Takamizawa Y. (1977) Trial of tamoxifen for the induction of ovulation. *Acta Obstet Gynecol Jpn* **29**:57–64.
41. Ruiz-Velasco V, Rosas-Arceo J, Mature MM. (1979) Chemical inducers of ovulation: comparative results. *Int J Fertil* **24**:61–64.
42. Macourt DC. (1974) A new synthetic agent for the induction of ovulation. *Med J Aust* **1**:631–632.
43. Gerhard I, Runnebaum N. (1979) Comparison between tamoxifen and clomiphene therapy in women with anovulation. *Arch Gynecol* **227**:279–288.
44. Steiner AZ, Terplan M, Paulson PJ. (2005) Comparison of tamoxifen and clomiphene citrate for ovulation induction: a meta-analysis. *Hum Reprod* **20**:1511–1515.
45. Messinis IE, Nillius SJ. (1982) Comparison between tamoxifen and clomiphene for induction of ovulation. *Acta Ostet Gynecol Scand* **61**:377–379.
46. Vegetti W, Riccaboni A, Colombo M, Baroni E, Diaferia D, Ragni G, Crosignani PG. (1999) Randomized study of induction of ovulation by two different molecules with antiestrogenic effects, in patients with chronic anovulation disorders. *Fertil Steril* **72**:S234–S235.
47. Boostanfar R, Jain JK, Mishell DR Jr, Paulson RJ. (2001) A prospective randomized trial comparing clomiphene citrate with tamoxifen citrate for ovulation induction. *Fertil Steril* **75**:1024–1026.
48. Elstein M, Fawcett GM. (1984) Effects of the anti-oestrogens, clomiphene and tamoxifen, on the cervical factor in female infertility. *Ciba Foundn Symp* **109**:173–179.
49. Buvat J, Buvat-Herbaut M, Marcolin G, Ardaens-Boulier K. (1987) Antiestrogens as treatment of female and male infertilities. *Horm Res* **28**:219–229.
50. Brown J, Farquhar C, Beck J, Boothroyd C, Hughes E. (2009) Clomiphene and anti-oestrogens for ovulation induction in PCOS. *Cochrane Database Syst Rev* **7**(4):CD002249.
51. Meirow D, Nugent D. (2001) The effects of radiotherapy and chemotherapy on female reproduction. *Hum Reprod Update* **7**:535–543.
52. Bines J, Oleske DM, Cobleigh MA. (1996) Ovarian function in premenopausal women treated with adjuvant chemotherapy for breast cancer. *J Clin Oncol* **14**:1718–1729.
53. Schattman GL, Navarro J. (2008) Breast cancer and fertility preservation. *Placenta* **29**:S147–S151.
54. Oktay K, Buyuk E. (2002) The potential of ovarian tissue transplant to preserve fertility. *Expert Opin Biol Ther* **2**:361–370.

55. Sonmezer M, Oktay K. (2006) Fertility preservation in young women undergoing breast cancer therapy. *Oncologist* **11**:422–434
56. Oktay K, Buyuk E, Davis O, *et al.* (2003) Fertility preservation in breast cancer patients: IVF and embryo cryopreservation after ovarian stimulation with tamoxifen. *Hum Reprod* **18**:90–95.
57. Jordan VC, Fritz NF, Langan-Fahey S, Thompson M, Tormey DC. (1991) Alteration of endocrine parameters in premenopausal women with breast cancer during long-term adjuvant therapy with tamoxifen as the single agent. *J Natl Cancer Inst* **83**:1488–1491.
58. Early Breast Cancer Trialists' Collaborative Group (EBCTCG). (2005) Effects of chemotherapy and hormonal therapy for early breast cancer on recurrence and 15-year survival: an overview of the randomised trials. *Lancet* **365**:1687–1717.
59. Fisher B, Costantino JP , Wickerham DL, Cecchini RS, Cronin WM, Robidoux A, Bevers TB, Kavanah MT, Atkins JN, Margolese RG, Runowicz CD, James JM, Ford LG, Wolmark N. (2005)Tamoxifen for the prevention of breast cancer: current status of the National Surgical Adjuvant Breast and Bowel Project P-1 Study. *JNCI* **97**(22):1652–1662.
60. Oktay K, Buyuk E, Libertella N, Akar M, Rosenwaks Z. (2005) Fertility preservation in breast cancer patients: a prospective controlled comparison of ovarian stimulation with tamoxifen and letrozole for embryo cryopreservation. *J Clin Oncol* **23**:4347–4353.

6

CLUES ABOUT ESTROGEN AND ANTIESTROGEN ACTION FROM X-RAY CRYSTALLOGRAPHY

Geoffrey L. Greene and V. Craig Jordan

Abstract

During the 1980s the structure-function relationship of nonsteroidal estrogens and antiestrogens using an a prolactin gene target in primary cultures of rat pituitary cells created basic models for the estrogen receptor (ER)-mediated modulation of estrogen-induced prolactin synthesis. However, the cloning and sequencing of the human ER, made possible by the creation of monoclonal antibodies to the ER, created a new dimension in our understanding of the mechanics of estrogen and antiestrogen action. Although the whole human ER has not been crystallized with ligands that binds to the ligand-binding domain (LBD), the crystallization and resolution of estrogen and antiestrogen LBD complexes are consistent and informative. Simply stated, an array of different complexes are formed with selective ER modulators (SERMs) and a pure antiestrogen causes considerable structural disruption of the LBD. This chapter tells the story of how the shape of the ligand–ER complex programs the interaction with coactivators for estrogen actions or becomes the signal for the destruction of the complex.

Keywords: Structure Function; X-Ray Crystallography; LBD; SERMs; Homodimer; Coactivators; AF-2; LxxLL; Hydrophobic Cleft; Helix 12; Molecular Switch; LxxML; Corepressor; Steric Exclusion; Indirect Antagonism; SERD; Molecular Switch.

INTRODUCTION

The description of the selective binding of [^{3}H] estradiol in the estrogen target tissues of the immature rat (uterus, vagina)[1] and the subsequent isolation of the estrogen receptor protein (ER) as an extractable protein from the rat uterus[2,3] was not only an advance in molecular endocrinology, but also an advance that would improve the therapeutics of breast cancer. From that point in time in the 1960s, it was possible to study and understand the binding of ligands to the ER and perhaps gain an insight into the mechanism of action of estrogens and antiestrogens. Early studies of the competitive binding of estrogens and antiestrogens with [^{3}H] estradiol for the ER *in vitro*[4,5] were unable to distinguish between estrogens and antiestrogens biologically. All that could be concluded was that antiestrogens had low binding affinity for the ER and this was why such large doses were necessary for blocking estrogen action.[4] Also, it was concluded that the low affinity of antiestrogens for the ER was part of their mechanism of action: the ligand would not remain long enough bound to the receptor to activate estrogen action.[4] This was all to change with the discovery of the pharmacological properties of 4-hydroxytamoxifen, a metabolite of tamoxifen then thought to be the principal metabolite of tamoxifen.[6] 4-Hydroxytamoxifen has a binding affinity for the ER equivalent to estradiol, so if it was possible to have high affinity antiestrogens, then low affinity was not the mechanism of antiestrogen action. The shape of the resulting complex was the key to efficacy and the subsequent modulation of signal transduction. 4-Hydroxytamoxifen was subsequently adopted as the standard laboratory antiestrogen in cell culture, and 20 years later it was used as an antiestrogenic ligand to be crystallized with the ligand-binding domain of the human ER.[7]

In the 1970s, what was needed was a model cell system for studying the structure–function relationships of ligands that bind to the ER. In this way, the intrinsic efficiency of the ligand–ER complex could be deciphered, without concerns about pharmacokinetics and metabolism. The ER positive breast cancer cell line MCF-7 had been described[8] but the fact that the cells apparently grew spontaneously in culture and would not respond proliferatively to estradiol but would when inoculated into athymic mice[9] led to considerable controversy in the field. Maybe estrogen was acting indirectly to promote breast cancer growth? Nevertheless, tamoxifen did block the spontaneous growth of MCF-7 cells and this blockade could be reversed with estradiol.[10] Interestingly enough, ERs isolated from MCF-7 cells would prove to be essential for creating the first monoclonal antibodies to the human ER[11] and would subsequently be the critical tool necessary for cloning and sequencing the human ER cDNA.[12]

The first cell system used to study the modulation of the ligand–ER complex *in vitro* was primary cultures of the immature rat pituitary gland. The target for the ER was the prolactin gene.[13,14] The first publication validated the mechanism of action of nonsteroidal antiestrogens to regulate estrogen-induced gene transcription as competitive inhibition of estradiol binding to the ER and that it was an advantage, but not a requirement, for an antiestrogen to be metabolically activated.[13] As with other drug–receptor interactions, affinity and the intrinsic efficacy of the drug–receptor complex are not interconnected for drug action. Numerous studies described the structure–function relationships of triphenylethylenes to modulate the ER complex between the extremes of estrogenic intrinsic efficacy and complete antiestrogen action.[15-19] Structure–activity relationship[20] studies permitted the creation of a map of the hypothetical folding of the ER complex (see Fig. 1). However, it was the serendipitous advance in deciphering breast cancer cell replication *in vitro* that was to enhance the interpretation of all future laboratory studies.

In the mid-1980s, the Katzenellenbogen laboratory[21] made the critical discovery that ER positive breast cancer cells had all been cultured in media containing high concentrations of a pH indicator, phenol red, which contained a contaminant that was an estrogen.[22,23] Removal of phenol red from media now permitted the structure–activity relationship studies of nonsteroidal antiestrogens to be extrapolated from prolactin gene modulation to the replication of breast cancer cell lines.[24,25]

However, the critical question to be addressed in molecular pharmacology was: what is the essential interaction of the antiestrogenic side chain with the ER that modulates estrogen-like and antiestrogen action? An antiestrogenic region (AER) that interacts with the

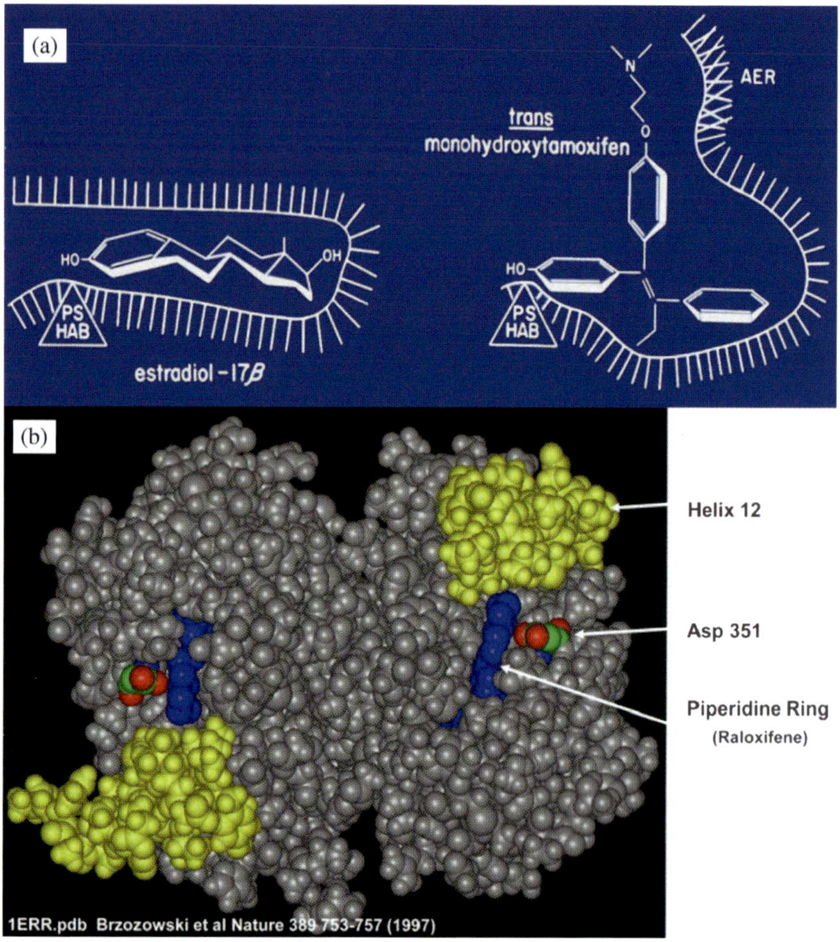

Figure 1. **(a)** Hypothetical model of estrogen and antiestrogen action at the ligand-binding domain (LBD) of the ER. Estradiol (left) binds with high-affinity binding (HAB) via the 3-phenolic hydroxyl of estradiol and then is sealed within the LBD to trigger Prolactin transcription as part of estrogen action. The hydroxylated metabolite of tamoxifen (known as monohydroxy tamoxifen originally, but now called 4-hydroxy tamoxifen) binds with HAB through the phenolic hydroxyl, but the angular shape of the ligand and the bulky antiestrogenic side chain prevents the ligand from effectively being sealed within the LBD. The antiestrogenic side chain was hypothesized to interact with an anti-estrogenic region (AEN) the ER. [This research was originally published in the *Journal of Biological Chemistry*. Lieberman ME, Gorski J, Jordan VC. (1983) An estrogen receptor model to describe the regulation of prolactin synthesis

appropriately positioned alkylaminoethoxy side chain on the ligand backbone had been proposed previously[14,20,26] but how to find it? Several advances were necessary before progress could occur. A model of acquired drug-resistance to tamoxifen in athymic mice needed to be developed, the ER needed to be sequenced, the sequence of the ER needed to be screened in drug-resistant MCF-7 breast tumors, ER needed to be stably transfected into ER-negative breast cancer cells and a suitable gene target modulated. All this was done to propose a hypothetical modulation of the antiestrogen ER complex prior to the crystallization of the ligand-binding domain with estradiol and raloxifene.[27] A biological clue was found in the human ER that would complement the structural knowledge of the ER ligand-binding domain complex with functional information at a transforming growth factor alpha (TGFα) gene target.

A mutation, asp351tyr, was noted in one MCF-7 tumor cell line variant with acquired resistance to tamoxifen.[28] The first transfection of the wild-type ER into the ER-negative breast cancer cell line MDA MB231[29] eventually allowed other mutant ERs to be transfected. The response of transfected ERs to different ligands could now be monitored at the estrogen-responsive TGFα gene. Simply stated, the 4-hydroxytamoxifen ER complex was an estrogen with wild-type and asp351tyr ER, while raloxifene was an antiestrogen with wild-type ER, but estrogenic with the asp351tyr ER. This observation was the first time a natural mutation (in the unlikely location of the ligand-binding domain) was observed to convert an antiestrogen to an estrogen.[30,31] The crystal structures of the ER ligand-binding domain bound to 4OH Tam and raloxifene[7,27] indicated that while raloxifene's side chain

Figure 1. (*Continued*) by antiestrogens *in vitro*. *J Biol Chem* **258**: 4741–4745.] **(b)** More than a decade later, with the resolution of the x-ray crystallography of the estradiol or raloxifene LBD complex (Brzozowski, *Nature*, 1997), the actual shape of a complex could be visualized. The figure shows a view from the outside of the raloxifene LBD dimer complex. Helix 12 (in yellow) is pushed back by the bulky antiestrogen side chain. The presumed AER turns out to be aspartic acid 351, which is neutralized and shielded by the piperidine ring of raloxidene to create a nonestrogen complex. Coactivators cannot bind because helix 12 cannot seal the anti-estrogenic ligand into the LBD.

shielded and possibly neutralized asp351, the side chain of tamoxifen is significantly shorter and barely interacted with asp351. The hypothesis that the side chain prevented the interaction of asp351 with AF-1 was used to interrogate the ligand–ER complex with mutations of asp351 and structural derivation of raloxifene,[32-35] and it was concluded that this amino acid is important for altering surface interactions with other coregulators of hormone action. With this historical background, we will now focus upon the advances and insights which x-ray crystallography has provided for our knowledge of the molecular modulation of the ER complex.

THE COMPLEXITY OF CRYSTALLIZING THE ER

A number of laboratories have attempted to express and crystallize full-length steroid receptors as well as other nuclear receptors. Thus far, the only intact receptors to be crystallized are the peroxisome-proliferator-activated receptor gamma (PPARγ) and the retinoid X alpha (RXRα) receptors, as a single heterodimeric complex bound to a cognate DNA response element (PPRE).[36] The major reason for this lack of success is primarily an inability to express, purify and concentrate receptor proteins at the levels needed to produce homogeneous crystals. While such efforts are still underway and will be discussed later in this chapter, the bulk of our own efforts and the efforts of others have focused on the ligand-binding domains (LBDs), which are soluble, well structured, and capable of binding ligands with essentially identical kinetics and selectivity as full-length receptors, due to the modular nature of nuclear receptors. In addition, the LBD contains one of two well-described transcription-activating functions (AF-2) associated with nuclear receptors. AF-2 behavior in response to diverse ligands is largely faithfully replicated in the LBD compared to the full-length protein.

ERα AND ERβ LBD CRYSTAL STRUCTURES

The first ER LBD structures to be solved and published were of ERα LBDs bound to estradiol and raloxifene (Figs. 1b and 2).[27] These

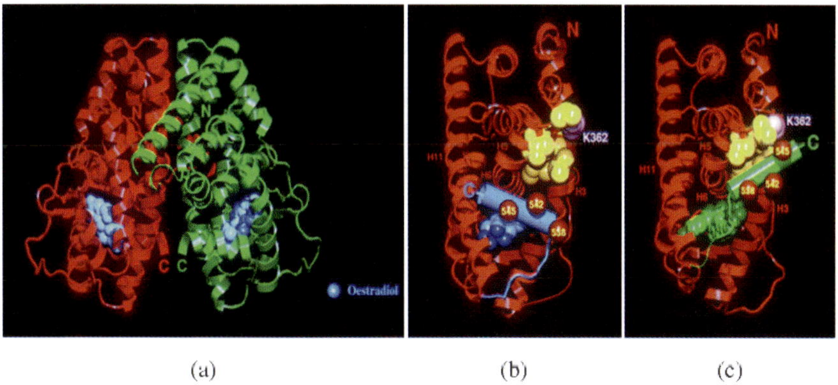

Figure 2. Structures of human ERα LBD bound to estradiol (blue) or to raloxifene (green). (**a**) Homodimer complex showing each monomer (red and green) interacting across an extensive symmetrical dimer interface made up primarily of helix 11, in a head-to-head arrangement. (**b**) Position of H12 in the presence of estradiol (blue cyclinder and blue ligand). (**c**) Position of H12 in the presence of raloxifene (green cyclinder and green ligand). Hydrophobic residues in the AF-2 cleft (H3 and H5; yellow) and Lys 362 (K362, pink) are also shown. Asp 538, Glu 542 and Asp 545 are shown as brown spheres. (Adapted from Ref. 27).

structures were very informative, because they showed for the first time how steroid receptor LBDs are organized into highly structured but conformationally flexible domains that could respond allosterically to naturally occurring hormones as well as synthetic agonists and antagonists. In particular, these structures confirmed that estrogen agonists and antagonists (e.g. SERMs) bind in the same hydrophobic pocket in the LBD core and that both classes of ligands are entirely contained within the LBD pocket, suggesting that entry of the ligand involves cooperative interaction with a malleable region of the LBD that is able to ultimately surround the ligand to give a fairly stable conformation. In agreement with extensive *in vitro* data,[37] the ERα LBD formed a homodimer in these crystal structures, with a symmetrical 'head-to-head' dimer interface. Also noted by the Hubbard group[27] were the distinct differences in the LBD conformations stabilized by estradiol and raloxifene, which resulted in the reorientation of helix 12 (H12) (Fig. 2c), suggesting a possible mechanism by which SERMs could interfere with the binding of known coactivators to the AF-2 function[38] located in the LBD.

In 1998, Shiau *et al.*[7] published the crystal structures of ERα LBDs bound to DES and 4-hydroxytamoxifen (OHT), which clearly revealed the nature and functional consequences of SERM versus estrogen agonist binding to the ERα LBD (Fig. 3). These structures, one of which (DES-ERα LBD) also included a peptide that represented the known NR interaction domain of the P160 coactivator GRIP-1 (SRC-2), revealed the basic mechanism by which this coactivator and other closely related coactivators (SRC-1 and SRC-3) recognize the ER LBD and how SERMs preclude this interaction, acting as AF-2, but not necessarily AF-1, antagonists. As described earlier, the ER LBD contains a second activation function (AF-2) that maps to a surface-exposed hydrophobic pocket, providing a docking site for coregulatory proteins.[39] Coactivators, including mediator and members of the p160 family (SRCs1–3), bind to the AF-2 surface via the amino acid motif LxxLL, in which the leucine residues dock into the hydrophobic cleft (Fig. 3).[7,40] Binding specificity is added by oppositely charged amino acids at either end of the ER hydrophobic cleft, which form a charge clamp with the LxxLL peptide backbone. The mechanism of transcriptional activation by ERs is via recruitment of these coactivators, which mediate chromatin remodeling and also recruit the basal transcription apparatus.

The most carboxy-terminal helix — helix 12 — of the LBD acts as a molecular switch,[41] forming one side of the AF-2 surface by docking against helices 3 and 11 in the presence of agonist ligands like E2 and DES (Fig. 3).[42] The remainder of the AF-2 surface is formed by helices 3–5 for the ERs (ERα and ERβ), which is equivalent to helices 3–4 in the retinoid X receptor (RXR) family. Antagonists such as OHT and raloxifene obstruct AF-2 through a bulky side chain that protrudes into the AF-2 surface (Figs. 2 and 3), displacing helix 12 and preventing coactivator recruitment to the LBD.[7,43] Helix 12 is relocated into the hydrophobic cleft with some SERMs, an effect mediated by sequences in helix 12 (LxxML) homologous to the LxxLL motif. Thus, helix 12 provides allosteric control of transcription mediated by its dynamic localization.

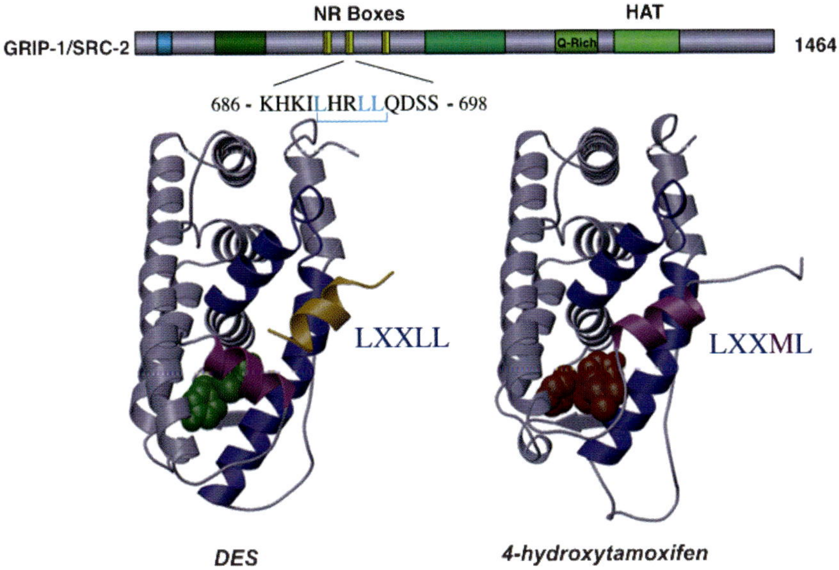

Figure 3. Structures of the DES-Erα LBD-GRIP-1 NR box II peptide complex and of the OHT-ERα LBD complex. Shown at the top is the overall schematic of GRIP-1/SRC-2, including the three NR interaction domains. The sequence of the GRP-1/NR box II peptide, bound to the AF-2 cleft formed by H3, H4, H5 and H12 (purple) in the DES-ERα LBD structure, is also shown. DES is in green. In the OHT-ERα LBD structure, H12 (purple) occupies the AF-2 cleft. OHT is in red/brown. NR box peptides stabilize the agonist AF-2 conformation, and SERMs prevent the formation of an active AF-2. (Adapted from Ref. 7.)

The AF-2 surface regulates the specificity of ER signaling by differential recruitment of cofactors to the two ER subtypes, ERα and ERβ.[44–46] analogous to other nuclear receptors. The two ER subtypes have distinct tissue distributions and phenotypes,[47–49] making the development of subtype-specific ligands of widespread importance for the treatment and prevention of breast cancer and for other indications, especially hormone replacement therapy.

Importantly, helix 12 is the structural link between ligand, coactivator and corepressor. The details of this communication and its differential regulation by receptor, receptor subtype and dimer partner are still being explored. However, a number of structural and functional studies over the past 15 years have shed light on some of the details of these signaling pathways, facilitating improved understanding of how diverse, ER subtype-selective ligands mediate differential effects in hormone-sensitive tissues and cancers through the two ERs.

Progress toward the development of subtype-selective ligands was significantly advanced with the reports of crystal structures of ERα[27] and ERβ.[50] The ligand-binding pockets of the subtypes are similar, but not identical. The ERβ ligand-binding pocket is smaller (390 Å^3 vs. 490 Å^3 for ERα) and differs in two residues from ERα: Leu-384 and Met-421 in ERα are replaced by Met-336 and Ile-373, respectively, in ERβ.[50] Notably, these two substitutions give rise to the selectivity of many, but not all, ligands for ERα and ERβ. Although several ERβ-selective ligands have been described,[51] only a relatively small number of crystal structures for these ligands bound to ERβ[50,52–56] and ERα[55–57] have been reported. However, these structures have provided valuable insights into the molecular mechanisms of ERα versus ERβ selectivity.

One of the more interesting ERα/ERβ-selective ligands is 5,11-cis-diethyl-5,6,11,12-tetrahydrochrysene-2,8-diol (THC), which exerts opposite effects on the transcriptional activity of ERα and ERβ.[56] THC acts as an ERα agonist and as an ERβ antagonist. The two ER crystal structures confirmed that THC stabilizes a conformation of ERα LBD that permits coactivator recruitment and a conformation of ERβ LBD that prevents coactivator association (Fig. 4). A comparison of these structures, taken together with functional data, revealed that THC does not act on ERβ through the same steric exclusion mechanisms associated with other known ER antagonists. Instead, THC antagonizes ERα through a novel mechanism termed "indirect antagonism" to reflect the contributions of helix 11 and other LBD residues to helix 12 positioning in response to ligands with variable agonist/antagonist behaviors. It is very likely that other AF-2 antagonists function by a similar allosteric mechanism, rather than by the direct H12 interference observed for OHT, raloxifene and ICI 164,384, or the closely related ICI 182,780 (fulvestrant).

Yet another class of ER antagonists is those that are capable of destabilizing ER, also known as SERDs (selective estrogen receptor downregulators). The best-known SERD is fulvestrant (ICI 182,780),

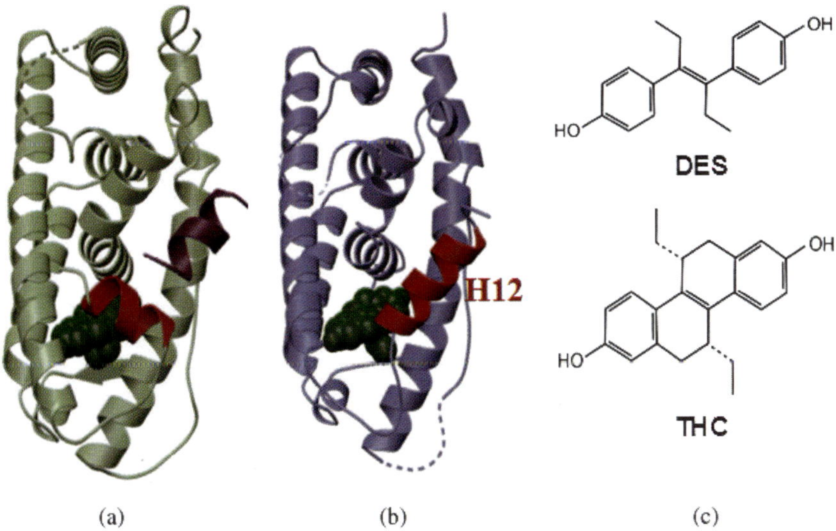

Figure 4. Structures of ERα **(a)** and ERβ **(b)** LBDs bound to THC. **(a)** THC–ERα LBD-GRIP-1 NR box II peptide. **(b)** THC-ERβ LBD. The two ER LBDs adopt distinct agonist **(a)** and antagonist **(b)** conformations when bound to THC. In both complexes, helix 12 is colored red, and THC (green) is shown in space-filling representation. In **(a)**, the GRIP-1 coactivator peptide is depicted as a purple ribbon, and H12 is a red ribbon in both structures. **(c)** Chemical structures of diethylstilbestrol (DES) and *R*,*R*-5,11-*cis*-diethyl-5,6,11,12-tetrahydrochrysene-2,8-diol (THC). DES is a full agonist on ERα and ERβ, whereas THC is an ERα agonist and ERβ antagonist. (Adapted from Ref. 56.)

frequently referred to as a "pure antiestrogen", although it is possible that this molecule does not act as a complete estrogen antagonist in all cell/tissue contexts. The only reported structure of an ER LBD bound to the closely related ICI 164,384 molecule was obtained by Pike *et al.* and reported in 2001.[58] What is most interesting about this crystal structure (Fig. 5), which consists of a somewhat distorted homodimeric complex between ICI 164,384 and the rat ERβ LBD, is the positioning of the 7α-alkyl extended side chain of estradiol, which extends into the bottom of the hydrophobic AF-2 cleft formed by helices 3 and 5, thereby preventing H12 from docking, which results in a disordered H12 that has no electron density and therefore cannot be seen in the structure. No other published ER LBD structure has this feature. Unfortunately, the comparable

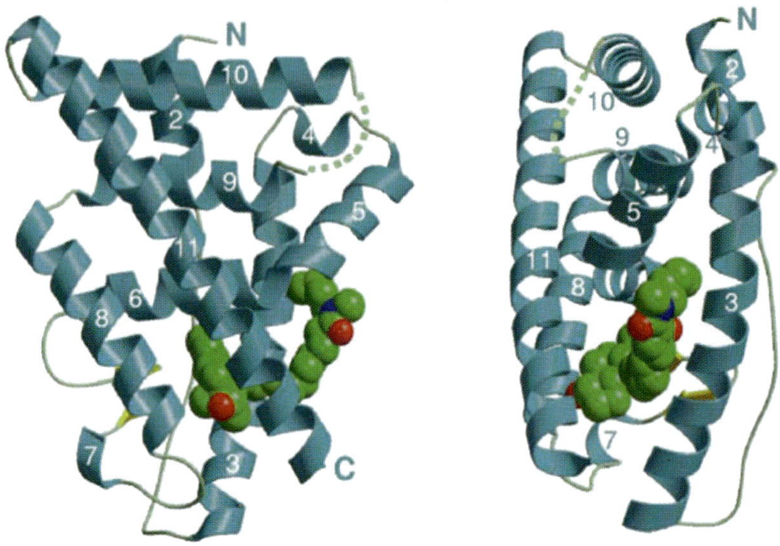

Figure 5. Structure of the ERβ LBD bound to ICI 164,384. Two rotated views of the rat ERβ LBD monomer occupied by ICI 164,384. Helices are numbered (H1-H11). Helix 12 is absent due to the insertion of the 7α alkyl amide side chain of estradiol into the lower portion of the AF-2 cleft formed by H3 and H5, with H4 at the top. (Reproduced from Ref. 50.)

structure of ERα LBD bound to either ICI analog has not been solved. Another interesting feature of the ICI-ERβ LBD structure is the 180° flipped orientation of estradiol, along the A–D ring axis formed by the two OH groups of 17β-estradiol, which allows the 7α-alkyl side chain to enter the "11β channel" which is also used by the side chains of OHT and raloxifene, although without the disruption of H12 docking.[58] As described below, the resulting exposure of a more hydrophobic surface on the ER LBD likely accounts for the SERD behavior of ICI.

GW5638/DPC974 is an example of a molecule that displays both SERM and SERD behavior.[59] The crystal structure of ERα LBD bound to GW5638[60] reveals an unexpected conformation in which H12 is repositioned by direct contacts between the carboxyl side chain of GW5638 and the N terminus of H12 (Fig. 6). The associated decrease

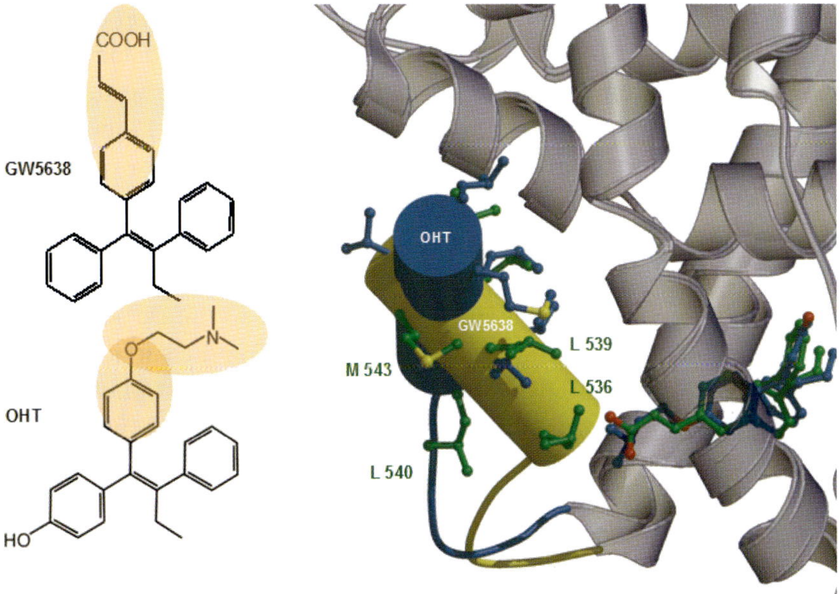

Figure 6. Overlay of GW5638-ERα and OHT-ERα LBD structures. H12 for OHT-ERα is shown as a blue cylinder and H12 for GW-ERα as a gold cylinder. The chemical structures for the two ligands are also shown on the left. Shaded areas represent the extended side chains. In the LBD pocket, OHT is in blue and GW5638 is in green. (Adapted from Ref. 60.)

in ERα stability may account for the ability of GW5638 to inhibit tamoxifen-resistant MCF-7 breast tumor explants.[61]

The acrylate side chain of GW5638/7604, a tamoxifen analog, interacts with Leu-536 and Tyr-537 of the ERα LBD to induce capping of H12 and an unexpected conformation for this AF-2 molecular switch. The resulting increase in ERα surface hydrophobicity correlates with a decrease in receptor stability, in sharp contrast to the increased stability observed for OHT-ERα.[60] Thus, the GW5638-ERα structure adds another level of complexity to the observed conformational flexibility of H12 by showing that it not only controls the recruitment of cofactors but also influences the stability of ERα. Overall, our data suggest that antagonist-mediated increases in surface hydrophobicity can contribute to ERα instability.

In total, the GW5638-ERα LBD structure and supporting data may explain how GW5638/7604 is an effective inhibitor of tamoxifen-resistant MCF-7 tumor explants. Clearly, relatively subtle ligand modifications can significantly alter the conformation of the H12 molecular switch. In addition to preventing coactivator recruitment by occlusion of the AF-2 cleft, similar to OHT and RAL, GW5638/7604 also destabilizes ERα, although less so than the more potent ER antagonists ICI 164,384 and 182,780. This effect is associated with a rotation of H12, induced by the tethering of Leu-536 and Tyr-537 to the carboxyl moiety on GW5638/7604, which leads to an increase in the surface hydrophobicity of the ERα LBD. This increased surface hydrophobicity is associated with a decrease in ERα stability. Therefore, GW5638/7604 is a more potent growth inhibitor than OHT and RAL because it also downregulates ERα protein. The distinct pharmacologies of tamoxifen, GW5638 and ICI are due, at least in part, to important differences in the respective ERα AF-2 conformations (Figs. 5 and 6).

CHALLENGES FOR THE FUTURE

It is clear that we have learned much about SERM action and ER subtype-selective ligands from the many ERα and ERβ LBD structures that have been published. There is undoubtedly much more mechanistic and predictive information to be gained from additional structures of ER LBDs bound to novel SERMs as well as existing SERMs whose behaviors are diverse, complex and not fully understood. However, structural information about how the different ER functional domains communicate with each other and how better to exploit these interactions to manage or prevent ER-dependent breast cancers or menopausal symptoms remains elusive. The expression and crystallization of multidomain nuclear receptor polypeptides has proven to be very difficult. In fact, only one such structure has been reported to date. In 2008 Rastinejad *et al.*[36] published the crystal structure of the peroxisome-proliferator-activated receptor gamma (PPARγ) bound to a consensus DNA response element (PPRE) as a heterodimer with the retinoid X receptor alpha (RXRα). Included were the cognate receptor ligands rosiglitazone, for PPARγ, and

9-*cis*-retinoic acid for RXRα, as well as a peptide that represents an interaction motif for the P160 coactivators (SRC family). Importantly, PPARγ and RXRα form a nonsymmetrical complex that allows the LBD of PPARγ to contact multiple domains in both proteins, but especially the DBD of RXRα. Three interacting surfaces link PPARγ and RXRα, including some that depend on the presence of DNA (PPRE). The PPARγ LBD cooperates with both DBDs to stabilize PPRE binding. This structure was surprising, for a few reasons. First, it was not previously realized that the LBD and DBD regions physically interact to such an extent when bound to DNA. Second, while the aminoterminal (NTD) region is unstructured for both receptors, it is accommodated within the crystal packing, a fortuitous outcome. Third, the three different PPARγ ligands that were studied did not significantly influence the overall structure of the complex or the interface between the interacting domains, in contrast to the many altered LBD conformations that have been observed for a number of nuclear receptor LBDs, including ERα and ERβ LBDs, when bound to SERMs with diverse chemical scaffolds and biological properties.[62]

Although low-resolution cryo-EM and/or other solution structure information about the PPARγ/RXR complex as well as full-length RXR/VDR bound to DNA has been published,[63,64] no additional crystal structures have been reported and no structure information about the organization of a multidomain ER/ERE complex has been reported. This information will be essential for designing new classes of SERMs that can target surfaces that are important for ER action in breast cancer treatment and prevention even when known SERMs are ineffective.

REFERENCES

1. Jensen EV, Jacobson HI. (1962) Basic guides to the mechanism of estrogen action. *Recent Prog Horm Res* **18**:387–414.
2. Toft D, Gorski J. (1966) A receptor molecule for estrogens: isolation from the rat uterus and preliminary characterization. *Proc Nat Acad Sci USA* **55**(6):1574–1581.
3. Toft D, Shyamala G, Gorski J. (1967) A receptor molecule for estrogens: studies using a cell-free system. *Proc Nat Acad Sci USA* **57**(6):1740–1743.

4. Korenman SG. (1970) Relation between estrogen inhibitory activity and binding to cytosol of rabbit and human uterus. *Endocrinolgy* **87**:1119–1123.
5. Skidmore J, Walpole AL, Woodburn J. (1972) Effect of some triphenylethylenes on oestradiol binding *in vitro* to macromolecules from uterus and anterior pituitary. *J Endocrinol* **52**(2):289–298.
6. Jordan VC, Collins MM, Rowsby L, Prestwich G. (1977) A monohydroxylated metabolite of tamoxifen with potent antioestrogenic activity. *J Endocrinol* **75**(2):305–316.
7. Shiau AK *et al.* (1998) The structural basis of estrogen receptor/coactivator recognition and the antagonism of this interaction by tamoxifen. *Cell* **95**(7):927–937.
8. Soule HD, Vazguez J, Long A, Albert S, Brennan M. (1973) A human cell line from a pleural effusion derived from a breast carcinoma. *J Natl Cancer Inst* **51**(5):1409–1416.
9. Shafie SM. (1980) Estrogen and the growth of breast cancer: new evidence suggests indirect action. *Science* **209**(4457):701–702.
10. Lippman ME, Bolan G. (1975) Oestrogen-responsive human breast cancer in long term tissue culture. *Nature* **256**(5518):592–593.
11. Greene GL, Fitch FW, Jensen EV. (1980) Monoclonal antibodies to estrophilin: probes for the study of estrogen receptors. *Proc Nat Acad Sci USA* **77**(1):157–161.
12. Greene GL *et al.* (1986) Sequence and expression of human estrogen receptor complementary DNA. *Science* **231**(4742):1150–1154.
13. Lieberman ME, Jordan VC, Fritsch M, Santos MA, Gorski J. (1983) Direct and reversible inhibition of estradiol-stimulated prolactin synthesis by antiestrogens *in vitro*. *J Biol Chem* **258**(8):4734–4740.
14. Lieberman ME, Gorski J, Jordan VC. (1983) An estrogen receptor model to describe the regulation of prolactin synthesis by antiestrogens *in vitro*. *J Biolo Chem* **258**(8):4741–4745.
15. Jordan VC *et al.* (1984) Structural requirements for the pharmacological activity of nonsteroidal antiestrogens *in vitro*. *Mol Pharmacol* **26**(2):272–278.
16. Jordan VC, Lieberman ME. (1984) Estrogen-stimulated prolactin synthesis *in vitro*. Classification of agonist, partial agonist, and antagonist actions based on structure. *Mol Pharmacol* **26**(2):279–285.
17. Jordan VC, Koch R, Mittal S, Schneider MR. (1986) Oestrogenic and antioestrogenic actions in a series of triphenylbut-1-enes: modulation of prolactin synthesis *in vitro*. *Br J Pharmacol* **87**(1):217–223.
18. Jordan VC, Koch R, Langan S, McCague R. (1988) Ligand interaction at the estrogen receptor to program antiestrogen action: a study with nonsteroidal compounds *in vitro*. *Endocrinology* **122**(4):1449–1454.
19. Jordan VC, Koch R. (1989) Regulation of prolactin synthesis *in vitro* by estrogenic and antiestrogenic derivatives of estradiol and estrone. *Endocrinology* **124**(4):1717–1726.

20. Jordan VC. (1984) Biochemical pharmacology of antiestrogen action. *Pharmacol Rev* **36**(4):245–276.
21. Berthois Y, Katzenellenbogen JA, Katzenellenbogen BS. (1986) Phenol red in tissue culture media is a weak estrogen: implications concerning the study of estrogen-responsive cells in culture. *Proc Nat Acad Sci USA* **83**(8):2496–2500.
22. Bindal RD, Carlson KE, Katzenellenbogen BS, Katzenellenbogen JA. (1988) Lipophilic impurities, not phenolsulfonphthalein, account for the estrogenic activity in commercial preparations of phenol red. *J Steroid Biochem* **31**(3):287–293.
23. Bindal RD, Katzenellenbogen JA. (1988) Bis(4-hydroxyphenyl)[2-(phenoxysulfonyl)phenyl]methane: isolation and structure elucidation of a novel estrogen from commercial preparations of phenol red (phenolsulfonphthalein). *J Med Chem* **31**(10):1978–1983.
24. Murphy CS, Langan-Fahey SM, McCague R, Jordan VC. (1990) Structure-function relationships of hydroxylated metabolites of tamoxifen that control the proliferation of estrogen-responsive T47D breast cancer cells *in vitro*. *Mol Pharmacol* **38**(5):737–743.
25. Murphy CS, Parker CJ, McCague R, Jordan VC. (1991) Structure-activity relationships of nonisomerizable derivatives of tamoxifen: importance of hydroxyl group and side chain positioning for biological activity. *Mol Pharmacol* **39**(3):421–428.
26. Jordan VC. (1987) Laboratory models of breast cancer to aid the elucidation of antiestrogen action. *J Lab Clin Med* **109**(3):267–277.
27. Brzozowski AM *et al.* (1997) Molecular basis of agonism and antagonism in the oestrogen receptor. *Nature* **389**(6652):753–758.
28. Wolf DM, Jordan VC. (1994) The estrogen receptor from a tamoxifen-stimulated MCF-7 tumor variant contains a point mutation in the ligand binding domain. *Breast Cancer Res Treat* **31**(1):129–138.
29. Jiang SY, Jordan VC. (1992) Growth regulation of estrogen receptor-negative breast cancer cells transfected with complementary DNAs for estrogen receptor. *J Natl Cancer Inst* **84**(8):580–591.
30. Levenson AS, Catherino WH, Jordan VC. (1997) Estrogenic activity is increased for an antiestrogen by a natural mutation of the estrogen receptor. *J Steroid Biochem Mol Biol* **60**(5–6):261–268.
31. Levenson AS, Jordan VC. (1998) The key to the antiestrogenic mechanism of raloxifene is amino acid 351 (aspartate) in the estrogen receptor. *Cancer Research* **58**(9):1872–1875.
32. Schafer JI, Liu H, Tonetti DA, Jordan VC. (1999) The interaction of raloxifene and the active metabolite of the antiestrogen EM-800 (SC 5705) with the human estrogen receptor. *Cancer Res* **59**(17):4308–4313.
33. MacGregor Schafer J, Liu H, Bentrem DJ, Zapf JW, Jordan VC. (2000) Allosteric silencing of activating function 1 in the 4-hydroxytamoxifen estrogen receptor

complex is induced by substituting glycine for aspartate at amino acid 351. *Cancer Res* **60**(18):5097–5105.

34. Liu H, Lee ES, Deb Los Reyes A, Zapf JW, Jordan VC. (2001) Silencing and reactivation of the selective estrogen receptor modulator–estrogen receptor alpha complex. *Cancer Res* **61**(9):3632–3639.
35. Liu H, *et al.* (2002) Structure–function relationships of the raloxifene-estrogen receptor-alpha complex for regulating transforming growth factor-alpha expression in breast cancer cells. *J Biol Chem* **277**(11):9189–9198.
36. Chandra V, *et al.* (2008) Structure of the intact PPAR-gamma–RXR–nuclear receptor complex on DNA. *Nature* **456**(7220):350–356.
37. Notides AC, Hamilton DE, Auer HE. (1975) A kinetic analysis of the estrogen receptor transformation. *J Biol Chem* **250**(10):3945–3950.
38. Brou C, *et al.* (1993) Different TBP-associated factors are required for mediating the stimulation of transcription *in vitro* by the acidic transactivator GAL-VP16 and the two nonacidic activation functions of the estrogen receptor. *Nucleic Acids Res* **21**(1):5–12.
39. McKenna NJ, O'Malley BW. (2002) Combinatorial control of gene expression by nuclear receptors and coregulators. *Cell* **108**(4):465–474.
40. Darimont BD, *et al.* (1998) Structure and specificity of nuclear receptor-coactivator interactions. *Genes Dev* **12**(21):3343–3356.
41. Schulman IG, Juguilon H, Evans RM. (1996) Activation and repression by nuclear hormone receptors: hormone modulates an equilibrium between active and repressive states. *Mole Cell Biol* **16**(7):3807–3813.
42. Renaud JP, *et al.* (1995) Crystal structure of the RAR-gamma ligand-binding domain bound to all-trans retinoic acid. *Nature* **378**(6558):681–689.
43. Bourguet W, *et al.* (2000) Crystal structure of a heterodimeric complex of RAR and RXR ligand-binding domains. *Molecular Cell* **5**(2):289–298.
44. Bramlett KS, Wu Y, Burris TP. (2001) Ligands specify coactivator nuclear receptor (NR) box affinity for estrogen receptor subtypes. *Mol Endocrinol (Baltimore, Md)* **15**(6):909–922.
45. Wong CW, Komm B, Cheskis BJ. (2001) Structure-function evaluation of ER alpha and beta interplay with SRC family coactivators. ER selective ligands. *Biochemistry* **40**(23):6756–6765.
46. Warnmark A, Almlof T, Leers J, Gustafsson JA, Treuter E. (2001) Differential recruitment of the mammalian mediator subunit TRAP220 by estrogen receptors ERalpha and ERbeta. *J Biol Chem* **276**(26):23397–23404.
47. Matt N, Ghyselinck NB, Wendling O, Chambon P, Mark M. (2003) Retinoic acid-induced developmental defects are mediated by RARbeta/RXR heterodimers in the pharyngeal endoderm. *Development* **130**(10):2083–2093.
48. Wang YX, *et al.* (2003) Peroxisome-proliferator-activated receptor delta activates fat metabolism to prevent obesity. *Cell* **113**(2):159–170.

49. Mueller SO, Korach KS. (2001) Estrogen receptors and endocrine diseases: lessons from estrogen receptor knockout mice. *Curr Opin Pharmacol* **1**(6):613–619.
50. Pike AC, *et al.* (1999) Structure of the ligand-binding domain of oestrogen receptor beta in the presence of a partial agonist and a full antagonist. *EMBO J* **18**(17):4608–4618.
51. Veeneman GH. (2005) Non-steroidal subtype selective estrogens. *Curr Med Chem* **12**(9):1077–1136.
52. Mewshaw RE, *et al.* (2005) ERbeta ligands. 3. Exploiting two binding orientations of the 2-phenylnaphthalene scaffold to achieve ERbeta selectivity. *J Med Chem* **48**(12):3953–3979.
53. Manas ES, *et al.* (2004) Structure-based design of estrogen receptor-beta selective ligands. *J Am Chem Soc* **126**(46):15106–15119.
54. Henke BR, *et al.* (2002) A new series of estrogen receptor modulators that display selectivity for estrogen receptor beta. *J Med Chem* **45**(25):5492–5505.
55. Malamas MS *et al.* (2004) Design and synthesis of aryl diphenolic azoles as potent and selective estrogen receptor-beta ligands. *J Med Chem* **47**(21):5021–5040.
56. Shiau AK *et al.* (2002) Structural characterization of a subtype-selective ligand reveals a novel mode of estrogen receptor antagonism. *Nat Struct Biol* **9**(5):359–364.
57. Manas ES, Xu ZB, Unwalla RJ, Somers WS. (2004) Understanding the selectivity of genistein for human estrogen receptor-beta using X-ray crystallography and computational methods. *Structure (Camb)* **12**(12):2197–2207.
58. Pike AC, *et al.* (2001) Structural insights into the mode of action of a pure antiestrogen. *Structure* **9**(2):145–153.
59. Wittmann BM, Sherk A, McDonnell DP. (2007) Definition of functionally important mechanistic differences among selective estrogen receptor down-regulators. *Cancer Res* **67**(19):9549–9560.
60. Wu YL, *et al.* (2005) Structural basis for an unexpected mode of SERM-mediated ER antagonism. *Mol Cell* **18**(4):413–424.
61. Connor CE, *et al.* (2001) Circumventing tamoxifen resistance in breast cancers using antiestrogens that induce unique conformational changes in the estrogen receptor. *Cancer Res* **61**(7):2917–2922.
62. Nettles KW, *et al.* (2008) NFkappaB selectivity of estrogen receptor ligands revealed by comparative crystallographic analyses. *Nat Chem Biol* **4**(4):241–247.
63. Orlov I, Rochel N, Moras D, Klaholz BP. (2012) Structure of the full human RXR/VDR nuclear receptor heterodimer complex with its DR3 target DNA. *EMBO J* **31**(2):291–300.
64. Brelivet Y, Rochel N, Moras D. (2012) Structural analysis of nuclear receptors: from isolated domains to integral proteins. *Mol Cell Endocrinol* **348**(2):466–473.

7

COREGULATORS AND THEIR ROLE IN SELECTIVE ESTROGEN RECEPTOR MODULATOR ACTION

Christopher D. Chien, Anton Wellstein and Anna Riegel

Abstract

The majority of breast cancer cases rely on estrogen receptor signaling and hence treatment strategies have focused in large part on blocking signaling through the estrogen receptor. Direct antagonists have undesirable side effects since they have antagonist effects in all tissues of the body. The discovery and development of selective estrogen receptor modulators (SERMs) has revolutionized the treatment of breast cancer as they have antagonist properties in the breast and yet agonist properties in other tissues lessening the severity of side effects of estrogen signaling blockade. Resistance can occur in patients treated with SERMs and this is thought to be due to changes in the balance of recruitment of coregulator molecules to estrogen-regulated transcriptional complexes.The effects of this shift may alter SERM function in the breast, moving it from antagonist to agonist. In this chapter, we will present the evidence that coregulator recruitment can alter the response to SERMs.

Keywords: Diethylstilbestrol (DES); Clear Cell Adenocarcinoma (CCAC); Charles; SERMs; Breast Cancer; Estrogen Receptor; Coregulator.

INTRODUCTION

Breast cancer is the most commonly diagnosed cancer in women, with an estimated 226,870 new cases, and the second leading cause of cancer mortality in the United States, with an estimated 39,510 deaths annually.[1] Globally, breast cancer is the most prevalent type of cancer in women, with an estimated 1,383,500 new cases. Additionally, it is the leading cause of cancer-related death in women, with an estimated 458,400 deaths.[2] Sixty percent of these breast cancer cases show expression of estrogen receptor α (ERα),[3,4] which is believed to be a key driver of breast cancer proliferation. Therapies for the treatment of ERα-positive breast cancer focus on the disruption of ERα activity.

Tamoxifen, a selective estrogen receptor modulator (SERM), is the standard option for the treatment of ER-positive breast cancer.[5] SERMs are classified as selective due to their differential agonist/antagonist effects in a cell- or tissue-specific context. In the breast, tamoxifen acts as an antagonist and competes with estrogen for binding to ER downregulating signaling through ERα. In the skeletal bone and uterus, it acts as an agonist upregulating ERα activity. This tissue-specific response from the same receptor–ligand interaction suggests that there is another mechanism driving the outcome from the same stimulus in different organs. Our current understanding indicates that steroid receptor coregulators primarily mediate this difference. These factors determine the transcriptional outcome of SERMs based on their levels and recruitment to sites of ligand-mediated transcriptional output.

The existence of transcriptional coregulators was first proposed in studies looking at transcriptional "squelching." The yeast Gal4 transcription factor was able to negatively regulate the expression of genes without the need to bind to DNA or for the Gal4 target sequence.[6] A similar effect was seen when ERα was shown to prevent the activation of progesterone and glucocorticoid receptor on promoters which did not contain any estrogen response elements (EREs).[7] These data suggested that the steroid receptors competed for a rate-limiting pool of factors responsible for the decrease in their transcriptional activity. This work preceded the identification of the numerous factors shown to bind to and regulate the function of steroid receptors, in particular the ER. These factors are known as steroid receptor coregulators, which can be divided into coactivators, which facilitate the enhancement of transcription, and corepressors, which dampen transcription. Based on *in vitro* studies, it is now believed that the major determinant in the agonist/antagonist properties of SERMs lies in the balance in recruitment of coactivators and corepressors.[8,9] For example, in tamoxifen-sensitive cells, the tamoxifen-bound ER recruits corepressors and prevents the activation of estrogen-regulated genes. Conversely, in tamoxifen-resistant cells, the tamoxifen-bound ER recruits coactivators and allows transcription of estrogen-regulated genes (Fig. 1). Although numerous steroid receptor coregulators have been identified, in this chapter we will cover only the most-studied coregulators, as they relate to the activity of SERMs, with most of the focus on tamoxifen and its effects on the ER. This is not to say that other coregulators do not play significant roles as yet unreported in different subtypes of breast cancer and their response to therapy.

STEROID RECEPTOR COACTIVATORS IN SERM ACTION

Steroid receptor coactivators are named for the fact that they increase the transcriptional activity of ligand-bound steroid receptors. The most-studied family of coactivators in relation to SERM activity is the p160 steroid receptor coactivator (SRC) family. This family consists of three members: *SRC-1* (NCOA1), *TIF-2* (GRIP-1, SRC-2, NCOA2), and

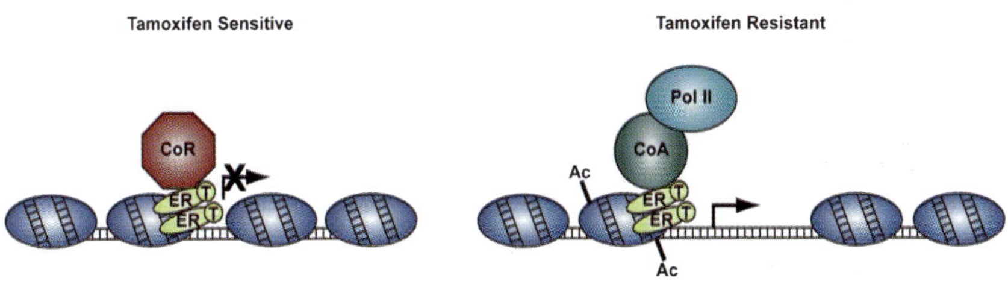

Figure 1. Recruitment of coregulators by SERMs determines their transcriptional activity. SERMs such as tamoxifen (T) bind to estrogen receptors (ER) in the cytoplasm and translocate to the nucleus, where they bind to estrogen-responsive elements in DNA. In a tamoxifen-sensitive cell, tamoxifen-bound the ER recruits corepressors (CoR), which do not allow active transcription. In contrast, cells that are tamoxifen-resistant recruit coactivators (CoA), which recruit other proteins involved in transcription such as histone acetyltransferases that can acetylate histones (Ac) to change the conformation of DNA and proteins from the basal transcription machinery such as RNA polymerase II (Pol II).

AIB1 (SRC-3, ACTR, p/CIP, RAC3, TRAM-1, NCOA3). They all contain a highly conserved basic helix loop helix (bHLH), Per Arnt Sim (PAS) A and B domain, nuclear receptor interaction domain (RID), CBP/p300 interaction domain (CID), and polygluatamine region (poly Q) (Fig. 2). They are all proposed to have inherent histone acetyltransferase (HAT) activity, yet histone acetyltransferase domains have only been confirmed in SRC-1 and AIB1 (Fig. 2). The predominant function of these proteins is to bind to activated nuclear receptors and recruit basal transcription machinery and other molecules involved in chromatin remodeling, such as CBP/p300[10] and CARM-1.[11] Their general effect is an enhancement of transcription due to modifications which allow a more relaxed conformation of the chromatin in the area of activated steroid receptor binding to DNA.[12] Numerous studies have indicated the involvement of the p160 SRC family in breast cancer.[13] In this section we will cover both the preclinical and clinical studies that demonstrate a role for the steroid receptor coactivator family members in regulating SERM activity.

SRC-1

SRC-1 has been shown to be involved in breast cancer tumorigenesis. The founding member of the p160 SRC family was cloned in the

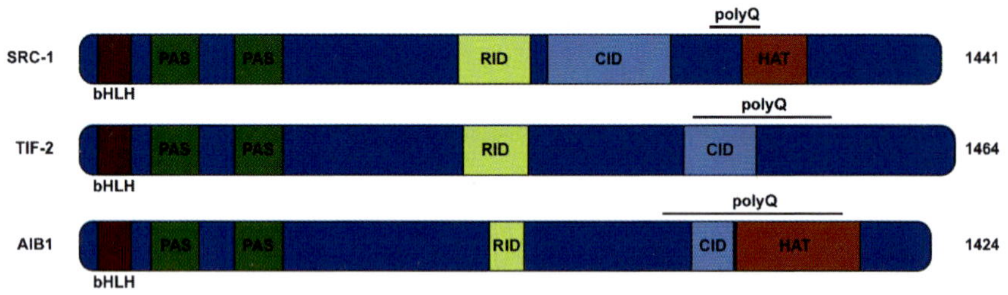

Figure 2. p160 steroid receptor coactivator family. The three members of the p160 steroid receptor coactivator family are SRC-1, TIF-2, and AIB1. They all contain basic helix loop helix (bHLH), Per Arnt Sim (PAS), nuclear receptor interaction domain (RID), CBP/p300 interaction domain (CID), and glutamine rich regions (poly Q). They all have been proposed to have inherent histone acetyltransferase activity, but only SRC-1 and AIB1 have confirmed histone acetyltransferase domains (HAT).

laboratory of Dr. Bert O'Malley in 1995.[14] In this same study SRC-1 was shown to associate with both ERα and progesterone receptor in the presence of agonist and this association was inhibited by ERα antagonist. More recently, SRC-1 has been implicated to play a role in breast cancer. It has been implicated in the invasiveness of breast cancer cells.[15-17] Interestingly, in one of these studies knockout of the SRC-1 gene had no effect on primary tumor formation, yet decreased the incidence of metastatic lesions in mice.[17] The expression of SRC-1 is correlated with human epidermal growth factor 2 (HER2) expression and poor prognosis for patients with breast cancer.[18]

Early studies with this coactivator would prove to be critical for understanding the modulation of SERM activity by coactivators. The first indications that a coactivator might modulate the activity of SERMs were in studies using transient transfection assays to determine the effect of SRC-1 expression on transcriptional activity stimulated by tamoxifen. The first study in 1997, at the laboratory of Dr. O'Malley, used HepG2 cells, in which tamoxifen acts as an agonist. In these studies SRC-1 was shown to potentiate the transcription induced by both estrogen and tamoxifen.[8] Intriguingly, overexpression of SRC-1 in HeLa cells where tamoxifen acts as an antagonist also led to an increase in transcription with estrogen and tamoxifen treatment, suggesting that levels of the coactivator determine the agonist or antagonist response to tamoxifen. A second study, by Dr. Myles Brown's group,[19] took advantage of the fact that two SERMs, tamoxifen and raloxifene, have antiestrogenic effects in the mammary gland and yet opposing effects in uterine tissue. In the uterus tamoxifen acts as an agonist and raloxifene acts as an antagonist. When looking at recruitment of coactivators to EREs in estrogen-stimulated genes after estrogen and SERM treatment, they observed a difference in the recruitment of SRC-1 and other coactivators such as AIB1, CBP, and acetylated histones, which are markers of transcriptional activation. SRC-1 and the other coactivators were recruited in response to estrogen and tamoxifen in Ishikawa endometrial carcinoma cells but not in the mammary carcinoma cell line MCF-7. They postulated that one of the coactivators may be responsible for the effect of tamoxifen acting like an agonist in the Ishikawa cells and observed that the only difference in expression of coregulators between the MCF-7 and Ishikawa cells was the level of expression of SRC-1. When they overexpressed SRC-1 in MCF-7 cells, which have relatively low levels of SRC-1, tamoxifen then worked as an agonist to stimulate the expression of c-Myc and insulin-like growth factor (IGF-1). Conversely, when they knocked down the levels of SRC-1 in Ishikawa cells they found that the agonist activity of tamoxifen was decreased and was no longer able to stimulate expression of c-Myc and IGF-1. Subsequent studies also showed that the difference in response to estrogen and raloxifene in human ovarian cancer cell lines was due to the difference in recruitment of SRC-1 and other coregulators to the promoters of these same genes.[20] Interestingly, high SRC-1 levels were predictive of response to tamoxifen in a small study on breast cancer patients with recurrent disease given tamoxifen after relapse; however, the study was done with a small number of patient samples, did not measure the levels of other coregulators or growth factors, and relied on mRNA expression to form their conclusions.[21]

Other studies have also shown that SRC-1 is important in altering the cellular response to tamoxifen. SRC-1 is recruited to EREs in response to estrogen and not tamoxifen in cultured primary breast cancer cells sensitive to tamoxifen.[22] Breast cancer patients with high levels SRC-1 and HER2 were more likely to have recurrent disease after tamoxifen treatment than patients with tumors expressing high levels of HER2 without SRC-1 expression.[23] Other mechanisms of tamoxifen resistance have been proposed to act through protein kinase A phosphorylation of ERα; however, SRC-1 is recruited in response to this phosphorylation, suggesting that the recruitment of the coactivator is critical in switching tamoxifen from an antagonist to an agonist in the breast.[24] Coassociation of p160 SRC family members with ERα was also seen in tamoxifen-resistant LY2 cells as opposed to MCF-7 tamoxifen-sensitive cells after tamoxifen treatment.[18] In this same study, using a tissue microarray from 560 patients, they observed that SRC-1 expression correlated with reduced disease-free survival and coassociation of SRC-1 with ERα in patients who relapse relative to patients who do not have recurrent disease. Taken together these studies indicate that SRC-1 plays a role in the modulation of SERM activity from an antagonist to an agonist. Recently, overexpression of SRC-1 has also been associated with resistance to aromatase inhibitors suggesting a broader role for SRC-1 in anti-estrogen therapy.[25]

TIF-2

To date there have been limited studies indicating that TIF-2 modulates response to tamoxifen or other SERMs. In relation to tumorigenesis there is one study associating increased TIF-2 mRNA expression with breast cancer.[26] However, in contrast to TIF-2 mRNA, TIF-2 protein expression levels were not shown to be different between normal and malignant breast tissue when measured by immunohistochemistry (IHC).[27] But there does appear to be a role for TIF-2 in mammary gland morphogenesis. Conditional knockout of TIF-2 in PR-expressing tissues led to a significant decrease of ductal side branching and alveologenesis in hormone-treated mammary glands, suggesting a role for TIF-2 in altering cell proliferation downstream of PR signaling.[28] It is possible that TIF-2 may play a role in subtypes of breast cancer that have yet to be identified.

AIB1

The most highly studied coactivator in relation to modulating the response to SERMs and in particular tamoxifen is the amplified in breast cancer 1 (AIB1) protein. AIB1 has been identified as an oncogene on the basis of a mouse model where overexpression of AIB1 directed to the mammary gland led to the formation of tumors.[29] The AIB1 gene was identified to be amplified in 2–10% of breast cancer samples[30–34] and this was associated with a significant increase in AIB1 mRNA levels in breast tumor samples.[13] In addition, the levels of AIB1 mRNA have been reported to be elevated in 13–64% of breast cancer samples.[31,35–37] It has been noted that the protein levels of AIB1 in tumor samples do not always correlate with the AIB1 mRNA levels,[37] though this could be due to differences in the sensitivity and linearity of quantitative PCR methods versus semiquantitative IHC methods. An added complication is that AIB1 protein can be detected at high levels in the nucleus and/or the cytoplasm.[34] This may be due to the existence of a splice variant of AIB1 discovered by the Riegel laboratory[38] which produces a mainly cytoplasmic shorter isoform of the AIB1 protein, AIB1-Δ4, which is overexpressed in breast cancer cell lines and tumor samples relative to normal breast tissue.[38,39] Overexpression of AIB1-Δ4 is associated with increased metastatic potential of breast cancer cells.[39,40] Irrespective of this additional complexity, higher levels of AIB1 mRNA or AIB1 gene amplification in breast cancers are generally predictive of worse disease outcome, higher grade disease, shorter disease-free interval, and reduced overall survival.[13,27,41,42]

AIB1 has been widely studied for its role in resistance to SERM therapy in breast cancer both preclinically and clinically. Most studies have focused on the coexpression of AIB1 with HER2. The earliest evidence that AIB1 and HER2 might have a role in tamoxifen resistance is from *in vitro* experiments at Dr. C. Kent Osborne's lab.[43] They created the MCF-7/HER2-18 breast cancer cell subline that expressed the HER2 receptor at 45 times the level expressed in the parental MCF-7 cell line. These parental and MCF-7/HER2-18 cells were allowed to form tumors in ovariectomized mice supplemented with a slow release estrogen pellet. When the tumors reached 500 mm^2 the estrogen pellet was removed and the animals were treated with tamoxifen. Strikingly, the MCF-7/HER2-18 tumors grew in response to tamoxifen treatment while the parental MCF-7 tumors did not grow. This implicated HER2 in acquisition of tamoxifen resistance but not necessarily AIB1. However, MCF-7 cells also have greater than 20-fold expression of the AIB1 gene and have high levels of AIB1 mRNA,[31] suggesting that there may be a role for AIB1 in tamoxifen resistance. Later data would show that in tamoxifen-resistant HER2 overexpressing BT474 breast cancer cells, depletion of AIB1 sensitized these cells to the growth inhibitory effects of tamoxifen compared to the parental cells.[44] This indicated a role for AIB1 in modulating the response to tamoxifen. In addition, overexpression of the AIB1-Δ4 isoform in breast cancer and endometrial carcinoma cells *in vitro* increased the agonist properties of tamoxifen.[45]

Confirmation of the crosstalk between AIB1 and HER2 in the development of tamoxifen resistance was illustrated in two separate papers. The first paper described a correlation between AIB1 and HER2 with worse outcomes after treatment with tamoxifen.[46] In this report the authors measured the levels of AIB1 and HER2 in samples from breast cancer patients who either did or did not receive adjuvant tamoxifen by Western blot. Patients with high levels of AIB1, irrespective of HER2 expression, who were treated with tamoxifen had a worse disease-free survival (DFS). However, the most significant finding from this paper was that patients who had high AIB1 and high HER2 expression treated with tamoxifen had worse DFS than all other patients, indicating that tamoxifen resistance developed in these patients. The second report of a link between AIB1 and HER2 relied on IHC to measure levels of AIB1 and HER family members, as well as fluorescence *in situ* hybridization (FISH) to measure amplification of the AIB1 gene.[47] Breast cancer patients in this study were all ERα-positive, and they received tamoxifen for a median of 5 years and were followed for a median of 6.45 years. Patients who had high levels of AIB1 and HER2 had an increased risk of relapse after treatment with tamoxifen. These studies suggest that high expression of AIB1 in conjunction with high levels of HER2 is a predictor of patients who might develop tamoxifen resistance.

These studies were followed up with *in vitro* work to explore the possibility of crosstalk between the estrogen and growth factor signaling pathways.[48] Subsequent work reported that in the tamoxifen-resistant MCF-7/HER2-18 breast cancer cell line described previously,[43] several molecules downstream of the human epidermal growth factor receptors — epidermal growth factor receptor (EGFR) and HER2 — and molecules associated with — estrogen signaling were activated/phosphorylated upon stimulation with growth factors — epidermal growth factor (EGF) and heregulin (HRG) — or estrogenic compounds (estrogen or tamoxifen). Activation of these same molecules and other signaling molecules was reduced when MCF-7/HER2-18 cells were treated with gefitinib, a small molecule inhibitor targeted specifically for the tyrosine kinase domain of EGFR. The authors also described that AIB1 is recruited after treatment with tamoxifen in the tamoxifen-resistant MCF-7/HER2-18 cells but not in tamoxifen-sensitive parental MCF-7 cells to an ERE in the pS2 promoter, indicating that the switch of tamoxifen from an antagonist to an agonist is in part due to the recruitment of AIB1 to estrogen-responsive genes. Additionally, validation in a preclinical *in vivo* model indicated that the growth response to tamoxifen of tumors derived from tamoxifen-resistant MCF-7/HER2-18 cells could be completely abrogated by treatment with gefitinib. Interestingly, in this same experiment estrogen- and gefitinib-treated tumors grew just as well as tamoxifen-treated MCF-7/HER2-18 cells, which the authors suggested was due to estrogen stimulation of growth that was not completely dependent on human epidermal growth factor (HER) signaling. In contrast, the growth stimulation due to tamoxifen was dependent on HER signaling. The authors also saw that the activation/phosphorylation of AIB1 was increased by HRG, estrogen, and tamoxifen in the MCF-7/HER2-18 cells relative to the parental MCF-7 cells and this activation was reversed by gefitinib treatment. Phosphorylation of AIB1 in extracts from the tumors in the *in vivo* experiment was found as well. These data suggest a role for AIB1 in the regulation of tamoxifen response through a mechanism that involves growth factor signaling activation of AIB1. Consistent with this correlation is that the development of hormone independence and SERM resistance in breast tumors is widely associated with increases in growth factor signaling through receptor tyrosine kinases such as EGFR, HER2, and HER3.[48,49] In addition, the Riegel lab has also shown that AIB1 can potentiate HER family signaling *in vitro* and *in vivo*.[38,50,51] Thus, it is perhaps not surprising that high levels of AIB1 coupled with high EGFR or HER2 are associated with increased relapse after tamoxifen therapy.[46,47,52]

In terms of a mechanism of AIB1- and HER2-induced tamoxifen resistance, one model is that the increases in HER2 and AIB1 lead directly to enhancement of agonist properties of tamoxifen in breast cancer cells.[53] Alternatively, the unliganded ERα activated by HER kinase pathways (such as mitogen-activated protein kinase, also known, as MAPK)[48] or other kinase, pathways that activate AIB1[54] could be capable of activating a mitogenic AP-1–dependent gene set[55] that possibly bypasses tamoxifen-repressive effects. It is worth noting that AIB1 can also coactivate AP-1-dependent genes and may be directly involved in potentiating the ER signaling at AP-1-dependent promoters.[38,56] In addition, it has been reported that the HER2 gene is itself regulated by estrogen and HER2 expression levels are determined by the balance in the levels of recruitment of AIB1 and PAX2.[57] In a tamoxifen-resistant cell line derived from MCF-7 cells (Tam-R),[58] the total levels of HER2 were increased and the total levels of PAX2 were decreased. Tamoxifen treatment of Tam-R cells led to the recruitment of AIB1 and no recruitment of PAX2 to the ERE in an intron of HER2, and in tamoxifen-sensitive parental MCF-7 cells the opposite was observed. This suggests that AIB1 regulation of HER2 levels is critical to the modulation in the SERM activity of tamoxifen.

Tamoxifen also induces the expression of AIB1 *in vitro*,[59,60] suggesting a positive feedback loop that would increase AIB1-mediated signaling in tumors. In agreement with this, tamoxifen induces AIB1 in breast tumors of patients even at low doses,[61] suggesting that this positive feedback also occurs *in vivo*. Overall, the *in vitro* and clinical data support a sequential and powerful positive feedback model of antiestrogen resistance whereby tamoxifen treatment initially leads to an increase in AIB1 that potentiates the agonist properties of tamoxifen, as well as any estrogen-liganded ER. This leads to increased levels and activation of HER2 and subsequent stimulation of kinase cascades. HER family signaling is also potentiated directly by AIB1.[50,51] The signaling cascades can further activate AIB1 and the unliganded ER, leading to hormone-independent stimulation of proliferation and endocrine resistance. Also worth noting is that the mechanisms of resistance to aromatase inhibitors reported to date seem to also involve increases in coactivator activity and crosstalk of HER2 and AIB1.[62-64]

STEROID RECEPTOR COREPRESSORS IN SERM ACTION

Most of the work that has been done in regard to the modulation of SERM activity in breast cancer has focused primarily on the steroid receptor coactivators. However, there has also been substantial work on

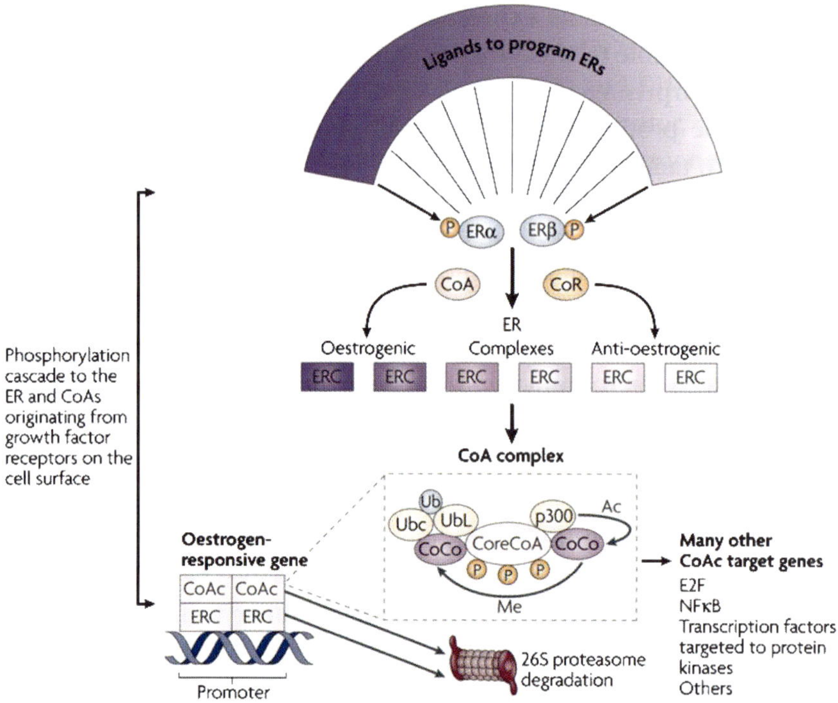

Figure 3. Molecular networks potentially influence the expression of SERM action in a target tissue. The shape of the ligands that bind to the ERs α and β programs the complex to become an estrogenic or antiestrogenic signal. The context of the ER complex (ERC) can influence the expression of the response through the numbers of corepressors (CoR) or coactivators (CoA). In simple terms, a site with few CoAs or high levels of CoRs might be a dominant antiestrogenic site. However, the expression of estrogenic action is not simply the binding of the receptor complex to the promoter of the estrogen-responsive gene, but a dynamic process of CoA complex assembly and destruction. A core CoA, such as, steroid receptor coactivator protein 3 (SRC-3), and the ERC are influenced by phosphorylation cascades that phosphorylate target sites on both complexes. The core CoA then assembles an activated multiprotein complex containing specific co-coactivators (CoCo) that might include p300, each of which has a specific enzymatic activity to be activated later. The CoA complex (CoAc) binds to the ERC at the estrogen-responsive gene promoter to switch on transcription. The CoCo proteins then perform methylation (Me) or acetylation (Ac) to activate dissociation of the complex. Simultaneously, ubiquitylation by the bound ubiquitin-conjugating enzyme (Ubc) targets ubiquitin ligase (UbL) destruction of protein members of the complex through the 26S proteasome. The ERs are also ubiquitylated and destroyed in the 26S proteasome.

level (through regulation of transcription or through micro-RNA regulation) or posttranslational level (through phosphorylation, acetylation, methylation, and ubiquitination) (reviewed in Refs. 9, 12, and 13). Peptidyl-prolyl cis/trans isomerase NIMA-interacting 1 (Pin1) regulates the expression of SMRT.[77] NCoR expression has been shown to be regulated through the proteasome.[78] Even with these various types of regulation of the coregulators, the most-studied coregulators in relation to SERM activity have broad expression in many tissue types.[14,65,79] An overview of how molecular networks impinging on coactivators and repressors might influence SERM signaling is shown in Fig.3.

FUTURE DIRECTIONS

From the data discussed in this review, a reasonable therapeutic strategy for endocrine-resistant breast cancer would be to interrupt coactivator crosstalk with both nuclear receptor and growth factor signaling. To produce this effect *in vivo,* we could reduce the cellular levels of these molecules by increased degradation. We could inhibit coactivator activation by preventing phosphorylation or other activating posttranslational modifications, or we could interrupt coactivator interaction with steroid receptors or other factors in the transcription complex, such as CBP/p300. Some of these approaches have already been tried; for instance, ligand-induced interruption of receptor–coactivator interaction has already been investigated[81] and new therapies may emerge from these approaches. In the long term,

Figure 3. (*Continued*) Therefore, a regimented cycle of assembly, activation, and destruction occurs on the basis of the preprogrammed ER complex. However, the coactivator, specifically SRC3, has ubiquitous action and can further modulate or amplify the ligand-activated trigger through many modulating genes that can consolidate and increase the stimulatory response of the ERC in a tissue. Therefore, the target tissue is programmed to express a spectrum of responses between full estrogen action and antiestrogen action on the basis of the shape of the ligand and the sophistication of the tissue-modulating network. NFκB, nuclear factor κB. Figure and legend reproduced with permission from *Nature Reviews Cancer*.[80]

perhaps the simplest, effective approach would be to design a ligand that would cause selective degradation of the coactivators. Interestingly, 17β-estradiol can cause degradation of AIB1[59] and perhaps we can develop a SERM that will stimulate coactivator degradation without stimulating breast proliferation. Such an antiproliferative compound would reduce endocrine resistance by inhibiting coactivator potentiation of HER family signaling. The availability of an expanding generation of new SERM reagents holds promise for novel compounds with these properties.

REFERENCES

1. Siegel R, Naishadham D, Jemal A. (2012) Cancer statistics, 2012. *CA Cancer J Clin* **62**(1):10–29.
2. Jemal A *et al.* (2011) Global cancer statistics. *CA Cancer J Clin* **61**:69–90.
3. M Stierer *et al.* (1993) Immunohistochemical and biochemical measurement of estrogen and progesterone receptors in primary breast cancer. Correlation of histopathology and prognostic factors. *Ann Surg* **218**:13.
4. Andersen J, Poulsen HS. (1989) Immunohistochemical estrogen receptor determination in paraffin-embedded tissue. Prediction of response to hormonal treatment in advanced breast cancer. *Cancer* **64**:1901–1908.
5. Jordan VC. (2003) Tamoxifen: a most unlikely pioneering medicine. *Nat Rev Drug Discov* **2**(3):205–213.
6. Gill G, Ptashne M. (1988) Negative effect of the transcriptional activator GAL4. *Nature* **334**:721–724.
7. Meyer ME *et al.* (1989) Steroid hormone receptors compete for factors that mediate their enhancer function. *Cell* **57**:433–442.
8. Smith CL *et al.* (1997) Coactivator and corepressor regulation of the agonist/antagonist activity of the mixed antiestrogen, 4-hydroxytamoxifen. *Mol Endocrinol* **11**:657–666.
9. Smith CL *et al.* (2004) Coregulator function: a key to understanding tissue specificity of selective receptor modulators. *Endocr Rev.* **25**:45–71.
10. Chen H *et al.* (1997) Nuclear receptor coactivator ACTR is a novel histone acetyltransferase and forms a multimeric activation complex with P/CAF and CBP/p300. *Cell* **90**:569–580.
11. Chen D *et al.* (1999) Regulation of transcription by a protein methyltransferase. *Science* **284**:2174–2177.
12. Xu J, Wu R-C, O'Malley BW. (2009) Normal and cancer-related functions of the p160 steroid receptor co-activator (SRC) family. *Nat Rev Cancer* **9**:615–630.

13. Lahusen T, Henke RT, Kagan BL, Wellstein A, Riegel AT. (2009) The role and regulation of the nuclear receptor co-activator AIB1 in breast cancer. *Breast Cancer Res Treat* **116**:225–237.
14. Oñate SA, Tsai SY, Tsai M-J, O'malley BW. (1995) Sequence and characterization of a coactivator for the steroid hormone receptor superfamily. *Science* **270**:1354–1357.
15. Qin L *et al.* (2011) Steroid receptor coactivator-1 upregulates integrin 5 expression to promote breast cancer cell adhesion and migration. *Cancer Res* **71**:1742–1751.
16. Qin L, Liu Z, Chen H, Xu J. (2009) The steroid receptor coactivator-1 regulates twist expression and promotes breast cancer metastasis. *Cancer Res* **69**.3819–3827.
17. Wang S *et al.* (2009) Disruption of the SRC-1 gene in mice suppresses breast cancer metastasis without affecting primary tumor formation. *Proc Natl Acad Sci USA* **106**:151–156.
18. Redmond AM *et al.* (2009) Coassociation of estrogen receptor and p160 proteins predicts resistance to endocrine treatment; SRC-1 is an independent predictor of breast cancer recurrence. *Clin Cancer Res* **15**:2098–2106.
19. Shang Y *et al.* (2002) Molecular determinants for the tissue specificity of SERMs. *Science* **295**:2465–2468.
20. Sasaki H *et al.* (2008) Difference between genomic actions of estrogen versus raloxifene in human ovarian cancer cell lines. *Oncogene* **27**:2737–2745.
21. Berns EM, van Staveren IL, Klijn JG, Foekens JA. (1998) Predictive value of SRC-1 for tamoxifen response of recurrent breast cancer. *Breast Cancer Res Treat* **48**:87–92.
22. Fleming FJ *et al.* (2004) Differential recruitment of coregulator proteins steroid receptor coactivator-1 and silencing mediator for retinoid and thyroid receptors to the estrogen receptor–estrogen response element by -estradiol and 4-hydroxy-tamoxifen in human breast cancer. *J Clin Endocrinol Metab* **89**:375–383.
23. Fleming FJ *et al.* (2004) Expression of SRC-1, AIB1, and PEA3 in HER2-mediated endocrine-resistant breast cancer: a predictive role for SRC-1. *J Clin Pathol* **57**:1069–1074.
24. Zwart W *et al.* (2007) PKA-induced resistance to tamoxifen is associated with an altered orientation of ERalpha towards co-activator SRC-1. *EMBO J* **26**:3534–3544.
25. McBryan J *et al.* (2012) Metastatic progression with resistance to aromatase inhibitors is driven by the steroid receptor coactivator SRC-1. *Cancer Res* **72**:548–559.
26. Kurebayashi J *et al.* (2000) Expression levels of estrogen receptor-alpha, estrogen receptor-beta, coactivators, and corepressors in breast cancer. *Clin Cancer Res* **6**:512–518.

27. Hudelist G *et al.* (2003) Expression of sex steroid receptors and their co-factors in normal and malignant breast tissue: AIB1 is a carcinoma-specific co-activator. *Breast Cancer Res Treat* **78**:193–204.
28. Mukherjee A *et al.* (2006) Steroid receptor coactivator 2 is critical for progesterone-dependent uterine function and mammary morphogenesis in the mouse. *Mol Cell Biol* **17**:6571–6583.
29. Torres-Arzayus MI *et al.* (2004) High tumor incidence and activation of the PI3K/AKT pathway in transgenic mice define AIB1 as an oncogene. *Cancer Cell* **6**:263–274.
30. Guan X-Y *et al.* (1996) Hybrid selection of transcribed sequences from microdissected DNA: isolation of genes within amplified region at 20q11–q13.2 in breast cancer. *Cancer Res* **56**:3446–3450.
31. Anzick SL. (1997) AIB1, a steroid receptor coactivator amplified in breast and ovarian cancer. *Science* **277**:965–968.
32. Cuny M *et al.* (2000) Relating genotype and phenotype in breast cancer: an analysis of the prognostic significance of amplification at eight different genes or loci and of p53 mutations. *Cancer Res* **60**:1077–1083.
33. Shibata A *et al.* (2001) Somatic gene alteration of AIB1 gene in patients with breast cancer. *Endocr J* **48**:199–204.
34. Iwase H *et al.* (2003) Clinical significance of AIB1 expression in human breast cancer. *Breast Cancer Res Treat* **80**:339–345.
35. Glaeser M, Floetotto T, Hanstein B, Beckmann MW, Niederacher D. (2001) Gene amplification and expression of the steroid receptor coactivator SRC3 (AIB1) in sporadic breast and endometrial carcinomas. *Horm Metab Res* **33**:121–126.
36. Bouras T, Southey MC, Venter DJ. (2001) Overexpression of the steroid receptor coactivator AIB1 in breast cancer correlates with the absence of estrogen and progesterone receptors and positivity for p53 and HER2/neu. *Cancer Res* **61**:903–907.
37. List H, Reiter R, Singh B, Wellstein A, Riegel A. (2001) Expression of the nuclear coactivator AIB1 in normal and malignant breast tissue. *Breast Cancer Res Treat* **68:**21–28.
38. Reiter R, Wellstein A, Riegel AT. (2001) An isoform of the coactivator AIB1 that increases hormone and growth factor sensitivity is overexpressed in breast cancer. *J Biol Chem* **276**:39736–39741.
39. Long W *et al.* (2010) SRC-3Δ4 mediates the interaction of EGFR with FAK to promote cell migration. *Mol Cell* **37**:321–332.
40. Chien CD *et al.* (2011) Role of the nuclear receptor coactivator AIB1-Δ4 splice variant in the control of gene transcription. *J Biol Chem* **286**:26813–26827.
41. Zhao C *et al.* (2003) Elevated expression levels of NCOA3, TOP1, and TFAP2C in breast tumors as predictors of poor prognosis. *Cancer* **98**:18–23.

42. Harigopal M *et al.* (2009) Estrogen receptor co-activator (AIB1) protein expression by automated quantitative analysis (AQUA) in a breast cancer tissue microarray and association with patient outcome. *Breast Cancer Res Treat* **115**:77–85.
43. Benz CC *et al.* (1992) Estrogen-dependent, tamoxifen-resistant tumorigenic growth of MCF-7 cells transfected with HER2/neu. *Breast Cancer Res Treat* **24**:85–95.
44. Su Q *et al.* (2008) Role of AIB1 for Tamoxifen Resistance in Estrogen Receptor–Positive Breast Cancer Cells. *Oncology* **75**:159–168.
45. Reiter R, Oh AS, Wellstein A, Riegel AT. (2004) Impact of the nuclear receptor coactivator AIB1 isoform AIB1-Δ3 on estrogenic ligands with different intrinsic activity. *Oncogene* **23**:403–409.
46. Osborne CK *et al.* (2003) Role of the estrogen receptor coactivator AIB1 (SRC-3) and HER-2/neu in tamoxifen resistance in breast cancer. *J Natl Cancer Inst* **95**:353–361.
47. Kirkegaard T *et al.* (2007) Amplified in breast cancer 1 in human epidermal growth factor receptor–positive tumors of tamoxifen-treated breast cancer patients. *Clin Cancer Res* **13**:1405–1411.
48. Shou J *et al.* (2004) Mechanisms of tamoxifen resistance: increased estrogen receptor–HER2/neu cross-talk in ER/HER2-positive breast cancer. *J Natl Cancer Inst* **96**:926–935.
49. Silva CM, Shupnik MA. (2007) Integration of steroid and growth factor pathways in breast cancer: focus on signal transducers and activators of transcription and their potential role in resistance. *Mol Endocrinol* **21**:1499.
50. Lahusen T, Fereshteh M, Oh A, Wellstein A, Riegel AT. (2007) Epidermal growth factor receptor tyrosine phosphorylation and signaling controlled by a nuclear receptor coactivator, amplified in breast cancer 1. *Cancer Res* **67**:7256–7265.
51. Fereshteh MP *et al.* (2008) The nuclear receptor coactivator amplified in breast cancer-1 is required for Neu (ErbB2/HER2) activation, signaling, and mammary tumorigenesis in mice. *Cancer Res* **68**:3697–3706.
52. Dihge L *et al.* (2007) Epidermal growth factor receptor (EGFR) and the estrogen receptor modulator amplified in breast cancer (AIB1) for predicting clinical outcome after adjuvant tamoxifen in breast cancer. *Breast Cancer Res Treat* **109**:255–262.
53. Massarweh S, Schiff R. (2007) Unraveling the mechanisms of endocrine resistance in breast cancer: new therapeutic opportunities. *Clin Cancer Res* **13**:1950–1954.
54. Wu R-C *et al.* (2004) Selective phosphorylations of the SRC-3/AIB1 coactivator integrate genomic responses to multiple cellular signaling pathways. *Mol Cell* **15**:937–949.
55. Lupien M *et al.* (2010) Growth factor stimulation induces a distinct ER cistrome underlying breast cancer endocrine resistance. *Genes Dev* **24**:2219–2227.

56. Oh AS *et al.* (2008) Tyrosine phosphorylation of the nuclear receptor coactivator AIB1/SRC-3 is enhanced by Abl kinase and is required for its activity in cancer cells. *Mol Cel Biol* **28**:6580–6593.
57. Hurtado A *et al.* (2008) Regulation of ERBB2 by oestrogen receptor-PAX2 determines response to tamoxifen. *Nature* **456**:663–666.
58. Knowlden JM *et al.* (2003) Elevated levels of epidermal growth factor receptor/c-erbB2 heterodimers mediate an autocrine growth regulatory pathway in tamoxifen-resistant MCF-7 cells. *Endocrinology* **144**:1032–1044.
59. Lauritsen KJ, List H-J, Reiter R, Wellstein A, Riegel AT. (2002) A role for TGF-beta in estrogen and retinoid mediated regulation of the nuclear receptor coactivator AIB1 in MCF-7 breast cancer cells. *Oncogene* **21**:7147–7155.
60. Mc Ilroy M, Fleming FJ, Buggy Y, Hill ADK, Young LS. (2006) Tamoxifen-induced ER-alpha–SRC-3 interaction in HER2-positive human breast cancer: a possible mechanism for ER isoform-specific recurrence. *Endocr Relat Cancer* **13**:1135–1145.
61. Haugan Moi LL *et al.* (2010) Effect of low-dose tamoxifen on steroid receptor coactivator 3/amplified in breast cancer 1 in normal and malignant human breast tissue. *Clin Cancer Res* **16**:2176–2186.
62. Belosay A, Brodie AMH, Njar VCO. (2006) Effects of novel retinoic acid metabolism blocking agent (VN/14-1) on letrozole-insensitive breast cancer cells. *Cancer Res* **66**:11485–11493.
63. Shin I, Miller T, Arteaga CL. (2006) ErbB receptor signaling and therapeutic resistance to aromatase inhibitors. *Clin Cancer Res* **12**:1008s–1012s.
64. Flågeng MH *et al.* (2009) Nuclear receptor co-activators and HER-2/neu are upregulated in breast cancer patients during neo-adjuvant treatment with aromatase inhibitors. *Br J Cancer* **101**:1253–1260.
65. Hörlein AJ *et al.* (1995) Ligand-independent repression by the thyroid hormone receptor mediated by a nuclear receptor co-repressor. *Nature* **377**:397–404.
66. Chen JD, Evans RM. (1995) A transcriptional co-repressor that interacts with nuclear hormone receptors. *Nature* **377**:454–457.
67. Perissi V, Jepsen K, Glass CK, Rosenfeld MG. (2010) Deconstructing repression: evolving models of co-repressor action. *Nat Rev Genet* **11**:109–123.
68. Liu XF, Bagchi MK. (2004) Recruitment of distinct chromatin-modifying complexes by tamoxifen-complexed estrogen receptor at natural target gene promoters *in vivo*. *J Biol Chem* **279**:15050–15058.
69. Kurokawa H, Lenferink A, Simpson J. (2000) Inhibition of HER2/neu (erbB-2) and mitogen-activated protein kinases enhances tamoxifen action against HER2-overexpressing, tamoxifen-resistant breast cancer cells. *Cancer Res*.
70. Lavinsky RM *et al.* (1998) Diverse signaling pathways modulate nuclear receptor recruitment of N-CoR and SMRT complexes. *Proc Natl Acad Sci USA* **95**:2920–2925.

71. Girault I *et al.* (2003) Expression analysis of estrogen receptor alpha coregulators in breast carcinoma: evidence that NCOR1 expression is predictive of the response to tamoxifen. *Clin Cancer Res* **9**:1259–1266.
72. Morrison AJ, Herrera RE, Heinsohn EC, Schiff R, Osborne CK. (2003) Dominant negative N-CoR relieves transcriptional inhibition of retinoic acid receptor but does not alter the agonist/antagonist activities of the tamoxifen-bound estrogen receptor. *Mol Endocrinol* **17**:200101441–1554.
73. Keeton EK, Brown M. (2005) Cell cycle progression stimulated by tamoxifen-bound estrogen receptor-α and promoter-specific effects in breast cancer cells deficient in N-CoR and SMRT. *Mol Endocrinol* **19**:1543.
74. Cheng X, Kao HY. (2009) G protein pathway suppressor 2 (GPS2) is a transcriptional corepressor important for estrogen receptor-mediated transcriptional regulation. *J Biol Chem* **284**:36395–36404.
75. Sarvilinna N, Eronen H, Miettinen S, Vienonen A, Ylikomi T. (2005) Steroid hormone receptors and coregulators in endocrine-resistant and estrogen-independent breast cancer cells. *Int J Cancer* **118**:832–840.
76. Peterson TJ, Karmakar S, Pace MC, Gao T, Smith CL. (2007) The silencing mediator of retinoic acid and thyroid hormone receptor (SMRT) corepressor is required for full estrogen receptor transcriptional activity. *Mol Cel Biol* **27**:5933–5948.
77. Ryo A, Wulf G, Lee TH, Lu KP. (2009) Pinning down HER2-ER crosstalk in SMRT regulation. *Trends Biochem Sci* **34**(4):162–165.
78. Zhang J, Guenther MG, Carthew RW, Lazar MA. (1998) Proteasomal regulation of nuclear receptor corepressor-mediated repression. *Genes Dev* **12**:1775–1780.
79. Misiti S, Schomburg L, Yen PM, Chin WW. (1998) Expression and hormonal regulation of coactivator and corepressor genes. *Endocrinology* **139**:2493–2500.
80. Jordan VC. (2007) Chemoprevention of breast cancer with selective oestrogen-receptor modulators. *Nat Rev Cancer* **7**:46–53.
81. Hall JM, McDonnell DP. (2005) Coregulators in nuclear estrogen receptor action: from concept to therapeutic targeting. *Mol Interv* **5**:343–357.

8

ROLE OF ESTROGEN IN FAMILIAL BREAST CANCER

Eliot Rosen and Claudine Isaacs

Abstract

Breast cancers that occur in women who carry mutations of the breast cancer susceptibility gene 1 (*BRCA1*) are mostly estrogen receptor (ER)-negative; while those that occur in *BRCA2* mutation carriers are usually ER-positive, similar to sporadic breast cancers. Despite these phenotypic differences, the evidence suggests that hormonal factors affect the risk of breast cancer in both *BRCA1* and *BRCA2* mutation carriers. In particular, prophylactic oophorectomy reduces breast cancer risk and the antiestrogen tamoxifen reduces the occurrence of contralateral breast cancer in both *BRCA1* and *BRCA2* carriers. Interestingly, while both *BRCA1* and *BRCA2* exert important functions in DNA repair and the maintenance of genomic integrity, cell-biologic and animal studies indicate that *BRCA1* exerts an additional function in the regulation of ER activity that may, in part, explain why *BRCA1* mutation carriers develop hormonally influenced tumor types. Here, *BRCA1* binds directly to the ER and inhibits its transcriptional activity, through a specific posttranslational modification (monoubiquitination). *BRCA1* also binds to the progesterone receptor (PR) and inhibits its activity; and, in mouse models, *BRCA1* loss-of-function mutations confer enhanced proliferative responses to estrogen and progesterone and an increased incidence of ER-driven mammary cancers. Finally, recent studies suggest that *BRCA1* may function in mammary epithelial cell differentiation by promoting the conversion of ER-negative stem cells to ER-positive cells. Together, these experimental findings are beginning to shed light on the puzzling clinical observations that hormonal manipulations can alter breast cancer risk in *BRCA1* mutation carriers even though a large majority of *BRCA1*-related breast cancers are ER-negative.

Keywords: *BRCA1*; *BRCA2*; Hormone Replacement Therapy; Oophorectomy; Tamoxifen; Estrogen Receptor.

INTRODUCTION

The familial component of breast cancer has long been recognized. In 1866, the French surgeon Paul Broca first reported on the characteristics of familial breast cancer when he described the pattern of cancers seen in his wife's family.[1] About a century later, in the early 1970s, Henry Lynch characterized hereditary breast and ovarian cancer syndrome.[2] The identification of *BRCA1* in 1994 and of *BRCA2* in 1995 allowed the introduction of genetic testing for breast cancer susceptibility into clinical practice and for a growing understanding of the biology of these genes.

In this chapter we will review the evidence that hormonal factors contribute to the etiology of *BRCA1*- and *BRCA2*-related breast cancers, studies suggesting that the *BRCA1* gene normally functions, in part, to regulate the response of mammary epithelial cells to estrogen, and the underlying molecular mechanisms.

CURRENT CLINICAL SITUATION

Cancer Risks in *BRCA1* and *BRCA2* Carriers

A large body of literature has reported on the cancer risks, or penetrance, associated with deleterious mutations in *BRCA1* and *BRCA2*. Available data are derived from population-based studies of women with breast or ovarian cancer unselected for family history as well as multigeneration kindreds containing several women with breast and ovarian cancer, particularly of early onset. A meta-analysis[3] of 10 studies found that the mean cumulative breast cancer risk for *BRCA1* carriers to age 70 was 57% [95% confidence interval (CI) = 47%–66%], and 49% (95% CI = 40%–57%) for *BRCA2* carriers. For ovarian cancer, the risk to age 70 was 40% (95% CI = 35%–46%) for *BRCA1* carriers and 18%

(95% CI = 13%–23%) for *BRCA2* carriers. In the U.S. general population, the comparable risk of breast and ovarian cancer is 7% and less than 2%, respectively. *BRCA1* carriers have a higher rate of premenopausal breast cancer, but for *BRCA1* and *BRCA2* carriers the risk of ovarian cancer before age 40 is low, with most women diagnosed after age 50.[3] Additionally, breast cancers that develop in *BRCA1* carriers have specific histopathologic features. These tumors have consistently been noted to be more frequently high-grade[4,5] and have medullary or atypical medullary features.[4,5] Estrogen receptors (ERs) are often negative, with studies demonstrating that approximately 70%–90% of *BRCA1*-related breast cancers, are ER-negative.[6,7] Additionally, amongst *BRCA1* carriers, 57% of breast cancers, were triple-negative [ER-, progesterone receptor (PR), and HER2-negative]. While it had been thought that ductal carcinoma *in situ* (DCIS) occurred infrequently in *BRCA1* carriers, more recent studies indicate that DCIS should be considered part of the tumor spectrum in both *BRCA1* and *BRCA2* carriers. For example, in a study examining the rate of DCIS (with or without associated invasive breast cancer) amongst women presenting for genetic counseling and testing, DCIS was seen in 37% of *BRCA1/2* carriers compared to 34% of noncarriers.[8] On molecular analyses, *BRCA1*-associated breast cancers showed an increased incidence of basal-like phenotype and p53 mutations[7,9] but a decreased incidence of overexpression of HER2.[7,10,11] Studies examining *BRCA2*-associated breast cancers have demonstrated that these appear to be similar to sporadic breast cancer with respect to hormone receptor status.[6,12]

Additionally, mutation carriers face significantly elevated risk of contralateral breast cancer. At 10 years postdiagnosis, one study reported these risks to be approximately 43% in *BRCA1* carriers and 35% in *BRCA2* carriers,[13] compared to a general population risk of contralateral breast cancer of 5%–10%. However, these risks do not account for tamoxifen use or oophorectomy.

Endocrine Factors Influencing Breast Cancer Risk

Cancer risks associated with mutations in *BRCA1* and *BRCA2* are affected by many endogenous and exogenous factors, and several are modifiable. The following discussion briefly summarizes the current literature on risk modifiers in mutation carriers, including: (1) reproductive factors; (2) lifestyle factors; (3) oophorectomy; (4) hormonal drugs for breast cancer treatment or risk reduction (e.g. tamoxifen); and (5) exogenous hormones such as hormone replacement.

Reproductive Risk Factors

Reproductive risk factors for sporadic breast cancer, such as a late age at menarche, early age at first birth, and prolonged breast feeding, are well established. It is hypothesized that these reproductive factors reduce the risk of breast cancer by decreasing the levels of estrogen and progesterone, inducing breast differentiation, and postponing resumption of ovulatory cycles.[14,15] Less definitive information on the impact of these factors is available in mutation carriers.

Generally, in keeping with the findings in sporadic breast cancer, studies suggest that early menarche confers slightly elevated risks for breast cancer among *BRCA1/BRCA2* carriers.[16,17] However, the data, with respect to the impact of breastfeeding and parity on cancer risk in mutation carriers are less consistent.[18] In a population-based study of over 1400 newly diagnosed breast cancer cases matched to controls, *BRCA1/2* testing was performed in cases and identified 94 carriers.[19] In both sporadic and hereditary breast cancer, the number of full-term pregnancies was inversely associated with breast cancer risk, but breastfeeding offered protection only in sporadic breast cancer. A second retrospective cohort study of 1601 mutation carriers from the International *BRCA1/2* Carrier Study Cohort[20] confirmed that increasing numbers of full-term pregnancies were associated with a statistically significant reduction in the risk of breast cancer for both *BRCA1* and *BRCA2* carriers (14% reduction for each birth; 95% CI = 6%–22%) and breastfeeding was not associated with any reduction in breast cancer risk for either *BRCA1* or *BRCA2* carriers. Other studies, however, suggest that increasing parity increased the risk of breast cancer among *BRCA1* carriers[16] and *BRCA2* carriers.[17,21]

Lifestyle Factors Influencing Risk

Studies have examined the effects of exogenous factors on risk in *BRCA1/2* carriers. These exposures may affect estrogen metabolism, thereby influencing breast cancer risk. In the course of genetic counseling, many women inquire about the role of these factors, particularly diet and exercise, as these behaviors have the potential to be modifiable. There is a relative paucity of data on the significance of these factors in mutation carriers.

Diet and exercise. In a study of 1008 women with newly diagnosed breast cancer which included 104 *BRCA1/2* carriers, physical activity as an adolescent and nonobese body weight at menarche and young adulthood were shown to delay the onset of breast cancer in *BRCA1* and *BRCA2* carriers.[22] It is postulated that these factors may impact breast cancer risk by reducing levels of estrogen. There is a scarcity of data on the impact of diet on breast cancer risk in mutation carriers.

Alcohol. Although epidemiologic studies in the general population have shown a positive association between alcohol use and breast cancer risk, a small international multi-institutional case control study involving 323 *BRCA1* and *BRCA2* carriers did not reveal such an association.[23]

Cigarette smoking. Carcinogens associated with cigarette smoke may increase the risk of breast cancer due to their DNA damaging properties; however, antiestrogenic effects of smoking may be associated with a decreased risk of breast cancer. The largest study to date reported on an international multi-institutional matched case control study of 2538 cases of breast cancer among *BRCA1/2* carriers.[24] Although current smoking in carriers was not associated with an increase in breast cancer risk, *BRCA1* carriers with a past history of smoking may have a somewhat increased risk (OR 1.27, 95% CI 1.06–1.50). Noting that more information is needed about dose, length, and age of cigarette exposure, the authors speculated that early estrogen exposure may be concerning, whereas postmenopausal exposure may be associated with a decreased breast cancer risk in *BRCA1* carriers.

Coffee consumption. An international multi-institutional matched case control analysis of 1690 *BRCA1/2* carriers showed that high levels of caffeinated coffee consumption (4–5 cups, but especially 6 or more cups per day) were associated with a reduced risk of breast cancer (OR 0.75, 95% CI 0.47–1.19; OR 0.31, 95% CI 0.13–0.71, respectively).[25] The authors posited that caffeine may upregulate CYP1A2, which in turn affects estrogen metabolism. In addition, coffee is a known phytoestrogen, and as such binds estrogen, leading to lower bioavailable estrogen.

Bilateral Oophorectomy

Risk-reducing bilateral salpingo-oophorectomy (RRSO) has been demonstrated to reduce the risk of both breast and ovarian cancer in mutation carriers. The protective effect on breast cancer risk is felt to be mediated by its hormonal effect in inducing early menopause. A multi-institution prospective study of 597 *BRCA1/2* carriers who had not undergone risk-reducing mastectomy compared the incidence of breast cancer in those undergoing RRSO with the incidence in those choosing surveillance for ovarian cancer.[26] With a median follow-up of three years, RRSO was associated with about a halving in the incidence of breast cancer [hazard ratio (HR) 0.53, 95% CI 0.29–0.96]. The benefit was more pronounced in *BRCA2* (HR 0.23, 95% CI 0.08–0.92) than in *BRCA1* carriers, for whom it did not achieve statistical significance (HR 0.61, 95% CI 0.3–1.22) (Fig. 1). Additionally, in this study, RRSO reduced the risk of ER-positive (HR 0.22, 95% CI 0.05–1.05) but not ER-negative disease (HR 1.1, 95% CI 0.48–2.51). While the numbers in this study are small, and overall limit the conclusions that can be drawn, these findings were postulated to support the notion that hormonal ablation with oophorectomy was not as effective in reducing breast cancer risk in *BRCA1* carriers as in *BRCA2* carriers. Other studies, however, have found similar effects of RRSO in both *BRCA1* and *BRCA2* carriers. In retrospective international case control study of 1439 *BRCA1/2* carriers with breast cancer matched to 1866 carriers without breast cancer, a prior history of oophorectomy was found to be associated with significant and slightly higher reduction in risk of

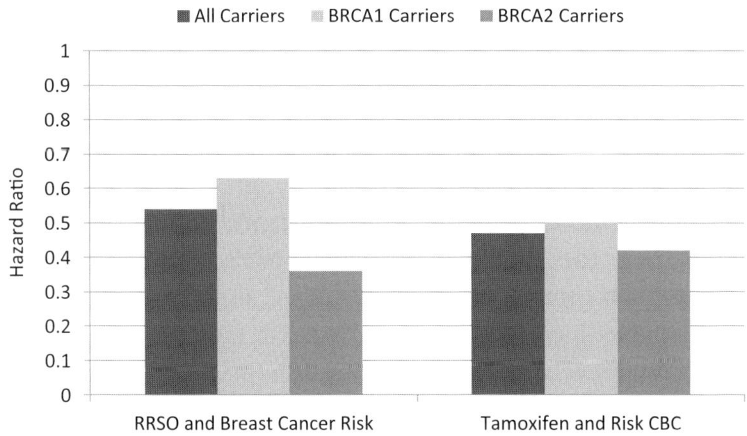

Figure 1. Impact of oophorectomy and tamoxifen on breast cancer risk in *BRCA* carriers.

[1] Domchek *et al* Journal of the Am Med Association 2010;
[2] Gronwald J *et al*, Int J Cancer 118:2281–4, 2006
RRSO: risk reducing salpingo-oophorectomy;
CBC: contralateral breast cancer.

breast cancer in *BRCA1* carriers than in *BRCA2* carriers [56% reduction in *BRCA1* carriers (OR 0.44, 95% CI 0.29–0.66) versus 46% reduction in *BRCA2* carriers (OR 0.57, 95% CI 0.28–1.15)].[27] Additionally, a recent prospective multicenter international study which included 2466 mutation carriers[28] demonstrated that while *BRCA2* carriers experienced a greater reduction in risk (HR 0.36, 95% CI 0.16–0.82), *BRCA1* carriers also derived significant benefit (HR 0.63, 95% CI 0.41–0.96). Thus, the bulk of the data indicate that oophorectomy, which acts through endocrine effects, reduced the risk of breast cancer to a similar degree in *BRCA1* and *BRCA2* carriers.

Additionally, a small pilot study on *BRCA1* carriers recently reported on the impact of hormonal chemoprevention with a gonadotropin-releasing hormone agonist (GnRHA) with low dose add-back estradiol and testosterone. Following 12 months of therapy, mammographic density in mutation carriers was significantly decreased, suggesting that the hypoestrogenic state induced by this regimen could potentially result in decreased breast cancer risk in mutation carriers.[29]

Antiestrogens for Breast Cancer Treatment or Risk Reduction

The role of tamoxifen in mutation carriers has been examined in two settings — the first in reducing the risk of contralateral breast cancer in mutation carriers with breast cancer and the second as a primary chemoprevention agent. There is very little direct data on the latter.

Role in reducing contralateral breast cancer. The use of tamoxifen to reduce the risk of contralateral breast cancer in sporadic patients is well established.[30] In *BRCA1/2* mutation carriers, retrospective studies have demonstrated similar findings (Fig. 1). In the largest study published to date, Gronwald *et al.* studied 1036 *BRCA1* and *BRCA2* carriers: 285 women had bilateral breast cancer and 751 matched controls had unilateral breast cancer.[31] Subjects were ascertained from 49 centers internationally. When prescribed as treatment for an initial breast cancer, tamoxifen was found to reduce the odds of contralateral breast cancer by 50% in *BRCA1* carriers (OR 0.50, 95% CI 0.30–0.85) and 58% for *BRCA2* carriers (OR 0.42, 95% CI 0.17–1.02). Thus, this report suggests that tamoxifen reduces contralateral breast cancer risk in patients with ER-positive disease regardless of *BRCA1/2* status. Additionally, in a prospective study, Metcalfe and colleagues studied 491 *BRCA1/2* carriers with early stage breast cancer.[13] Although they identified a 41% reduction in contralateral breast cancer risk associated with tamoxifen use (HR 0.59, 95% CI 0.35–1.01, $p = 0.05$), this trend was not significant on multivariate analysis.

Role as chemoprevention agent. Data from three large studies — the NSABP P-1 Breast Cancer Prevention Trial (NSABP P-1)[32], the International Breast Cancer Intervention Study (IBIS-I).[33], and the Study of Tamoxifen and Raloxifene (STAR P2)[34] — showed that use of the selective estrogen receptor modulators (SERMs) tamoxifen and raloxifene for five years reduced the risk of ER-positive breast cancer in healthy women by roughly 30%–45%.[35-37] Subjects were at increased risk for breast cancer by virtue of their age, family history, and/or the presence of conditions such as lobular cancer *in situ* (LCIS) or atypical hyperplasia. The majority of women in these families were not *BRCA1/2* mutation carriers, and thus data are limited in this subgroup. For example, in the NSABP P-1 trial, in which genotyping was performed in the 288 women who developed breast cancer after entry into the trial, only 19 *BRCA1/2* carriers were identified.[38] Eight had received tamoxifen and 11 had received placebo. Tamoxifen use was therefore found to be associated with a 62% reduced incidence of breast cancer in *BRCA2* carriers (RR 0.38, 95% CI 0.06–1.56), but no reduction in *BRCA1* carriers (RR 1.67, 95% CI 0.32–10.70). As evidenced by the wide confidence intervals, the study was not adequately powered to address the effects of tamoxifen. However, it has been widely cited, in part because the findings are consistent with the hypothesis that tamoxifen is likely to be more effective in *BRCA2* carriers, who are more likely to develop hormone-receptor-positive disease, whereas *BRCA1* carriers are more likely to develop hormone-receptor-negative breast cancer.[6,12,38]

Data regarding whether the effect of oophorectomy and tamoxifen either in reducing contralateral breast cancer risk is greater than for intervention alone are inconsistent. In the study by Gronwald *et al.* described above,[31] tamoxifen had no protective effect in women who had an oophorectomy (OR 0.83, 95% CI 0.24–2.89), but this subgroup was small (26 cases and 123 controls). In contradistinction, Metcalfe *et al.* found that both tamoxifen and oophorectomy reduced the risk of contralateral breast cancer but the combination of the two was even more effective particularly among young women (HR 0.09; 95% CI 0.01–0.68).[13] In both of these studies, the degree of protection observed was felt to be surprising, given that the majority of the tumors in *BRCA1* carriers would be expected to be ER-negative.

There are no data on the role of raloxifene or aromatase inhibitors in mutation carriers.

Hormone Replacement Therapy

In the general population, findings from the Women's Health Initiative (WHI) trial demonstrated that the use of postmenopausal combined therapy with estrogen and progesterone resulted in an increased breast cancer risk.[39] The use of conjugated equine estrogen alone in postmenopausal women who had undergone a hysterectomy was not associated with an increased incidence of breast cancer.[40] In a

cells.[52,55] The inhibition of estrogen-stimulated cell proliferation by *BRCA1* may be due, in part, to *BRCA1*-mediated inhibition of signaling through ER localized at the cell membrane.[55,56]

Mechanisms Underlying BRCA1 Inhibition of ER Activity

Subsequent studies suggested a molecular basis for this inhibition through a direct physical interaction between the *BRCA1* and ER proteins.[51,57,58] A detailed structural analysis has identified the interacting regions on each protein, including contact points within the amino terminus of *BRCA1* (amino acids 67–134) and within the carboxyl terminal transcriptional activation domain (amino acids 338–379), allowing the creation of a partial three-dimensional model of the *BRCA1*: the ER protein complex.[58] While the binding of *BRCA1* to the ER is necessary in order to inhibit ER activity, it is not sufficient, as evidenced by the finding that certain mutant *BRCA1* proteins bind but do not inhibit the ER.[51,58] This observation was subsequently explained, in part, by the findings that *BRCA1* regulates the posttranslational modification of the ER, including its phosphorylation, acetylation, and ubiquitination.[59-61]

While *BRCA1* does not directly regulate ER phosphorylation, it was found that knockdown of *BRCA1* led to increased phosphorylation of the ER at several sites within its amino terminal activation domain (serine-118 and serine-167), through a mechanism that involves inhibition of protein phosphatase 2A activity and stimulation of the activity of Akt, a serine–threonine kinase that is known to phosphorylate the ER at these sites.[59] It was suggested that *BRCA1* regulation of Akt signaling modulates ER activity through serine phosphorylation events within the amino terminal activation domain of the ER.

Acetylation of the ER lysines 302 and 303 is mediated by the coactivator protein p300 and increases its sensitivity to low concentrations of estrogen.[62,63] Interestingly, it has been found that *BRCA1* overexpression inhibits p300 expression and inhibits ER acetylation, while *BRCA1* underexpression causes increased p300 expression and increased ER acetylation[61,64] (Fig. 2). Furthermore, mutations of the ER at several acetylation sites did not affect ER binding to *BRCA1* but did

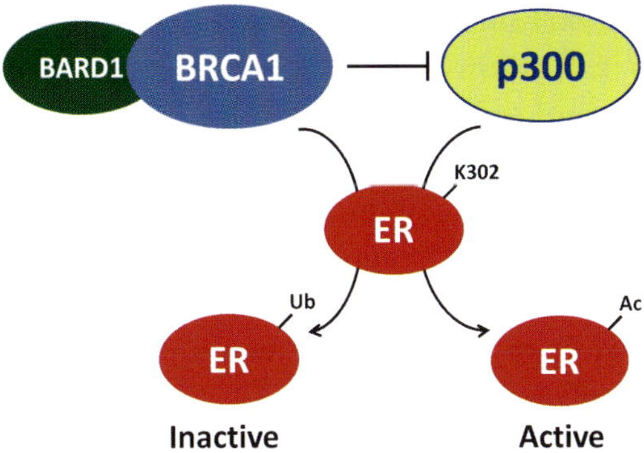

Figure 2. Mechanistic basis of *BRCA1* inhibition of ER activity. *BRCA1* ubiquitinates ER and inhibits the acetylation of ER on lysine-302, thus promoting an inactive configuration of ER. Abbreviations: ER, estrogen receptor; K302, lysine-302; Ub, ubiquitin; Ac, acetyl.

render the ER protein resistant to repression by *BRCA1*, suggesting that *BRCA1* may indirectly regulate ER activity by regulating the levels of the coactivator p300.[64]

With regard to ubiquitination, a complex of *BRCA1* and *BARD1* (a *BRCA1*-associated RING domain-containing protein) directly monoubiquitinates the ER at the lysine 302,[60] a site that is also a target for acetylation. Monoubiquitination of the ER does not cause its degradation but does appear to be required for *BRCA1* inhibition of ER activity, since mutations of *BRCA1* that selectively inactivate its ubiquitin ligase activity did not alter *BRCA1* binding to the ER but did abrogate its ability to inhibit ER activity. Conversely, mutations of the ER at several ubiquitination sites rendered it resistant to inhibition by *BRCA1*. These findings suggest that *BRCA1* regulates ER activity, in part, by regulating the relative degree of its acetylation versus ubiquitination.

BRCA1 Inhibits Ligand-Independent Activation of ER

As we noted earlier, the absence of *BRCA1* or knockdown of *BRCA1* causes activation of the ER in the absence of its ligand, estradiol.[49,50,59] These findings suggest that one function of *BRCA1* may be to prevent its activation in the absence of estrogen. In this regard, our findings suggest that the so-called "ligand-independent" activation of the ER due to *BRCA1* knockdown could be attributed, at least in part, to a phosphatidylinositol-3 kinase (PI3K)/Akt signaling pathway that results in phosphorylation of several key serine residues (serine 118 and serine 167) within the amino terminal transcriptional activation domain (AF-1) of the ER.[59] In agreement with this hypothesis, several PI3K inhibitors blocked ER activation due to *BRCA1* knockdown, and ectopic expression of Akt by itself was sufficient to cause ligand-independent activation of the ER.[59]

BRCA1 Regulation of Estrogen Action in the Mouse Model

A commonly studied mouse model of *BRCA1*-deficient breast cancer features a conditional mammary epithelial cell–targeted homozygous deletion of *BRCA1* exon 11 (which codes for more than 60% of the *BRCA1* protein) coupled to a heterozygous *p53* mutation.[65] About 35%–50% of these mice develop mammary cancer by the age of 12 months. In this model, the mammary tissue showed evidence for increased endogenous estrogen effect.[66] In addition, exogenous estrogen (17β-estradiol) caused abnormally sustained mammary epithelial cell proliferation, mammary hyperplasia, and the appearance of mammary preneoplasia. When the mammary epithelial cell–targeted deletion of *BRCA1* exon 11 was combined with both *p53* heterozygosity and mammary-targeted conditional ER overexpression, the result was an increased incidence of mammary preneoplasia as well as invasive mammary adenocarcinomas.[66]

In agreement with the finding that bilateral oophorectomy reduces breast cancer risk in women who carry germ-line *BRCA1* mutations, bilateral oophorectomy conferred a reduced incidence of mammary cancers in mice with mammary epithelial cell–targeted *BRCA1* deficiency coupled to a heterozygous *p53* mutation.[67] The reduction in the mammary cancer incidence was observed only at times later than 135 days postoophorectomy. These results suggest that in mice, as in humans, ovarian steroids contribute to the early stages of *BRCA1*-related mammary carcinogenesis and removal of the ovaries can reduce the risk of cancer development if performed early enough.

BRCA1 and Tamoxifen

Several studies suggest that *BRCA1* regulates the response of mammary epithelial cells to tamoxifen both in cultured cells and in the mouse. Thus, in ER positive MCF-7 human breast cancer cells, knockdown of *BRCA1* caused a relative increase in the ER agonist activity and reduction of the ER antagonist activity of tamoxifen.[50] At a molecular level, it appears that *BRCA1* knockdown causes the recruitment of coactivators rather than corepressors to the tamoxifen-bound ER complex at ER-regulated gene promoters.[68] Interestingly, in the above-described mouse model of *BRCA1/p53*-deficient mammary tumorigenesis, exposure to tamoxifen not only failed to inhibit tumorigenesis but also increased the incidence of mammary carcinomas by the age of 12 months.[50] Tamoxifen also caused an increase in the rate of mammary epithelial cell proliferation and the incidence of mammary hyperplasia at the age of 6 months, in agreement with the idea that *BRCA1* can influence the response of the ER to tamoxifen *in vivo*. It should be noted that in the mouse model tested both copies of *BRCA1* were disrupted along with one copy of p53. Pending studies (Furth, personal communication) are addressing the impact of reducing rather than ablating *BRCA1* function in the presence of intact p53 the genetic situation women carrying *BRCA1* mutation are in.

BRCA1 and Aromatase

As inherited mutations of *BRCA1* confer an increased risk for both premenopausal and postmenopausal breast cancer, it is relevant to consider additional mechanisms through which *BRCA1* may regulate estrogen action in the absence of ovarian estrogen production. Estrogens derived from adrenal androgens through the action of the enzyme aromatase (CYP19A1) are thought to contribute significantly to the development of breast cancer in postmenopausal women. In this regard, several studies suggest that *BRCA1* acts to inhibit aromatase expression in ovarian granulosa cells, adipocytes, mammary fibroblasts, and breast cancer cells.[69-71] In addition, a study of breast and ovarian tissues from *BRCA1* mutation carriers showed increased aromatase levels.[72] These findings suggest that in addition to increased ER activity, inactivation of *BRCA1* might also cause increased estrogen synthesis via aromatization.

BRCA1 and Progesterone Action

Since the progesterone receptor (PR) is an important transcriptional target of the ER and since the PR may also play a role in breast cancer development, we will briefly consider the functional interaction of *BRCA1* and the PR. It was recently reported that *BRCA1* binds to the PR and inhibits its activity.[73] Conversely, knockdown of *BRCA1* caused an increase in progesterone-stimulated PR activity. In agreement with these findings, administration of progesterone caused an exaggerated proliferative response in the mammary glands of *BRCA1*-deficient mice, which was characterized by an increase in mammary epithelial cell density and an increase in tertiary side branching similar to that observed during pregnancy.[73] In detailed mechanistic studies, it was found that *BRCA1* inhibits progestin-stimulated PR activity by preventing the PR from binding to the progesterone response element in target genes and by promoting the formation of a corepressor rather than a coactivator complex.[74]

In a mouse model for *BRCA1*-related mammary tumorigenesis characterized by a mammary epithelial cell–targeted deletion of *BRCA1* exon 11 coupled to a homozygous deletion of *p53*, treatment with a progesterone antagonist, mifepristone (RU-486, the "morning after" pill), completely prevented mammary tumorigenesis.[75] In this model, the mammary epithelial cells exhibited increased PR expression due to a defect in proteasomal degradation of the PR. However, we were unable to adduce any evidence that *BRCA1* regulates the degradation of the PR in human mammary epithelial cells (unpublished data). These studies suggest that antiprogestin treatment might be useful in preventing the development of *BRCA1*-mutant breast cancer, although this remains to be proven.

Finally, it has also been documented that *BRCA1* can associate with the androgen receptor (AR) and stimulate its transcriptional activity.[76,77] The significance of this observation for breast cancer development is uncertain, although it is known that androgens can block the stimulation of mammary epithelial cell proliferation by estrogen and progesterone.[78]

Loss of the ER in BRCA1-Related Breast Cancer

Most human breast cancers (about 70%) are ER-positive at diagnosis. However, as noted above,[6,7] most *BRCA1*-mutant breast cancers (70%–90%) are ER-negative at diagnosis. Thus, any model for the molecular pathogenesis of breast cancers that occur in *BRCA1* mutation carriers must account for the observations that hormonal factors can influence the risk of *BRCA1*-related breast cancers in humans and mice, but the cancers that develop are mostly ER-negative. In this regard, most *BRCA1*-mutant breast cancers are also PR-negative and negative for *HER/Neu* gene amplification, and thus fall into the "triple-negative" phenotype, which portends a biologically aggressive cancer. These observations are consistent with two models: one in which the tumor cells start out as ER-positive and become ER-negative later during the multistep process of carcinogenesis, and the other in which hormonal factors stimulate the growth of ER-negative tumor cells via a paracrine mechanism.

In the above-described mouse model of *BRCA1*-related tumorigenesis, both tamoxifen and estradiol induced the formation of mammary hyperplasias in mice with a mammary-targeted homozygous deletion of the full-length *BRCA1* isoform.[50,66] In these mice, the preneoplasias and cancers that developed were consistently ER-negative. These findings suggest that during the development of these cancers, the "loss" of ER expression is an early event that is integral to their pathogenesis and not a late or random event. While the mechanism(s) by which ER expression is lost in *BRCA1*-mutant breast cancers is unknown, it has been reported that ER-negative human *BRCA1*-mutant breast cancers often exhibit increased CpG methylation of the ER gene.[79] In another study, a mechanism was described whereby *BRCA1* is recruited to the ER gene promoter by the octamer-binding transcription factor 1 (OCT1); and both *BRCA1* and OCT1 are required for ER expression.[80] In this model, the absence of functional *BRCA1* causes loss of ER expression at the transcriptional level.

BRCA1 Regulation of ER Activity: Role in Tumor Suppression

A major unresolved issue regarding *BRCA1*'s role as a tumor suppressor is the question of which function(s) are required to block breast cancer development: the DNA repair functions, the endocrine regulatory functions, and/or other functions. This is a complex issue, in part because *BRCA1* exerts multiple functional activities within each category. The "other functions" include the regulation of a variety of transcriptional pathways as well as a function for *BRCA1* in promoting mammary epithelial cell differentiation that is not well understood at a molecular level;[81-83] but a recent report suggests that *BRCA1* may play a role in the determination of mammary stem cell fate, in particular the conversion of ER-negative stem/progenitor cells to ER-positive luminal cells.[84] To confound matters, another recent report suggests that the target cell population for *BRCA1*-mutant mammary carcinogenesis is abnormal luminal progenitor cells from tumors with a basal-like triple-negative phenotype develop.[85] Fig. 3 shows a summary of *BRCA1* molecular functions that may contribute to breast cancer tumor suppression.

In regard to *BRCA1* tumor suppressor function, it is interesting that a recent study suggests that the ubiquitin ligase function of *BRCA1* is not required for certain key functions of *BRCA1* related to

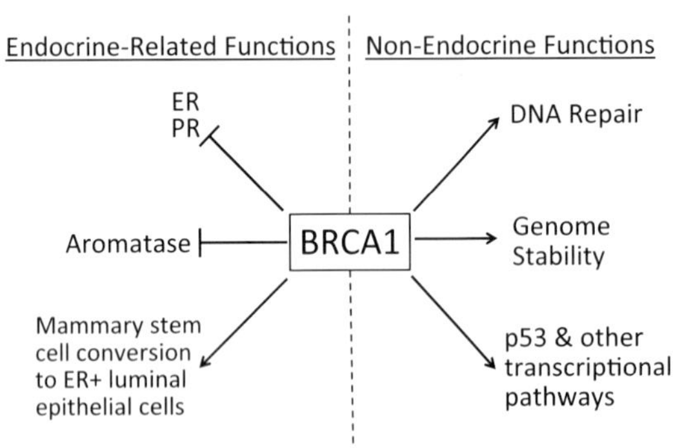

Figure 3. Potential breast cancer-related tumor suppressor functions of *BRCA1*. The loss of these functions due to inherited mutations of *BRCA1* may explain, in part, the contribution of endocrine factors to *BRCA1*-mutant mammary carcinogenesis and why most *BRCA1*-mutant breast cancers are hormone receptor negative.

the maintenance of genomic stability, including the repair of double-strand DNA breaks by homology-directed DNA repair (HDR) and the formation of Rad51-containing nuclear foci following exposure to ionizing radiation.[86] The same *BRCA1* mutation that selectively ablated its ubiquitin ligase activity but allowed retention of several, but not all, DNA-repair-related functions (^{26}isoleucine → alanine) also abolished the ability of *BRCA1* to both monoubiquitinate the ER and inhibit ER activity.[61] However, the effect of this mutation of *BRCA1*-related mammary tumorigenesis has not been described. Furthermore, a breast-cancer-associated point mutant *BRCA1* protein deficient for ubiquitin ligase activity (^{61}cysteine → glycine) also failed to inhibit ER activity and to rescue the hypersensitivity of *BRCA1*-deficient cells to ionizing radiation, which kills cells primarily by causing double-strand breaks of the sugar-phosphate backbone of DNA.[51,87] Finally, some studies suggest that the *BRCA1/BARD1* heterodimer, which is required for the *BRCA1* ubiqituitin ligase function, is also required for tumor suppression.[88]

If the ubiquitin ligase function of *BRCA1* were required for tumor suppression, it would suggest that the ability of *BRCA1* to mediate HDR of double-strand DNA breaks is not sufficient for tumor suppression. It would further suggest that ubiquitination of the ER and/or another substrate of the *BRCA1/BARD1* ubiquitin ligase contributes to breast cancer suppression. These issues still remain to be sorted out.

SUMMARY

Various clinical/epidemiological studies suggest that hormonal factors alter the risk of breast cancer development in *BRCA1* mutation carriers and, therefore, contribute to the etiology of *BRCA1*-related breast cancers. In agreement with these observations, both *in vitro* and *in vivo* experimental studies indicate that *BRCA1* regulates the response of mammary epithelial cells to estrogen and progesterone, in addition to its well-established roles in DNA repair. The remaining clinical questions are whether antiestrogens (tamoxifen or raloxifene) can prevent breast cancer development in *BRCA1* mutation carriers as they do in other high risk populations of women, and whether there is a role for antiprogestins in the prevention of *BRCA1* mutation-related breast cancers. The research issues include the mechanism(s) by which most *BRCA1*-related cancers evolve to a hormone-receptor-negative status and the role of different *BRCA1* functions (regulation of hormone action, mammary differentiation, and DNA repair pathways) in suppressing breast cancer formation.

ACKNOWLEDGMENT

Dr. Isaacs received support from the Familial Cancer registry of Georgetown University (NIH/NCI grant p30 CA 051-00811, the Fisher Center for Familiar Research and Survivors for the Cure).

Dr. Rosen has been supported, in part, by research grants from the USPHS (R01-CA150646), Susan G. Komen for the Cure (KG110580), and Living in Pink. Dr. Rosen has also received support from the World Class University program funded by the Ministry of Education, Science and Technology through the National Research Foundation of Korea (R31-10069).

REFERENCES

1. Broca P. (1866) Traite des tumeurs. *Paris: Asselin* **1**(2).
2. Lynch HT, Krush AJ. (1971) Carcinoma of the breast and ovary in three families. *Surg Gynecol Obstet* **133**(4):644–648.
3. Chen S, Parmigiani G. (2007) Meta-analysis of BRCA1 and BRCA2 penetrance. *J Clin Oncol* **25**(11):1329–1333.
4. Anonymous. (1997) Pathology of familial breast cancer: differences between breast cancers in carriers of BRCA1 or BRCA2 mutations and sporadic cases. Breast Cancer Linkage Consortium. *Lancet* **349**(9064):1505–1510.
5. Brekelmans CT *et al.* (2006) Survival and prognostic factors in BRCA1-associated breast cancer. *Ann Oncol* **17**(3):391–400.
6. Atchley DP *et al.* (2008) Clinical and pathologic characteristics of patients with BRCA-positive and BRCA-negative breast cancer. *J Clin Oncol* **26**(26): 4282–4288.
7. Lakhani SR *et al.* (2002) The pathology of familial breast cancer: predictive value of immunohistochemical markers estrogen receptor, progesterone receptor, HER-2, and p53 in patients with mutations in BRCA1 and BRCA2. *J Clin Oncol* **20**(9):2310–2318.

8. Hwang ES *et al.* (2007) Ductal carcinoma *in situ* in BRCA mutation carriers. *J Clin Oncol* **25**(6):642–647.
9. Phillips KA *et al.* (1999) Frequency of p53 mutations in breast carcinomas from Ashkenazi Jewish carriers of BRCA1 mutations. *J Natl Cancer Inst* **91**(5): 469–473.
10. Phillips KA, Andrulis IL, Goodwin PJ. (1999) Breast carcinomas arising in carriers of mutations in BRCA1 or BRCA2: are they prognostically different? *J Clin Oncol* **17**(11):3653–3663.
11. Quenneville LA *et al.* (2002) HER-2/neu status and tumor morphology of invasive breast carcinomas in Ashkenazi women with known BRCA1 mutation status in the Ontario Familial Breast Cancer Registry. *Cancer* **95**(10):2068–2075.
12. Rennert G *et al.* (2007) Clinical outcomes of breast cancer in carriers of BRCA1 and BRCA2 mutations. *N Engl J Med* **357**(2):115–123.
13. Metcalfe K *et al.* (2004) Contralateral breast cancer in BRCA1 and BRCA2 mutation carriers. *J Clin Oncol* **22**(12):2328–2335.
14. Jordan VC. (2009) A century of deciphering the control mechanisms of sex steroid action in breast and prostate cancer: the origins of targeted therapy and chemoprevention. *Cancer Res* **69**(4):1243–1254 (in Eng.).
15. Russo J *et al.* (2005) Breast differentiation and its implication in cancer prevention. *Clin Cancer Res* **11**(2 Pt 2):931s–936s.
16. Gronwald J *et al.* (2006) Influence of selected lifestyle factors on breast and ovarian cancer risk in BRCA1 mutation carriers from Poland. *Breast Cancer Res Treat* **95**(2):105–109.
17. Kotsopoulos J *et al.* (2007) Age at first birth and the risk of breast cancer in BRCA1 and BRCA2 mutation carriers. *Breast Cancer Res Treat* **105**(2):221–228.
18. Jernstrom H *et al.* (2004) Breast-feeding and the risk of breast cancer in BRCA1 and BRCA2 mutation carriers. *J Natl Cancer Inst* **96**(14):1094–1098.
19. Lee E *et al.* (2008) Effect of reproductive factors and oral contraceptives on breast cancer risk in BRCA1/2 mutation carriers and noncarriers: results from a population-based study. *Cancer Epidemiol Biomarkers Prev* **17**(11): 3170–3178.
20. Andrieu N *et al.* (2006) Pregnancies, breast-feeding, and breast cancer risk in the International BRCA1/2 Carrier Cohort Study (IBCCS). *J Natl Cancer Inst* **98**(8): 535–544.
21. Cullinane CA *et al.* (2005) Effect of pregnancy as a risk factor for breast cancer in BRCA1/BRCA2 mutation carriers. *Int J Cancer* **117**(6):988–991.
22. King MC, Marks JH, Mandell JB. (2003) Breast and ovarian cancer risks due to inherited mutations in BRCA1 and BRCA2. *Science* **302**(5645):643–646.
23. McGuire V *et al.* (2006) No increased risk of breast cancer associated with alcohol consumption among carriers of BRCA1 and BRCA2 mutations ages <50 years. *Cancer Epidemiol Biomarkers Prev* **15**(8):1565–1567.
24. Ginsburg O *et al.* (2009) Smoking and the risk of breast cancer in BRCA1 and BRCA2 carriers: an update. *Breast Cancer Res Treat* **114**(1):127–135.

25. Nkondjock A *et al.* (2006) Coffee consumption and breast cancer risk among BRCA1 and BRCA2 mutation carriers. *Int J Cancer* **118**(1):103–107.
26. Kauff ND *et al.* (2008) Risk-reducing salpingo-oophorectomy for the prevention of BRCA1- and BRCA2-associated breast and gynecologic cancer: a multicenter, prospective study. *J Clin Oncol* **26**(8):1331–1337.
27. Eisen A *et al.* (2005) Breast cancer risk following bilateral oophorectomy in BRCA1 and BRCA2 mutation carriers: an international case-control study. *J Clin Oncol* **23**(30):7491–7496.
28. Rebbeck TR, Kauff ND, Domchek SM. (2009) Meta-analysis of risk reduction estimates associated with risk-reducing salpingo-oophorectomy in BRCA1 or BRCA2 mutation carriers. *J Natl Cancer Inst* **101**(2):80–87.
29. Weitzel JN *et al.* (2007) Reduced mammographic density with use of a gonadotropin-releasing hormone agonist-based chemoprevention regimen in BRCA1 carriers. *Clin Cancer Res* **13**(2):654–658.
30. Anonymous. (2005) Effects of chemotherapy and hormonal therapy for early breast cancer on recurrence and 15-year survival: an overview of the randomised trials. *Lancet* **365**(9472):1687–1717.
31. Gronwald J *et al.* (2006) Tamoxifen and contralateral breast cancer in BRCA1 and BRCA2 carriers: an update. *Int J Cancer* **118**(9):2281–2284.
32. Fisher B *et al.* (1998) Tamoxifen for prevention of breast cancer: report of the National Surgical Adjuvant Breast and Bowel Project P-1 Study. *J Natl Cancer Inst* **90**(18):1371–1388.
33. Cuzick J *et al.* (2002) First results from the International Breast Cancer Intervention Study (IBIS-I): a randomised prevention trial. *Lancet* **360**(9336): 817–824.
34. Vogel VG *et al.* (2006) Effects of tamoxifen vs. raloxifene on the risk of developing invasive breast cancer and other disease outcomes: the NSABP Study of Tamoxifen and Raloxifene (STAR) P-2 trial. *JAMA* **295**(23):2727–2741.
35. Cuzick J *et al.* (2007) Long-term results of tamoxifen prophylaxis for breast cancer—96-month follow-up of the randomized IBIS-I trial. *J Natl Cancer Inst* **99**(4):272–282.
36. Fisher B *et al.* (2005) Tamoxifen for the prevention of breast cancer: current status of the National Surgical Adjuvant Breast and Bowel Project P-1 Study. *J Natl Cancer Inst* **97**(22):1652–1662.
37. Vogel VG *et al.* (2006) Effects of tamoxifen vs. raloxifene on the risk of developing invasive breast cancer and other disease outcomes: the NSABP Study of Tamoxifen and Raloxifene (STAR) P-2 Trial. *JAMA* **295**(23):2727–2741.
38. King MC *et al.* (2001) Tamoxifen and breast cancer incidence among women with inherited mutations in BRCA1 and BRCA2: National Surgical Adjuvant Breast and Bowel Project (NSABP-P1) Breast Cancer Prevention Trial. *JAMA* **286**(18): 2251–2256.

39. Chlebowski RT *et al.* (2003) Influence of estrogen plus progestin on breast cancer and mammography in healthy postmenopausal women: the Women's Health Initiative Randomized Trial. *JAMA* **289**(24):3243-3253.
40. Stefanick ML *et al.* (2006) Effects of conjugated equine estrogens on breast cancer and mammography screening in postmenopausal women with hysterectomy. *JAMA* **295**(14):1647-1657.
41. Eisen A *et al.* (2008) Hormone therapy and the risk of breast cancer in BRCA1 mutation carriers. *J Natl Cancer Inst* **100**(19):1361-1367.
42. Rebbeck TR *et al.* (2005) Effect of short-term hormone replacement therapy on breast cancer risk reduction after bilateral prophylactic oophorectomy in BRCA1 and BRCA2 mutation carriers: the PROSE Study Group. *J Clin Oncol* **23**(31):7804-7810.
40. Miki Y, Swensen J, Shattuck-Eidens D, *et al.* (1994) A strong candidate for the breast and ovarian cancer susceptibility gene. *Science* **266**:66-71.
41. Merajver SD, Frank TS, Xu J, *et al.* (1995) Germline BRCA1 mutations and loss of the wild-type allele in tumors from families with early onset breast and ovarian cancer. *Clin Cancer Res* **1**:539-544.
42. Wilson CA, Ramos L, Villasenor MR, *et al.* (1999) Localization of human BRCA1 and its loss in high-grade, non-inherited breast carcinomas. *Nat Genet* **21**:236-240.
43. Esteller M, Silva JM, Dominguez G, *et al.*(2000) Promoter hypermethylation and BRCA1 inactivation in sporadic breast and ovarian tumors. *J Natl Cancer Inst* **92**:564-569.
44. Staff S, Isola J, Tanner M. (2003) Haplo-insufficiency of BRCA1 in sporadic breast cancer. *Cancer Res* **63**:4978-4983.
45. Rosen EM, Fan S, Pestell RG, *et al.* (2003) BRCA1 gene in breast cancer. *J Cell Physiol* **196**:19-41.
46. Gudmundsdottir K, Ashworth A. (2006) The roles of BRCA1 and BRCA2 and associated proteins in the maintenance of genomic stability. *Oncogene* **25**: 5864-5874, Review.
47. Thompson D, Easton DF. (2002) Breast Cancer Linkage Consortium: cancer incidence in BRCA1 mutation carriers. *J Natl Cancer Inst* **94**:1358-1365.
48. Fan S, Wang J, Yuan R, *et al.* (1999) BRCA1 inhibition of estrogen receptor signaling in transfected cells. *Science* **284**:1354-1356.
49. Zheng L, Annab LA, Afshari CA, *et al.* (2001) BRCA1 mediates ligand-independent transcriptional repression of the estrogen receptor. *Proc Natl Acad Sci USA* **98**:9587-9592.
50. Jones LP, Li M, Halama ED, *et al.* (2005) Promotion of mammary cancer development by tamoxifen in a mouse model of Brca1-mutation related breast cancer. *Oncogene* **24**:3554-3562.
51. Fan S, Ma YX, Wang C, *et al.* (2001) Role of direct interaction in BRCA1 inhibition of estrogen receptor activity. *Oncogene* **20**:77-87.

52. Xu J, Fan S, Rosen EM. (2005) Regulation of the estrogen-inducible gene expression profile by the breast cancer susceptibility gene BRCA1. *Endocrinology* **146**:2031-2047.
53. Zheng L, Annab LA, Afshari CA, *et al.* (2001) BRCA1 mediates ligand-independent transcriptional repression of the estrogen receptor. *Proc Natl Acad Sci USA* **98**:9587-9592.
54. Wang C, Fan S, Li Z, *et al.* (2005) Cyclin D1 antagonizes BRCA1 repression of estrogen receptor alpha activity. *Cancer Res* **65**:6557-6567.
55. Razandi M, Pedram A, Rosen EM, *et al.* (2004) BRCA1 inhibits membrane estrogen and growth factor receptor signaling to cell proliferation in breast cancer. *Mol Cell Biol* **24**:5900-5913.
56. Pedram A, Razandi M, Levin ER. (2006) Nature of functional estrogen receptors at the plasma membrane. *Mol Endocrinol* **20**:1996-2009.
57. Kawai H, Li H, Chun P, *et al.* (2002) Direct interaction between BRCA1 and the estrogen receptor regulates vascular endothelial growth factor (VEGF) transcription and secretion in breast cancer cells. *Oncogene* **21**:7730-7739.
58. Ma YX, Tomita Y, Fan S, *et al.* (2005) Structural determinants of the BRCA1: estrogen receptor interaction. *Oncogene* **24**:1831-1846.
59. Ma Y, Hu C, Riegel AT, *et al.* (2003) Growth factor signaling pathways modulate BRCA1 repression of estrogen receptor-alpha activity. *Mol Endocrinol* **21**: 1905-1923.
60. Eakin CM, Maccoss MJ, Finney GL, *et al.* (2007) Estrogen receptor alpha is a putative substrate for the BRCA1 ubiquitin ligase. *Proc Natl Acad Sci USA* **104**:5794-5799.
61. Ma Y, Fan S, Hu C, *et al.* (2010) BRCA1 regulates acetylation and ubiquitination of estrogen receptor-alpha. *Mol Endocrinol* **24**:76-90.
62. Cui Y, Zhang M, Pestell R, *et al.* (2004) Phosphorylation of estrogen receptor alpha blocks its acetylation and regulates estrogen sensitivity. *Cancer Res* **64**:9199-9208.
63. Wang C, Fu M, Angeletti RH, *et al.* (2001) Direct acetylation of the estrogen receptor alpha hinge region by p300 regulates transactivation and hormone sensitivity. *J Biol Chem* **276**:18375-18383.
64. Fan S, Ma YX, Wang C, *et al.* (2002) p300 modulates the BRCA1 inhibition of estrogen receptor activity. *Cancer Res* **62**:141-151.
65. Xu X, Wagner K-U, Larson D, *et al.* (1999) Conditional mutation of Brca1 in mammary epithelial cells results in blunted ductal morphogenesis and tumour formation. *Nat Genet* **22**:37-43.
66. Jones LP, Tilli MT, Assefnia S, *et al.* (2008) Activation of estrogen signaling pathways collaborates with loss of Brca1 to promote development of ERalpha-negative and ERalpha-positive mammary preneoplasia and cancer. *Oncogene* **27**:794-802.

67. Bachelier R, Xu X, Li C, *et al.* (2005) Effect of bilateral oophorectomy on mammary tumor formation in BRCA1 mutant mice. *Oncol Rep* **14**:1117–1120.
68. Wen J, Li R, Lu Y, *et al.* (2009) Decreased BRCA1 confers tamoxifen resistance in breast cancer cells by altering estrogen receptor–coregulator interactions. *Oncogene* **28**:575–586.
69. Hu Y, Ghosh S, Amleh A, *et al.* (2005) Modulation of aromatase expression by BRCA1: a possible link to tissue-specific tumor suppression. *Oncogene* **24**:8343–8348.
70. Lu M, Chen D, Lin Z, *et al.* (2006) BRCA1 negatively regulates the cancer-associated aromatase promoters I.3 and II in breast adipose fibroblasts and malignant epithelial cells. *J Clin Endocrinol Metab* **91**:4514–4519.
71. Subbaramaiah K, Hudis C, Chang SH, *et al.* (2008) EP2 and EP4 receptors regulate aromatase expression in human adipocytes and breast cancer cells. Evidence of a BRCA1 and p300 exchange. *J Biol Chem* **283**:3433–3444.
72. Chand AL; kConFab, Simpson ER, *et al.* (2009) Aromatase expression is increased in BRCA1 mutation carriers. *BMC Cancer* **9**:148.
73. Ma Y, Katiyar P, Jones LP, *et al.* (2006) The breast cancer susceptibility gene BRCA1 regulates progesterone receptor signaling in mammary epithelial cells. *Mol Endocrinol* **20**:14–34.
74. Katiyar P, Ma Y, Riegel A, *et al.* (2009) Mechanism of BRCA1-mediated inhibition of progesterone receptor transcriptional activity. *Mol Endocrinol* **23**:1135–1146.
75. Poole AJ, Li Y, Kim Y, *et al.* (2006) Prevention of Brca1-mediated mammary tumorigenesis in mice by a progesterone antagonist. *Science* **314**:1467–1470.
76. Park JJ, Irvine RA, Buchanan G, *et al.* (2000) Breast cancer susceptibility gene 1 (BRCA1) is a coactivator of the androgen receptor. *Cancer Res* **60**:5946–5949.
77. Yeh S, Hu YC, Rahman M, *et al.* (2000) Increase of androgen-induced cell death and androgen receptor transactivation by BRCA1 in prostate cancer cells. *PNAS USA* **97**:11256–11261.
78. von Schoultz B. (2007) Androgens and the breast. *Maturitas* **57**:47–49, Review.
79. Archey WB, McEachern KA, Robson M, *et al.* (2002) Oncogene **21**:7034–7041.
80. Hosey AM, Gorski JJ, Murray MM, *et al.* (2007) Molecular basis for estrogen receptor alpha deficiency in BRCA1-linked breast cancer. *J Natl Cancer Inst* **99**:1683–1694.
81. Rajan JV, Wang M, Marquis ST, *et al.* (1996) Brca2 is coordinately regulated with Brca1 during proliferation and differentiation in mammalian epithelial cells. *PNAS USA* **93**:3078–3083.
82. Kubista M, Rosner M, Kubista E, *et al.* (2002) Brca1 regulates in vitro differentiation of mammary epithelial cells. *Oncogene* **21**:4747–4756.
83. Furuta S, Jiang X, Gu B, *et al.* (2005) Depletion of BRCA1 impairs differentiation but enhances proliferation of mammary epithelial cells. *PNAS USA* **102**:9176–9181.
84. Liu S, Ginestier C, Charafe-Jauffret E, *et al.* (2008) BRCA1 regulates human mammary stem/progenitor cell fate. *Proc Natl Acad Sci USA* **105**:1680–1685.

85. Lim E, Vaillant F, Wu D, *et al.* (2009) Aberrant luminal progenitors as the candidate target population for basal tumor development in BRCA1 mutation carriers. *Nat Med* **15**:907–913.
86. Reid LJ, Shakya R, Modi AP, *et al.* (2008) E3 ligase activity of BRCA1 is not essential for mammalian cell viability or homology-directed repair of double-strand DNA breaks. *Proc Natl Acad Sci USA* **105**:20876–20881.
87. Ruffner H, Joazeiro CA, Hemmati D, *et al.* (2001) Cancer-predisposing mutations within the RING domain of BRCA1: loss of ubiquitin protein ligase activity and protection from radiation hypersensitivity. *Proc Natl Acad Sci USA* **98**:5134–5139.
88. Shakya R, Szabolcs M, McCarthy E, *et al.* (2008) The basal-like mammary carcinomas induced by Brca1 or Bard1 inactivation implicate the BRCA1/BARD1 heterodimer in tumor suppression. *Proc Natl Acad Sci USA* **105**:7040–7045, 2008.

9

ADVANCES WITH AROMATASE INHIBITORS

Saranya Chumsri, Ting Bao, Gauri Sabnis and Angela Brodie

Abstract

Aromatase is the key enzyme in the biosynthesis of estrogen from precursor androgens via the process termed aromatization. Aromatase is highly expressed in the placenta and the granulosa cells of the ovary. Other tissues including adipose tissue, muscle the normal breast and breast cancer also express the enzyme but at much lower levels. In postmenopausal women these tissues serve as the main sources of estrogen production when the ovaries no longer function. Nevertheless the concentration of estrogens in the breast tissue is similar to levels in premenopausal women or as high as 4-6 times the serum concentration and up to 10 times in breast cancer tissue. Aromatase inhibitors reduce estrogen levels in postmenopausal patients by inhibiting aromataization. The clinical experience with the three FDA approved AIs, letrozole, anastrozole and exemestane is reviewed according to disease stage including metastatic disease and early stage breast cancer as well as the role of AIs in chemoprevention. Although AIs have proved to be the most effective treatment for hormone responsive breast cancer, the emergence of resistance to them can be a limitation on their sustained clinical benefit. Cross talk between ER and HER2 or MAPK as well as PI3/K signaling pathways appear to be key mechanisms which cause previously hormone-responsive tumors to become resistance to AIs. Phosphorylation of ER via activation of MAPK or PI3K/Akt pathways can lead to transcription of genes that regulate and promote tumor proliferation despite estrogen deprivation. Cotargeting the ER signaling pathway with AIs and other novel agents to disrupt cross talk appears to be a promising approach to prolong the benefits of AIs. Multiple therapies are currently under clinical development to target resistance and to individualize treatment for breast cancer patients.

Keywords: Breast Cancer; Aromatase Inhibitors.

INTRODUCTION

Aromatase is the key enzyme in the biosynthesis of estrogens from androgens via the process termed "aromatization".[1] This is a three-step reaction in which 1 mol of O_2 and 1 mol of NADPH are required for each step. Aromatase is a member of the cytochrome P450 superfamily and the product of the *CYP19A1* gene on chromosome 15q21.1.[2] While most of these enzymes are not highly specific, aromatase has an androgen-specific cleft and is only involved in the conversion of testosterone to estradiol, androstenedione to estrone, and hydroxytestosterone to estriol. It consists of a heme group and a polypeptide chain of 503 amino-acid residues.[3] Aromatase is highly expressed in the placenta and in the granulose cells of ovarian follicles. Other tissues, including subcutaneous fat, muscle, liver, brain, normal breast, and breast cancer, also express aromatase but at lower levels. In postmenopausal women, these tissues serve as a main source of estrogen production when ovaries no longer function. Peripherally synthesized estrogens exert their activity in an intracrine or paracrine fashion without being released into the circulation. Therefore, the concentration of estradiol in plasma of postmenopausal women is extremely low in the range of 7 pg per milliliter (25 pmol per liter), compared to 110 pg per milliliter (400 pmol per liter) in premenopausal women.[4] Nonetheless, the concentration of estrogens in breast tissue can be similar to levels in premenopausal women or as high as 4–6 times the serum concentration and up to 10 times in breast cancer tissue.[4]

Estrogens are the main driving force in the majority of breast cancers. It has been recognized for more than a century that oopharectomy can result in breast cancer growth reduction. Since the main source of estrogens in postmenopausal women is from the peripheral conversion of androgens via aromatization, inhibition of aromatase enzyme in postmenopausal women will lead to significant further reduction in estrogens. The discovery of aromatase inhibitors (AIs) in the early 1970s has truly revolutionized the treatment for hormone-receptor-positive postmenopausal breast cancer patients. They have created a novel aspect in drug development and enhanced therapeutic possibilities. AIs are now considered to be the standard of care for postmenopausal women with hormone receptor positive breast cancer.[5] Nevertheless, the emergence of resistance to AIs remains the key impediment in the treatment of hormone receptor positive breast cancer. The clinical development and experience, mechanisms of resistance to AIs, and novel approaches to overcome resistance are reviewed in this chapter.

CLINICAL DEVELOPMENT OF AROMATASE INHIBITORS

In 1896, Sir George Beatson was the first to demonstrate that the majority of breast cancers were dependent on ovarian "hormones" and that bilateral oophorectomy could induce breast cancer regression in premenopausal women with breast cancer.[6,7] Later, other surgical modalities aimed at reducing estrogen production, such as adrenalectomy and hypophysectomy, were described as therapeutic options for breast cancer patients.[7] In the early 1970s, the first selective estrogen receptor modulator (SERM), tamoxifen (at that time also known as ICI 46,474) was approved for the treatment of breast cancer in the U.K. (1973) and then in the U.S. (1977).[8] Even though tamoxifen is quite effective in treating hormone receptor positive breast cancer, it has unfavorable side effects due to its partially estrogenic effects in the uterus and vascular system resulting in an increased risk of endometrial cancer,[9] thromboembolism, and stroke.[10] Harry and Angela Brodie hypothesized that an alternative approach to block the production of estrogen without having partial estrogenic activity would be more effective, with a safer side effect profile. With that approach in mind, over 100 steroidal compounds were systematically examined as selective inhibitors of aromatase.[11,12] This led to the identification of 1,4,6-androstatreienedione[11] and 4-hydroxy-androstenedione (4-OH-A).[12,13] The lead compound 4-OH-A was demonstrated to significantly reduce estrogen concentrations in rodents and cause regression of rat mammary tumors more effectively compared to tamoxifen and without estrogenic effect on other tissues, including the uterus.[12] Around the same time, aminoglutethimide, which was initially developed as an antiepileptic drug, was found to inhibit multiple cytochrome P450 enzymes and resulted in adrenal insufficiency. Using aminoglutethimide initially as a medical adrenalectomy, Richard Santen and colleagues demonstrated that it was effective in treating breast cancer,in the late 1970s.[13] Subsequently, Santen *et al.* showed that aminoglutethimide inhibits the aromatase enzyme.[14] Given the fact that it also inhibits CYP11,[15] cortisol replacement was required in conjunction with aminoglutethimide. For this reason and because of other side effects such as drowsiness and rash, the general use of aminoglutethimide did not advance.[13] By this time, the first clinical trial with 4-OH-A as a selective AI for the treatment of breast cancer was launched at the Royal Marsden Hospital in London from the collaboration between Angela Brodie and two oncologists, Charles Coombes and Paul Goss, as well as a clinical chemist, Mitch Dowsett. This and subsequent clinical trials demonstrated significant benefit and responses to 4-OH-A even in patients with tamoxifen-resistant breast cancer.[3,16,17] In the mid-1980s, 4-OH-A was renamed "formestane" and became the first "selective aromatase inhibitor" used clinically for the treatment of breast cancer.[13] Later on, their discovery inspired other investigators to identify a wide variety of AIs. Subsequently, another compound, known as fadrozole, also demonstrated clinical efficacy. Nonetheless, due to the requirement for intramuscular injection of formestane and aldosterone suppression of fadrozole, these two agents were superseded by other inhibitors with better oral bioavailability and less side effects, respectively.[13]

The oral AIs that are widely used in the clinic nowadays can be further classified into two types. Type I or steroidal AIs are referred to as substrate analogs, due to their resemblance to androstenedione.

Many steroidal AIs bind to the substrate-binding site of the aromatase enzyme and are converted to a reactive intermediate that binds irreversibly and covalently, causing permanent inactivation of aromatase. Such inhibitors are also known as "inactivators" or "suicide inhibitors" because the enzyme is inactivated by its own mechanism of action.[18,19] It is possible that the steroidal AIs may exert androgenic effects but this has not been proven in the clinic.[20] Such AIs include formestane and exemestane (Aromasin®). Anastrozole (Arimidex®) and letrozole (Femara®) are considered to be type II AIs or nonsteroidal AIs. They bind noncovalently to the heme moiety of the aromatase enzyme and prevent binding of androgens by occupying the substrate-binding site. This process is reversible via competitive inhibition of the endogenous substrate like androstenedione.[1,20] Vorozole is another nonsteroidal AI, but it is not commercially available in the United States. It appeared to be as effective as megestrol acetate in postmenopausal women with tamoxifen-resistant metastatic breast cancer.[21] Currently, anastrozole, letrozole, and exemestane are approved by U.S. Food and Drug Administration (FDA) for postmenopausal women with hormone receptor positive breast cancer in the adjuvant setting, as well as for first- and second-line treatment in the metastatic setting. Letrozole is also approved for extended adjuvant therapy after prior adjuvant tamoxifen therapy. The classification of AIs according to the mechanisms of action and the development timeline are summarized in Table 1.

Table 1.

Generations	Type 1 steroidal inhibitors	Type 2 Non-steroidal inhibitors
Nonspecific inhibitor		Aminoglutethimide
Previous selective inhibitors not currently in clinical use	Formestane	Fadrozole Rogletimide Vorozole
Selective oral inhibitors currently in clinical use	Exemestane (Aromasin®)	Anastrozole (Arimidex®) Letrozole (Femara®)

CLINICAL EXPERIENCE WITH AROMATASE INHIBITORS

In contrast to tamoxifen, AIs markedly reduce estrogen levels in postmenopausal women by inhibiting aromatization. Due to the specificity of the aromatase enzyme, selective inhibition of aromatase would not interfere with the biosynthesis of other steroids and thus a corticosteroid supplement is not necessary. Oral AIs appear to have comparable efficacy across all three drugs. Even though there is evidence that letrozole is more potent in suppressing plasma estrogen levels than anastrozole,[22] the clinical significance of this finding remains unclear. AIs are generally well tolerated, with minimal side effects. The common adverse effects include hot flashes, vaginal dryness, and headache, which are typically mild. Unlike SERMs, AIs do not have the partially estrogenic effects, particularly in the uterus and cardiovascular systems. Long term follow-up of patients receiving AIs in the ATAC (arimidex, tamoxifen, alone or in combination) trial, which by far is the largest trial with the longest (100 months) follow-up in the adjuvant setting where there are no confounding symptoms from metastatic disease, showed no evidence of increased risk of uterine cancer and venous thromboembolism.[10,23-25] Compared to SERM, there is a significant increase in musculoskeletal symptoms, osteopenia, osteoporosis, and fracture rate [375 vs. 234 cases; incidence rate ratio (IRR) 1.55, $P < 0.0001$] in breast cancer patients treated with AIs.[24] Nonetheless, it is reassuring that the fracture risk appears to increase only while patients are on active treatment and there was no significant difference after treatment was completed (146 vs. 143 cases; IRR 1.03, $P = 0.79$).[23] A more detailed account of current adjuvant antihormone therapy is presented in a later chapter.

In summary, for first-line treatment of metastatic breast cancer, all oral AIs appear to be well-tolerated and superior to or at least as good as tamoxifen in response rate, median time to progression, and clinical benefit rate.[26-29] The largest study, involving 916 patients, demonstrated that letrozole is superior to tamoxifen, with a longer time to disease progression of 42 versus 23 weeks [hazard ratio (HR) 0.70,

$P = 0.0001$].[30] In the adjuvant setting, recent meta-analysis demonstrated that upfront monotherapy with AIs for 5 years confers an absolute 2.9% reduction in recurrence (9.6% vs. 12.6%; $P < 0.00001$) compared to tamoxifen. Moreover, sequential treatment with 2–3 years of tamoxifen followed by AIs is associated with an absolute 3.1% decrease in recurrence (5.0% vs. 8.1%; $P < 0.00001$). Nonetheless, there is no significant difference in overall survival but longer follow-up is required for further evaluation.[31] Important clinical trials of AIs in the metastatic and adjuvant setting are summarized in Table 2. Furthermore, AIs have also been evaluated in comparison with tamoxifen in other clinical settings, including the neoadjuvant or preoperative setting and chemoprevention.

We review below the clinical trial experience with AIs according to the disease stage (including metastatic and early stage disease) as well as the role of AIs in chemoprevention. Significant clinical trials of AIs are summarized in Table 2 according to the stage of the disease.

Metastatic Setting

For the first-line treatment, all oral AIs have been demonstrated to be well-tolerated and superior to or at least as good as tamoxifen in response rate, median time to progression, and clinical benefit rate.[26-29] The largest study, involving 916 patients, demonstrated that letrozole is superior to tamoxifen, with a longer time to disease progression of 42 versus 23 weeks (HR 0.70, $P = 0.0001$).[30] The subsequent two trials for anastrozole showed somewhat conflicting results. While the smaller North American study with 353 patients showed a significant improvement in time to disease progression (11.1 vs. 5.6 months, $P = 0.005$),[27] the similar design study with a larger population, the TARGET trial, failed to confirm these findings (8.2 vs. 8.3 months, HR 0.99, $P = 0.941$).[29] Thus, anastrozole appears to be at least as good as tamoxifen. A European phase III study comparing exemestane with tamoxifen (EORTC 10951) also demonstrated a significant improvement in progression free survival (PFS) of exemestane over tamoxifen (HR = 0.79, $P = 0.04$).[32] For the second-line treatment after progression on tamoxifen, all of the third-generation AIs

Table 2. Impoprtant clinical trails of AIs.

Key metastatic AI vs. tamoxifen as first-line therapy trials:

Trials	AIs	N	Time to progression
TARGET	A	668	8.2 vs. 8.3 months (HR 0.99, $P = 0.941$)
North American Study	A	353	11.1 vs. 5.6 months ($P = 0.005$)
International Letrozole study	L	916	42 vs. 23 weeks (HR 0.70, $P = 0.0001$)
EORTC 10951	E	382	HR 0.79, $P = 0.04$

Key adjuvant AI vs. tamoxifen trials:

Strategies	Trial	AIs	N	Disease-free survival
Upfront monotherapy (AIs 5 yrs vs. Tam 5 yrs)	ATAC	A	6241	HR 0.9, $P = 0.025$
	BIG 1-98	L	4922	HR 0.88, $P = 0.03$
	ABCSG-12	A	1803	HR 1.1, $P = 0.59$
Sequential (AIs after 2–3 yrs of Tam vs. Tam 5 yrs)	IES	T/E vs. T	4724	HR 0.76, $P = 0.0001$
	BIG 1-98	T/L vs. L	3094	HR 1.05 (CI 0.84–0.32)
		L/T vs. L	3086	HR 0.96 (CI 0.76–1.21)
	TEAM	T/E vs. E	9766	HR 0.97, $P = 0.604$
	N-SAS BC-03	T vs. T/A	696	HR 0.69, $P = 0.14$
	ABCSG8	T/A vs. T	3714	HR 0.85, $P = 0.067$
	ARNO95	T/A vs. T	979	HR 0.66, $P = 0.049$
	ITA	T/A vs. T	448	EFS HR 0.57, $P = 0.005$
Extended (Tam 5 yrs followed by AIs for 8–10 yrs total)	MA.17	L	5170	HR 0.68, $P < 0.0001$
	ABCSG6a	A	852	HR 0.62, $P = 0.031$
	NSABP B33	E	1562	HR 0.68, $P = 0.07$

Abbreviations: A = anastrozole, L = letrozole, E = exemestane, Tam = tamoxifen, N = number of patients, HR = hazard ratio, CI = confidence interval, EFS = event-free survival.

seem to have a marginal benefit over megestrol acetate, with only a modest improvement in response, clinical benefit, and median time to progression.[21,33-36] Due to the rapid adoption of AIs into the adjuvant and first-line metastatic setting, the second-line treatment with AIs is currently less clinically relevant.[4]

Adjuvant Setting

Several strategies were employed to investigate the benefit of AIs as an adjuvant treatment for hormone receptor positive breast cancer. They included upfront treatment with AIs, switching therapy to AIs after 2–3 years of tamoxifen, and extended AI therapy after the completion of tamoxifen for 5 years (Table 2). For the upfront strategy, there were three phase III clinical trials, namely ATAC,[25] BIG 1-98,[37,38] and ABCSG-12,[39] comparing 5 years of anastrozole or letrozole to 5 years of tamoxifen. The HR for disease-free survival (DFS) of these studies ranged from 0.88 to 1.10, with no significant improvement in overall survival (OS). The lack of OS benefit may have been related to considerable subsequent crossover after these trials had been reported. It is worth noting that the combination arm of anastrozole concurrently with tamoxifen in the ATAC trial not only show no additional benefit of the combination but also that the benefit of anastrozole was lost.[40]

Since prolonged exposure to tamoxifen can lead to acquired resistance, several clinical trials were launched to evaluate the switching approach after a few years of tamoxifen to non-cross-resistant agents like AIs (Table 2). Five clinical trials (ABCSG-8,[41] BIG 1-98,[38] ARNO 95,[42] ITA,[43] and IES[44]) comparing 5 years of tamoxifen to sequential treatment of AI after 2–3 years of tamoxifen demonstrated a significant benefit of the switching strategy over tamoxifen alone, with the HR for DFS ranging from 0.57 to 0.76. There were only two trials comparing 5 years of AIs to the sequential therapy which included the BIG 1-98[38] and TEAM[45] trials. Both of the trials did not show significant differences between the AI-alone group and the tamoxifen-sequenced-with-AI group as well as the AI-sequenced-with-tamoxifen group in BIG 1-98. However, there appear to be more early relapses in the tamoxifen-followed-by-letrozole group compared to the upfront AI group, particularly in women with lymph node involvement.[38]

Given the prolonged relapse prospect of hormone receptor positive breast cancer, an extended course of endocrine therapy has been evaluated in multiple clinical trials. Currently, extended therapy of tamoxifen beyond 5 years is still controversial. Although there appears to be some marginal benefit from extended tamoxifen, this is offset by its toxicities, particularly increased risks of endometrial cancer and thromboembolism.[46–48] In contrast, three large randomized clinical trials (ABCSG 6a,[49] MA.17,[50] and NSABP B-33[51]) have demonstrated that extending the duration of endocrine therapy with AIs after 5 years of tamoxifen can be beneficial.

Neoadjuvant Setting

Neoadjuvant therapy refers to any treatment — including chemotherapy, endocrine therapy or targeted therapies — that is given prior to surgery. This can be as a mean for downstaging the tumor to improve the opportunity for breast conserving surgery (BCS)[52] and, more attractively, providing an *in vivo* measurement of tumor response.[4,53] Two clinical trials (IMPACT[54] and PROACT[55]) evaluated the use of tamoxifen and anastrozole for three months prior to surgery. There was no significant difference in overall response and it appears that anastrozole is at least as effective as tamoxifen. In contrast, the PO24[56] study showed a significant increase in the overall response rate and BCS conversion rate with letrozole for four months. A smaller Russian trial also showed a significant improvement in the overall response rate and BCS conversion rate with exemestane for 12 weeks. Nonetheless, a head-to-head comparison between all three AIs in the neoadjuvant setting (ACOSOG Z1031) demonstrated no significant difference in the response rate between all three AIs (anastrozole 66.7% vs. letrozole 70.9% vs. exemestane 60.5%).

Chemoprevention

Based on these promising results of the superiority of AIs over tamoxifen and the approximately 50% reduction in the risk of contralateral breast cancer,[25] AIs are now being investigated as chemopreventive agents. Two large phase III trials are ongoing to evaluate this aspect, including the IBIS II trial in Europe and the MAP-3 trial in Canada.

RESISTANCE TO AROMATASE INHIBITORS

Aromatase inhibition is one of the first specific targeted therapies in oncology and continues to be one of the most successful treatments in hormone responsive breast cancer. Nevertheless, the benefit of AIs is often limited by the development of resistance, which can occur in a significant number of patients in the adjuvant setting and is often inevitable in the metastatic setting. Multiple small phase II studies demonstrated that switching between steroidal and non-steroidal AIs regardless of the sequence can produce 0%–26% objective response rates.[20,57,58] Although the objective response rates are small, the clinical benefit rate, which includes stable disease of longer than 6 months, can be as high as 50%–62%.[20,59–62] The larger randomized, double-blinded, placebo-controled phase III trial comparing fulvestrant and exemestane (EFECT trial) in postmenopausal women with hormone receptor positive breast cancer refractory to a nonsteroidal AI showed a comparable overall response rate of 7.4% versus 6.7% ($P = 0.736$) and a clinical benefit rate of 32.2% versus 31.5% ($P = 0$ .853), respectively. However, the median time to progression was merely 3.7 months in both groups.[63] Given that switching between steroidal and nonsteroidal AIs results in some but modest clinical benefit, this highlights the partial noncross-resistance between different classes of AIs.[20]

Multiple resistant mechanisms of AIs have been described in the past. Overactive growth factor receptor signaling pathways and cross-talks between different pathways appear to be the key feature of the resistant phenotype. Manipulation and interfering with these other signaling pathways may provide an attractive strategy for circumventing the resistance to AI therapy in breast cancer. An intratumoral aromatase breast cancer model was previously developed in the Brodie laboratory. This model mimics hormone receptor positive breast cancer in postmenopausal women, and has proven useful for comparing the antitumor activity of AIs and other endocrine therapies, as well as for evaluating mechanisms of resistance to different types of AIs.[64] By collecting MCF-7 overexpressing aromatase (MCF-7Ca) xenograft tumors treated with letrozole at different time

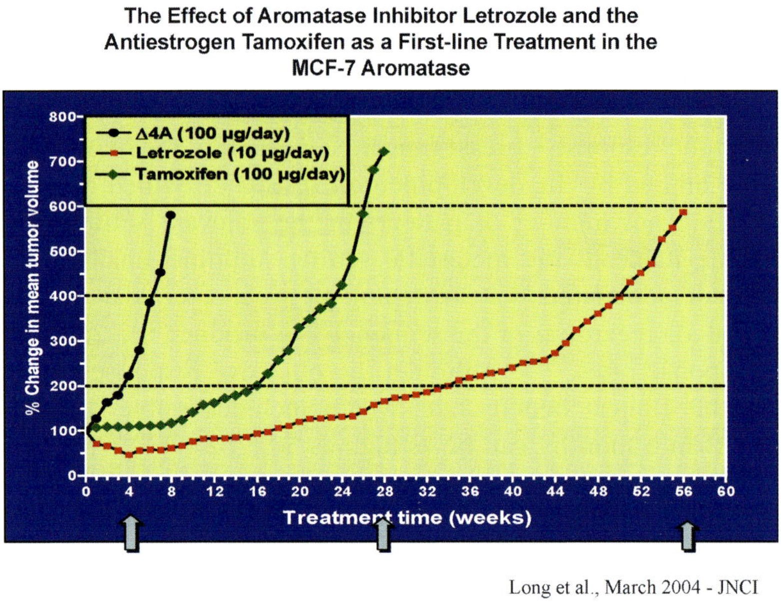

Figure 1A. The effect of letrozole and tamoxifen as a first-line treatment on the growth of MCF-7Ca xenografts. Animals were inoculated with MCF-7Ca cells at two sites on each flank and were supplemented with androstenedione (100 μg/d) for the duration of the experiment. When the tumors reached a measurable size (~300 mm^3), the animals were assigned to three groups (n = 20 per group) and injected s.c. daily with vehicle (control), or tamoxifen (100 μg/d), or letrozole (10 μg/d). Tumor volumes were measured weekly and were expressed as the percent change relative to the initial tumor volume. Two mice per group were sacrificed and tumors were collected for analysis at 4, 28, and 56 weeks.

points (Fig. 1A), we found that the resistant state corresponds with the activation of the human epidermal growth factor receptor 2 (HER2)/mitogen-activated protein kinase (MAPK) signaling pathway. The expressions of HER2, p-Shc, Grb-2, p-Raf, p-Mek1/2 and p-MAPK were all significantly increased in the treatment group compared to control. The effect on Her2 could be observed as early as 4 weeks even when tumors were still responding to letrozole.[65] Downstream targets of MAPK like p90 ribosomal S6 kinase (p90RSK) and the ETS-domain containing protein (Elk) were also found to be activated by phosphorylation (Fig. 1B). Nevertheless, total estrogen receptor α (ERα)

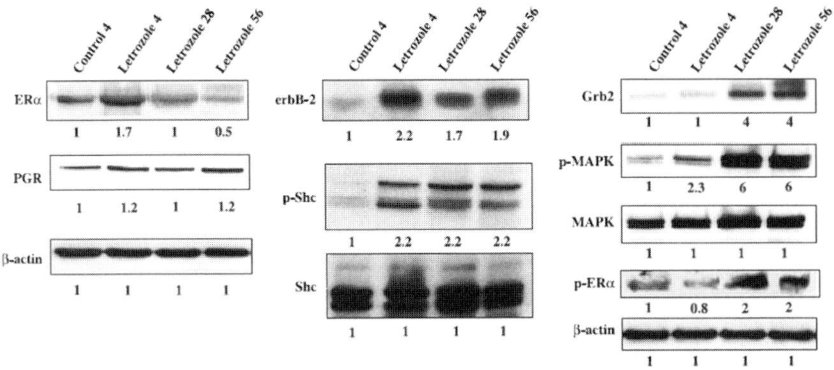

Figure 1B. The effect of letrozole treatment on Her-2, p-Shc, Shc, Grb2, p-MAPK, MAPK, ER-α, p-ER-α, and PR expression in MCF-7Ca tumor xenografts. Letrozole-treated tumors collected at 4, 28, and 56 weeks (A), were analyzed by western immunoblotting, and compared to vehicle-treated tumors collected at week 4.[65]

expression was significantly decreased by 28 weeks, when the tumors became resistant to letrozole compared to before letrozole was started. While the total amount of ERα was decreased, the levels of phosphorylated ERα (p-ERα) in letrozole-resistant tumors were significantly higher than the p-ERα level in tumors without treatment. Moreover, the level of the progesterone receptor (PR), which is a downstream target of activated ERα, remained unchanged compared to parental MCF-7Ca, suggesting a constitutive activation of the ERα signaling pathway despite a reduction in ERα total protein (Fig. 1B).[65] In contrast to letrozole resistance, using the same treatment model, we found that resistance to anastrozole is associated with upregulations of the insulin-like growth factor 1 (IGF-1) receptor and the phosphatidylinositol 3-kinase (PI3K)/Akt pathway.[66] Tumors treated with anastrozole were collected at 14 weeks, when they became resistant to anastrozole and subjected to western blotting. While there seems to be a reduction in total MAPK and p-MAPK, the expressions of IGF-1R and the mammalian target of rapamycin (mTOR) as well as p-mTOR, a critical downstream effecter of Akt, were increased compared to the vehicle control group (Fig. 2B). The constitutive activation of ERα in this ligand-independent manner may be explained by

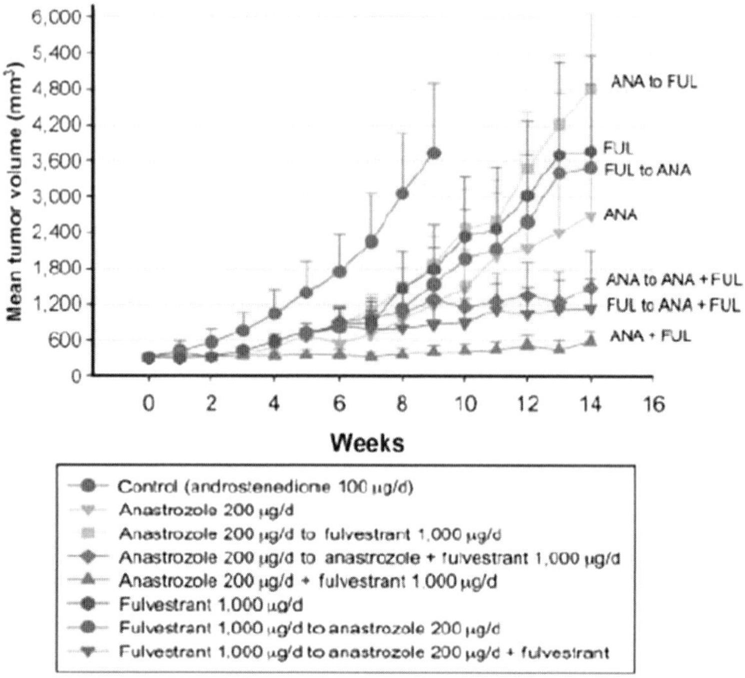

Figure 2A. Effect of anastrozole and fulvestrant alone, in sequence, or in combination on the growth of MCF-7Ca xenografts. MCF-7Ca xenografts were grown in female ovariectomized nude mice. When the tumors reached a measurable size (~300 mm^3), animals were assigned to the treatment groups for the duration of the experiment. The treatment groups were injected with anastrozole (200 μg/d; n = 30; ANA), anastrozole (200 μg/d) plus fulvestrant (1,000 μg/d; n = 10; ANA + FUL), and fulvestrant (1,000 μg/d; n = 20; FUL) in addition to androstenedione (100 μg/d) 5 d/wk. The control group received androstenedione (100 μg/d). At week 5, mice in the anastrozole group were assigned to the following treatment groups: one group continued on anastrozole treatment (200 μg/d; n = 10; ANA); another group was given a combination of anastrozole (200 μg/d) and fulvestrant (1,000 μg/d; n = 12; ANA to ANA + FUL); and a third group switched to fulvestrant (1,000 μg/d; n = 8; ANA to FUL). Animals (n = 20) treated with fulvestrant (1,000 μg/d) were assigned to three treatment groups: one group continued treatment on fulvestrant (1,000 μg/d; n = 6; FUL); another group was given a combination of anastrozole (200 μg/d) and fulvestrant (1,000 μg/d; n = 7; FUL to ANA + FUL); and a third group switched to anastrozole (200 Ag/d; n = 7; FUL to ANA). Tumors were measured weekly and tumor volumes were calculated. On week 6, the differences in growth rate of control versus ANA (P = 0.001), ANA + FUL ($P < 0.0001$), and FUL (P = 0.012) were statistically significant. The differences in growth rate of ANA versus ANA + FUL (P = 0.037) and FUL versus

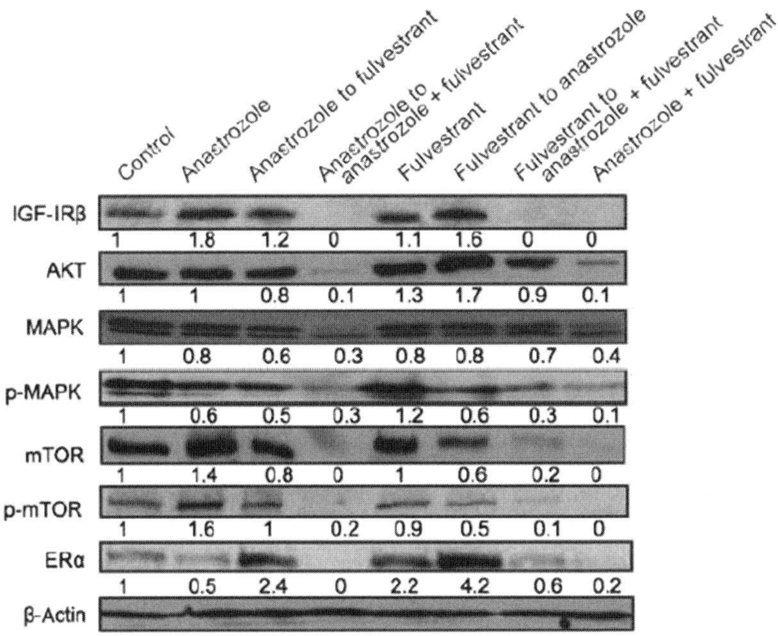

Figure 2B. Effect of anastrozole and fulvestrant alone, in sequence, or in combination on tumoral signaling transduction proteins. The tumors of mice treated with anastrozole, anastrozole + fulvestrant, fulvestrant, anastrozole to fulvestrant, anastrozole to anastrozole + fulvestrant, fulvestrant to anastrozole, and fulvestrant to anastrozole + fulvestrant were collected after 14 wk, analyzed by Western blot, and compared with vehicle-treated tumors (Control) collected at 9 wk. Blots show IGF-IRβ at 95 kDa, AKT at 60 kDa, MAPK and p-MAPK at 42 to 44 kDa, mTOR and p-mTOR at 289 kDa, and ERα at 66 kDa. The blots were stripped and reprobed for β-actin (45 kDa) to verify equal loading. Numbers below the blots represent fold change in protein expression compared with control obtained by densitometric analysis. Representative of at least two independent experiments.

Figure 2A. (*Continued*) ANA + FUL ($P = 0.011$) were statistically significant. On week 9, the differences in growth rate of control versus ANA ($P = 0.013$) and ANA + FUL ($P < 0.0001$) were statistically significant whereas the difference between control and fulvestrant had a marginal statistical significance ($P = 0.056$). At week 14, the differences in growth rate of ANA versus ANA + FUL ($P = 0.0003$) and ANA to ANA + FUL ($P = 0.019$) were statistically significant. The differences in growth rate between FUL versus ANA + FUL ($P = 0.009$) and FUL to ANA + FUL ($P = 0.047$) were considered statistically significant. Points, mean tumor volume (mm3); bars, SE.

the crosstalk activation between the ERα and other signaling pathways like the MAPK and PI3K/Akt pathways. Previous studies demonstrated that MAPK can phosphorylate ERα directly or indirectly via Elk-1 and p90RSK and result in the transcription of genes involved in growth regulation and tumor progression.[67–69] This crosstalk between the ERα and other signaling pathways may occur by both genomic and nongenomic mechanisms. At the genomic level, the crosstalk may occur via direct phosphorylation of ERα and/or its associated coactivator like AIB1 (see Chapter 7).[70,71] Both MAPK and Akt can directly phosphorylate ERα within the AF-1 domain at serine 118 and serine 167, respectively.[72] Nongenomic interaction between ERα and the epidermal growth factor receptor (EGFR) signaling pathway at the inner cell membrane, which has been described as membrane-initiated steroid signaling, was also described in tamoxifen resistance.[73] However, it remains unclear whether this phenomenon also occurs in AI resistance.

Further studies with the above xenograft model demonstrated that disrupting these signaling pathways by means of specific inhibitors and discontinuing treatment can avert resistance and prolong the benefit of AIs. Treatment with trastuzumab, which is a monoclonal antibody against Her2, can reverse letrozole-resistance. Thus, treatment with trastuzumab after letrozole resistance can restore ERα expression and resensitize the letrozole resistant cells (LTLT-Ca) and tumors to endocrine therapy by downregulating the Her2/MAPK pathway and upregulating ERα. Trastuzumab by itself did not appear to have antitumor activity in the parental endocrine-responsive MCF-7Ca. Nonetheless, the combination of trastuzumab and letrozole upfront failed to prevent the emergence of resistance and to prolong the response to letrozole.[74] In addition, inhibition of the PI3K/Akt pathway by wortmannin together with fulvestrant, an estrogen selective receptor downregulator, resulted in more effective tumor regression compared to either of the agents alone in the long-term estrogen-deprived aromatase-transfected ER-positive breast cancer model.[75] Furthermore, the crosstalk between ERα and these two major signaling cascades appears to be dynamic interactions and can be reversed by interrupting the treatment. Interestingly, resuming letrozole after withdrawing it for 4–6 weeks resulted in marked tumor regression with the second course of letrozole treatment (Fig. 2A).[76,77] Analysis of tumors after stopping letrozole treatment demonstrated a significant reduction in phosphorylated MAPK and increases in ERα and aromatase expression levels which were comparable to the vehicle-treated control MCF-7Ca (Fig. 2B). However, in letrozole responsive tumors, intermittent treatment with letrozole (6 weeks on and 6 weeks off) appeared to be inferior to continuous treatment. It was observed that the MCF-7Ca tumors treated with a cyclic schedule could acquire resistance quicker than those in the continuous letrozole treatment group.[77] The dynamic alteration in Her2/MAPK activation appeared to be secondary to the change in the stability of Her2 protein and not to Her2 gene amplification.[74]

Multiple emerging clinical data support our observations that there is an upregulation of the Her2/MAPK pathway corresponding to AI resistance and that dual targeting of the ER signaling pathway with AIs and other growth factor receptor signaling cascades may confer a better clinical outcome. Lipton *et al.* demonstrated that approximately 26% of patients treated with letrozole converted from serum Her2 negative to positive at the time of disease progression.[78] Serum Her2, which is the extracellular domain of Her2 protein that is shed into serum, has been shown to correlate with the overexpression of Her2 protein in tumor cells and can be used to predict the response to trastuzumab.[79] The dynamic interaction between the ER and the Her2/MAPK signaling pathway was illustrated in a small clinical study of 10 patients with ER-negative/HER2-positive advanced breast cancer treated with trastuzumab and chemotherapy. Intriguingly, the tumors of three patients on this trial converted from ER-negative to ER-positive after the treatment with trastuzumab and subsequently responded to letrozole.[80] For dual inhibition with AIs and Her2 inhibitors, three clinical trials have demonstrated a better clinical outcome when combining Her2 targeted therapy with AIs. A phase II trial of letrozole and trastuzumab in ER-positive/Her2-negative metastatic breast cancer patients demonstrated that the combination was well tolerated, with a clinical benefit rate of 50%.[81] Another randomized phase III trial (TANDEM trial) of anastrozole ± trastuzumab showed a significant improvement in PFS with an addition of trastuzumab to anastrozole

(3.8 vs. 5.6 months, $P = 0.0059$).[82] Furthermore, lapatinib, which is a dual tyrosine kinase inhibitor of Her2 and EGFR, also showed superiority when combining with letrozole in a randomized phase III trial of ER-positive metastatic breast cancer. In this trial, there was a significant improvement in PFS from 3.0 months to 8.2 months with an HR of 0.71 ($P = 0.019$) with addition of lapatinib to letrozole in patients with Her2-positive breast cancer.[83] There was no significant improvement in PFS for patients with Her2-negative tumors. Interestingly, the preplanned Cox regression analysis of patients with Her2-negative tumors who relapsed less than 6 months after tamoxifen discontinuation demonstrated a nonsignificant trend toward improvement in PFS for the combination (HR 0.78, $P = 0.117$).[83] This result highlights the significance of dual inhibition of the Her2 and ER signaling pathways in tumors that have been resistant to endocrine therapy even though the initial tumors may lack Her2 overexpression. Furthermore, given that there was no benefit from combining letrozole and lapatinib upfront in ER-positive and Her2-negative tumors, this implies that blockade of Her2 and estrogen deprivation from the beginning may not delay the emergence of resistance[72] because inhibiting HER2 raises ER levels. This finding mirrors the preclinical model discussed previously. In addition, cotargeting the downstream signaling of the PI3K/Akt pathway with an mTOR inhibitor has also been under investigation in multiple clinical trials. A recent randomized phase II study of letrozole ± everolimus for 4 months in the preoperative setting demonstrated a significant increase in tumor shrinkage as determined by ultrasound (58% vs. 47%, $P = 0.03$) and a greater reduction in the cell proliferation index, Ki67 (57% vs. 30%).[84]

CONCLUSIONS AND FUTURE DIRECTIONS

AIs are effective treatment for postmenopausal women with hormone receptor positive breast cancer. Due to the specificity of the aromatase enzyme, selective AIs are well tolerated, with minimal side effects. They are considered as a standard of care in both the adjuvant and the metastatic setting. While AIs have been proven to be very potent in hormone responsive breast cancer, the emergence of resistance to them is the major setback of their sustained clinical benefit. Our preclinical work and several clinical trials suggest that the crosstalk between ER and MAPK as well as the PI3K/Akt signaling pathway serves as the key escape mechanisms which allow these previously hormone responsive tumors to become resistant to AIs. Constitutive activation of the ER by phosphorylation via activation of the MAPK or PI3K/Akt pathway can lead to transcription of genes that regulate and promote tumor proliferation despite estrogen deprivation. Furthermore, inhibition of the Her2/MAPK pathway by specific inhibitors like trastuzumab and lapatinib together with aromatase inhibition may reverse the resistance of these tumors to AIs. It is worth noting that the benefits of this combination can be achieved only when the tumors are already resistant to AIs. Cotargeting the ER signaling pathway with AIs and other novel agents to disrupt the crosstalk between the MAPK and PI3K/Akt signaling cascade appears to be a promising approach to enhance and prolong the efficacy of AIs. Currently, there are multiple novel targeted therapies under clinical development that can also be used to target the MAPK and PI3K/Akt signaling pathway besides trastuzumab and lapatinib. These agents include MEK inhibitors, Raf inhibitors, PI3K inhibitors, mTOR inhibitors, and Akt inhibitors. Lastly, biological analysis from clinical trials with AIs is crucial for determining the clinically relevant mechanisms of AI resistance in individual patients. This will provide a smarter way to target the resistance and to individualize therapy for breast cancer patients.

REFERENCES

1. Meyer AS. (1955) Conversion of 19-hydroxy-delta 4-androstene-3,17-dione to estrone by endocrine tissue. *Biochim Biophys Acta* **17**(3):441–442.
2. Evans CT, Ledesma DB, Schulz TZ, Simpson ER, Mendelson CR. (1986) Isolation and characterization of a complementary DNA specific for human aromatase-system cytochrome P-450 mRNA. *Proc Natl Acad Sci USA* **83**(17):6387–6391.
3. Ghosh D, Griswold J, Erman M, Pangborn W. (2009) Structural basis for androgen specificity and oestrogen synthesis in human aromatase. *Nature* **457**(7226):219–223.
4. Smith IE, Dowsett M. (2003) Aromatase inhibitors in breast cancer, *N Engl J Med* **348**(24):2431–2442.

5. Burstein HJ, Prestrud AA, Seidenfeld J, Anderson H, Buchholz TA, Davidson NE, Gelmon KE, Giordano SH, Hudis CA, Malin J, Mamounas EP, Rowden D, Solky AJ, Sowers MR, Stearns V, Winer EP, Somerfield MR, Griggs JJ. (2010) American Society of Clinical Oncology clinical practice guideline: update on adjuvant endocrine therapy for women with hormone receptor-positive breast cancer. *J Clin Oncol* **28**(23):3784–3796.
6. Beatson G. (1896) On the treatment of inoperable cases of carcinoma of the-mamma. Suggestions for a new method of treatment with illustrative cases. *Lancet II* 104–107.
7. Love RR, Philips J. (2002) Oophorectomy for breast cancer: history revisited. *J Natl Cancer Inst* **94**(19):1433–1434.
8. Jordan VC, Brodie AM. (2007) Development and evolution of therapies targeted to the estrogen receptor for the treatment and prevention of breast cancer. *Steroids* **72**(1):7–25.
9. Fisher B, Costantino JP, Redmond CK, Fisher ER, Wickerham DL, Cronin WM. (1994) Endometrial cancer in tamoxifen-treated breast cancer patients: findings from the National Surgical Adjuvant Breast and Bowel Project (NSABP) B-14. *J Natl Cancer Inst* **86**(7):527–537.
10. Jordan VC. (1995) Tamoxifen: toxicities and drug resistance during the treatment and prevention of breast cancer. *Annu Rev Pharmacol Toxicol* **35**:195–211.
11. Schwarzel WC, Kruggel WG, Brodie HJ. (1973) Studies on the mechanism of estrogen biosynthesis. 8. The development of inhibitors of the enzyme system in human placenta. *Endocrinology* **92**(3):866–880.
12. Brodie AM, Schwarzel WC, Shaikh AA, Brodie HJ. (1977) The effect of an aromatase inhibitor, 4-hydroxy-4-androstene-3,17-dione, on estrogen-dependent processes in reproduction and breast cancer. *Endocrinology* **100**(6):1684–1695.
13. Santen RJ, Brodie H, Simpson ER, Siiteri PK, Brodie A. (2009) History of aromatase: saga of an important biological mediator and therapeutic target. *Endocr Rev* **30**(4):343–375.
14. Santen RJ, Santner S, Davis B, Veldhuis J, Samojlik E, Ruby E. (1978) Aminoglutethimide inhibits extraglandular estrogen production in postmenopausal women with breast carcinoma. *J Clin Endocrinol & Metab* **47**(6):1257–1265.
15. Cash R, Brough AJ, Cohen MN, Satoh PS. (1967) Aminoglutethimide (Elipten-Ciba) as an inhibitor of adrenal steroidogenesis: mechanism of action and therapeutic trial. *J Clin Endocrinol & Metab* **27**(9):1239–1248.
16. Coombes CR, Goss P, Dowsett M, Gazet JC, Brodie A. (1984) 4-hydroxyandostenedione in treatment of postmenopausal patients with advanced breast cancer. *Lancet*, **1**:1237–1239.
17. Goss PE, Coombes RL, Powles TJ, Dowsett M, Brodie AMH. (1986) Treatment of advanced postmenopausal breast cancer with aromatase inhibitor, 4-hydroxyandrostenedione-phase 2 report. *Cancer Res* **46**:4823–4826.

18. Brodie AMH, Garrett WM, Hendrickson JR, Marcotte PA, Robinson CH. (1981) Inactivation of aromatase activity in placental and ovarian microsomes by 4-hydroxyandrostene-3,17-dione and 4-acetoxyandrostene-3,17-dione. *Steroids* **38**:693–702.
19. Hong Y, Yu B, Sherman M, Yuan YC, Zhou D, Chen S. (2007) Molecular basis for the aromatization reaction and exemestane-mediated irreversible inhibition of human aromatase. *Mol Endocrinol* **21**(2):401–414.
20. Miller WR, Bartlett J, Brodie AM, Brueggemeier RW, di Salle E, Lonning PE, Llombart A, Maass N, Maudelonde T, Sasano H, Goss PE. (2008) Aromatase inhibitors: are there differences between steroidal and nonsteroidal aromatase inhibitors and do they matter? *Oncologist* **13**(8):829–837.
21. Goss PE, Winer EP, Tannock IF, Schwartz LH. (1999) Randomized phase III trial comparing the new potent and selective third-generation aromatase inhibitor vorozole with megestrol acetate in postmenopausal advanced breast cancer patients. North American Vorozole Study Group. *J Clin Oncol* **17**(1):52–63.
22. Geisler J, Haynes B, Anker G, Dowsett M, Lonning PE. (2002) Influence of letrozole and anastrozole on total body aromatization and plasma estrogen levels in postmenopausal breast cancer patients evaluated in a randomized, cross-over study. *J Clin Oncol* **20**(3):751–757.
23. Howell A, Cuzick J, Baum M, Buzdar A, Dowsett M, Forbes JF, Hoctin-Boes G, Houghton J, Locker GY, Tobias JS. (2005) Results of the ATAC (Arimidex, Tamoxifen, Alone or in Combination) trial after completion of 5 years' adjuvant treatment for breast cancer. *Lancet* **365**(9453):60–62.
24. Baum M, Budzar AU, Cuzick J, Forbes J, Houghton JH, Klijn JG, Sahmoud T. (2002) Anastrozole alone or in combination with tamoxifen versus tamoxifen alone for adjuvant treatment of postmenopausal women with early breast cancer: first results of the ATAC randomised trial. *Lancet* **359**(9324):2131–2139.
25. Forbes JF, Cuzick J, Buzdar A, Howell A, Tobias JS, Baum M. (2008) Effect of anastrozole and tamoxifen as adjuvant treatment for early-stage breast cancer: 100-month analysis of the ATAC trial. *Lancet Oncol* **9**(1): 45–53.
26. Bonneterre J, Buzdar A, Nabholtz JM, Robertson JF, Thurlimann B, von Euler M, Sahmoud T, (2001) Webster A., Steinberg M., Anastrozole is superior to tamoxifen as first-line therapy in hormone receptor positive advanced breast carcinoma. *Cancer* **92**(9):2247–2258.
27. Nabholtz JM, Buzdar A, Pollak M, Harwin W, Burton G, Mangalik A, Steinberg M, Webster A, von Euler M. (2000) Anastrozole is superior to tamoxifen as first-line therapy for advanced breast cancer in postmenopausal women: results of a North American multicenter randomized trial. Arimidex Study Group. *J Clin Oncol* **18**(22):3758–3767.
28. Mouridsen HT. (2004) Aromatase inhibitors in advanced breast cancer. *Semin Oncol* **31**(6 Suppl 12):3–8.

29. Bonneterre J, Thurlimann B, Robertson JF, Krzakowski M, Mauriac L, Koralewski P, Vergote I, Webster A, Steinberg M, von Euler M. (2000) Anastrozole versus tamoxifen as first-line therapy for advanced breast cancer in 668 postmenopausal women: results of the tamoxifen or arimidex randomized group efficacy and Tolerability study. *J Clin Oncol* **18**(22):3748–3757.
30. Mouridsen H, Gershanovich M, Sun Y, Perez-Carrion R, Boni C, Monnier A, Apffelstaedt J, Smith R, Sleeboom HP, Janicke F, Pluzanska A, Dank M, Becquart D, Bapsy PP, Salminen E, Snyder R, Lassus M, Verbeek JA, Staffler B, Chaudri-Ross HA, Dugan M. (2001) Superior efficacy of letrozole versus tamoxifen as first-line therapy for postmenopausal women with advanced breast cancer: results of a phase III study of the International Letrozole Breast Cancer Group. *J Clin Oncol* **19**(10):2596–2606.
31. Dowsett M, Cuzick J, Ingle J, Coates A, Forbes J, Bliss J, Buyse M, Baum M, Buzdar A, Colleoni M, Coombes C, Snowdon C, Gnant M, Jakesz R, Kaufmann M, Boccardo F, Godwin J, Davies C, Peto R. (2010) Meta-analysis of breast cancer outcomes in adjuvant trials of aromatase inhibitors versus tamoxifen. *J Clin Oncol* **28**(3):509–518.
32. Paridaens TP Dirix L. (2004) First line hormonal treatment for metastatic breast cancer with exemestane or tamoxifen in postmenopausal patients: randomized phase III trial of the EORTC Breast Group. *Proc Am Soc Clin Oncol* p. 6.
33. Buzdar A, Douma J, Davidson N, Elledge R, Morgan M, Smith R, Porter L, Nabholtz J, Xiang X, Brady C. (2001) Phase III, multicenter, double-blind, a randomized study of letrozole, an aromatase inhibitor, for advanced breast cancer versus megestrol acetate. *J Clin Oncol* **19**(14):3357–3366.
34. Buzdar A, Jonat W, Howell A, Jones SE, Blomqvist C, Vogel CL, Eiermann W, Wolter JM, Azab M, Webster A, Plourde PV. (1996) Anastrozole, a potent and selective aromatase inhibitor, versus megestrol acetate in postmenopausal women with advanced breast cancer: results of overview analysis of two phase III trials. Arimidex Study Group. *J Clin Oncol* **14**(7):2000–2011.
35. Dombernowsky P, Smith I, Falkson G, Leonard R, Panasci L, Bellmunt J, Bezwoda W, Gardin G, Gudgeon A, Morgan M, Fornasiero A, Hoffmann W, Michel J, Hatschek T, Tjabbes T, Chaudri HA, Hornberger U, Trunet PF. (1998) Letrozole, a new oral aromatase inhibitor for advanced breast cancer: double-blind randomized trial showing a dose effect and improved efficacy and tolerability compared with megestrol acetate. *J Clin Oncol* **16**(2):453–461.
36. Kaufmann M, Bajetta E, Dirix LY, Fein LE, Jones SE, Zilembo N, Dugardyn JL, Nasurdi C, Mennel RG, Cervek J, Fowst C, Polli A, di Salle E, Arkhipov A, Piscitelli G, Miller LL, Massimini G. (2000) Exemestane is superior to megestrol acetate after tamoxifen failure in postmenopausal women with advanced breast cancer: results of a phase III randomized double-blind trial. The Exemestane Study Group. *J Clin Oncol* **18**(7):1399–1411.

37. Thurlimann B, Keshaviah A, Coates AS, Mouridsen H, Mauriac L, Forbes JF, Paridaens R, Castiglione-Gertsch M, Gelber RD, Rabaglio M, Smith I, Wardley A, Price KN, Goldhirsch A. (2005) A comparison of letrozole and tamoxifen in postmenopausal women with early breast cancer. *N Engl J Med* **353**(26):2747–2757.
38. Mouridsen H, Giobbie-Hurder A, Goldhirsch A, Thurlimann B, Paridaens R, Smith I, Mauriac L, Forbes JF, Price KN, Regan MM, Gelber RD, Coates AS. (2009) Letrozole therapy alone or in sequence with tamoxifen in women with breast cancer. *N Engl J Med* **361**(8):766–776.
39. Gnant M, Mlineritsch B, Schippinger W, Luschin-Ebengreuth G, Postlberger S, Menzel C, Jakesz R, Seifert M, Hubalek M, Bjelic-Radisic V, Samonigg H, Tausch C, Eidtmann H, Steger G, Kwasny W, Dubsky P, Fridrik M, Fitzal F, Stierer M, Rucklinger E, Greil R, Marth C. (2009) Endocrine therapy plus zoledronic acid in premenopausal breast cancer. *N Engl J Med* **360**(7):679–691.
40. Lin NU, Winer EP. (2008) Advances in adjuvant endocrine therapy for postmenopausal women. *J Clin Oncol* **26**(5):798–805.
41. Jakesz R, Gnant M, Greil R. (2005) The benefits of sequencing adjuvant tamoxifen and anastrozole in postmenopausal women with hormone responsive early breast cancer: 5-year analysis of ABCSG Trial 8. 28th Annual San Antonio Breast Cancer Symposium, San Antonio, TX, 2005.
42. Kaufmann M, Jonat W, Hilfrich J, Eidtmann H, Gademann G, Zuna I, von Minckwitz G. (2007) Improved overall survival in postmenopausal women with early breast cancer after anastrozole initiated after treatment with tamoxifen compared with continued tamoxifen: the ARNO 95 Study. *J Clin Oncol* **25**(19):2664–2670.
43. Boccardo F, Rubagotti A, Guglielmini P, Fini A, Paladini G, Mesiti M, Rinaldini M, Scali S, Porpiglia M, Benedetto C, Restuccia N, Buzzi F, Franchi R, Massidda B, Distante V, Amadori D, Sismondi P. (2006) Switching to anastrozole versus continued tamoxifen treatment of early breast cancer. Updated results of the Italian tamoxifen anastrozole (ITA) trial. *Ann Oncol* (17 Suppl) 7:vii,10–14.
44. Coombes RC, Kilburn LS, Snowdon CF, Paridaens R, Coleman RE, Jones SE, Jassem J, Van de Velde CJ, Delozier T, Alvarez I, Del Mastro L, Ortmann O, Diedrich K, Coates AS, Bajetta E, Holmberg SB, Dodwell D, Mickiewicz E, Andersen J, Lonning PE, Cocconi G, Forbes J, Castiglione M, Stuart N, Stewart A, Fallowfield LJ, Bertelli G, Hall E, Bogle RG, Carpentieri M, Colajori E, Subar M, Ireland E, Bliss JM. (2007) Survival and safety of exemestane versus tamoxifen after 2–3 years' tamoxifen treatment (Intergroup Exemestane Study): a randomised controlled trial. *Lancet* **369**(9561):559–570.
45. Rea D, Hasenburg A, Seynaeve C. (2009) Five years of exemestane as initial therapy compared to 5 years of tamoxifen followed by exemestane: the TEAM Trial, a prospective, randomized, phase III trial in postmenopausal women with hormone-sensitive early breast cancer. 32nd Annual San Antonio Breast Cancer Symposium, San Antonio, TX, 2009.

46. Stewart HJ, Prescott RJ, Forrest AP. (2001) Scottish adjuvant tamoxifen trial: a randomized study updated to 15 years. *J Natl Cancer Inst* **93**(6):456–462.
47. Fisher B, Dignam J, Bryant J, Wolmark N. (2001) Five versus more than five years of tamoxifen for lymph node–negative breast cancer: updated findings from the National Surgical Adjuvant Breast and Bowel Project B-14 randomized trial. *J Natl Cancer Inst* **93**(9):684–690.
48. Peto R, Davies C. (2007) ATLAS (Adjuvant Tamoxifen, Longer Against Shorter): international randomized trial of 10 versus 5 years of adjuvant tamoxifen among 11,500 women — preliminary results. 30th Annual San Antonio Breast Cancer Symposium, San Antonio, TX, 2007.
49. Jakesz R, Greil R, Gnant M, Schmid M, Kwasny W, Kubista E, Mlineritsch B, Tausch C, Stierer M, Hofbauer F, Renner K, Dadak C, Rucklinger E, Samonigg H. (2007) Extended adjuvant therapy with anastrozole among postmenopausal breast cancer patients: results from the randomized Austrian Breast and Colorectal Cancer Study Group Trial 6a. *J Natl Cancer Inst* **99**(24):1845–1853.
50. Goss PE, Ingle JN, Martino S, Robert NJ, Muss HB, Piccart MJ, Castiglione M, Tu D, Shepherd LE, Pritchard KI, Livingston RB, Davidson NE, Norton L, Perez EA, Abrams JS, Cameron DA, Palmer MJ, Pater JL. (2007) Efficacy of letrozole extended adjuvant therapy according to estrogen receptor and progesterone receptor status of the primary tumor: National Cancer Institute of Canada Clinical Trials Group MA.17. *J Clin Oncol* **25**(15):2006–2011.
51. Mamounas EP, Jeong JH, Wickerham DL, Smith RE, Ganz PA, Land SR, Eisen A, Fehrenbacher L, Farrar WB, Atkins JN, Pajon ER, Vogel VG, Kroener JF, Hutchins LF, Robidoux A, Hoehn JL, Ingle JN, Geyer CE, Jr, Costantino JP, Wolmark N. (2008) Benefit from exemestane as extended adjuvant therapy after 5 years of adjuvant tamoxifen: intention-to-treat analysis of the National Surgical Adjuvant Breast And Bowel Project B-33 trial. *J Clin Oncol* **26**(12):1965–1971.
52. Smith IE, Lipton L. (2001) Preoperative/neoadjuvant medical therapy for early breast cancer. *Lancet Oncol* **2**(9):561–570.
53. Forrest AP, Levack PA, Chetty U, Hawkins RA, Miller WR, Smyth JF, Anderson TJ. (1986) A human tumour model. *Lancet* **2**(8511):840–842.
54. Smith IE, Dowsett M, Ebbs SR, Dixon JM, Skene A, Blohmer JU, Ashley SE, Francis S, Boeddinghaus I, Walsh G. (2005) Neoadjuvant treatment of postmenopausal breast cancer with anastrozole, tamoxifen, or both in combination: the Immediate Preoperative Anastrozole, Tamoxifen, or Combined with Tamoxifen (IMPACT) multicenter double-blind randomized trial. *J Clin Oncol* **23**(22):5108–5116.
55. Cataliotti L, Buzdar AU, Noguchi S, Bines J, Takatsuka Y, Petrakova K, Dube P, de Oliveira CT. (2006) Comparison of anastrozole versus tamoxifen as preoperative therapy in postmenopausal women with hormone receptor–positive breast cancer: the Pre-Operative "Arimidex" Compared to Tamoxifen (PROACT) trial. *Cancer* **106**(10):2095–2103.

56. Eiermann W, Paepke S, Appfelstaedt J, Llombart-Cussac A, Eremin J, Vinholes J, Mauriac L, Ellis M, Lassus M, Chaudri-Ross HA, Dugan M, Borgs M. (2001) Preoperative treatment of postmenopausal breast cancer patients with letrozole: a randomized double-blind multicenter study. *Ann Oncol* **12**(11):1527–1532.
57 Thurlimann B, Paridaens R, Serin D, Bonneterre J, Roche H, Murray R, di Salle E, Lanzalone S, Zurlo MG, Piscitelli G. (1997) Third-line hormonal treatment with exemestane in postmenopausal patients with advanced breast cancer progressing on aminoglutethimide: a phase II multicentre multinational study. Exemestane Study Group. *Eur J Cancer* **33**(11):1767–1773.
58. Carlini P, Frassoldati A, De Marco S, Casali A, Ruggeri EM, Nardi M, Papaldo P, Fabi A, Paoloni F, Cognetti F. (2001) Formestane, a steroidal aromatase inhibitor after failure of non steroidal aromatase inhibitors (anastrozole and letrozole): is a clinical benefit still achievable? *Ann Oncol* **12**(11):1539–1543.
59. Geisler J, Johannessen DC, Anker G, Lonning PE. (1996) Treatment with formestane alone and in combination with aminoglutethimide in heavily pretreated breast cancer patients: clinical and endocrine effects. *Eur J Cancer* **32**A(5):789–792.
60. Harper-Wynne C, English J, Meyer L, Bower M, Archer C, Sinnett HD, Lowdell C, Coombes RC. (1999) Randomized trial to compare the efficacy and toxicity of cyclophosphamide, methotrexate and 5-fluorouracil (CMF) with methotrexate mitoxantrone (MM) in advanced carcinoma of the breast. *Br J Cancer* **81**(2):316–322.
61. Bertelli G, Garrone O, Merlano M, Occelli M, Bertolotti L, Castiglione F, Pepi F, Fusco O, Del Mastro L, Leonard RC. (2005) Sequential treatment with exemestane and non-steroidal aromatase inhibitors in advanced breast cancer. *Oncology* **69**(6):471–477.
62. Zilembo N, E Bajetta, Bichisao E, Martinetti A, La Torre I, Bidoli P, Longarini R, Portale T, Seregni E, Bombardieri E. (2004) The estrogen suppression after sequential treatment with formestane in advanced breast cancer patients. *Biomed Pharmacother* **58**(4):255–259.
63. Chia S, Gradishar W, Mauriac L, Bines J, Amant F, Federico M, Fein L, Romieu G, Buzdar A, Robertson JF, Brufsky A, Possinger K, Rennie P, Sapunar F, Lowe E, Piccart M. (2008) Double-blind, randomized placebo-controlled trial of fulvestrant compared with exemestane after prior nonsteroidal aromatase inhibitor therapy in postmenopausal women with hormone receptor-positive, advanced breast cancer: results from EFECT. *J Clin Oncol* **26**(10):1664–1670.
64. Macedo LF, Sabnis G, Brodie A. (2008) Preclinical modeling of endocrine response and resistance: focus on aromatase inhibitors. *Cancer* **112**(3 Suppl):679–688.
65. Jelovac D, Sabnis G, Long BJ, Macedo L, Goloubeva OG, Brodie AM. (2005) Activation of mitogen-activated protein kinase in xenografts and cells during prolonged treatment with aromatase inhibitor letrozole. *Cancer Res* **65**(12):5380–5389.

66. Macedo LF, Sabnis GJ, Goloubeva OG, Brodie A. (2008) Combination of anastrozole with fulvestrant in the intratumoral aromatase xenograft model. *Cancer Res* **68**(9):3516–3522.
67. Kato S, Endoh H, Masuhiro Y, Kitamoto T, Uchiyama S, Sasaki H, Masushige S, Gotoh Y, Nishida E, Kawashima H, Metzger D, Chambon P. (1995) Activation of the estrogen receptor through phosphorylation by mitogen-activated protein kinase. *Science* **270**(5241):1491–1494.
68. Joel PB, Traish AM, Lannigan DA. (1998) Estradiol-induced phosphorylation of serine 118 in the estrogen receptor is independent of p42/p44 mitogen-activated protein kinase. *J Biol Chem* **273**(21):13317–13323.
69. Chen D, Washbrook E, Sarwar N, Bates GJ, Pace PE, Thirunuvakkarasu V, Taylor J, Epstein RJ, Fuller-Pace FV, Egly JM, Coombes RC, Ali S. (2002) Phosphorylation of human estrogen receptor alpha at serine 118 by two distinct signal transduction pathways revealed by phosphorylation-specific antisera. *Oncogene* **21**(32):4921–4931.
70. Wu RC, Qin J, Yi P, Wong J, Tsai SY, Tsai MJ, O'Malley BW. (2004) Selective phosphorylations of the SRC-3/AIB1 coactivator integrate genomic responses to multiple cellular signaling pathways. *Mol Cell* **15**(6):937–949.
71. Font de Mora J, Brown M. (2000) AIB1 is a conduit for kinase-mediated growth factor signaling to the estrogen receptor. *Mol Cell Biol* **20**(14):5041–5047.
72. Johnston SR. (2009) Enhancing the efficacy of hormonal agents with selected targeted agents. *Clin Breast Cancer* **9** (Suppl 1):28–36.
73. Fan P, Wang J, Santen RJ, Yue W. (2007) Long-term treatment with tamoxifen facilitates translocation of estrogen receptor alpha out of the nucleus and enhances its interaction with EGFR in MCF-7 breast cancer cells. *Cancer Res* **67**(3):1352–1360.
74. Sabnis G, Schayowitz A, Goloubeva O, Macedo L, Brodie A. (2009) Trastuzumab reverses letrozole resistance and amplifies the sensitivity of breast cancer cells to estrogen. *Cancer Res* **69**(4):1416–1428.
75. Sabnis G, Goloubeva O, Jelovac D, Schayowitz A, Brodie A. (2007) Inhibition of the phosphatidylinositol 3-kinase/Akt pathway improves response of long-term estrogen-deprived breast cancer xenografts to antiestrogens. *Clin Cancer Res* **13**(9):2751–2757.
76. Sabnis GJ, Macedo LF, Goloubeva O, Schayowitz A, Brodie AM. (2008) Stopping treatment can reverse acquired resistance to letrozole. *Cancer Res* **68**(12):4518–4524.
77. Sabnis G, Goloubeva O, Gilani R, Macedo L, Brodie A. (2010) Sensitivity to the aromatase inhibitor letrozole is prolonged after a "break" in treatment. *Mol Cancer Ther* **9**(1):46–56.
78. Lipton A, Leitzel K, Ali SM, Demers L, Harvey HA, Chaudri-Ross HA, Evans D, Lang R, Hackl W, Hamer P, Carney W. (2005) Serum HER-2/neu conversion to positive at the time of disease progression in patients with breast carcinoma on hormone therapy. *Cancer* **104**(2):257–263.

79. Lipton A, Ali S, Leitzel K. (2008) HER-2 protein expression predicts response to trastuzumab in FISH-positive patients with metastatic breast cancer. 31st AACR-CTRC San Antonio Breast Cancer Symposium, San Antonio, TX, 2008.
80. Munzone E, Curigliano G, Rocca A, Bonizzi G, Renne G, Goldhirsch A, Nole F. (2006) Reverting estrogen-receptor-negative phenotype in HER-2-overexpressing advanced breast cancer patients exposed to trastuzumab plus chemotherapy. *Breast Cancer Res* **8**(1):R4.
81. Marcom PK, Isaacs C, Harris L, Wong ZW, Kommarreddy A, Novielli N, Mann G, Tao Y, Ellis MJ. (2007) The combination of letrozole and trastuzumab as first or second-line biological therapy produces durable responses in a subset of HER2 positive and ER-positive advanced breast cancers. *Breast Cancer Res Treat* **102**(1):43–49.
82. Mackey J, Kaufman B, Clemens M. (2006) Trastuzumab prolongs progression-free survival in hormone-dependent and HER2-positive metastatic breast cancer. *Breast Cancer Res Treat* **100** (suppl 1):abstract 3.
83. Johnston S, Pippen J, Jr, Pivot X, Lichinitser M, Sadeghi S, Dieras V, Gomez HL, Romieu G, Manikhas A, Kennedy MJ, Press MF, Maltzman J, Florance A, O'Rourke L, Oliva C, Stein S, Pegram M. (2009) Lapatinib combined with letrozole versus letrozole and placebo as first-line therapy for postmenopausal hormone receptor–positive metastatic breast cancer. *J Clin Oncol* **27**(33):5538–5546.
84. Baselga J, Semiglazov V, van Dam P, Manikhas A, Bellet M, Mayordomo J, Campone M, Kubista E, Greil R, Bianchi G, Steinseifer J, Molloy B, Tokaji E, Gardner H, Phillips P, Stumm M, Lane HA, Dixon JM, Jonat W, Rugo HS. (2009) Phase II randomized study of neoadjuvant everolimus plus letrozole compared with placebo plus letrozole in patients with estrogen receptor–positive breast cancer. *J Clin Oncol* **27**(16):2630–2637.

10

ADJUVANT ANTIHORMONE THERAPY

Anthony Howell and V. Craig Jordan

Abstract

The development of effective, life-saving strategies for the adjuvant antihormone treatment of estrogen receptor (ER)-positive breast cancer is a major advance in medicine. The effectiveness of the long-term adjuvant tamoxifen therapy has saved perhaps millions of lives worldwide during the past 30 years. However, tamoxifen therapy for the postmenopausal patient has the concern of a small but significant increase in endometrial cancer and thromboembolic events. The innovation of blocking endogenous estrogen synthesis with aromatase inhibitors has not only reduced the side effects of tamoxifen with endometrial cancer and thromboembolic disorders but also enhanced survivorship. Tamoxifen remains the endocrine treatment of choice for the premenopausal patient.

Keywords: Overview Analysis; Tamoxifen; Antiestrogens; Decreased Mortality.

INTRODUCTION

One hundred years ago, Dr. George Beatson[1] made the astonishing decision to treat advanced breast cancer in a premenopausal woman by removing her ovaries. Beatson was aware that farmers in Scotland could affect lactation in their animals by removal of the ovaries, so after completing preliminary studies on rabbits, he reasoned that the "ovarian irritation" of the mammary glands — breast cancer — in a woman might be influenced by the removal of her ovaries. He initially reported that one of his patients had a dramatic response to the procedure, and by 1900 Stanley Boyd[2] at Charing Cross Hospital in London had collected the records of 46 premenopausal women (excluding 8 Edinburgh cases) with advanced breast cancer to document their responses to oophorectomy.

Beatson observed, however, that only 1 in 3 patients (actually 37%) responded to the procedure. The reason for this was to remain obscure for the next 60 years.

What was remarkable about Bestson's original report in 1896 was that the endocrine role of the ovaries was completely unknown. In 1923, Drs. Allen and Doisy[3] in St. Louis described their finding of an "estrus-stimulating principle" in the follicular fluid of pig ovaries. The ovaries were known to control reproductive functions, and it was also known that female animals came into heat and accepted males only at certain times in their reproductive cycles.

Allen and Doisy's studies illustrated the close relationship that was necessary between the laboratory and the clinic in order to unlock the mysteries of human physiology. They devised a test system for examining follicular fluid by first ovariectomizing mice and eliminating the reproductive cycle. They discovered that the vaginal epithelium underwent the same cellular changes (called cornification) in response to injections of ovarian follicular fluid as the vagina did when the unovariectomized mouse went into heat or "estrus" (from the Latin word meaning "frenzy"). The chemicals produced by the ovaries — that were subsequently identified as circulating messengers — were named "estrogens". It was thus discovered that these estrogens could selectively activate estrogen target tissues (uterus, vagina, and breast) in a woman. The concept of endocrine control of an organ by a distant organ became an established fact.

Now, we should return to the original question regarding why some breast tumors responded to the removal of ovarian influence whereas other did not. In 1959, [^{3}H] hexestrol was synthesized and shown to bind to estrogen target tissues in animals.[4] Early attempts to translate these data to discover whether injected [^{3}H] hexestrol bound to some breast cancers were successful but the technique was not pursued.[5] In 1962, Drs. Elwood Jensen and Herb Jacobson synthesized radioactive estradiol — the most potent of the natural estrogens — in a woman's body.[6] With this breakthrough, these investigators were able to address the question of differing tissue responses to estrogen. Although natural estrogens circulate throughout the female body, only target tissues become activated. Jensen and Jacobson injected radioactive estradiol into immature female rats (i.e. animals that have no estrogen because their ovaries are inactive) and killed the animals at different times over the next 24 hr to find out where the estrogen had gone. They found that the radioactivity was initially bound to all tissues, but was only retained in the estrogen target tissues (uterus and vagina). They reasoned that there must be a mechanism, a receptor perhaps, that enabled the target tissues to retain estrogen.

THE ESTROGEN RECEPTOR HYPOTHESIS

The estrogen receptor (ER) protein was subsequently extracted and isolated from the rat uterus in 1966 in parallel research ventures by Dr. Jack Gorski's group at the University of Illinois and by Jensen's group at the University of Chicago.[7,8] The work rapidly translated to the clinic with the finding that [^{3}H] estradiol with the binding specifically to the ER in breast biopsies and subsequently work established the concept of a pattern of estrogen target tissue containing the ER throughout a woman's body.

Jensen's work, however, did not stop here. He reasoned that since estrogen controls estrogen action via the ER, then perhaps breast cancer growth was regulated by the same mechanism. In the laboratory, he established that some breast cancers did indeed contain the ER. He conducted clinical studies that illustrated the principle that patients whose tumor contained the ER would respond to an endocrine maneuver, but that those whose tumor was "ER-negative" would be unlikely to respond.[9] The generality of the principle was established in Bethesda, Maryland in 1974 at the National Cancer Institute-sponsored meeting. Investigators from around the world were invited to pool their clinical data on the usefulness of the ER assay to predict hormone-responsive breast

Table 1. Comparison of early clinical experience with antiestrogens as a treatment for advanced breast cancer.

Antiestrogen	Daily dose	Year	Response rate	Toxicity
Ethamoxytriphetol	500–4500 mg	1960	25%	acute psychotic episodes
Clomiphene	100–300 mg	1964–1974	34%	fear of cataracts
Nafoxidine	180–240 mg	1976	31%	cataracts, ichthyosis, photophobia
Tamoxifen	20–40 mg	1971–1973	31%	transient thrombocytopenia*

*"The particular advantage of this drug is the low incidence of troublesome side effects." Cole et al: *Br J Cancer* **25**:270–275, 1971.
"Side effects were usually trivial." Ward: *Br Med J* **1**:13–14, 1973.

found to be potent, long-acting estrogenic drugs. ICI Pharmaceuticals (now Astra Zeneca) had already provided the triphenylethylenes to Sir Alexander Haddow for his pioneering studies[14] on the value of high-dose estrogen therapy to treat advanced breast cancer in postmenopausal women and to treat prostate cancer in men. This breakthrough in 1944 established a somewhat paradoxical, but effective, therapeutic option that is still used today (see Chapter 2).

What interested the scientists at ICI was the fact that the new antiestrogens were derivatives of the estrogen triphenylethylene; but it was clear that improvements in the toxicity profiles and potencies had to be made before any clinically useful agent could be developed for long-term therapy.

The breakthrough came in the mid-1960s, when Drs. Harper and Walpole discovered that the *trans* isomer of a triphenylethylene, ICI 46,474, was the antiestrogenic compound of choice, whereas it *cis* geometric isomer, ICI 47,699, was an estrogen in all tests.[15] Although the option as a fertility regulator was kept open — in fact, tamoxifen, as it came to be called, was marketed during the 1970s in Great Britain and Ireland for the induction of ovulation (see Chapter 5) — Walpole suggested using tamoxifen as a treatment for metastatic breast cancer.

Earlier in his career, Walpole had established a solid reputation as a cancer researcher, and had actually conducted clinical studies with triphenylethylenes in the treatment of breast cancer in the 1940s.[16] ICI 46,474 had antiestrogenic activity in rats and primates, but was paradoxically an estrogen in mice. The question Walpole wished to address was whether ICI 46,474 would work as an antitumor agent in women with breast cancer, but had fewer estrogen-like side effects than high-dose estrogen.

Dr. Mary Cole and coworkers at the Christie Hospital in Manchester, England conducted the first preliminary evaluation of ICI 46,474 for the palliative treatment of breast cancer. Initially, 46 postmenopausal patients with late-stage disease were treated with the drug; 10 patients responded.[17] The pioneering team at the Christie Hospital subsequently placed hundreds of patients on palliative tamoxifen treatment. Their results were subsequently confirmed in a small dose-response study by Dr. Harold Ward in Birmingham, England.[18]

TAMOXIFEN WENT FORWARD ALONE

The late Dr. Arthur Walpole (Fig. 2) deserves the credit for the application of tamoxifen as a potential palliative treatment for advanced breast cancer in postmenopausal women.[19] Tamoxifen, however, was only one of numerous compounds being tested by pharmaceutical companies during the 1960s. Table 1 compares and contrasts the different compounds that underwent evaluation.

ICI Pharmaceuticals had the good fortune to discover a potent antiestrogen with a low profile of toxic side effects. At the onset, it was clear that tamoxifen had several advantages compared to the other contenders. Tamoxifen had higher antitumor potency (i.e. a lower dose could be used) than other compounds being evaluated at the time. Moreover, while side effects represented a major problem with the other antiestrogens, tamoxifen had virtually no troublesome side effects. In fact, tamoxifen seemed to have none of the troublesome side effects observed with high-dose estrogen therapy or androgen therapy, which were the standard treatments of the time.[17] Based

Figure 2. Dr. Arthur Walpole, who died suddenly on July 2, 1977. At the time of his death, he had retired as the Head of the Fertility Control Program at ICI's Pharmaceuticals Division, Alderly Park, near Macclesfield, Cheshire; but he had continued to work as a consultant on the Joint Research Scheme between ICI Pharmaceuticals Division and the Department of Pharmacology at the University of Leeds, England.

on experience with drugs, such as clomiphene, investigators were also concerned about the potential of long-term therapy leading to cataract development. Even during the 1960s, however, the ICI scientists believed that tamoxifen would not produce cataracts based on animal studies and less reported circulating desmosterol.[20] We now know this is only partially correct. When tamoxifen was compared with placebos in the U.S. P-1 prevention study,[21] there was a significant rise in the diagnosis of cataracts and surgeries in women taking tamoxifen.

Back in the 1960s and '70s, ICI was the only company that had chosen to develop an antiestrogenic drug for the treatment of advanced breast cancer. Scientists at ICI did not conduct any studies with the drug as an antitumor agent. Indeed, in late 1972, all of the data with ICI 46,474 were reviewed and the Research Director terminated clinical trials and stopped the development project. The Marketing Department had decided that a treatment for metastatic breast cancer was not going to generate sufficient revenue. Arthur Walpole was toward the end of his career and chose to take early retirement, but agreed to remain an employee only if funds could be given to a young man he had met, Craig Jordan, who (as he did) wanted to turn ICI 46,474 into a drug for treating breast cancer. They subsequently worked together on an ICI/University joint research scheme when Jordan returned as Lecturer in the Department of Pharmacology at the University of Leeds in September 1974. Walpole died suddenly on June 2, 1977, before he could witness the success of the laboratory strategy for the treatment and prevention of breast cancer. The story of the early development of tamoxifen has been told in the refereed literature[22-24] and Foundations of a Career.

However, there was little enthusiasm from the clinical community about this innovation — all hopes seemed pinned instead on the discovery and application of new cytotoxic chemotherapeutic agents. Thirty years ago, clinical researchers reasoned that their only hope of curing breast cancer lay in the development of better and stronger cytotoxics. That perspective would eventually change when tamoxifen was successfully developed as an ER-targeted treatment for all stages of breast cancer and as the first drug to reduce the incidence of breast cancer in high-risk women.

FROM SCIENTIFIC PRINCIPLES TO EVALUATION-APPROPRIATE APPLICATION OF TAMOXIFEN FOR BREAST CANCER THERAPY

In 1973, Nolvadex, the ICI Pharmaceuticals Division (now Astra Zeneca) brand of tamoxifen (as its citrate salt), was approved by the Committee on the Safety of Medicines in the United Kingdom for the treatment of breast cancer. Similarly, approval was given in the United States for the treatment of advanced disease in postmenopausal women by the Food and Drug Administration on December 30, 1977. Tamoxifen is available in more than 110 countries as endocrine therapy for the treatment of breast cancer. To mark this achievement, ICI Pharmaceuticals was presented with the Queen's Award for Technological Achievement by the Lord Lieutenant of Cheshire, Viscount Leverhulme, on July 6, 1978. One of us (V. C. J.) was the only person from outside ICI Pharmaceuticals to be invited to the celebration that day. The remarkable success of tamoxifen encouraged a closer examination of its pharmacology and toxicology with a view to further development and wider applications. Tamoxifen was the pioneer precisely because no other competitive medicines were available initially and, in fact, no competing nonsteroidal antiestrogens were even serious competitors to replace it. This is a unique situation in therapeutics, with one agent available for clinical care for over 30 years with no improvements. The development of aromatase inhibitors would change that for the treatment of postmenopausal patients. However, principles for the targeted antihormone treatment of breast cancer were established in the laboratory and subsequently tested in clinical trials in the late 1970s.[25]

ADJUVANT TAMOXIFEN: FROM LABORATORY STUDIES TO CLINICAL TRIALS

Throughout the 1970s and '80s, the development of tamoxifen became a conversation between the laboratory and the clinic. The initial success of adjuvant monotherapy with L-phenylalanine mustard or combination chemotherapy in the mid-1970s[26,27] to delay the recurrence of node-positive breast cancer helped encourage the investigation of other, perhaps less toxic, therapies. Laboratory studies using the DMBA-induced rat mammary carcinoma model were first used to explore whether tamoxifen would be effective as an adjuvant therapy and whether the drug produced a tumoristatic or tumoricidal effect *in vivo*.

Previous *in vitro* studies by Marc Lippman had indicated that tamoxifen could be a tumoricidal drug[28] but the results from the *in vivo* DMBA studies (first reported at a breast cancer symposium at King's College, Cambridge, England in September 1977; see Fig. 3) demonstrated that a one-month course of tamoxifen therapy given one month after the carcinogenic insult only delayed the appearance

Figure 3. Participants at a Breast Cancer Symposium in September 1977 at King's College, Cambridge, England. The concept of extended adjuvant tamoxifen treatment

of mammary tumors; continuous therapy for six months, on the other hand, resulted in 90% of the animals remaining tumor-free (Fig. 4). Indeed, tumors appeared whenever tamoxifen therapy was stopped.[29] Thus, tamoxifen was shown to have a tumoristatic component to its mode of action, and the laboratory results indicated that long-term (up to five years) or indefinite therapy might be the best clinical strategy for adjuvant tamoxifen treatment.[29,30]

Several trials of tamoxifen monotherapy as an adjuvant to mastectomy were initiated in the mid- to late 1970s. The majority of clinical trial organizations selected a conservative course of one year of adjuvant tamoxifen therapy. This decision was based on a number of reasonable concerns. Patients with advanced disease usually respond to tamoxifen for one year, and it was expected that ER-negative disease would be encouraged to grow prematurely during adjuvant therapy. If this growth occurred, then the physician would have already used a valuable palliative drug and would have only combination chemotherapy to slow the relentless growth of recurrent disease. A related argument involved the changing strategy for the application of adjuvant combination chemotherapy. Recurrent treatment cycles (two years) were found to be of no long-term benefit for patients. In contrast, it seemed that aggressive, short-term courses of treatment (six months) with the most active cytotoxic drugs had the best chance to kill tumor cells before the premature development of drug

Figure 3. (*Continued*) was first proposed at this meeting. Clinical studies of one-year adjuvant tamoxifen were in place; regrettably, a decade later, the approach was shown to produce little survival benefit for patients. Insets: (top) Craig Jordan, who presented the new concept; (bottom left) Professor Michael Baum, the session chairman, who was about to launch the Nolvadex Adjuvant Trial Organization's (NATO) two-year adjuvant tamoxifen trial; and (bottom right) Dr. Helen Stewart, who was a participant at the conference. Dr. Stewart would initiate a pilot trial in 1978 and, led by Sir Patrick Forest, would later guide the full randomized Scottish Trial of five years' adjuvant tamoxifen treatment versus control in the 1980s. Both clinical trials were later proven to produce survival advantages for patients. The concept of longer tamoxifen treatment producing more survival benefits for patients was eventually established indirectly by Oxford Overview analysis in 1992 and directly by the Swedish group led by Dr. Lars Rutqvist.

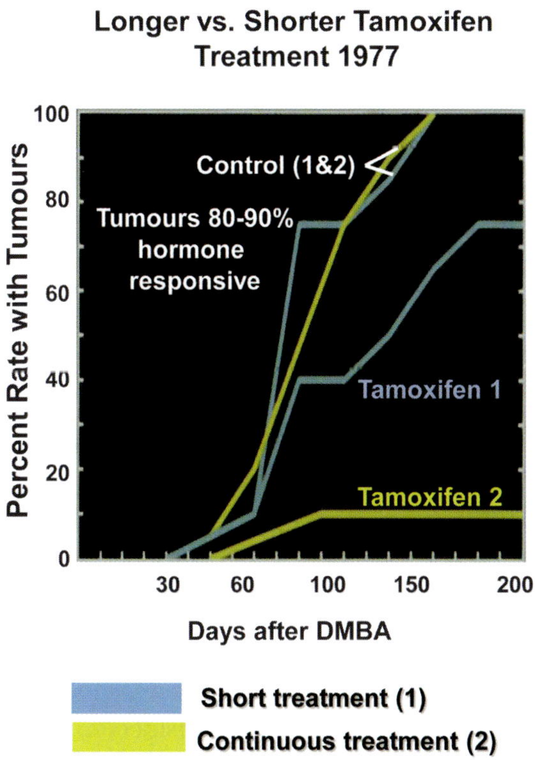

Figure 4. The effectiveness of long-term tamoxifen treatment in the dimethylbenzanthracene (DMBA)-induced rat mammary carcinoma model. The administration of 20 mg DMBA by gavage to 50-day-old female Sprague Dawley rats resulted in all animals developing mammary tumors 160 days later. The short-term (30 days) administration of different daily doses (12.5–800 μg) of tamoxifen between days 30 and 60 after DMBA resulted in a delay of tumor formation. However, not all animals were protected from the carcinogen. In contrast, the daily administration of a clinically relevant dose (50 μg daily — 0.25 mg/kg in rats or 20 mg daily to a 70 kg woman) of tamoxifen continuously, starting 30 days after DMBA, resulted in 90% of animals remaining tumor-free.

resistance. The same intuitive sense that "longer might not be better" contributed to a reluctance on the part of the clinical researchers to start with long-term tamoxifen therapy.

Finally, there were sincere concerns about the side effects of adjuvant therapy and the ethical issues of treating patients who might never have recurrent disease. Although this argument primarily focused on chemotherapy and node-negative patients, it is fair to say that few women in the mid-1970s had received extended therapy with tamoxifen — therefore, long-term side effects were largely unknown. Most tamoxifen-treated patients had only received about two years of treatment for advanced disease before drug resistance developed. Potential side effects, such as thrombosis and osteoporosis, were only of secondary importance.

The use of long-term adjuvant tamoxifen in the disease-free patient, however, would change that perspective.

SUMMARY OF SCIENTIFIC PRINCIPLES FOR THE TREATMENT AND PREVENTION OF BREAST CANCER WITH TAMOXIFEN AND AROMATASE INHIBITORS

In the 1970s, a cluster of papers were published in the *European Journal of Cancer* [summarized in *Eur J. Cancer* (2008), **44**:30–38] describing the idea of targeting tamoxifen to patients with ER-positive breast tumors,[31] and proposed the strategic value of using long-term tamoxifen in an adjuvant setting with consideration of the antitumor properties of the hydroxylated metabolites of tamoxifen.[30] The relevance of the metabolic activation of tamoxifen remains controversial and a focus of research, and this is presented in Chapter 11.[32] At the time, these laboratory results and principles were slow to be embraced by the clinical community. Nevertheless, through methodical studies and the inspirational overview of clinical trials by the Early Breast Cancer Trialists Collaborative Group conducted at intervals at Oxford, the antitumor power of adjuvant tamoxifen monotherapy was demonstrated repeatedly. Now, it is clear that hundreds of thousands (perhaps millions) of women worldwide have benefitted from adjuvant tamoxifen and continue to do so.

Not surprisingly, in the 1970s and early 1980s, tamoxifen was noted to exhibit some predictable side effects, such as menopausal symptoms, (antiestrogenic action), increases in blood clots, vaginal hyperplasia and increased vaginal discharge,[33] but compared with high-dose adjuvant cytotoxic chemotherapy regimens containing Adriamycin, the safety profile of tamoxifen was trivial. Or so it was thought.

In the mid-1980s, laboratory experiments described the target-site-specific effects of tamoxifen in the rodent uterus, human endometrial cancer and human breast cancer inoculated into athymic (immune—deficient) ovariectomized mice. Tamoxifen's well-documented estrogen-like actions in the mouse uterus[20] translated to the growth of human endometrial tumors but not the growth of breast tumors.[34] Clearly, tamoxifen could either switch on or switch off target sites specifically with the tamoxifen–ER complex. These laboratory results[34] also carried a warning for women's health. "However, the fact that tamoxifen has been successfully used to treat endometrial cancer does not exclude the possibility of an increasing incidence of endometrial tumors during prolonged tamoxifen therapy for breast cancer. A larger cohort of patients under long-term tamoxifen therapy (>5 yr) needs to be monitored for the occurrence of tamoxifen-stimulated endometrial tumors." Unfortunately, this concern was not supported by evidence from adjuvant clinical trials, and so no action was taken, until Dr. Tommy Fornander and colleagues in Sweden examined their clinical trial database.[35] Contralateral breast cancer was reduced with tamoxifen [first noted by J. Cuzick and M. Baum; *Lancet* (1985) **2**:282] but endometrial cancer incidence was increased by up to five times by long-term (five years) adjuvant therapy treatment. These data were subsequently confirmed and extended in the NSABP B14 study.[21] Clearly, a less "estrogenlike" strategy for adjuvant antihormone therapy was necessary and this strategy was already positioned through translation work conducted by Angela Brodie (Chapter 9). Before we consider the current comparisons of tamoxifen and aromatase inhibitors in adjuvant trials in postmenopausal patients, it is perhaps important to summarize the important findings with tamoxifen obtained by Oxford Overview analysis over the past two decades. The database with the application of a "nonsteroidal" adjuvant therapy tamoxifen has become the foundation and benchmark for all aromatase inhibitors and any other reversible anticancer agents applied as adjuvant therapies.

Oxford Overview Analysis

The first overview analysis of adjuvant therapy for breast cancer was conducted in 1984 by Richard Peto, Rory Collins and Richard Gray, leading the team for the Clinical Trials Unit of Oxford University. Analysis of clinical trials results pertaining to tamoxifen demonstrated not only a decrease in recurrence-free survival for postmenopausal women receiving tamoxifen but also an increase in overall survival. These data were refined, checked and presented again at the National Cancer Institute Consensus Conference in Bethesda, Maryland in 1985, where the panel concluded that adjuvant tamoxifen should be the standard of care for all postmenopausal women with ER-positive primary tumor and positive nodes.[36]

As an aside, this was the year that ICI Pharmaceutical Division (Zeneca) was awarded the start of their "use patent" for tamoxifen originally submitted and denied from 1965 onward (25 years!). The patent would now extend into the 21st century, creating the resources to advance chemoprevention with tamoxifen in the United States and the major clinical trial of anastrozole, in their aromatase inhibitor. Anastrozole versus Tamoxifen and the Combination (ATAC), then the single largest adjuvant endocrine clinical trial, and became pivotal in leading progress with breast cancer therapy.

The overview of the clinical trials with tamoxifen has been published in 1998 and 2005.[37–39] The 1998 and 2005 had three major therapeutic conclusions:

(1) Tamoxifen was only effective as an adjuvant therapy in patients with an ER-positive breast tumor.
(2) Longer was better than short adjuvant therapy in the treatment of ER-positive breast cancer. The power of this principle was best illustrated in premenopausal women receiving tamoxifen monotherapy: one year of adjuvant tamoxifen was completely ineffective at improving either recurrence rates or mortality but five years of it decreased recurrence by 50% and mortality by 30%. The scientific principles published before any of the trials started to recruit patients had veracity.[25]
(3) The concern that the increased incidence of endometrial cancer during long-term adjuvant tamoxifen therapy might significantly reduce the value of tamoxifen as a cheap and effective life-saving medicine was calculated to be incorrect.[37,38]

We will now summarize the 2011 report on the relevance of breast cancer hormone receptors to the efficacy of adjuvant tamoxifen. The meta-analysis of data was derived from 20 randomized clinical trials (n = 21,457) of adjuvant tamoxifen employing a five-year treatment duration (80% compliance). Again, the continuing evaluation of adjuvant tamoxifen demonstrates the veracity of science in "the real world":

(1) The ER-positive disease (n = 10,645) tamoxifen reduced recurrence rates during the first 10 years but thereafter there was no gain or loss out to 15 years.
(2) Marginal ER-positive disease (10–19 femtommoles/mg cytosol protein — from assays no longer used or quantitation employed) recurrence rates were substantial and significant.
(3) The progesterone receptor was of no value in predicting responsiveness to tamoxifen.
(4) Breast cancer mortality was reduced by a third for the first 15 years.
(5) All-cause mortality was substantially reduced despite small increases in thromboembolic and uterine cancer deaths (only women over 55 years of age) in women taking tamoxifen.

Clinical trials will now be summarized to clarify the state of knowledge with the use of aromatase inhibitors versus tamoxifen.

SUMMARY OF ADJUVANT AROMATASE INHIBITOR CLINICAL TRIALS IN POSTMENOPAUSAL PATIENTS WITH ER-POSITIVE BREAST CANCER

Response of advanced breast cancer after surgical removal of the adrenal glands in postmenopausal women indicated that adrenal precursors of estrogen were important stimulators of tumor growth when, as we now know, they are converted to estrogen by the enzyme aromatase in peripheral tissues, including the breast. The chance finding that the antiepileptic drug aminoglutethomide caused adrenal suppression led to its introduction for the treatment of advanced breast cancer after Richard Santen presented the results of a pioneering randomized trial showing that it was equivalent to adrenalectomy, which made this type of extensive surgery obsolete.[40] Subsequently, a number of agents able to reduce postmenopausal estrogen were developed and compared in clinical trials, leading to studies which indicated that the most potent agents which inhibited aromatase, anastrozole, letrozole and exemestone, were the agents of choice and improved survival compared with older adrenal suppressants, aromatase inhibitors and other endocrine therapies such as progestogens and antiestrogens in patients with advanced breast cancer.[41]

Because of several positive features of the modern AIS, such as ease of administration (once a day), relatively low toxicity and positive data from trials in advanced disease, a number of major international and national collaborative trials were initiated to test whether the AIs were superior to tamoxifen and whether they could be integrated with tamoxifen as adjuvant therapy.[42] The first major adjuvant study, initiated by Michael Baum and his colleagues in 1996, was the ATAC trial, which compared anastrozole, tamoxifen and a combination of the two drugs given for five years, with over 3000 patients entered in each arm.[43] The 10-year results of the ATAC trial indicated that the time to breast cancer recurrence was greater from anastrozole compared with tamoxifen in patients with hormone-receptor-positive tumours (HR 0.79, 95% CI 0.70–0.89, $p = 0.0002$) and, interestingly, the absolute differences in time to recurrence increased over time (2.7% at 5 years and 4.3% at 10 years) even though active treatment was for five years only.[44] The combined arm of the trial was stopped when it appeared to be less effective than even tamoxifen — further indication that, with rare exceptions, endocrine therapies given together are inferior to sequential use.[45]

In another major randomized trial, with over 8000 patients, the effectiveness of sequential letrozole and tamoxifen was compared with continuous use of each agent for five years (BIG 1-98 Trial). The results showed that letrozole was significantly superior to tamoxifen with respect to disease-free (HR 0.82, 95% CI 0.74–0.92) and overall survival (HR 0.79, 95% CI 0.69–0.90). The investigators concluded that single-agent letrozole was the treatment of choice, but that either sequence of therapy could be used if clinically indicated since this was not associated with an appreciable decrement in effect.[46] In comparative studies as neoadjuvant therapy there appears to be little clinical difference between the three potent AIs to date,[47] but the results of a randomized clinical trial of a head-to-head comparison of letrozole and anastrozole as adjuvant therapy are awaited and will determine which AI to use in preference to tamoxifen.[48]

THE PLACE OF TAMOXIFEN

Currently, the major indication for tamoxifen is for the endocrine treatment of premenopausal women. Five years of tamoxifen in this group of patients with tumors which are hormone-receptor-positive reduces the risk of recurrence by about one third irrespective of whether chemotherapy is given in addition and the size and ancillary nodal status of the tumor.[39] There is continued debate as to whether ovarian ablation (OA) by surgery or the use of LHRH analogs adds to the effectiveness of tamoxifen, or whether AIs and OA together is superior to tamoxifen. An overview analysis of clinical studies of OA, tamoxifen and OA + tamoxifen was inconclusive[49] and the clinical community is awaiting the results of several trials which are testing a comparison of OA + tamoxifen or either alone or an AI + OA.[50] A randomized trial of OA + tamoxifen vs. OA + anastrozole showed, if anything, that OA + tamoxifen was superior.[51] In view of all these uncertainties, tamoxifen alone remains the endocrine treatment of choice for premenopausal women with estrogen-receptor-positive tumors.

Tamoxifen may also be used in postmenopausal women who are intolerant of AI therapy, and also if a sequential regimen is indicated. For example, if it is unclear whether a woman aged 50 is through the menopause or not, clinically it is wise to start adjuvant treatment with tamoxifen and then change to an AI after 2–3 years, since such sequential treatment is superior to continuing tamoxifen for five years.

Tamoxifen has been off-patent for many years and, since it is now relatively inexpensive, may be used as an effective therapy in developing countries or even in developed countries with fiscal restraint. AIs are also coming off-patent but it remains to be seen how their costs will decline. Ideally, we should confirm by clinical trial to test whether tamoxifen or AIs are effective in populations where they have not been used. Such an approach was used by Richard Love, who confirmed the value of adjuvant OA and tamoxifen in a randomized trial in Vietnam.[52]

Endocrine therapy and chemotherapy are both active in ER-positive breast cancer. The gains for both are similar and each reduces mortality by about one third.[53] Chemotherapy will reduce risk by one third after the benefit of endocrine therapy is calculated. In women with ER-positive disease and a good initial prognosis, the absolute benefits of additional chemotherapy may be very small. Algorithms to assess the benefits of adding chemotherapy include Adjuvant Online,[54] gene arrays[55] and a compendium of clinical and immunophenotypic markers known as ICH4.[56] At present it is not clear which of these (or others) is optimal for prediction.

CAN WE IMPROVE ADJUVANT ENDOCRINE THERAPY?

Although the advances in adjuvant therapy over the past 20 years have been considerable, there is clearly great room for improvement when we consider that mortality is reduced by endocrine therapy by one third. This suggests that the remainder of tumors are resistant to therapy either de novo or acquired after a period of responsiveness.

It is important to understand the differences in outcome between women with ER-positive and ER-negative tumors. ER-negative tumors tend to relapse within the first seven years after primary tumor diagnosis and later relapse is unusual, so that after this time patients can be considered cured. However, the pattern of relapse in ER-positive tumors is quite different and occurs at a rate of 1%–2% per year continuously. This continued relapse rate has led to trials of extending endocrine therapy beyond five years. We await the results of the ATLAS trial comparing 5 with 10 years of tamoxifen. We know that giving 3–4 years of letrozole after 5 years of tamoxifen reduces relapses and improves survival compared with no letrozole[57] but, as yet, we do not have data on extending AIs beyond 5 years, which is clearly an important research question.[58] Another approach to improving endocrine therapy is to attempt to reverse the resistance in tumor cells, a phenomenon which is of major importance in reducing the potential benefits of treatment. Resistance may be intrinsic to the tumor epithelial cell or induced by the activity of adjacent stromal cells. A major cause of intrinsic resistance is activation of growth factor signal transduction pathways within the tumor cell, which is associated with, for example, increased PI3 kinase and mTOR activity. Recent trials with endocrine therapy and mTOR inhibitors together have shown particular promise in reversing resistance in advanced breast cancer.[59] Recently, resistance to tamoxifen was demonstrated in cocultures of stoma and mammary tumor epithelium which could be reversed by agents such as metformin and the Src inhibitor, dasatinib.[60] Clinical trials are currently in progress to test the benefit of adjuvant metformin.[61] It is known that many endocrine agents can change with time from being tumor antagonists to tumor stimulators — translation of these observations into treatment algorithms is currently being explored.[62]

Another approach is to search for endocrine agents more potent than tamoxifen or AIs. Recent studies indicate that the pure antiestrogen fulvestrant given at the appropriate dose is more effective than anastrozole, for advanced breast cancer, and could potentially be used as adjuvant therapy.[63]

An equally important cause of "resistance" is failure of the patient to take medication for the prescribed five or more years. The rates of nonadherence for women prescribed either tamoxifen or anastrozole reported in studies range from 11% to 85%. Lack of compliance is related to side effects and a number of other reasons, including fear of side effects, lack of interest and a dislike of taking medication. It is of the utmost importance to improve adherence, which must involve strategies to improve interactions with patients.[64] This may also involve lifestyle measures, since there have been several reports of a relative reduction in efficacy of AIs in overweight women.[65]

CONCLUSION

Long-term adjuvant endocrine therapy targeted to the ER is a major advance in cancer therapeutics. Nevertheless, the knowledge of the evolution of endocrine resistance has now revealed new targets whose vulnerability can be exploited to decrease mortality further. The fashion of "personalized medicine" to pick the correct agent for the appropriately vulnerable tumor holds significant promise for future adjuvant clinical trials.

ACKNOWLEDGEMENTS

This work (V.C.J.) was supported by the Department of Defense Breast Program under award No. W81XWH-06-1-0590 Center of Excellence; subcontract under SU2C (AACR) grant No. SU2C-AACR-DT0409; the Susan G. Komen for the Cure Foundation under award No. SAC100009; GHUCCTS CTSA (Grant #UL1RR031975) and the Lombardi Comprehensive Cancer Center Support Grant (CCSG) Core Grant NIH P30 CA051008. The views and opinions of the author(s) do not reflect those of the U.S. Army or the Department of Defense.

REFERENCES

1. Beatson G. (1896) On treatment of inoperable cases of carcinoma of the mamma: suggestions for a new method of treatment with illustrative cases. *Lancet* **2**:104–107.
2. Boyd S. (1900) On oophorectomy in cancer of the breast. *Br Med J* **2**:1161–1167.
3. Allen E, Doisy EA. (1923) An ovarian hormone: preliminary repeats on its location, extraction and partial purification and action in test animals. *JAMA* **81**:819–821.
4. Glascock RF, Hoekstra WG. (1959) Selective accumulation of tritium-labelled hexoestrol by the reproductive organs of immature female goats and sheep. *Biochem J* **72**:673–682.
5. Folca PJ, Glascock RF, Irvine WT. (1961) Studies with tritium-labelled hexoestrol in advanced breast cancer. Comparison of tissue accumulation of hexoestrol with response to bilateral adrenalectomy and oophorectomy. *Lancet* **2**(7206):796–798.
6. Jensen EV, Jacobson HI. (1962) Basic guides to the mechanism of estrogen action. *Recent Prog. Horm Res* **18**:387–414.

7. Toft D, Gorski J. (1966) A receptor molecule for estrogens: isolation from the rat uterus and preliminary characterization. *Proc Natl Acad Sci USA* **55**(6): 1574–1581.
8. Jensen EV, *et al.* (1968) A two-step mechanism for the interaction of estradiol with rat uterus. *Proc Natl Acad Sci USA* **59**(2):632–638.
9. Jensen EV, Block GE, Smith S, Kyser K, DeSombre ER. (1971) Estrogen receptors and breast cancer response to adrenalectomy. *Natl Cancer Inst Monogr* **34**:55–70.
10. McGuire WL, Carbone PP, Vollmer EP. (1975) *Estrogen Receptors in Human Breast Cancer*, Raven.
11. Legha SS, Slavik M, Carter SK. (1976) Nafoxidine — an antiestrogen for the treatment of breast cancer. *Cancer* **38**(4):1535–1541.
12. Herbst AL, Griffiths CT, Kistner RW. (1964) Clomiphene citrate (Nsc-35770) in disseminated mammary carcinoma. *Cancer Chemother Rep* **43**:39–41.
13. Company WSM. (1967) Official literature on new drugs: clomiphene citrate (Clomid). *Clin Pharmacol Ther* **8**:891–897.
14. Haddow A, Watkinson JM, Paterson E, Koller PC. (1944) Influence of synthetic oestrogens on advanced malignant sisease. *Br Med J* **2**(4368):393–398.
15. Harper MJ, Walpole AL. (1966) Contrasting endocrine activities of cis and trans isomers in a series of substituted triphenylethylenes. *Nature* **212**(5057):87.
16. Walpole AL, Paterson E. (1949) Synthetic oestrogens in mammary cancer. *Lancet* **2**(6583):783–786.
17. Cole MP, Jones CT, Todd ID. (1971) A new anti-oestrogenic agent in late breast cancer. An early clinical appraisal of ICI46474. *Br J Cancer* **25**(2):270–275.
18. Ward HW. (1973) Anti-oestrogen therapy for breast cancer: a trial of tamoxifen at two dose levels. *Br Med J* **1**(5844):13–14.
19. Jordan VC. (1988) The development of tamoxifen for breast cancer therapy: a tribute to the late Arthur L. Walpole. *Breast Cancer Res Treat* **11**(3):197–209.
20. Harper MJ, Walpole AL. (1967) A new derivative of triphenylethylene: effect on implantation and mode of action in rats. *J Reprod Fertil* **13**(1):101–119.
21. Fisher B, *et al.* (1998) Tamoxifen for prevention of breast cancer: report of the National Surgical Adjuvant Breast and Bowel Project P-1 Study. *J Natl Cancer Inst* **90**(18):1371–1388.
22. Jordan VC. (2003) Tamoxifen: a most unlikely pioneering medicine. *Nat Rev Drug Discov* **2**(3):205–213.
23. Jordan VC. (2006) Tamoxifen (ICI46,474) as a targeted therapy to treat and prevent breast cancer. *Br J Pharmacol* **147** (Suppl 1) :S269–S276.
24. Poirot M. (2011) Four decades of discovery in breast cancer research and treatment: an interview with V Craig Jordan. *Int J Dev Biol* **55**:703–712.
25. Jordan VC. (2008) Tamoxifen: catalyst for the change to targeted therapy. *Eur J Cancer* **44**(1):30–38.

26. Fisher B, *et al.* (1975) 1-phenylalanine mustard (L-PAM) in the management of primary breast cancer: a report of early findings. *N Engl J Med* **292**(3):117–122.
27. Bonadonna G *et al.* (1976) Combination chemotherapy as an adjuvant treatment in operable breast cancer. *N Engl J Med* **294**(8):405–410.
28. Lippman ME, Bolan G. (1975) Oestrogen-responsive human breast cancer in long term tissue culture. *Nature* **256**(5518):592–593.
29. Jordan V, Dix CJ, Allen KE. (1979) The effectiveness of long term tamoxifen treatment in a laboratory model for adjuvant hormone therapy of breast cancer. *Adjuvant Ther Cancer* **2**:19–26.
30. Jordan VC, Allen KE. (1980) Evaluation of the antitumour activity of the non-steroidal antioestrogen monohydroxytamoxifen in the DMBA-induced rat mammary carcinoma model. *Eur J Cancer* **16**(2):239–251.
31. Jordan VC, Koerner S. (1975) Tamoxifen (ICI 46,474) and the human carcinoma 8S oestrogen receptor. *Eur J Cancer* **11**(3):205–206.
32. Brauch H, Jordan VC. (2009) Targeting of tamoxifen to enhance antitumour action for the treatment and prevention of breast cancer: the "personalised" approach? *Eur J Cancer* **45**(13):2274–2283.
33. Furr BJ, Jordan VC. (1984) The pharmacology and clinical uses of tamoxifen. *Pharmacol Ther* **25**(2):127–205.
34. Gottardis MM, Robinson SP, Satyaswaroop PG, Jordan VC. (1988) Contrasting actions of tamoxifen on endometrial and breast tumor growth in the athymic mouse. *Cancer Res* **48**(4):812–815.
35. Fornander T, *et al.* (1989) Adjuvant tamoxifen in early breast cancer — occurrence of new primary cancers. *Lancet* **1**(8630):117–120.
36. Anonymous. (1985) Consensus conference. Adjuvant chemotherapy for breast cancer. *JAMA* **254**(24):3461–3463.
37. EBCTCG. (1998) Tamoxifen for early breast cancer: an overview of the randomised trials. *Lancet* **351**(9114):1451–1467.
38. EBCTCG. (2005) Effects of chemotherapy and hormonal therapy for early breast cancer on recurrence and 15-year survival: an overview of the randomised trials. *Lancet* **365**(9472):1687–1717.
39. Davies C, *et al.* (2011) Relevance of breast cancer hormone receptors and other factors to the efficacy of adjuvant tamoxifen: patient-level meta-analysis of randomised trials. *Lancet* **378**(9793):771–784.
40. Santen RJ, *et al.* (1981) A randomized trial comparing surgical adrenalectomy with aminoglutethimide plus hydrocortisone in women with advanced breast cancer. *N Engl J Med* **305**(10):545–551.
41. Mauri D, Pavlidis N, Polyzos NP, Ioannidis JP. (2006) Survival with aromatase inhibitors and inactivators versus standard hormonal therapy in advanced breast cancer: meta-analysis. *J Natl Cancer Inst* **98**(18):1285–1291.
42. Dowsett M, *et al.* (2010) Meta-analysis of breast cancer outcomes in adjuvant trials of aromatase inhibitors versus tamoxifen. *J Clin Oncol* **28**(3):509–518.

43. Howell A, *et al.* (2005) Results of the ATAC (Arimidex, Tamoxifen, Alone or in Combination) trial after completion of 5 years' adjuvant treatment for breast cancer. *Lancet* **365**(9453):60–62.
44. Forbes JF, *et al.* (2008) Effect of anastrozole and tamoxifen as adjuvant treatment for early-stage breast cancer: 100-month analysis of the ATAC trial. *Lancet Oncol* **9**(1):45–53.
45. Dowsett M, Howell A. (2002) Breast cancer: aromatase inhibitors take on tamoxifen. *Nat Med* **8**(12):1341–1344.
46. Regan MM, *et al.* (2011) Assessment of letrozole and tamoxifen alone and in sequence for postmenopausal women with steroid hormone receptor-positive breast cancer: the BIG 1-98 randomised clinical trial at 8.1 years median follow-up. *Lancet Oncol* **12**(12):1101–1108.
47. Ellis MJ, *et al.* (2011) Randomized phase II neoadjuvant comparison between letrozole, anastrozole, and exemestane for postmenopausal women with estrogen receptor-rich stage 2 to 3 breast cancer: clinical and biomarker outcomes and predictive value of the baseline PAM50-based intrinsic subtype — ACOSOG Z1031. *J Clin Oncol* **29**(17):2342–2349.
48. O'Shaughnessy J. (2007) A decade of letrozole: FACE. *Breast Cancer Res Treat* **105** (Suppl 1):67–74.
49. Cuzick J, *et al.* (2007) Use of luteinising-hormone-releasing hormone agonists as adjuvant treatment in premenopausal patients with hormone-receptor-positive breast cancer: a meta-analysis of individual patient data from randomised adjuvant trials. *Lancet* **369**(9574):1711–1723.
50. Griggs JJ, *et al.* (2011) American Society of Clinical Oncology endorsement of the cancer care Ontario practice guideline on adjuvant ovarian ablation in the treatment of premenopausal women with early-stage invasive breast cancer. *J Clin Oncol* **29**(29):3939–3942.
51. Gnant M, *et al.* (2011) Adjuvant endocrine therapy plus zoledronic acid in premenopausal women with early-stage breast cancer: 62-month follow-up from the ABCSG-12 randomised trial. *Lancet Oncol* **12**(7):631–641.
52. Love RR, *et al.* (2002) Oophorectomy and tamoxifen adjuvant therapy in premenopausal Vietnamese and Chinese women with operable breast cancer. *J Clin Oncol* **20**(10):2559–2566.
53. Peto R, *et al.* (2012) Comparisons between different polychemotherapy regimens for early breast cancer: meta-analyses of long-term outcome among 100,000 women in 123 randomised trials. *Lancet* **379**(9814):432–444.
54. Hajage D, *et al.* (2011) External validation of Adjuvant! Online breast cancer prognosis tool. Prioritising recommendations for improvement. *PLoS One* **6**(11):e27446.
55. Rutgers E, *et al.* (2011) The EORTC 10041/BIG 03-04 MINDACT trial is feasible: results of the pilot phase. *Eur J Cancer* **47**(18):2742–2749.

56. Cuzick J, *et al.* (2011) Prognostic value of a combined estrogen receptor, progesterone receptor, Ki-67, and human epidermal growth factor receptor 2 immunohistochemical score and comparison with the Genomic Health recurrence score in early breast cancer. *J Clin Oncol* **29**(32):4273–4278.
57. Jin H, Tu D, Zhao N, Shepherd LE, Goss PE. (2012) Longer-term outcomes of letrozole versus placebo after 5 years of tamoxifen in the NCIC CTG MA.17 trial: analyses adjusting for treatment crossover. *J Clin Oncol* **30**(7):718–721.
58. Jordan VC, *et al.* (2011) The St. Gallen Prize Lecture 2011: evolution of long-term adjuvant anti-hormone therapy: consequences and opportunities. *Breast* **20** (Suppl 3):S1–S11.
59. Baselga J, *et al.* (2012) Everolimus in postmenopausal hormone-receptor-positive advanced breast cancer. *N Engl J Med* **366**(6):520–529.
60. Martinez-Outschoorn UE, *et al.* (2011) Anti-estrogen resistance in breast cancer is induced by the tumor microenvironment and can be overcome by inhibiting mitochondrial function in epithelial cancer cells. *Cancer Biol Ther* **12**(10):924–938.
61. Goodwin PJ, Stambolic V. (2011) Obesity and insulin resistance in breast cancer—chemoprevention strategies with a focus on metformin. *Breast* **20** (Suppl 3):S31–S35.
62. Jordan VC, Ford LG. (2011) Paradoxical clinical effect of estrogen on breast cancer risk: a "new" biology of estrogen-induced apoptosis. *Cancer Prev Res (Phila)* **4**(5):633–637.
63. Howell A, Sapunar F. (2011) Fulvestrant revisited: efficacy and safety of the 500-mg dose. *Clin Breast Cancer* **11**(4):204–210.
64. Banning M. (2012) Adherence to adjuvant therapy in post-menopausal breast cancer patients: a review. *Eur J Cancer Care (Engl)* **21**(1):10–19.
65. Sestak I, *et al.* (2010) Effect of body mass index on recurrences in tamoxifen and anastrozole-treated women: an exploratory analysis from the ATAC trial. *J Clin Oncol* **28**(21):3411–3415.

NOTE ADDED IN PROOF

The Adjuvant Tamoxifen: Long against Short (ATLAS) trial comparing 5 years vs. 10 years of tamoxifen demonstrates superiority for 10 years of tamoxifen in the decade after tamoxifen is stopped (Davis *et al.* 2012 *Lancet*). It is estimated that there is a 50% decrease in mortality in the second decade following diagnosis and after 10 years of adjuvant tamoxifen compared to no treatment.

11

PHARMACOGENOMICS OF TAMOXIFEN

Hiltrud B. Brauch and V. Craig Jordan

Abstract

The discovery of tamoxifen as an anticancer drug more than four decades ago and the constant gain of knowledge about its metabolism and mechanistic action together with long term clinical experience could finally yield strategies for individualized endocrine treatment of breast cancer. This notion evolves from tamoxifen efficacy being subject to genetic predisposition, which explains the interindividual variability of outcome in adjuvant postmenopausal breast cancer treatment. This chapter introduces the evolution of knowledge that tamoxifen's efficacy depends on the formation of clinically active metabolites 4-hydroxytamoxifen and endoxifen, which have a greater affinity to the estrogen receptor and ability to control cell proliferation as compared to the parent drug. It explains why the cytochrome P450 (CYP) 2D6 enzyme plays a key role in this biotransformation and why lack of tamoxifen efficacy can be linked to low CYP2D6 activity. There is now considerable mechanistic, pharmacologic and clinical pharmacogenetic evidence in support of the notion that CYP2D6 genetic variants and phenocopying effects through drug interaction by CYP2D6 inhibitors influence plasma concentrations of active tamoxifen metabolites and negatively impact tamoxifen outcome. These interrelations are particularly critical for patients with nonfunctional (poor metabolizer) and severely impaired (intermediate metabolizer) CYP2D6 variants, and also for patients in need of comedications such as serotonin reuptake inhibitors to control adverse effects like hot flashes and other menopausal symptoms. A personalized approach for an optimal tamoxifen benefit should therefore consider a CYP2D6 genotype-guided adjuvant endocrine treatment strategy and avoid nonadherence as well as strong CYP2D6 inhibitors as comedications.

By introducing the evolution of knowledge about tamoxifen metabolism and the key role of the drug metabolizing enzyme CYP2D6, the chapter explains the current understanding of the metabolic activation of tamoxifen to its putative active agent, endoxifen. It follows that tamoxifen resistance can be explained at the level of genetic predisposition of variable drug metabolism, a rationale underlying the pharmacogenomic principle that explains the interindividual variability of outcome in adjuvant postmenopausal breast cancer tamoxifen treatment. The focus is on the veracity of the hypothesis of CYP2D6 deficiency affecting the pharmacokinetics of tamoxifen and the interpretation of the available clinical evidence of the effects of CYP2D6 polymorphisms on tamoxifen outcome and the relevance of phenocopying effects through comedications such as selective serotonin reuptake inhibitors (SSRIs). Finally, we summarize the actions necessary for selecting appropriate long-term endocrine treatment for postmenopausal breast cancer patients with ER positive disease, and for improving the value of tamoxifen as a "personalized targeted treatment" within the context of improving women's health.

Keywords: CYP2D6; Metabolic Activation; Tamoxifen Metabolites; 4-Hydroxytamoxifen; Endoxifen; Selective Serotonin Reuptake Inhibitors.

INTRODUCTION

Tamoxifen, a nonsteroidal antiestrogen,[1] is used for the treatment of all stages of breast cancer[2-4] and in the U.S. it is available for reducing the incidence of breast cancer in both pre- and postmenopausal women at elevated risk.[5-7] It is important to remember that during early clinical studies tamoxifen did not show any improvement in efficacy over standard hormonal treatments (high-dose estrogen or androgen) for metastatic breast cancer.[2,8] The only advantage of tamoxifen was a reduced incidence of side effects for those 30% of patients who responded for about one year. However, laboratory studies to target the tumor estrogen receptor (ER)[9] employ long-term adjuvant therapy,[10] and consider that the chemoprevention of breast cancer[11,12] transformed and reinvented tamoxifen from an orphan drug to the gold standard for the endocrine treatment of breast cancer between 1984 and 2004. The targeting of tamoxifen to block estrogen-stimulated breast tumor growth with long-term (five years) adjuvant tamoxifen therapy[13] resulted in a major improvement in patient survivorship and has contributed significantly to the reduction in national death rates from breast cancer.[14,15] The recent development of aromatase inhibitors (AIs) as an effective treatment for breast cancer in postmenopausal patients[16] has improved disease-free survival and reduced the side effects of endometrial cancer and blood clots noted with tamoxifen.[17-20] However, AIs are not universally available in national healthcare systems worldwide because of significant financial constraints. Tamoxifen remains a cheap, life-saving, targeted therapy for both pre- and postmenopausal patients with breast cancer.

The application of the ER as a tumor target to treat breast cancer patients appropriately provided a valuable, but admittedly not perfect, test to increase the probability of tumor growth control during long-term adjuvant treatment. Tamoxifen does not enhance either disease-free or overall survival in patients with ER-negative tumors.[14,15]

At present, there are no universally accepted tumor markers to improve response rates for patients with ER-positive tumors. However, tamoxifen presents an early example in oncology research to demonstrate a relationship between genome variation and drug treatment outcome toward the individualization of medicine. The background is that, as with all major classes of drugs given at standard doses, a substantial portion of patients do not respond, respond only partially or experience adverse drug reactions. Drug concentrations in plasma can vary more than 600 times between two individuals of the same weight on the same drug dosage. Such variations can be of genetic, physiological, pathophysiological or environmental origin.[21] Pharmacogenomics employs the genome variability (polymorphisms) as an important contributor to the variability in drug response and addresses the question why a drug may work well in one person but have no effect in another and, moreover, why one person may tolerate a drug well while another suffers from severe side effects.[22] Pharmacogenomic testing by way of DNA and/or drug metabolite analyses aims for the prediction of treatment outcome. Its integration into clinical practice will require proof of patient benefit toward guided treatment strategies.

TAMOXIFEN METABOLISM IN HUMANS

Evolution of Knowledge About Metabolism

All of the antiestrogens that were discovered before 1977 had a low affinity for the ER.[23,24] Indeed, it was believed that nonsteroidal antiestrogens needed to have a low affinity for the ER in order to be antiestrogens. In other words, an effective complex could not be maintained to initiate biological effects. The discovery of 4-hydroxytamoxifen (4-OH Tam) changed this perspective, as this metabolite had a high affinity for the ER, and it was clear from structure–function relationships that this was an advantage, but not a requirement for metabolic activation to produce antiestrogenic activity.[25–27] The twists and turns of the story that deciphered tamoxifen metabolim have recently been reviewed.[28] Based upon the discovery of high-affinity antiestrogens, this subsequently changed the approach of medicinal chemists to the discovery of new compounds that became known as selective estrogen receptor modulators (SERMs).[29,30]

Tamoxifen, orally administered as tamoxifen citrate, is a prodrug that requires activation by hepatic drug-metabolizing enzymes to exert its therapeutic effect. The complex metabolism involves CYP enzymes, i.e. CYP3A4/5, CYP2D6, CYP2B6, CYP2C9 and CYP2C19, for the conversion of tamoxifen into primary metabolites N-desmethyltamoxifen and 4-OH Tam[31] (Fig. 1). Oxidation of the abundant N-desmethyltamoxifen results in 4-hydroxy-N-desmethyltamoxifen, also called endoxifen, with CYP2D6 being a leading enzyme (Fig. 1). 4-hydroxylation of tamoxifen increases the affinity for the ER[26,27,32,33] and it has been shown that 4-OH Tam and endoxifen are equipotent with respect to ER binding and inhibition of 17β-estradiol-induced cell proliferation. Recent gene array analyses of the spectrum of genes activated or suppressed by 4-OH Tam and endoxifen in MCF-7 breast cancer cells showed that both metabolites induce similar changes on global gene expression patterns.[34] More than 4000 genes are either up- or downregulated by estradiol, and in the copresence of 4-OH Tam or endoxifen the numbers are 2444 and 2390, respectively. Due to overlapping effects on 1365-estradiol-sensitive-genes, it has been concluded that the effects of 4-OH Tam

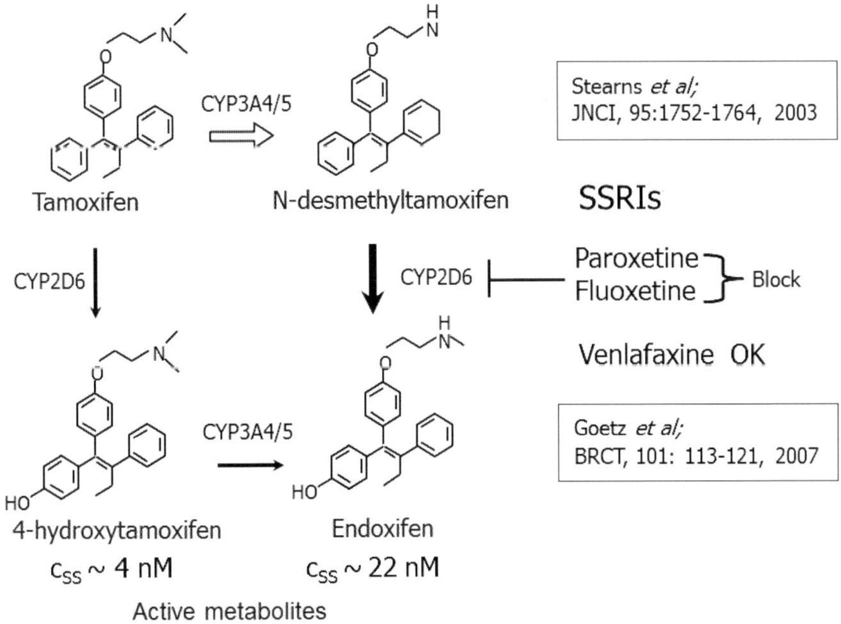

Figure 1. Tamoxifen metabolism and formation of the antiestrogenic metabolites 4-hydroxytamoxifen and endoxifen. The cytochrome P450 (CYP) enzyme CYP2D6 is the key enzyme in the formation of endoxifen, which is considered the prominent active metabolite due to higher steady-state plasma concentrations (C_{SS} ~ 22 nM) as compared to 4-hydroxytamoxifen (C_{SS} ~ 4 nM). The strength of the arrows shows that the formation of endoxifen mainly depends on the conversion of N-desmethyltamoxifen by CYP2D6. This metabolic step can be blocked by selective serotonine reuptake inhibitors (SSRIs, i.e. paroxetine and fluoxetine) due to CYP2D6 inhibition.[61,80]

and endoxifen are almost identical.[34] However, endoxifen has been found at more than sixfold higher concentrations in the plasma of tamoxifen-treated patients as compared to 4-OH Tam. Considering the identical ER binding profile and antiproliferative action of both metabolites as well as similar gene expression patterns[32] but higher circulating endoxifen levels than that of 4-OH Tam in patients,[35,36] and based on the law of mass action, it is currently anticipated that endoxifen is the principal metabolite in blocking the binding of estradiol at the tumor ER in breast cancer patients treated at the 20 mg daily dose of tamoxifen.[35]

An intriguing aspect of the investigations into the molecular pharmacology of endoxifen is the possibility that the antiestrogen may target ERα for rapid destruction in breast cancer cells.[37] The implication is that the shape of the endoxifen–ERα complex is perturbed significantly for rapid proteasomal degradation. Profound structural perturbations of the ER are noted with the pure antiestrogen ICI 164384[38] and the SERM GW5638,[39] and both compounds cause the rapid destruction of the ER. In contrast, the structure of endoxifen is almost identical to that of the related metabolite 4-OH Tam (Fig. 1), which causes accumulation of the ER. The structure of the 4-OH Tam–ER complex has been resolved,[40] and thus crystallization of the endoxifen–ER complex could provide further insight into the actions of endoxifen at the ER.

The Role of Cytochrome P450 2D6 (CYP2D6)

CYP2D6 is involved in the metabolism of many clinically important drugs, including β-blockers, antiarrythmics, antihypertensives, antipsychotics, antidepressants, opioids and others.[41,42] A recent analysis of the routes of elimination for the "top 200 drugs" in the U.S. (http://www.rxlist.com; most frequently prescribed drugs, April 2008) showed that of these, 15% were CYP2D6 substrates. Importantly, CYP2D6 is also at the heart of tamoxifen action due to its prominent role in the conversion of the nonactive N-desmethyltamoxifen into clinically active endoxifen (Fig. 1). There is considerable variability in the CYP2D6 drug-metabolizing capacity, which is regarded to be responsible for the clinically observed interindividual variability of drug treatment outcomes. To understand tamoxifen efficacy beyond known effects related to tumor ER expression,[15,43–45] it is important to understand CYP2D6 phenotypes and their underlying genetic variation, which differ by alleles, and their frequencies across ethnic groups. A first CYP2D6 phenotype distinct from an extensive metabolizer (EM) phenotype was identified by way of clinical observation more than 30 years ago (i.e. sparteine/debrisoquine polymorphism) and was termed a "poor metabolizer" (PM).[46–48] Since then, based on drug oxidation capacity, the four different CYP2D6 phenotypes EM, intermediate metabolizer (IM), PM and ultrarapid metabolizer (UM) have been identified, and their frequencies and global distributions have been established.[41,49-51] Not all CYP2D6 phenotypic variations can be attributed to underlying genetic variations; however, 10% of the more than 100 known variants (http:www.cypalleles.ki.se) have clinical relevance.

The PM phenotype is present in approximately 8% of the European population, with PM individuals carrying two nonfunctional (null) alleles leading to a loss of enzyme function (Table 1). Of the more than 20 known null alleles, CYP2D6 *3, *4, *5, *6 and *7 are prevalent in populations of European descent, with *4 being present in 80%–90% of all PMs. PM status, i.e. lack of catalytic function, can be deduced with greater than 99% certainty from the presence of two nonfunctional alleles, and therefore can be accurately predicted from the patient's genotype without the need to phenotype.[42,49,51] Another class of alleles accounts for severely impaired CYP2D6 expression and function, i.e. IM, due to the presence of CYP2D6 *9, *10 and *41 alleles.[42,51-53] Approximately 10% of Europeans are IM, being either homozygous or compound-heterozygous for an IM allele in combination with one null allele (Table 1). The EM phenotype results from the presence of one or two alleles with normal expression level and catalytic function, such as CYP2D6 *1, *2 and *35, and represents the most frequent phenotype within the European population, accounting for approximately 80%. The large group of EMs can be separated into homozygous or heterozygous EMs, depending on whether they carry two (EM/EM) or one (hetEM) functional allele (Table 1). HetEMs, carrying one *1 or *2 or *35 allele in combination with an IM or PM allele have a somewhat impaired enzyme expression and function, a reason why they have been classified as "intermediate metabolizers," assuming a gene dose effect. However, due to the substantial overlap, both in enzyme content and activity between EM/EM and hetEMs, this is not correct and therefore the predictive value of the hetEM genotype is rather poor. Importantly, the IM is a phenotype and genotype distinct from the hetEM based on the presence of *9, *10 and *41 and/or nonfunctional alleles.[51-54] The UM phenotype is present in approximately 10% of the European population and a gene

Table 1. Nomenclature of CYP2D6 phenotypes and frequencies in Germany*.

CYP2D6 phenotype	Definition	Frequency	Prevalent variants	Variant	Allelic frequency (%)
PM	Lacking active enzyme function, i.e. carriers of two non function (null) alleles (PM/PM)	8.3%	*3	2549Adel	1.9
			*4	1846A	22.1
			*5	Gene deletion	1.8
			*6	1707Tdel	0.7
			*7	2935C	0.2
IM	Impaired enzyme function, i.e.	9.7%	*9	2613-2615delA	2.5
	a) carriers of any two impaired function alleles (IM/IM)		*10	100T	2.0
	b) carriers of one impaired function and one PM allele (IM/PM)		*41	2988A*** −1584C, 2850T, 4180C	9.0
EM	Full catalytic function, i.e. homozygous carriers of full function alleles (EM/EM)	37.2%	*1	—	35.7
			*2		17.1
			*35		6.3
UM	Increased enzyme activity: i.e. carriers of duplicated functional alleles (EM/EMxN)	1.2%	*1/*1×N	Gene duplication	0.6**
			*2 /*1×N	Gene duplication	0.1**
			*1/*2×N	Gene duplication	0.4**
Het EM	Impaired enzyme function, i.e. carriers of one EM in combination with one impaired function or loss of function allele, respectively (EM/IM or EM/PM)	43.5%	PM, IM, EM alleles see above		

*Schroth *et al. Clin Cancer Res* 2010 (frequencies have been established with Roche AmpliChip CYP450 Test).
**The prevalence for UM is based on genotypes.
***Based on Raimundo *et al. Clin Pharmacol Ther* 2004 (not included in commercial AmpliChip).

duplication with up to 13 gene copies involving *1 and *2 alleles has been identified as an underlying molecular mechanism. Such an increase in enzyme activity can have profound consequences for the plasma concentrations of drug metabolites;[55,56] however, only 20%–30% of the UM phenotypes in the Caucasian population are accessible through genotyping and therefore the predictive value is rather low.[41,49,51,57] Nomenclature and definition of CYP2D6 phenotypes, i.e. PM, IM, EM, UM, hetEM as well as their expected frequencies in European populations (Table 1), were obtained from the genotyping of 33 CYP2D6 alleles in 492 German individuals (964 chromosomes).[58] It follows that tamoxifen pharmacogenomics in patients of European descent must primarily focus on PM and IM to identify those patients with impaired and/or lack of CYP2D6 function, but also on UM to identify patients with increased enzyme activity. In Asian populations, where PMs play a minor role but the *10 allele is prevalent (i.e. 57% in Han Chinese), tamoxifen pharmacogenomics requires a focus on IM. In northeastern African populations, where gene duplication is prevalent (e.g. 29% in Ethiopia and 21% in Saudi Arabia as compared to 1%–5% in Europeans), tamoxifen pharmacogenomics must focus on UM.[41,42,50,51]

CYP2D6 POLYMORPHISM AND TAMOXIFEN EFFICACY

Given the complexity of CYP2D6 phenotypic variability within and across populations, a thorough investigation of the CYP2D6 tamoxifen pharmacogenomics would require the correlation of the patient's tamoxifen steady-state metabolite plasma levels with the corresponding CYP2D6 genotypes. Moreover, both genotype and metabolite levels should be correlated with clinical outcomes to assess any potential for outcome prediction. As of today, such a genotype–phenotype–outcome correlation derived from the same study cohort is not available. A reason may be that clinical trials for the investigation of tamoxifen outcome were initiated a long time ago and no biological materials for the later conduct of pharmacogenomic studies were collected. In contrast, more recent clinical trials may have addressed this issue; however, their clinical follow-up time, as yet, may not be sufficient to yield meaningful results. However, the increasing awareness of the importance of the issue has sparked a number of investigations for the prediction of tamoxifen efficacy and outcome in early breast cancer. Owing to the lack of "perfect" patient cohorts, the currently available patient-based evidence comes from a number of retrospective studies, which investigated the association between concentrations of active tamoxifen metabolites and CYP2D6 genotypes or the association between CYP2D6 genotype and clinical outcome. It is worth noting that drug–drug interactions and tamoxifen adherence have been recognized to be critical confounders.

CYP2D6 Genotype and Endoxifen Production

The underlying rationale of the tamoxifen pharmacogenomic principle requires that CYP2D6 variants encoding an enzyme with reduced or absent enzyme function may be associated with lower plasma concentrations of active tamoxifen metabolites, therefore impacting the efficacy of tamoxifen. An initial study of 12 breast cancer patients with adjuvant tamoxifen showed lower baseline endoxifen concentrations for patients with variant CYP2D6.[36] Larger pharmacokinetic studies substantiated the role of the CYP2D6 genotype in the formation of endoxifen from tamoxifen *in vivo*. In a study of 80 patients with newly diagnosed breast cancer, Jin and colleagues measured tamoxifen and endoxifen plasma concentrations following initiation of adjuvant therapy.[35] They found a CYP2D6 gene–dose effect on steady-state endoxifen plasma concentrations in that PM/PM patients had the lowest, EM/EM patients the highest, and EM/PM patients intermediate concentrations ($P = 0.003$). Borghes *et al.* extended this study toward a quantitative assessment in 158 patients genotyped at 33 CYP2D6 alleles and confirmed these observations for PM, IM and UM ($P < 0.001$).[59] In both studies, coadministered antidepressants (e.g. paroxetine and fluoxetine) known to be potent CYP2D6 inhibitors (Fig. 1) significantly reduced the formation of endoxifen in EM patients. Studies from Norway (151 pre- and postmenopausal patients; $P < 0.005$[60]) and Germany (236 postmenopausal patients; $P = 10^{-14}$)

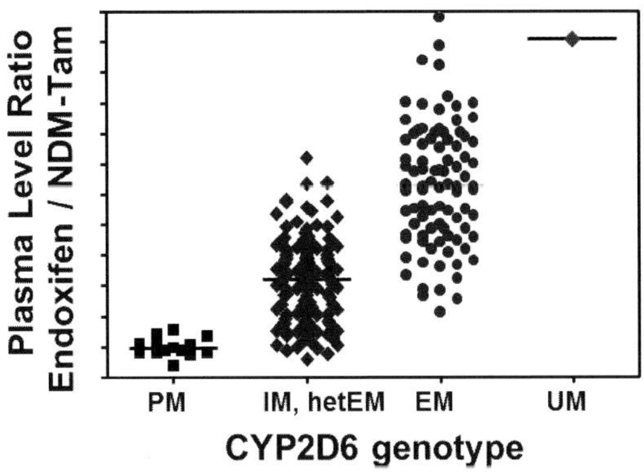

Figure 2. Gene-dose effect of steady-state endoxifen derived from 236 patients with early breast cancer treated with 20 mg/day adjuvant tamoxifen. Patients with no or low function alleles (PM) have the lowest, patients with impaired enzyme activity (IM and hetEM) have intermediate, and patients with two or more full function alleles (EM and UM) have the highest endoxifen plasma levels expressed as a metabolic ratio of the precursor metabolite N-desmethyltamoxifen (NDM-Tam).[61]

confirmed the gene-dose effect on the formation of endoxifen[61] (Fig. 2). Interestingly, the Norwegian study reported a similar effect for 4-OH Tam, but this was not the case in the German study. Similar studies from East Asia, where the CYP2D6*10 reduced function allele is prevalent, provided supportive evidence for this gene-dose relationship. In a prospective Korean study of early metastatic breast cancer, an association between the IM CYP2D6*10 homozygous variant and lower steady-state plasma concentrations was shown for 4-OH Tam and endoxifen (202 patients; $P < 0.0001$, respectively).[62] For 4-OH Tam this has also been shown in a small Chinese study with pre- and postmenopausal patients (37 patients; $P = 0.04$).[63]

CYP2D6 Genotype and Clinical Outcome

The variation of CYP2D6 activity determined by genotype or pharmacokinetics is expected to translate into effects on clinical outcome, and this has been the subject of intensive research. More than one and half a dozen original epidemiologic studies[58,62–78] and numerous reviews[79-87] have addressed this issue in the past few years. The majority of studies provided evidence in favor of this relationship, but few showed no or counter evidence. The wide interstudy heterogeneity mainly attributable to differences in study size and patient inclusion criteria, nonuniform outcome measures and limited CYP2D6 allele coverage makes an interstudy comparison difficult and hampers conclusions.

The first evidence linking CYP2D6 variants with treatment response came from a U.S. prospective randomized phase III trial of postmenopausal women with primary ER-positive breast cancer (North Central Cancer Treatment Group adjuvant breast cancer trial 89-30-52) for the investigation of the effect of adding the androgen fluoxymestrone for 1 year to the standard 5-year adjuvant tamoxifen (20 mg/day). Goetz *et al.*[67] isolated genomic DNA from paraffin-embedded tissue specimens of patients treated with 20 mg/daily adjuvant tamoxifen (n = 223 of 256 eligible) and the genotyped CYP2D6 variants *4 and *6. Statistical analysis was based on 190 patients for whom CYP2D6 *4 genotyping was possible [137 (72.1%) had wt/wt, 40 (21.1%) wt/*4 and 13 (6.8%) *4/*4 genotype]. After a median follow-up of 11.4 years, the CYP2D6 *4/*4 was associated with worse relapse-free (P = 0.023) and disease-free survival (P = 0.012). No association with overall survival was observed. The authors confirmed these findings in an extended study of 256 patients.[88] In the same year, a robust association between CYP2D6 genotypes and treatment outcome was published by Schroth *et al.*[73] They compared 206 breast cancer patients treated with adjuvant tamoxifen monotherapy (standard dose) and 280 patients without tamoxifen. A comprehensive genotyping of the CYP2D6 variants *4, *5, *10 and *41 in DNA derived from formalin-fixed paraffin-embedded normal breast tissues covered the vast majority of expected PM and IM genotypes (e.g. 95% and 90%, respectively). At a median follow-up of 71 months, carriers of the CYP2D6 *4, *5 *10 and *41 alleles had significantly more breast cancer recurrences, shorter relapse-free time (P = 0.02) and worse event-free survival (P = 0.02) compared to carriers of functional alleles. These associations were not observed in patients without tamoxifen.[73]

Other European studies apparently supportive of these findings differed with respect to inclusion criteria such as menopausal or high-risk status. For example, a British study of hereditary breast cancer patients suggested a relationship between CYP2D6 PM status and worse survival for *BRCA2* but not *BRCA1* carriers. However, the numbers of patients in this study were small and ER-negative patients had been included, making it impossible to distinguish a poor prognosis from a pharmacogenetic effect.[70] Similarly, a small nested cohort study of the Italian breast cancer tamoxifen chemoprevention trial reported that the frequency of the CYP2D6 *4/*4 genotype was significantly higher in women who developed breast cancer compared to those who did not (i.e. 3:1). Yet, there were only four patients with this genotype and, moreover, the chemoprevention setting may not compare with the adjuvant postmenopausal treatment setting — reasons why conclusions must be regarded as premature.[65] Furthermore, results of studies from The Netherlands[64] and Spain[66] may also be regarded as inconclusive due to small numbers of study participants and divergent study designs such as inclusion of high proportions of metastatic patients and patients treated with chemotherapy. Altogether, the results from these four European studies cannot contribute to clarifying the validity of CYP2D6-tamoxifen pharmacogenomics.

Three studies from Asia, i.e. Japan, China and Korea, where the CYP2D6*10 (IM) allele is prevalent, found that patients with IM/IM genotypes have a poor outcome of tamoxifen. Kiyotani *et al.* investigated 282 pre- and postmenopausal patients (an extension of their previous study[69] and reported significant risks of recurrence for carriers of one variant and for homozygous variant carriers, in support of a gene-dose effect ($P = 0.000036$).[78] A study of 152 Chinese women treated with adjuvant tamoxifen also found that patients with a CYP2D6 *10/*10 genotype had a significantly worse disease-free survival than a control group of 141 patients without tamoxifen ($P = 0.005$).[63] For metastatic breast cancer, a small Korean study of 21 patients showed that the median time to progression for *CYP2D6**10/*10 patients was significantly shorter than for patients with either *CYP2D6* wt/*10 or wt/wt genotypes ($P = 0.0032$). All six patients with progressive or stable disease lasting less than 24 weeks carried the *10/*10 genotype (P = 0.0186).[62] Although the sample sizes of some Asian studies were small and endpoints as well as patient characteristics and treatment settings varied, their findings of an implication of CYP2D6 genotypes predictive for tamoxifen outcome are in line with the large studies from Europe and the U.S.

No favorable association, i.e. no association between CYP2D6 *4 and tamoxifen response or breast cancer prognosis, was reported by a U.S. study.[71] Inverse findings were reported by a Swedish study showing a decrease of recurrences for carriers of the CYP2D6 *4 variant, which was then confirmed, although not significant, in their larger study (P = 0.055).[75,76] It is worth noting that these studies were limited to the analysis of CYP2D6 *4, which according to the recent findings by Schroth *et al.*[58] considerably limits the power to detect the CYP2D6–tamoxifen pharmacogenomic relationship (see below). Moreover, the study inclusion criteria differed with respect to tamoxifen dose and duration as well as concomitant chemotherapy, which may be regarded as additional confounders.[77] Similar arguments may apply to a study from Japan[72] reporting on a negative CYP2D6–tamoxifen association. In that study, however, only 42% of the participants were first-time tamoxifen users and the median follow-up time was below five years.

These uncertainties and seeming contradictions prompted Schroth *et al.* to perform additional studies[58,74] to scrutinize the importance of the study design, both at the level of clinical inclusion criteria and comprehensive CYP2D6 allele coverage for accurate CYP2D6 phenotype assignment. They were the first to employ a power calculation which, based on the previously observed strongest effect between PM and EM, required more than 1250 patients (e.g. 166 events) to detect a hazard ratio (HR) of 1.85 (originally reported by Goetz *et al.*[67] with a statistical power of 90% and one-sided α level. Thus, in their new study, they analyzed 1325 postmenopausal, hormone-receptor-positive patients from Germany and the U.S., all treated with adjuvant monotamoxifen (20 mg/day intended for 5 years). This number exceeded that of most other studies by 5–16 times, and it is worth noting that the quality of this study has been rated "good" in a recent meta-analysis of 11 tamoxifen–CYP2D6 outcome studies, with 6 studies rated "poor" and 3 "fair".[87] The preponderance of the study, despite its 20% overlap with previous studies,[67,68,73] is given by an extended clinical follow-up for the documentation of breast events and a more comprehensive CYP2D6 allele coverage for accurate phenotype assignments. Biological materials for the CYP2D6 genotyping included in 44% whole blood, 7% fresh frozen tissue and 49% formalin-fixed paraffin-embedded tissue samples. The results provide strong evidence in favor of patients with loss (PM) and impaired CYP2D6 activity (hetEM/IM) having higher time to recurrence and event-free survival rates ($P < 0.001$ and $P = 0.0029$, respectively; Fig. 3). The PM risk for breast cancer recurrence was close to twofold and this risk, albeit somewhat lower, was also observed in the group of patients who were carriers of any CYP2D6 deficiency or impaired allele. The observed linear increase in recurrence (EM 14.9%, hetEM/IM 20.9%, PM 29.0%) and mortality rates (EM 16.7%, hetEM/IM 18.0%, PM 29.0%) for increasing numbers of CYP2D6

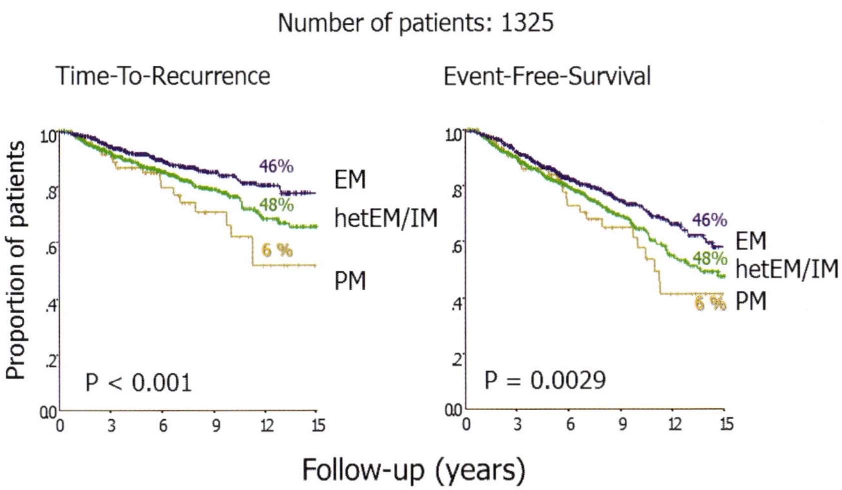

Figure 3. Kaplan-Meier estimates of time to recurrence and event-free survival show that *CYP2D6* polymorphism is a determinant of tamoxifen efficacy in early breast cancer. Among 1325 patients, there were 46% EM patients with two functional *CYP2D6* alleles, including patients with ultrarapid metabolism; 48% hetEM/IM patients with intermediate or one poor metabolism (PM) allele; and 6% PM patients homozygous for poor metabolism alleles. Curves were truncated at 15 years after diagnosis, and calculations included overall time (median follow-up 6.3 years).[74]

nonfunction/deficiency alleles stresses the allele dose effect in support of the CYP2D6-tamoxifen pharmacogenomic relationship. The HR for recurrence after adjustment for nodal status and tumor size was close to twofold for PM and almost 1.5-fold for het/EM as compared to EM, respectively. A comparison of PM with all patients (not stratified by CYP2D6 genotype) at 9 years' follow-up showed an 11.5% increase of recurrences, while for EM a 2% decrease was shown (Fig. 4). There was no association with overall survival.[74]

Schroth *et al.*[58] then investigated whether comprehensive CYP2D6 allele coverage is needed to unanimously detect the CYP2D6-tamoxifen pharmacogenomic relationship, and whether a limited coverage may provide a clue to explain previous studies with counterevidence. Results based on the in-depth genotyping of 33 CYP2D6 alleles (AmpliChip CYP450 Test, Roche Diagnostics, Branchburg) of the patient subcohort, for which blood-derived DNA

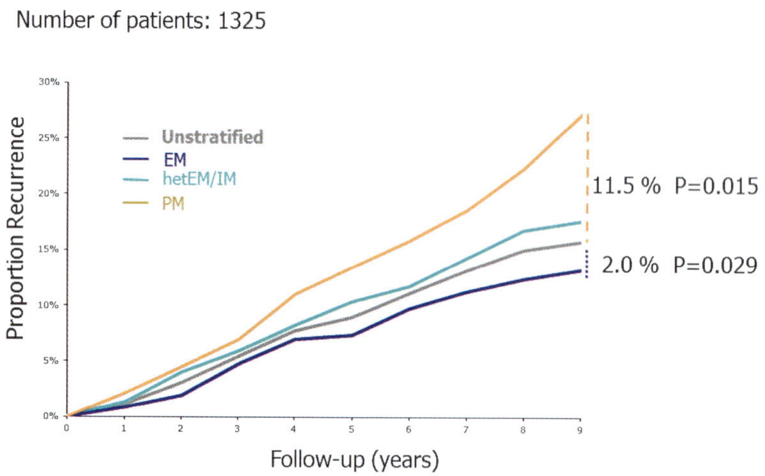

Figure 4. CYP2D6 polymorphism and recurrence probabilities upon adjuvant tamoxifen treatment. At nine years' follow-up the recurrence rate for patients homozygous for poor metabolism alleles (PM) was 11.5% higher compared to the entire patient group not stratified by the *CYP2D6* genotype ($P = 0.015$). In contrast, EM patients with two functional *CYP2D6* alleles (including patients with ultrarapid metabolism) had a 2% better recurrence rate ($P = 0.029$). For hetEM/IM patients (carriers of at least one impaired function or null function allele), the difference was less pronounced.[117]

was available (492 patients), underscored the importance of comprehensive CYP2D6 allele coverage for accurate phenotype assignment and subsequent recurrence risk assessment. Only now did it become clear that about one third of patients will be misclassified when the analysis is limited to the *4 allele. This analysis provided results reminiscent of some of the studies reporting no association. In contrast, the inclusion of all CYP2D6 reduced function alleles increased the PM-associated HR from 1.33 ($P = 0.58$) to 2.87 ($P = 0.006$) and the power from 7.8% to 63.2% (Fig. 5).[58] While the combined studies of Schroth *et al.*[58,73,74] provide important solutions toward the interpretation of the CYP2D6-tamoxifen pharmacogenomic relationship, a recent study by Kiyotani *et al.*[77] contributed evidence for chemotherapy being a confounder in CYP2D6-tamoxifen pharmacogenomics. The authors investigated the influence of concomitant treatment such as chemotherapy, LH-RH agonists or AIs and showed no significant association between CYP2D6 genotype and recurrence-free survival in a cohort of 167 patients with these comedications.[77] These findings underscore the necessity of an expedient study design of monotamoxifen treatment for the investigation of the CYP2D6-tamoxifen pharmacogenomic relationship as it has been continuously employed in the German[58,73] and German-U.S. studies.[74]

In the future it will be interesting to see whether the small difference in the outcome benefit of AI as compared to tamoxifen recently reported in the landmark trials BIG1-98[89] and ATAC[19,44] can be attributed to the lack of CYP2D6 genotype stratification. This possibility should be considered particularly in the light of insights from a biomathematical modeling exercise of the estimated benefit of tamoxifen according to CYP2D6 status. Using the BIG 1-98 information on recurrence probabilities and assuming that AI metabolism was CYP2D6-independent, it has been suggested that the benefit of five years of adjuvant tamoxifen may even exceed that of upfront AI treatment in postmenopausal CYP2D6 wt/wt patients.[90] Based on a more comprehensive CYP2D6 phenotype definition, another modelling by Schroth *et al.*[74] superimposed a hypothetical AI survival curve based on the reported HR of 0.76 for time to recurrence

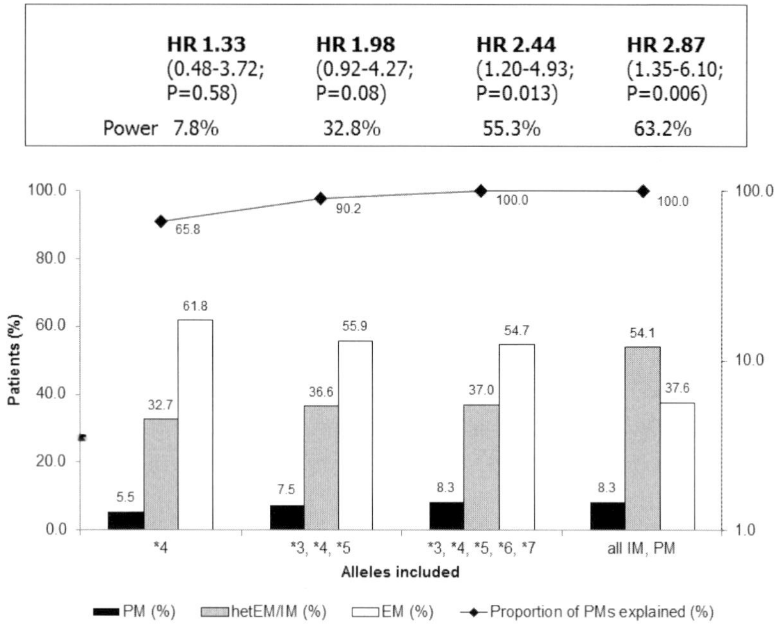

Figure 5. CYP2D6 phenotype predictability and increase of power to predict breast cancer recurrence by increasing numbers of *CYP2D6* variants in the analysis. Bars correspond to the proportion of PM (black), hetEM/IM (gray) and EM (white) patients within the tamoxifen patient cohort as a function of an increasing number of PM alleles *3, *4, *5, *6 and *7, or inclusion of IM alleles. Diamonds indicate the probability of a PM detection based on allele inclusion (plotted on the log scale, right axis). Boxed numbers indicate nonadjusted hazard ratios (HR) and 95% confidence intervals for recurrence of PM versus EM patients together with corresponding *P* values from Cox regression, depending on selected CYP2D6 alleles. The associated power of these Cox models is given as a measure of sensitivity and has been calculated for a significance level of 5% and with HR, number of events and proportion of phenotype groups depending on variant inclusion. EM was defined by patients without variant allele, with 6 UM included in this group.[58]

between anastrozole and tamoxifen from the ATAC trial[44] onto the time-to-recurrence curve for tamoxifen obtained by Schroth *et al.*[74] This not only showed that EM patients have the capacity to benefit from tamoxifen but that this benefit is likely to be comparable to that of the AI (Fig. 6).

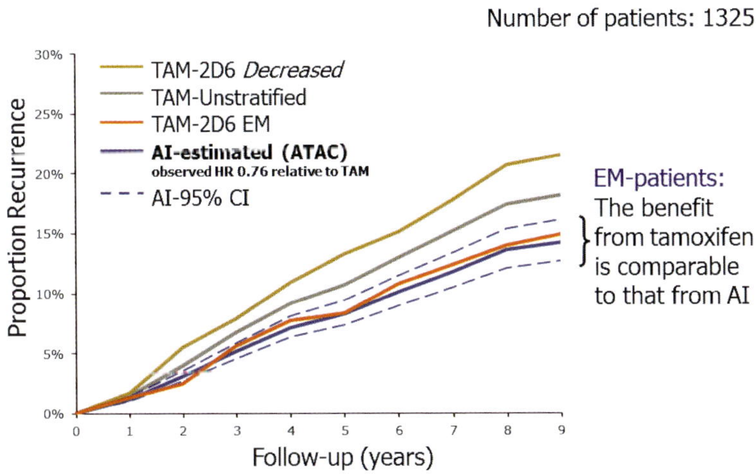

Figure 6. Kaplan-Meier estimates of time to recurrence comparing tamoxifen with a hypothetical AI curve. Nonadjusted, heterogeneity-corrected Kaplan-Meier estimates for the CYP2D6 phenotypes *Decreased* (gold line) and EM (blue line) as well as the entire tamoxifen cohort unselected by genotype (gray line); a 95% CI interval is shown for EM patients (blue dotted lines). EM patients have two functional CYP2D6 alleles; *Decreased* refers to patients with any intermediate or PM alleles. Making the Cox proportionality hazard assumption, a hypothetical AI survival curve (black line) was estimated based on a hazard ratio of 0.76 for anastrozole relative to tamoxifen[44] and the Kaplan-Meier estimate of the entire tamoxifen cohort.[74]

FACTORS UNDERMINING TAMOXIFEN EFFICACY AND OUTCOME

Side Effects, Adherence and Drug Interaction

It is clear that patients must complete a five-year course of tamoxifen, because five years of tamoxifen is superior to one or two years of adjuvant treatment. This principle is elegantly demonstrated in the overview analysis of clinical trials for premenopausal patients with ER-positive breast cancer (Fig. 7)[15,74] although, in general, the rates of tamoxifen adherence are higher than those observed for other medications, discontinuation of adjuvant tamoxifen in older women with ER-positive breast cancer has been evaluated. Five-year discontinuation rates were 23% in a randomized clinical trial of adjuvant

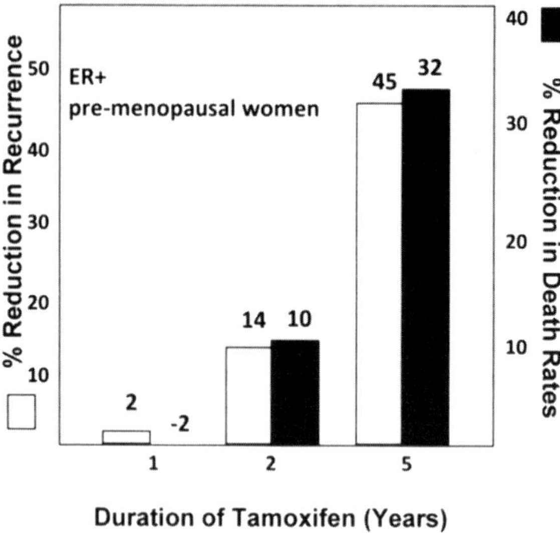

Figure 7. The influence of the durations of adjuvant tamoxifen therapy administered to premenopausal patients with estrogen receptor (ER)–positive (+) breast cancer.[15] The enhancement of a reduction of recurrences and a reduction of death rates in women taking only one year of adjuvant tamoxifen compared to five years serves to illustrate the benefits of the drug, the need for compliance, and the need to ensure that patients are not poor metabolizers by virtue of aberrations of CYP2D6 or phenocopying by taking SSRIs to reduce menopausal symptoms.[80]

therapy[3,74] and 24% as well as 40% in primary prevention trials.[91,92] In clinical practice, discontinuation rates range from 15% to 50%.[93-97] Healthcare-data-based analyses revealed that as many as half of patients stop their medication in the course of the five-year adjuvant treatment with tamoxifen and as many as 15% and 22% of patients stop taking tamoxifen during the first year.[98-100]

There is now evidence that low tamoxifen adherence has profound negative effects on patient outcome. A retrospective cohort study of 2080 incident breast cancer cases in the Tayside region of Scotland with 1633 patients on prescribed tamoxifen reported on the relationship between adherence and survival.[96] The median duration of tamoxifen use was 2.42 years and patients were usually highly adherent (93%). A longer tamoxifen use was associated with better survival (HR 0.85, 95% CI 0.83–0.87); however, a significant proportion of patients (19%) had low adherence levels of less than 80%, which is the equivalent of missing at least one tablet every five days, and these patients were at increased risk of death (HR 1.10, 95% CI 1.001–1.21). There was no difference with respect to social class, but the authors observed a trend for a higher proportion of younger women to have low adherence ($P < 0.001$).[96] Interestingly, a Dutch cohort of 1962 patients with early breast cancer treated with tamoxifen between 1994 and 2006 and investigated for the effect of concomitant CYP2D6 inhibitor use during tamoxifen treatment linked adherence with lower event-free time (HR 0.987, 95% CI 0.975–0.99, $P = 0.029$).[101] In this study, adherence was defined as the proportion of days covered by tamoxifen in the first year following initiation.

The main obstacle to compliance with tamoxifen is unacceptable side effects such as severe hot flashes (Fig. 8) and related menopausal symptoms.[93] Because hot flashes are an indicator of tamoxifen efficacy, the patient's lack of compliance imposes once more an obstacle to successful treatment. Supportive evidence came from the Women's Healthy Eating and Living trial (WHEL),[102] which enrolled 3088 primary breast cancer patients (aged 18–70) between 2 and 48 months after initial diagnosis to either dietary intervention ($n = 1537$) or observation ($n = 1551$) alone. Among the 864 women taking tamoxifen, 78% reported hot flashes, which in the vast majority (69%) were accompanied by night sweats. The remainder reported neither hot flashes nor night sweats (18%) or night sweats without hot flashes (4%). Patients reporting hot flashes at baseline were less likely to develop recurrent breast cancer than those who did not report hot flashes (12.9% vs. 21%, $P = 0.01$; 127 women had a confirmed breast cancer recurrence after 7.3 years' follow-up). Moreover, hot flashes were more predictive of outcome than age, grade, hormone receptor status or stage II cancer.[102] Goetz *et al.* in their study[67] showed that the incidence of hot flashes during adjuvant tamoxifen improved therapeutic outcomes and correlated with the CYP2D6 genotype. In particular, none of the patients with a CYP2D6 *4/*4 genotype (0/13) reported moderate or severe hot flashes compared to 20% (36/177) in the patient group with heterozygous (*4/wt) and homozygous wild-type genotype ($P = 0.064$). Accordingly, hot flashes can be attributed to higher

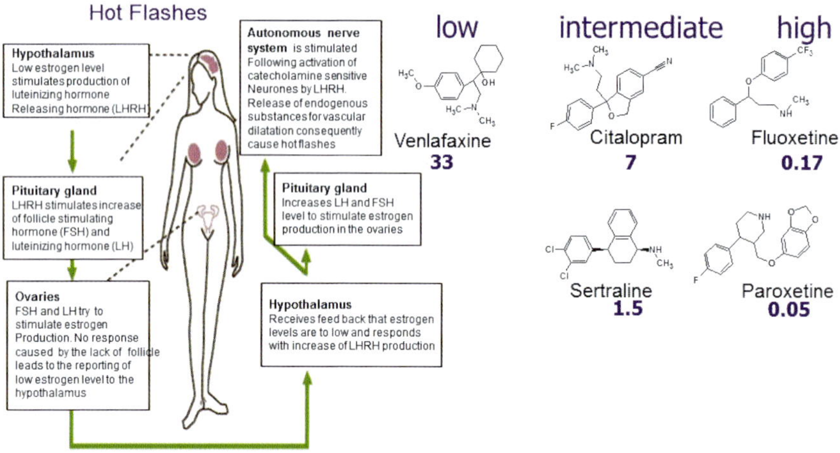

Figure 8. Selective serotonin reuptake inhibitors (SSRIs) and selective epinephrine reuptake inhibitor (SNRI; venlafaxine) used to ameliorate hot flushes and menopausal symptoms during tamoxifen therapy. SSRIs are CYP2D6 substrates and compete with N-desmethyltamoxifen for the CYP2D6 enzyme. They can be classified into high, intermediate and low binding substrates for the CYP2D6 enzyme. The inhibitor constants for venlafaxine (low), citalopram (intermediate) and sertraline (intermediate), fluoxetine and paroxetine (high) are 33, 7, 1.5, 0.17 and 0.05, respectively.[80]

tamoxifen metabolite levels in patients with functional CYP2D6 and drug efficacy. The additional link between the genotype and the occurrence of hot flashes under tamoxifen expands the knowledge about the genotype-based interindividual variability of the formation of steady-state endoxifen plasma concentrations[35,59,103] and both must be considered relevant to adjuvant tamoxifen outcome. It is of the utmost importance that patients experiencing hot flashes maintain and complete adjuvant tamoxifen (five years) despite the discomfort of adverse reaction.

To aid compliance patients are routinely prescribed selective serotonin reuptake inhibitors (SSRIs; Fig. 8), which reduce menopausal symptoms.[104–106] This, however, imposes a new challenge, because it is known that some SSRIs have a high affinity for the CYP2D6 enzyme[107,108] and therefore SSRIs can inhibit CYP2D6 activity and interfere with tamoxifen efficacy by blocking the conversion of tamoxifen to endoxifen (Fig. 1).The relative inhibitory concentrations of SSRIs for the CYP2D6 enzyme product are noted in Fig. 8. The effect of SSRIs on the plasma levels of endoxifen was initially reported by Stearns *et al.*[36] and was subsequently linked to the patients' CYP2D6 genotype by Jin *et al.*[35] Plasma concentrations after four months of tamoxifen therapy were significantly lower in patients with a CYP2D6 homozygous variant (20 nM, 95% CI 11.1–28.9 nM) or heterozygous genotype (43.1 nM, 95% CI 33.3–52.9 nM) than those with homozygous wild-type (78.0 nM, 95% CI 65.1–90.1 nM) (both $P = 0.003$). In their study, 24 of 78 patients took CYP2D6 inhibitors, including paroxetine, fluoxetine, sertraline, citalopram, amiodarone and metoclopramide. Among patients who carried a homozygous wild-type genotype, the mean plasma endoxifen concentration for patients using CYP2D6 inhibitors was 58% lower than that for patients not using SSRI comedication (38.6 nM vs. 91.4 nM; $P = 0.0025$), and in patients who were heterozygous for a nonfunctional CYP2D6 allele (wt/vt) this difference was 38% (31.0 nM vs. 51.7; $P = 0.08$). Moreover, women taking the weak CYP2D6 inhibitor venlafaxine (a serotonine norepinephrine reuptake inhibitor, SNRI) had slightly reduced plasma endoxifen concentrations compared to women taking the potent CYP2D6 inhibitor paroxetine.[35] Borges *et al.*[59] in their quantitative study of the influence of CYP2D6 PM, IM and UM genotypes on endoxifen plasma concentrations, reported no significant differences for tamoxifen, N-desmethyltamoxifen and 4-hydroxytamoxifen between users and nonusers of concomitant CYP2D6 inhibitors. However, the mean endoxifen plasma concentration was significantly lower in patients who took CYP2D6 inhibitors than in patients who did not (39.6 +/– 28.4 nmol/L vs. 71.5 +/– 41.2 nmol/L; $P < 0.01$).[59] When the authors divided the CYP2D6 inhibitors into potent (paroxetine, fluoxetine; $n = 19$) and weak [SSRs: sertraline and citalopram ($n = 14$) as well as celecoxib, diphenydramine and chlorpheniramine [$n = 13$)], they found a more pronounced decrease in mean endoxifen plasma concentrations with potent inhibitors than with weak inhibitors. Concomitant use of venlafaxine, which is considered the least potent inhibitor, did not show any significant effect. Taking into account CYP2D6 genotypes, the authors observed that the mean plasma endoxifen concentration was significantly lower in CYP2D6 EM patients who were taking potent CYP2D6 inhibitors than in patients who were not (23.5 +/− 9.5 nmol/L vs. 84.1 +/− 39.4 nmol/L; $P < 0.001$).[59] Thus, CYP2D6 genotype and concomitant potent CYP2D6 inhibitors are highly associated with plasma endoxifen concentrations and may substantially impact outcome during tamoxifen treatment by phenocopying effects, i.e. converting an EM into a PM phenotype status.

CYP2D6 Enzyme Activity and Clinical Outcomes

The CYP2D6 phenocopying effect of SSRIs on clinical outcome was explored by Goetz *et al.*[68] in their follow-up of the NCCTG trial. They investigated the role of CYP2D6 inhibitors in 256 patients that had been randomized to the tamoxifen-alone arm. Patients with the CYP2D6 wt/wt genotype who did not take CYP2D6 inhibitors were classified as EM ($n = 115$), whereas patients with either one or two *4 alleles or those taking a CYP2D6 inhibitor were classified as IM or PM ($n = 65$), depending on the strength of the inhibitor. Following these assignments, patients with decreased metabolism had shorter time to breast recurrence ($P = 0.015$) as well as shorter relapse free ($P = 0.007$) and disease-free ($P = 0.009$) survival than those with extensive CYP2D6 metabolism.[68] There was no significant effect on overall survival. The authors concluded that CYP2D6 metabolism, as measured by genetic variation and enzyme inhibition, is an independent predictor of breast cancer outcome in postmenopausal primary breast cancer patients receiving adjuvant tamoxifen. Accordingly, outcome during tamoxifen treatment may be influenced by its pharmacogenetics as well as coprescription of drugs interfering with the CYP2D6-mediated tamoxifen metabolism. Recently, a population-based, retrospective cohort study including 2430 breast cancer patients (66 years or older) in Ontario, Canada, treated with tamoxifen has been conducted on the influence of a coprescribed single SSRI antidepressant on the risk of death from breast cancer.[109] The following SSRIs were under investigation: paroxetine ($n = 630$; 25.9%), sertraline ($n = 541$, 22.3%), citalopram ($n = 467$; 19.2%), venlafaxine ($n = 365$; 15.0%), fluoxetine (n = 253; 10.4%) and fluvoxamine (n = 174; 7.2%). During follow-up (mean 2.38 years, SD 2.59 years), 374 patients (15.4%) died of breast cancer. Among all SSRIs, paroxetine use was identified to increase the risk of death from breast cancer, and this risk increased with the proportion of time of cotreatment. The proportion of time of tamoxifen with overlapping use of paroxetine of 25%, 50% and 75% increased the risk of death by 24%, 54% and 91% ($P < 0.05$ each). This study supports the hypothesis that paroxetine can reduce or abolish the benefit of tamoxifen in women with breast cancer.[109] In contrast, a Dutch study of 1962 early-stage breast cancer patients using tamoxifen, which investigated the effect of concomitant CYP2D6 inhibitor use and tamoxifen adherence on breast cancer recurrence, did not show an effect. The study included 150 patients (7.6%) with frequent use (equal to or more than 60 days) of CYP2D6 inhibitors. Among them, 110 patients used strong inhibitors (mainly fluoxetine, n = 25, or paroxetine, n = 82), 26 patients moderate ones and 14 patients weak ones. No effect of concomitant CYP2D6 inhibitor use has been observed during tamoxifen therapy for the endpoint event-free time. However, as the authors pointed out, their study may not have been sufficiently powered to detect such an effect. Rather, in this study, shorter event-free time (HR 0.987, 95% CI 0975–0.99, P = 0.029) has been linked with poor tamoxifen adherence.[101] With respect to the weak CYP2D6 inhibitor citalopram, a case control study nested in the population of female residents of Denmark involving patients with nonmetastatic ER-positive breast cancer (diagnosed between 1994 and 2001) treated at least one year with tamoxifen investigated the influence of citalopram on breast cancer recurrence (follow-up time 5–10 years).[110] Citalopram use was similar between cases with breast cancer recurrence (37/329 — 11.2%) and cases without recurrence (35/331 — 10.6%). The study provides clinical epidemiological support for the hypothesis that citalopram taken concurrently with tamoxifen does not reduce tamoxifen's protective effect against breast cancer recurrence. Similar negative associations for other, less frequently used SSRIs, such as fluoxetine, paroxetine, sertraline and escitalopram, were observed; however, numbers were small.[110] Currently, it appears that there might be differences in the observations of CYP2D6 inhibition by SSRIs and effect on outcomes between North American and European studies. Interestingly, a recent study from Israel reported no differences for overall recurrence rates between users and nonusers of CYP2D6-inhibiting drugs.[111] However, based on the recommendations of the proposals by the "Grading of Recommendations Assessment, Development and Evaluation Working Group," it is prudent to avoid strong CYP2D6 inhibitors in the setting of tamoxifen so as to avoid potential detrimental effects.[112]

CONCLUSIONS

Tamoxifen has been one of the highlights of mechanistic (targeted) cancer therapeutics over the past 40 years, saving the lives of many thousands of breast cancer patients annually. However, the variation of response to and the occurrence of side effects from tamoxifen have been recognized from the beginning and consequently led to intensive research on the underlying mechanisms. The increasing understanding of genetically based resistance to tamoxifen is a first step toward safer and more effective breast cancer therapy.

In summary, we can conclude that endoxifen is formed by the CYP2D6 enzyme,[31,36,113,114] and it is therefore anticipated that aberrant genotypes and other medicines that are metabolized by the same enzyme impair the actions of tamoxifen in patients and determine clinical outcomes.[35] This chapter has addressed the veracity of the hypothesis from the current literature to explore the possibility of targeting tamoxifen to improve women's health.

There is now strong evidence that hot flashes are indicators of tamoxifen efficacy and that tamoxifen efficacy and outcome depend on the drug's metabolism, which is subject to CYP2D6 genotype and pharmacointeractions. Data from numerous studies yield consistent results in linking plasma concentrations of active tamoxifen metabolites with genetically determined CYP2D6 metabolizer status, interference with strong CYP2D6 inhibitors as well as clinical outcome.[35,36,59,61,68] Most notably, considerable attention has been given to establishing a link between CYP2D6 genotype and tamoxifen outcome, because this could provide a powerful diagnostic tool for upfront outcome prediction should this association be proven. Most authors agree on this relationship[58,62,63,67–70,73,74,77,78] but some data initially appeared to be conflicting.[71,75,76] This has largely been resolved by recent new studies from Germany[58,74] and Japan[77,78] that provided powerful arguments for the lack of comparability between most published studies. As a consequence, the current insight into the CYP2D6–tamoxifen pharmacogenomic relationship is based on a few studies that implemented critical baseline issues. Disapproval of other studies refers to small study sizes and nonuniform patient inclusion criteria such as inclusion of mainly premenopausal patients or patients with metastatic diseases, insufficient CYP2D6 allele coverage and subsequent false phenotype assignments, additional chemotherapy being a strong confounder, and, importantly, inappropriate clinical endpoint analyses. Following a stringent stratification based on study quality, there is now strong support for the notion that genetic CYP2D6 variants negatively impact tamoxifen efficacy and treatment outcome[58,63,74,77,78] and that this negative impact can be phenocopied by CYP2D6 inhibition through prescribed comedications such as antidepressants.[68,112]

There are a number of potential clinical consequences from these emerging data. First of all, strict compliance with tamoxifen treatment is critical for efficacy and outcome, and therefore deviations from the prescribed course of adjuvant tamoxifen must be avoided even when side effects occur. Second, potent SSRIs such as paroxetine or fluoxetine should not be used for the relief of hot flashes in breast cancer patients receiving tamoxifen. Even though SSRIs are among the few evidence-based therapy options for menopausal vasomotor symptoms,[115] available data from the U.S. and Canada,[36,68,109,112] but not from Europe[101,110] and Israel,[111] indicate that they may compromise tamoxifen efficacy due to their interference with CYP2D6-dependent tamoxifen metabolism. Yet, this interference depends on the strength of the CYP2D6 inhibitor. If treatment of hot flashes is indicated, patients should be on the safe side with an SSRI such as citalopram or escitalopram or an SNRI such as venlafaxine, because these substances showed no significant inhibition of CYP2D6.[35,110,112] Third, the CYP2D6 genotype/phenotype treatment–outcome relationship points to the possible benefit of upfront CYP2D6 genotyping prior to the initiation of endocrine treatment. A comprehensive, robust, standardized and quality-controlled CYP2D6 genotyping test will need to analyze all relevant genetic variants that may affect tamoxifen metabolism, which should include PM alleles (i.e. *3, *4, *5, *6, *7; Fig. 5) and IM alleles, depending on the individual's ethnic origin (Table 1).[58] It is worth noting that *41 is the most frequent IM allele in Europeans, *17 is the principal IM allele in Africans and *10 dominates in Asians [*9 should also be considered (Table 1)].[41,42,50,51] Other areas of interest with respect to clinical application are the measurement of endoxifen plasma levels as a surrogate of the CYP2D6 phenotype and a possible dose increase of tamoxifen to overcome impaired CYP2D6 metabolism; however, the latter option requires further investigations before definite conclusions can be made. Note that no data and therefore no instructions for premenopausal and male breast cancer can be inferred.

In October 2006, the U.S. Food and Drug Administration (FDA) Clinical Pharmacology Subcommittee of the Advisory Committee for Pharmaceutical Sciences reviewed the relationship between tamoxifen and CYP2D6, and recommended that the package insert for tamoxifen be amended to warn postmenopausal women of the potential increased risk of treatment failure for patients with deficient CYP2D6 activity and that certain SSRIs can interfere with the bioactivation of tamoxifen.[84] Given alternative treatment options, i.e. tamoxifen versus AI, and considering the available scientific and clinical evidence, an individualized approach to endocrine treatment of postmenopausal breast cancer patients would be feasible. One might speculate that tamoxifen alone may be adequate for CYP2D6 EM/EM carriers whereas patients with variant CYP2D6 alleles (PM and possibly hetEM/IM) may fare better with upfront AI therapy. Indeed, the *in vivo* pharmacology and retrospective clinical trial data make a strong case for the use of the CYP2D6 genotype to guide selection of tamoxifen in postmenopausal breast cancer. However, there is still some resistance to formal recommendation within the clinical community and this may be explained by the lack of data available from prospective studies. Such a study has not been considered in the past owing to the lack of financial support and potentially low interest in the improved use of off-patent medications.The results of other large, retrospective tamoxifen pharmacogenetic trials comparing tamoxifen and AI outcome could, however, add to the predictive relevance of CYP2D6. It will be interesting to learn whether the small difference in the outcome benefit of AI as compared to tamoxifen recently reported from the landmark trials BIG 1-98[89] and ATAC[19,44] can be attributed to the lack of CYP2D6 genotype stratification. Such anticipated data may help the FDA and other regulatory agencies to reach a decision on the label warning of tamoxifen and, moreover, help to clarify the advice given to clinicians in favor of or against genotype-guided selection of antiestrogenic therapy for postmenopausal breast cancer patients.

Finally, the personalized approach to targeting tamoxifen in postmenopausal patients seems reasonable and should await timely translation into clinical practice. Indeed, the CYP2D6 genotype might be one of the first predictors of therapeutic response in cancer care. Because this approach is genome-based by utilizing CYP2D6 genotyping for the prediction of a patient's metabolizer phenotype, ethical issues need to be sufficiently addressed. In the light of acceptable alternatives for postmenopausal patients with early-stage breast cancer, an informed choice about adjuvant endocrine treatment and, most importantly, avoidance of a therapy that might potentially lack efficacy must be of prime interest. It will therefore be important to make patients and their caretakers aware of these issues and also initiate discussions with regulatory authorities. Because no alternatives exist for premenopausal patients and patients with male breast cancer, CYP2D6 genotyping currently cannot be considered for these tamoxifen settings.

FINAL THOUGHTS AND REFLECTIONS

The reason for this chapter being substantially longer than the others is that the topic of the pharmacogenomics of tamoxifen is particularly controversial and there is intense interest both in the scientific community and in the patient community. We felt that the case for and against genotyping needed to be made in full. However, there are other important advances that have occurred through the focused investigation of tamoxifen metabolites that ran in parallel with the clinical development of tamoxifen to become the gold standard for the treatment of breast cancer for the two decades 1980–2000. The discovery of 4-OH Tam[26] not only gave an insight into the metabolic activation of tamoxifen, but also provided the scientific community with the "standard" nonsteroidal antiestrogen to be used in all laboratory studies *in vitro* for the next four decades. The foundation for all that we know about the mechanism of action of nonsteroidal antiestrogens really flowed from this early work. The x-ray crystallography of the 4-OH Tam ER ligand-binding domain was crystallized by Shiau *et al.*[40] (Chapter 6).

Perhaps most importantly for a book entitled *Estrogen Action, Selective Estrogen Receptor Modulators (SERMs) and Women's Health*, the subsequent explosion of interest in SERMs not only with high affinity for the ER, but also a rapid excretion rate, has really led to the development of the whole new drug group that addresses the treatment and prevention of multiple diseases in women's health. Indeed, if the metabolism of tamoxifen had not been deciphered and linked to biology in the mid 1970s, keoxifene (raloxifene) would not have been synthesized for development as an anticancer agent with less estrogen-like properties than tamoxifen. Keoxifene's failure and the subsequent finding that it preserved bone density in oophorectomized animals[116] encouraged confirmation and propelled keoxifene into the clinic as raloxifene for the treatment and prevention of osteoporesis, whilst reducing the risk of breast cancer and giving no concerns about endometrial safety.

REFERENCES

1. Jordan VC. (2003) Tamoxifen: a most unlikely pioneering medicine. *Nat Rev Drug Discov* **2**:205–213.
2. Cole MP, Jones CT, Todd ID. (1971) A new anti-oestrogenic agent in late breast cancer. An early clinical appraisal of ICI 46474. *Br J Cancer* **25**:270–275.
3. Fisher B, Costantino J, Redmond C, Poisson R, Bowman D, Couture J, *et al.* (1989) A randomized clinical trial evaluating tamoxifen in the treatment of patients

with node-negative breast cancer who have estrogen-receptor-positive tumors. *N Engl J Med* **320**:479–484.

4. NATO. (1983) Controlled trial of tamoxifen as adjuvant agent in management of early breast cancer. Interim analysis at four years by Nolvadex Adjuvant Trial Organisation. *Lancet* **1**:257–261.
5. Cuzick J, Powles T, Veronesi U, Forbes J, Edwards R, Ashley S, *et al.* (2003) Overview of the main outcomes in breast-cancer prevention trials. *Lancet* **361**:296–300.
6. Fisher B, Costantino JP, Wickerham DL, Cecchini RS, Cronin WM, Robidoux A, *et al.* (2005) Tamoxifen for the prevention of breast cancer: current status of the National Surgical Adjuvant Breast and Bowel Project P-1 study. *J Natl Cancer Inst* **97**:1652 1662.
7. Powles TJ, Ashley S, Tidy A, Smith IE, Dowsett M. (2007) Twenty-year follow-up of the Royal Marsden randomized, double-blinded tamoxifen breast cancer prevention trial. *J Natl Cancer Inst* **99**:283–290.
8. Ingle JN, Ahmann DL, Green SJ, Edmonson JH, Bisel HF, Kvols LK, *et al.* (1981) Randomized clinical trial of diethylstilbestrol versus tamoxifen in postmenopausal women with advanced breast cancer. *N Engl J Med* **304**:16–21.
9.. Jordan VC, Koerner S. (1975) Tamoxifen (ICI 46,474) and the human carcinoma 8S oestrogen receptor. *Eur J Cancer* **11**:205–206.
10. Jordan VC, Allen KE. (1980) Evaluation of the antitumour activity of the non-steroidal antioestrogen monohydroxytamoxifen in the DMBA-induced rat mammary carcinoma model. *Eur J Cancer* **16**:239–251.
11. Jordan VC. (1976) Effect of tamoxifen (ICI 46,474) on initiation and growth of DMBA-induced rat mammary carcinomata. *Eur J Cancer* **12**:419–424.
12. Powles TJ, Hardy JR, Ashley SE, Farrington GM, Cosgrove D, Davey JB, *et al.* (1989) A pilot trial to evaluate the acute toxicity and feasibility of tamoxifen for prevention of breast cancer. *Br J Cancer* **60**:126–131.
13. Jordan VC. (2008) Tamoxifen: catalyst for the change to targeted therapy. *Eur J Cancer* **44**:30–38.
14. EBCTCG. (1998) Tamoxifen for early breast cancer: an overview of the randomised trials. Early Breast Cancer Trialists' Collaborative Group. *Lancet* **351**:1451–1467.
15. EBCTCG. (2005) Effects of chemotherapy and hormonal therapy for early breast cancer on recurrence and 15-year survival: an overview of the randomised trials. *Lancet* **365**:1687–1717.
16. Jordan VC, Brodie AM. (2007) Development and evolution of therapies targeted to the estrogen receptor for the treatment and prevention of breast cancer. *Steroids* **72**:7–25.
17. Coombes RC, Hall E, Gibson LJ, Paridaens R, Jassem J, Delozier T, *et al.* (2004) A randomized trial of exemestane after two to three years of tamoxifen therapy in postmenopausal women with primary breast cancer. *N Engl J Med* **350**:1081–1092.

18. Goss PE, Ingle JN, Martino S, Robert NJ, Muss HB, Piccart MJ, *et al.* (2003) A randomized trial of letrozole in postmenopausal women after five years of tamoxifen therapy for early-stage breast cancer. *N Engl J Med* **349**:1793–1802.
19. Howell A, Cuzick J, Baum M, Buzdar A, Dowsett M, Forbes JF, *et al.* (2005) Results of the ATAC (Arimidex, Tamoxifen, Alone or in Combination) trial after completion of 5 years' adjuvant treatment for breast cancer. *Lancet* **365**:60–62.
20. Thurlimann B, Keshaviah A, Coates AS, Mouridsen H, Mauriac L, Forbes JF, *et al.* (2005) A comparison of letrozole and tamoxifen in postmenopausal women with early breast cancer. *N Engl J Med* **353**:2747–2757.
21. Eichelbaum M, Ingelman-Sundberg M, Evans WE. (2006) Pharmacogenomics and individualized drug therapy. *Annu Rev Med* **57**:119–137.
22. Evans WE, Relling MV. (2004) Moving towards individualized medicine with pharmacogenomics. *Nature* **429**:464–468.
23. Korenman SG. (1970) Relation between estrogen inhibitory activity and binding to cytosol of rabbit and human uterus. *Endocrinology* **87**:1119–1923.
24. Skidmore J, Walpole AL, Woodburn J. (1972) Effect of some triphenylethylenes on oestradiol binding *in vitro* to macromolecules from uterus and anterior pituitary. *J Endocrinol* **52**:289–298.
25. Lieberman ME, Jordan VC, Fritsch M, Santos MA, Gorski J. (1983) Direct and reversible inhibition of estradiol-stimulated prolactin synthesis by antiestrogens *in vitro*. *J Biol Chem* **258**:4734–4740.
26. Jordan VC, Collins MM, Rowsby L, Prestwich G. (1977) A monohydroxylated metabolite of tamoxifen with potent antioestrogenic activity. *J Endocrinol* **75**:305–316.
27. Allen KE, Clark ER, Jordan VC. (1980) Evidence for the metabolic activation of nonsteroidal antioestrogens: a study of structure–activity relationships. *Br J Pharmacol* **71**:83–91.
28. Jordan VC. (2007) New insights into the metabolism of tamoxifen and its role in the treatment and prevention of breast cancer. *Steroids* **72**:829–842.
29. Jordan VC. (2003) Antiestrogens and selective estrogen receptor modulators as multifunctional medicines. 1. Receptor interactions. *J Med Chem* **46**:883–908.
30. Jordan VC. (2003) Antiestrogens and selective estrogen receptor modulators as multifunctional medicines. 2. Clinical considerations and new agents. *J Med Chem* **46**:1081–1111.
31. Desta Z, Ward BA, Soukhova NV, Flockhart DA. (2004) Comprehensive evaluation of tamoxifen sequential biotransformation by the human cytochrome P450 system *in vitro*: prominent roles for CYP3A and CYP2D6. *J Pharmacol Exp Ther* **310**:1062–1075.
32. Johnson MD, Zuo H, Lee KH, Trebley JP, Rae JM, Weatherman RV, *et al.* (2004) Pharmacological characterization of 4-hydroxy-N-desmethyl-tamoxifen, a novel active metabolite of tamoxifen. *Breast Cancer Res Treat* **85**:151–159.

33. Lim YC, Desta Z, Flockhart DA, Skaar TC. (2005) Endoxifen (4-hydroxy-N-desmethyl-tamoxifen) has anti-estrogenic effects in breast cancer cells with potency similar to 4-hydroxy-tamoxifen. *Cancer Chemother Pharmacol* **55**:471–478.
34. Lim YC, Li L, Desta Z, Zhao Q, Rae JM, Flockhart DA, *et al.* (2006) Endoxifen, a secondary metabolite of tamoxifen, and 4-OH-tamoxifen induce similar changes in global gene expression patterns in MCF-7 breast cancer cells. *J Pharmacol Exp Ther* **318**:503–512.
35. Jin Y, Desta Z, Stearns V, Ward B, Ho H, Lee KH, *et al.* (2005) CYP2D6 genotype, antidepressant use, and tamoxifen metabolism during adjuvant breast cancer treatment. *J Natl Cancer Inst* **97**:30–39.
36. Stearns V, Johnson MD, Rae JM, Morocho A, Novielli A, Bhargava P, *et al.* (2003) Active tamoxifen metabolite plasma concentrations after coadministration of tamoxifen and the selective serotonin reuptake inhibitor paroxetine. *J Natl Cancer Inst* **95**:1758–1764.
37. Wu X, Hawse JR, Subramaniam M, Goetz MP, Ingle JN, Spelsberg TC. (2009) The tamoxifen metabolite, endoxifen, is a potent antiestrogen that targets estrogen receptor alpha for degradation in breast cancer cells. *Cancer Res* **69**:1722–1727.
38. Pike AC, Brzozowski AM, Walton J, Hubbard RE, Thorsell AG, Li YL, *et al.* (2001) Structural insights into the mode of action of a pure antiestrogen. *Structure* **9**:145–153.
39. Wu YL, Yang X, Ren Z, McDonnell DP, Norris JD, Willson TM, *et al.* (2005) Structural basis for an unexpected mode of SERM-mediated ER antagonism. *Mol Cell* **18**:413–424.
40. Shiau AK, Barstad D, Loria PM, Cheng L, Kushner PJ, Agard DA, *et al.* (1998) The structural basis of estrogen receptor/coactivator recognition and the antagonism of this interaction by tamoxifen. *Cell* **95**:927–937.
41. Ingelman-Sundberg M. (2005) Genetic polymorphisms of cytochrome P450 2D6 (CYP2D6): clinical consequences, evolutionary aspects and functional diversity. *Pharmacogenomics J* **5**:6–13.
42. Zanger UM, Turpeinen M, Klein K, Schwab M. (2008) Functional pharmacogenetics/genomics of human cytochromes P450 involved in drug biotransformation. *Anal Bioanal Chem* **392**:1093–1108.
43. Carlson RW, Hudis CA, Pritchard KI. (2006) Adjuvant endocrine therapy in hormone receptor-positive postmenopausal breast cancer: evolution of NCCN, ASCO, and St Gallen recommendations. *J Natl Compr Canc Netw* **4**:971–979.
44. Forbes JF, Cuzick J, Buzdar A, Howell A, Tobias JS, Baum M. (2008) Effect of anastrozole and tamoxifen as adjuvant treatment for early-stage breast cancer: 100-month analysis of the ATAC trial. *Lancet Oncol* **9**:45–53.

45. Mauriac L, Keshaviah A, Debled M, Mouridsen H, Forbes JF, Thurlimann B, *et al.* (2007) Predictors of early relapse in postmenopausal women with hormone receptor-positive breast cancer in the BIG 1-98 trial. *Ann Oncol* **18**:859–867.
46. Eichelbaum M, Spannbrucker N, Dengler HJ. (1975) Proceedings: N-oxidation of sparteine in man and its interindividual differences. *Naunyn Schmiedebergs Arch Pharmacol* **287** (Suppl):R94.
47. Eichelbaum M, Spannbrucker N, Dengler HJ. (1979) Influence of the defective metabolism of sparteine on its pharmacokinetics. *Eur J Clin Pharmacol* **16**:189–194.
48. Mahgoub A, Idle JR, Dring LG, Lancaster R, Smith RL. (1977) Polymorphic hydroxylation of debrisoquine in man. *Lancet* **2**:584–586.
49. Griese EU, Zanger UM, Brudermanns U, Gaedigk A, Mikus G, Morike K, *et al.* (1998) Assessment of the predictive power of genotypes for the *in-vivo* catalytic function of CYP2D6 in a German population. *Pharmacogenetics* **8**:15–26.
50. Sistonen J, Sajantila A, Lao O, Corander J, Barbujani G, Fuselli S. (2007) CYP2D6 worldwide genetic variation shows high frequency of altered activity variants and no continental structure. *Pharmacogenet Genomics* **17**:93-101.
51. Zanger UM, Raimundo S, Eichelbaum M. (2004) Cytochrome P450 2D6: overview and update on pharmacology, genetics, biochemistry. *Naunyn Schmiedebergs Arch Pharmacol* **369**:23–37.
52. Raimundo S, Toscano C, Klein K, Fischer J, Griese EU, Eichelbaum M, *et al.* (2004) A novel intronic mutation, 2988G>A, with high predictivity for impaired function of cytochrome P450 2D6 in white subjects. *Clin Pharmacol Ther* **76**:128–138.
53. Toscano C, Klein K, Blievernicht J, Schaeffeler E, Saussele T, Raimundo S, *et al.* (2006) Impaired expression of CYP2D6 in intermediate metabolizers carrying the *41 allele caused by the intronic SNP 2988G>A: evidence for modulation of splicing events. *Pharmacogenet Genomics* **16**:755–766.
54. Zanger UM, Fischer J, Raimundo S, Stuven T, Evert BO, Schwab M, *et al.* (2001) Comprehensive analysis of the genetic factors determining expression and function of hepatic CYP2D6. *Pharmacogenetics* **11**:573–585.
55. Bertilsson L, Dahl ML, Sjoqvist F, berg-Wistedt A, Humble M, Johansson I, *et al.* (1993) Molecular basis for rational megaprescribing in ultrarapid hydroxylators of debrisoquine. *Lancet* **341**:63.
56. Dalen P, Dahl ML, Bernal Ruiz ML, Nordin J, Bertilsson L. (1998) 10-hydroxylation of nortriptyline in white persons with 0, 1, 2, 3, and 13 functional CYP2D6 genes. *Clin Pharmacol Ther* **63**:444–452.
57. Lovlie R, Daly AK, Matre GE, Molven A, Steen VM. (2001) Polymorphisms in CYP2D6 duplication-negative individuals with the ultrarapid metabolizer phenotype: a role for the CYP2D6*35 allele in ultrarapid metabolism? *Pharmacogenetics* **11**:45–55.
58. Schroth W, Hamann U, Fasching PA, Dauser S, Winter S, Eichelbaum M, *et al.* (2010) CYP2D6 polymorphisms as predictors of outcome in breast cancer

patients treated with tamoxifen: expanded polymorphism coverage improves risk stratification. *Clin Cancer Res* **16**:4468-4477.

59. Borges S, Desta Z, Li L, Skaar TC, Ward BA, Nguyen A, *et al.* (2006) Quantitative effect of CYP2D6 genotype and inhibitors on tamoxifen metabolism: implication for optimization of breast cancer treatment. *Clin Pharmacol Ther* **80**:61-74.
60. Gjerde J, Hauglid M, Breilid H, Lundgren S, Varhaug JE, Kisanga ER, *et al.* (2008) Effects of CYP2D6 and SULT1A1 genotypes including SULT1A1 gene copy number on tamoxifen metabolism. *Ann Oncol* **19**:56-61.
61. Murdter TE, Schroth W, Bacchus-Gerybadze L, Winter S, Heinkele G, Simon W, *et al.* (2011) Activity levels of tamoxifen metabolites at the estrogen receptor and the impact of genetic polymorphisms of phase I and II enzymes on their concentration levels in plasma. *Clin Pharmacol Ther* **89**: 708-717.
62. Lim HS, Ju LH, Seok LK, Sook LE, Jang IJ, Ro J. (2007) Clinical implications of CYP2D6 genotypes predictive of tamoxifen pharmacokinetics in metastatic breast cancer. *J Clin Oncol* **25**:3837-3845.
63. Xu Y, Sun Y, Yao L, Shi L, Wu Y, Ouyang T, *et al.* (2008) Association between CYP2D6 *10 genotype and survival of breast cancer patients receiving tamoxifen treatment. *Ann Oncol* **19**:1423-1429.
64. Bijl MJ, van Schaik RH, Lammers LA, Hofman A, Vulto AG, van GT, *et al.* (2009) The CYP2D6*4 polymorphism affects breast cancer survival in tamoxifen users. *Breast Cancer Res Treat* **118**:125-130.
65. Bonanni B, Macis D, Maisonneuve P, Johansson HA, Gucciardo G, Oliviero P, *et al.* (2006) Polymorphism in the CYP2D6 tamoxifen-metabolizing gene influences clinical effect but not hot flashes: data from the Italian Tamoxifen Trial. *J Clin Oncol* **24**:3708-3709.
66. Cajal T, Altes A, Pare L, del RE, Alonso C, Barnadas A, *et al.* (2010) Impact of CYP2D6 polymorphisms in tamoxifen adjuvant breast cancer treatment. *Breast Cancer Res Treat* **119**:33-38.
67. Goetz MP, Rae JM, Suman VJ, Safgren SL, Ames MM, Visscher DW, *et al.* (2005) Pharmacogenetics of tamoxifen biotransformation is associated with clinical outcomes of efficacy and hot flashes. *J Clin Oncol* **23**:9312-9318.
68. Goetz MP, Knox SK, Suman VJ, Rae JM, Safgren SL, Ames MM, *et al.* (2007) The impact of cytochrome P450 2D6 metabolism in women receiving adjuvant tamoxifen. *Breast Cancer Res Treat* **101**:113-121.
69. Kiyotani K, Mushiroda T, Sasa M, Bando Y, Sumitomo I, Hosono N, *et al.* (2008) Impact of CYP2D6*10 on recurrence-free survival in breast cancer patients receiving adjuvant tamoxifen therapy. *Cancer Sci* **99**:995-999.
70. Newman WG, Hadfield KD, Latif A, Roberts SA, Shenton A, McHague C, *et al.* (2008) Impaired tamoxifen metabolism reduces survival in familial breast cancer patients. *Clin Cancer Res* **14**:5913-5918.
71. Nowell SA, Ahn J, Rae JM, Scheys JO, Trovato A, Sweeney C, *et al.* (2005) Association of genetic variation in tamoxifen-metabolizing enzymes with

overall survival and recurrence of disease in breast cancer patients. *Breast Cancer Res Treat* **91**:249–258.

72. Okishiro M, Taguchi T, Jin KS, Shimazu K, Tamaki Y, Noguchi S. (2009) Genetic polymorphisms of CYP2D6 10 and CYP2C19 2, 3 are not associated with prognosis, endometrial thickness, or bone mineral density in Japanese breast cancer patients treated with adjuvant tamoxifen. *Cancer* **115**:952–961.
73. Schroth W, Antoniadou L, Fritz P, Schwab M, Muerdter T, Zanger UM, *et al.* (2007) Breast cancer treatment outcome with adjuvant tamoxifen relative to patient CYP2D6 and CYP2C19 genotypes. *J Clin Oncol* **25**:5187–5193.
74. Schroth W, Goetz MP, Hamann U, Fasching PA, Schmidt M, Winter S, *et al.* (2009) Association between CYP2D6 polymorphisms and outcomes among women with early stage breast cancer treated with tamoxifen. *Jama* **302**:1429–1436.
75. Wegman P, Vainikka L, Stal O, Nordenskjold B, Skoog L, Rutqvist LE, *et al.* (2005) Genotype of metabolic enzymes and the benefit of tamoxifen in postmenopausal breast cancer patients. *Breast Cancer Res* **7**:R284–R290.
76. Wegman P, Elingarami S, Carstensen J, Stal O, Nordenskjold B, Wingren S. (2007) Genetic variants of CYP3A5, CYP2D6, SULT1A1, UGT2B15 and tamoxifen response in postmenopausal patients with breast cancer. *Breast Cancer Res* **9**:R7.
77. Kiyotani K, Mushiroda T, Hosono N, Tsunoda T, Kubo M, Aki F, *et al.* (2010) Lessons for pharmacogenomics studies: association study between CYP2D6 genotype and tamoxifen response. *Pharmacogenet Genomics* **20**:565–568.
78. Kiyotani K, Mushiroda T, Imamura CK, Hosono N, Tsunoda T, Kubo M, *et al.* (2010) Significant effect of polymorphisms in CYP2D6 and ABCC2 on clinical outcomes of adjuvant tamoxifen therapy for breast cancer patients. *J Clin Oncol* **28**:1287–1293.
79. Brauch H, Schroth W, Eichelbaum M, Schwab M, Harbeck N. (2008) Clinical relevance of CYP2D6 genetics for tamoxifen response in breast cancer. *Breast Care (Basel)* **3**:43–50.
80. Brauch H, Jordan VC. (2009) Targeting of tamoxifen to enhance antitumour action for the treatment and prevention of breast cancer: the "personalised" approach? *Eur J Cancer* **45**:2274–2283.
81. Brauch H, Murdter TE, Eichelbaum M, Schwab M. (2009) Pharmacogenomics of tamoxifen therapy. *Clin Chem* **55**:1770–1782.
82. Dezentje VO, Guchelaar HJ, Nortier JW, Van d V, Gelderblom H. (2009) Clinical implications of CYP2D6 genotyping in tamoxifen treatment for breast cancer. *Clin Cancer Res* **15**:15–21.
83. Flockhart D. (2008) CYP2D6 genotyping and the pharmacogenetics of tamoxifen. *Clin Adv Hematol Oncol* **6**:493–494.
84. Hoskins JM, Carey LA, McLeod HL. (2009) CYP2D6 and tamoxifen: DNA matters in breast cancer. *Nat Rev Cancer* **9**:576–586.

85. Lash TL, Lien EA, Sorensen HT, Hamilton-Dutoit S. (2009) Genotype-guided tamoxifen therapy: time to pause for reflection? *Lancet Oncol* **10**:825–833.
86. Lash TL, Rosenberg CL. (2010) Evidence and practice regarding the role for CYP2D6 inhibition in decisions about tamoxifen therapy. *J Clin Oncol* **28**:1273–1275.
87. Seruga B, Amir E. (2010) Cytochrome P450 2D6 and outcomes of adjuvant tamoxifen therapy: results of a meta-analysis. *Breast Cancer Res Treat* **122**:609–617.
88. Goetz MP, Knox SK, Suman VJ, Rae JM, Safgren SL, Ames MM, *et al.* (2007) The impact of cytochrome P450 2D6 metabolism in women receiving adjuvant tamoxifen. *Breast Cancer Res Treat* **101**:113–121.
89. Coates AS, Keshaviah A, Thurlimann B, Mouridsen H, Mauriac L, Forbes JF, *et al.* (2007) Five years of letrozole compared with tamoxifen as initial adjuvant therapy for postmenopausal women with endocrine-responsive early breast cancer: update of study BIG 1-98. *J Clin Oncol* **25**:486–492.
90. Punglia RS, Burstein HJ, Winer EP, Weeks JC. (2008) Pharmacogenomic variation of CYP2D6 and the choice of optimal adjuvant endocrine therapy for postmenopausal breast cancer: a modeling analysis. *J Natl Cancer Inst* **100**:642–648.
91. Fisher B, Costantino JP, Wickerham DL, Redmond CK, Kavanah M, Cronin WM, *et al.* (1998) Tamoxifen for prevention of breast cancer: report of the National Surgical Adjuvant Breast and Bowel Project P-1 Study. *J Natl Cancer Inst* **90**:1371–1388.
92. Powles T, Eeles R, Ashley S, Easton D, Chang J, Dowsett M, *et al.* (1998) Interim analysis of the incidence of breast cancer in the Royal Marsden Hospital tamoxifen randomised chemoprevention trial. *Lancet* **352**:98–101.
93. Demissie S, Silliman RA, Lash TL. (2001) Adjuvant tamoxifen: predictors of use, side effects, and discontinuation in older women. *J Clin Oncol* **19**:322–328.
94. Fink AK, Gurwitz J, Rakowski W, Guadagnoli E, Silliman RA. (2004) Patient beliefs and tamoxifen discontinuance in older women with estrogen receptor–positive breast cancer. *J Clin Oncol* **22**:3309–3315.
95. Lash TL, Fox MP, Westrup JL, Fink AK, Silliman RA. (2006) Adherence to tamoxifen over the five-year course. *Breast Cancer Res Treat* **99**:215–220.
96. McCowan C, Shearer J, Donnan PT, Dewar JA, Crilly M, Thompson AM, *et al.* (2008) Cohort study examining tamoxifen adherence and its relationship to mortality in women with breast cancer. *Br J Cancer* **99**:1763–1768.
97. Partridge AH, Wang PS, Winer EP, Avorn J. (2003) Nonadherence to adjuvant tamoxifen therapy in women with primary breast cancer. *J Clin Oncol* **21**:602–606.
98. Baron AT, Lafky JM, Suman VJ, Hillman DW, Buenafe MC, Boardman CH, *et al.* (2001) A preliminary study of serum concentrations of soluble epidermal growth factor receptor (sErbB1), gonadotropins, and steroid hormones in healthy men and women. *Cancer Epidemiol Biomarkers Prev* **10**:1175–1185.

99. Fallowfield L. (2005) Acceptance of adjuvant therapy and quality of life issues. *Breast* **14**:612–616.
100. Owusu C, Buist DS, Field TS, Lash TL, Thwin SS, Geiger AM, *et al.* (2008) Predictors of tamoxifen discontinuation among older women with estrogen receptor-positive breast cancer. *J Clin Oncol* **26**:549–555.
101. Dezentje VO, van Blijderveen NJ, Gelderblom H, Putter H, van Herk-Sukel MP, Casparie MK, *et al.* (2010) Effect of concomitant CYP2D6 inhibitor use and tamoxifen adherence on breast cancer recurrence in early-stage breast cancer. *J Clin Oncol* **28**:2423–2429.
102. Mortimer JE, Flatt SW, Parker BA, Gold EB, Wasserman L, Natarajan L, *et al.* (2008) Tamoxifen, hot flashes and recurrence in breast cancer. *Breast Cancer Res Treat* **108**:421–426.
103. Stearns V, Davidson NE, Flockhart DA. (2004) Pharmacogenetics in the treatment of breast cancer. *Pharmacogenomics J* **4**:143–153.
104. Loprinzi CL, Sloan JA, Perez EA, Quella SK, Stella PJ, Mailliard JA, *et al.* (2002) Phase III evaluation of fluoxetine for treatment of hot flashes. *J Clin Oncol* **20**:1578–1583.
105. Loprinzi CL, Kugler JW, Sloan JA, Mailliard JA, LaVasseur BI, Barton DL, *et al.* (2000) Venlafaxine in management of hot flashes in survivors of breast cancer: a randomised controlled trial. *Lancet* **356**:2059–2063.
106. Stearns V, Beebe KL, Iyengar M, Dube E. (2003) Paroxetine controlled release in the treatment of menopausal hot flashes: a randomized controlled trial. *JAMA* **289**:2827–2834.
107. Crewe HK, Lennard MS, Tucker GT, Woods FR, Haddock RE. (1992) The effect of selective serotonin re-uptake inhibitors on cytochrome P4502D6 (CYP2D6) activity in human liver microsomes. *Br J Clin Pharmacol* **34**:262–265.
108. Jeppesen U, Gram LF, Vistisen K, Loft S, Poulsen HE, Brosen K. (1996) Dose-dependent inhibition of CYP1A2, CYP2C19 and CYP2D6 by citalopram, fluoxetine, fluvoxamine and paroxetine. *Eur J Clin Pharmacol* **51**:73–78.
109. Kelly CM, Juurlink DN, Gomes T, Duong-Hua M, Pritchard KI, Austin PC, *et al.* (2010) Selective serotonin reuptake inhibitors and breast cancer mortality in women receiving tamoxifen: a population-based cohort study. *BMJ* **340**:c693.
110. Lash TL, Cronin-Fenton D, Ahern TP, Rosenberg CL, Lunetta KL, Silliman RA, *et al.* (2010) Breast cancer recurrence risk related to concurrent use of SSRI antidepressants and tamoxifen. *Acta Oncol* **49**:305–312.
111. Siegelmann-Danieli N, Kurnik D, Lomnicky Y, Vesterman-Landes J, Katzir I, Bialik M, *et al.* (2011) Potent CYP2D6 inhibiting drugs do not increase relapse rate in early breast cancer patients treated with adjuvant tamoxifen. *Breast Cancer Res Treat* **125**:505–510.
112. Sideras K, Ingle JN, Ames MM, Loprinzi CL, Mrazek DP, Black JL, *et al.* (2010) Coprescription of tamoxifen and medications that inhibit CYP2D6. *J Clin Oncol* **28**:2768–2776.

113. Lien EA, Solheim E, Kvinnsland S, Ueland PM. (1988) Identification of 4-hydroxy-N-desmethyltamoxifen as a metabolite of tamoxifen in human bile. *Cancer Res* **48**:2304–2308.
114. Lien EA, Solheim E, Lea OA, Lundgren S, Kvinnsland S, Ueland PM. (1989) Distribution of 4-hydroxy-N-desmethyltamoxifen and other tamoxifen metabolites in human biological fluids during tamoxifen treatment. *Cancer Res* **49**:2175–2183.
115. Hickey M, Davis SR, Sturdee DW. (2005) Treatment of menopausal symptoms: what shall we do now? *Lancet* **366**:409–421.
116. Jordan VC, Phelps E, Lindgren JU. (1987) Effects of anti-estrogens on bone in castrated and intact female rats. *Breast Cancer Res Treat* **10**:31–35.
117. Brauch H, Schroth W. (2010) CYP2D6 tamoxifen. *Der Gynaecologe* **43**:579–585.

NOTE ADDED IN PROOF

The recent publication of limited sample sets from both the BIGI-98 (Regan *et al.* (2012) *Journal of National Cancer Institute* **104**:441–451) and ATAC (Rae *et al.* (2012) *Journal of National Cancer Institute* **104**:452–460) studies showing no correlation between tamoxifen efficacy and CYP2D6 genotyping has been controversial based on methodological considerations. These issues have been addressed in the refereed literature (Brauch *et al.* (2012) *Journal of Clinical Oncology* Oct 22 epub ahead of print).

12

ANTIHORMONE DRUG RESISTANCE

Robert Clarke and V. Craig Jordan

Abstract

The estrogen receptor (ER) has proved to be an excellent target for the treatment and prevention of breast cancer. Although the majority (80%) of breast cancer is ER positive, not all ER positive tumors respond to antihormone therapy. When an ER positive tumor does not respond at all to antihormone therapy it is described to have intrinsic resistance. This contrasts with acquired antihormone resistance where the tumor initially responds with regression but then cell populations expand that grow because of tamoxifen or despite estrogen deprivation. Based on laboratory studies using ER positive cell lines, the evolution of acquired antihormone resistance has been documented. Growth factor pathways expand and subvert the action of the ER at the genome. The new knowledge about molecular mechanisms of resistance has created new opportunities for combinations of antihormone therapies and inhibitors of growth factor pathways.

Keywords: Tamoxifen; Aromatase Inhibitors; Animal Models; Cell Cultures; Selective Estrogen Receptor Modulators.

INTRODUCTION

The innovation of targeting the tumor ER specifically using the nonsteroidal antiestrogen tamoxifen[1] ultimately changed the prognosis of women with breast cancer by proposing two new treatment strategies. The first strategy offered a new approach to therapy with long-term early adjuvant tamoxifen treatment following surgery, whereas the second established the possibility of using tamoxifen for chemoprevention.[2,3] In both cases the target would be the ER, to be blocked by tamoxifen (see Chapter 6).

In addition to blocking estrogen's binding to its receptor, another means of limiting estrogenic activity in breast tissue is by blocking the synthesis of estrogen. Aromatase inhibitors (AIs) block estrogen biosynthesis from its androgen precursors, thereby limiting the production of 17β-estradiol (see Chapter 9).[4,5] This approach has proven beneficial clinically with fewer side effects than tamoxifen and improvements in both recurrence and survival rates for postmenopausal patients.[6–9]

Breast cancer prevention trials were designed using the prior clinical experience with tamoxifen and demonstrated tamoxifen's efficacy in preventing ER-positive invasive breast cancer in women at high-risk (see Chapter 14).[10] However, many high-risk women do not benefit from population-based chemoprevention with tamoxifen yet they are exposed to side effects such as endometrial cancer and thromboembolic events.[11] As a result, a paradigm shift occurred with the finding that non-steroidal antiestrogens are, in fact, selective ER modulators (SERMs). The laboratory discovery that SERMs can maintain bone density but prevent mammary carcinogenesis led to the idea of treating osteoporosis while concurrently preventing breast cancer.[12–14] The laboratory finding that tamoxifen increases the growth of human endometrial cancer but stops the growth of breast cancer,[15] and its subsequent clinical confirmation,[16] clearly stressed the need to find a new chemopreventive medicine. Raloxifene is a drug similar in structure to tamoxifen and that is now prescribed indefinitely as a medicine to prevent osteoporosis, offering the beneficial side effect of ER-positive breast cancer prevention in postmenopausal women.[17,18] Additionally, raloxifene is FDA-approved as a prevention strategy to reduce the incidence of ER-positive breast cancer in at-risk postmenopausal women without increasing the incidence of endometrial cancer as occurs with tamoxifen.[19,20]

With this brief clinical background of progress in the quality of life and survivorship for women with breast cancer, and the practical progress in reducing the incidence of breast cancer, several principles emerged to focus laboratory efforts to enhance further advances.

The ER is present in about 80% of all breast cancers but only about half of these ER positive tumors respond to antihormone treatment. The tumors that are ER positive but do not respond to antihormone treatment have intrinsic resistance to therapy. These tumors remain a challenge despite intense efforts to devise new treatments. Early advances in the practical applications of new knowledge will be noted at the conclusion of this chapter. The tumors that respond initially with regression, but then experience antihormone therapy failure, are classified as developing acquired resistance.

The finding that five or more years of therapy can control recurrences of the growth of micrometastatic primary breast cancer, it is acquired resistance to antihormone therapy that must be addressed. Models must replicate clinical experience with the ER-positive tumor. The surviving cells whose growth is not blocked by antihormones have the plasticity to respond to treatment in what appears to be an adaptive Darwinian model of continued growth and replication in the presence of the drug.

To investigate the properties of acquired antihormone-resistant breast cancer cells, populations of MCF-7 cells have been created that have adapted to various antihormone environments. MCF-7 cells are well-suited for antihormone resistance studies since they are easily cultured and retain ER expression when treated with antihormones or estrogen withdrawal;[21-23] they are routinely used in the laboratory and have produced more data of practical knowledge for patient care than any other breast cancer cell line. Figure 1 illustrates the lineages of different subtypes of MCF-7 cells maintained in the laboratories at the Lombardi Comprehensive Cancer Center. The characteristics of several of these cell line models have recently been reviewed.[24]

Evolution of Acquired Antihormone Resistance

Based on laboratory evidence from both individual reports and studies of acquired resistance for over a decade, the evolution of acquired resistance to SERMs can now be described in distinct phases following long-term SERM treatment and long-term experiments *in vitro* and *in vivo* (Fig. 2).[22,25,26] The evolution (Figure 2) of acquired resistance

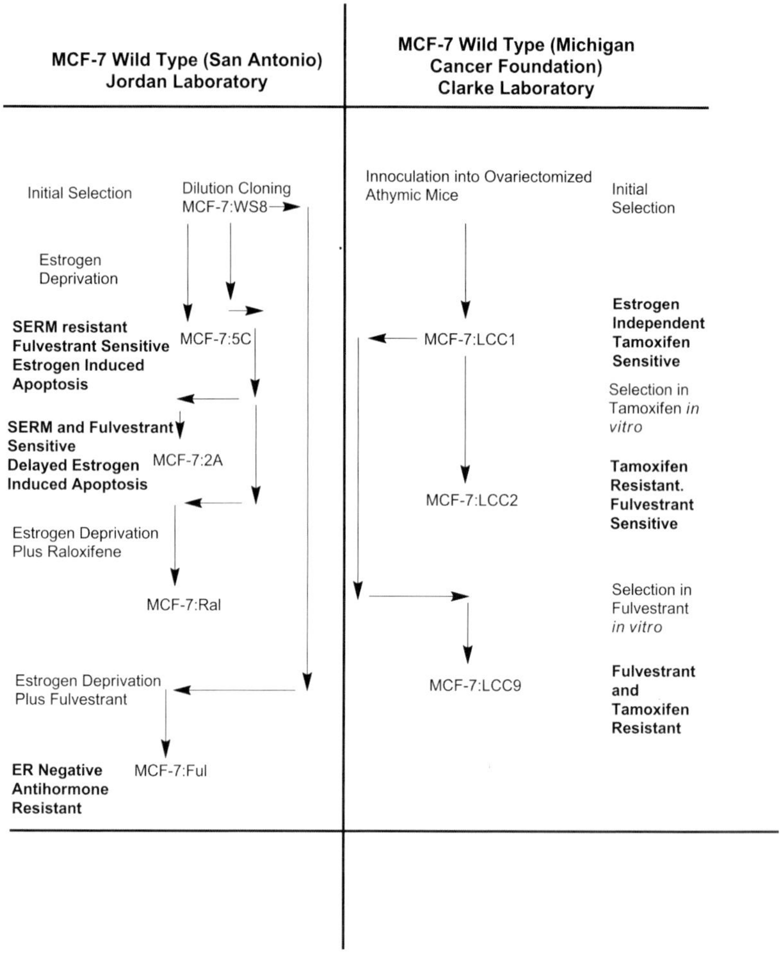

Figure 1. A flow diagram representation of the defined antihormone-resistant cell lines derived from MCF-7 cells. The Jordan laboratory obtained original "Soule" MCF-7 cells from the Michigan Cancer Foundation as a gift from Dr. Dean Edwards who was then at the University of Texas. The Clarke laboratory obtained MCF-7 cells as a gift from Dr. Marc Lippman at the NCI who obtained the original cells from Dr. Marvin Rich at the Michigan Cancer Foundation. All cells are genotyped by DNA fingerprinting.

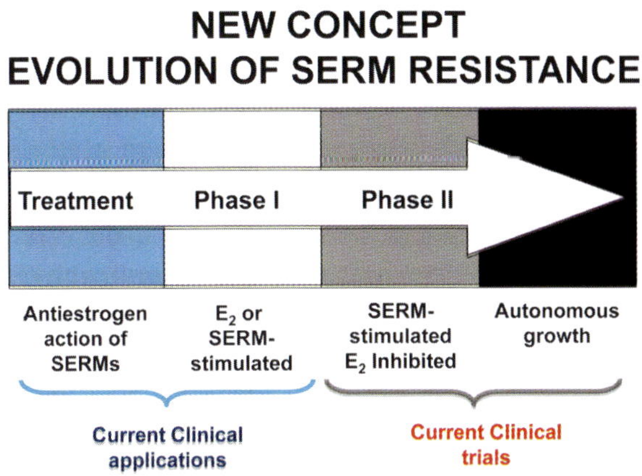

Figure 2. Evolution of acquired SERM resistance. After long-term treatment with SERMs (1–2 years *in vivo*), initially responsive ER-positive tumors become resistant to treatment and are stimulated by SERMs (Phase I of resistance) as well as by E_2. After long-term transplantation into SERM-treated animal (5+ years), breast tumor growth is inhibited by E_2, though still stimulated by SERMs (Phase II of resistance). This process with SERMs *in vivo* is replicated with estrogen deprivation with MCF-7 breast cancer cells *in vitro*; cells initially start to grow spontaneously but estrogen still induces growth (hypersensitivity). This is Phase I. Long-term estrogen deprivation causes spontaneous growth in culture but apoptosis with physiologic estrogens both *in vitro* and *in vivo* (Phase II).

occurs after an initial period of therapeutic success where antiestrogenic activity predominates and the SERMs are competitive inhibitors of estrogen-stimulated tumor growth in athymic mice.[27,28] The therapeutic phase of SERM action can be maintained for a year or two (at most) but eventually tumors start to grow despite continued tamoxifen;[29] these tumors can be re-transplanted into other tamoxifen-treated ovariectomized athymic mice.[30] Paradoxically, both physiologic estradiol and tamoxifen (there is cross-resistance with raloxifene and toremifene)[31] can then cause growth, indicating Phase I resistance. The pure antiestrogens ICI 164,382 and fulvestrant block Phase I growth with either tamoxifen or estradiol. These antiestrogens target the receptor for ubiquitination and degradation[32,33] and are generally called selective estrogen receptor down-regulators (SERDs). A similar form of acquired resistance to tamoxifen occurs with the T47D breast cancer cell line.[34,35] This type of acquired resistance is characteristic of resistance to tamoxifen during the treatment of metastatic ER-positive breast cancer. Consequently, treatment with either fulvestrant or an aromatase inhibitor are often effective second-line therapeutic agents in the clinic.[36,37] The laboratory principles are illustrated in Fig. 2.

These laboratory data are inconsistent with the successful adjuvant treatment of many node-positive and node-negative ER-positive breast cancers with five years of tamoxifen.[7] Not only is tamoxifen frequently effective during adjuvant therapy but it can maintain recurrence-free survival and reduce mortality by 30% during the 10 years following cessation of tamoxifen treatment. Laboratory studies have now provided an insight into this clinical advance (see also Chapter 2).

Repeated transplantation of tamoxifen-resistant tumors into subsequent generations of tamoxifen-treated athymic mice can change the clonal selection of tumor cells. Not only do the tumors remain tamoxifen-dependent for growth over a five-year period but the constant exposure to tamoxifen can change the tumor response to estradiol, which can switch from being a survival signal to an apoptotic trigger. Tumor regression occurs in response to physiologic estrogen, which may explain[38,39] the decreasing mortality of tamoxifen-treated patients following adjuvant tamoxifen. In other words, short-term adjuvant tamoxifen only pushes acquired resistance into Phase I resistance where estradiol is still a growth stimulator once tamoxifen is stopped. In contrast, longer tamoxifen treatment forces clonal selection into Phase II resistance where apoptosis occurs upon exposure to a woman's own estrogen. Indeed it was proposed that since tumors that regress and subsequently regrow in response to physiological estrogen can again respond to subsequent antihormone treatments, then this could be applied in the clinic.[39] This experiment has recently been reported in a small clinical study by Ellis *et al.*[40] (see Chapter 2)

The evolution of cell populations to long-term antihormone therapies has been replicated with raloxifene in a 10-year clinical study *in vivo.*[41] The study is important because raloxifene can be used indefinitely to prevent osteoporosis[18] and ER-positive breast cancer.[20] The same evolution of acquired resistance occurs with the development of

Phase I and Phase II raloxifene resistance, characterized by Phase I resistance with estradiol- or raloxifene-stimulated tumor growth and Phase II resistance characterized with estradiol-induced tumor regression. It is perhaps relevant to point out that MCF-7 cells exposed to both raloxifene and estrogen deprivation *in vitro* can rapidly advance to Phase II resistance with estradiol-induced apoptosis *in vivo*.[42]

Other clinically relevant points can be made regarding acquired SERM resistance in the laboratory. The T47D cell line advances to Phase I tamoxifen resistance but does not progress to Phase II. The fact that T47D cells have mutant p53 may be relevant as estrogen-induced apoptosis does not develop.

The pure antiestrogen fulvestrant is an excellent antiestrogen/antitumor agent in the laboratory and can be as effective as some AIs in the clinic.[37] Fulvestrant has begun to attract more attention with the recent successful use of twice the previously recommended dose.[43] Laboratory studies with Phase II tamoxifen-resistant tumors grown in athymic mice suggest that the second-line use of fulvestrant in an environment of physiologic estrogen may fail and, in fact, could cause enhanced tumor growth.[44] The reason for this is unknown. Since AIs are now the adjuvant treatment of choice for postmenopausal patients with ER-positive breast cancer, an examination of acquired resistance is mandatory. The principles first described for SERMs are true for aromatase inhibitors and the development of acquired resistance to estrogen deprivation *in vivo*[7,45,46] and *in vitro*.[21,47-50] The mechanism of estrogen-induced apoptosis in Phase II acquired resistance has been described in Chapter 2.

MECHANISMS OF ANTIHORMONE RESISTANCE

Multiple mechanisms of resistance have been proposed based on observations in experimental models and humans. These can be thought of broadly as reflecting pharmacologic/pharmacogenomic events that alter the pharmacokinetics of the drugs, or events more closely tied to alterations in the ability of endocrine therapies to modulate ERα-mediated cell death signaling. The former events are described briefly below; a more detailed discussion of the pharmacogenetic modulators of responsiveness is provided in Chapter 11. The latter events may or may not require continued ERα expression and function. Loss of detectable ERα protein expression confers resistance to drugs that directly or indirectly target its function and is the dominant *de novo* resistance mechanism. The Early Breast Cancer Trialist's meta analyses provide no compelling evidence that ER-negative primary breast tumors respond to ovariectomy or TAM.[51,52] Cases of such responses in ER-negative breast cancers are evident in the early literature but these may more closely reflect false negative determinations of ER expression. In marked contrast, loss of detectable ERα expression is not the primary receptor phenotype detected in breast tumors that acquire resistance — those that initially responded to an ER-targeted therapy but later recurred (reviewed in[53]).

The complexity of ER-regulated signaling, and how it controls cell fate decisions, is poorly understood. Many individual genes and signaling pathways have been implicated and are reviewed in detail elsewhere.[53-55] However, it has been evident for some time that a focus on pathway studies fails to capture the more complex underlying network structure of signaling in the context of antihormone resistance.[56,57] The network is a complex system with features of redundancy and degeneracy, where multiple smaller pathways interconnect and multiple signals likely converge. The sum of these various signals, however that is determined, is most likely to be the ultimate determinant of whether a cell chooses to proliferate (actively enter the cell cycle), arrest (exit the cell cycle temporarily or permanently), or die.[58,59] Both proliferation and survival appear central in the determination of responsiveness. For example, a profound G_0/G_1 cell cycle arrest is widely seen following treatment of sensitive cells with antiestrogens or E2 withdrawal. However, this alone does not explain the ability of these therapies to increase overall survival for some patients; clearly, cancer cells also must be dying.[54]

Pharmacological Mechanisms

The metabolism and distribution of tamoxifen has been extensively studied and reviewed.[60,61] While greater than 98% of tamoxifen and its major metabolites are bound to serum proteins, the average estimated intratumor concentrations of tamoxifen and its antiestrogenic metabolites can reach concentrations that exceed that of estradiol by several orders of magnitude.[53,60] Nonetheless, it remains conceivable that, in some cases, the generation of a weaker profile of antiestrogenic metabolites, perhaps due to pharmacogenetic effects, may produce a level of total intratumor antiestrogenicity that does not fully block endogenous estrogenicity. This outcome may be most relevant in those tumors that have relatively high intratumor estradiol concentrations. 17β-estradiol is the primary estrogen present in human breast tumors, although its intratumor concentration varies widely.[53] Breast tumors can both synthesize their own estradiol and take up estrogens from blood,[62] which is why systemic AIs are so effective. Nonetheless, tamoxifen exhibits excellent pharmacodynamic properties in most patients, explaining its proven clinical efficacy in over 70% of patients with tumors that coexpress both ERα and the progesterone receptor.[60] Some patients experience a tumor flare upon the initiation of tamoxifen therapy, indicated by a temporary worsening of pain.[63,64] This adverse response can be a good predictor of a beneficial response if the treatment can be maintained. The cause of this flare is poorly understood but it likely partly reflects the tissue specific agonist effects of tamoxifen, particularly when the flare is seen in bone and/or is accompanied by hypercalcemia.[55] An initial tamoxifen-induced increase in serum estrogens may also contribute to tumor flare.[65] Tamoxifen can stimulate the growth of some breast cancer models,[29,30] a phenotype that is likely also a reflection of the agonist activities of tamoxifen as mediated through its interactions with ERα. Whether this also accounts for progression of the disease during tamoxifen treatment is less clear. Flare upon therapy initiation and disease progression during therapy and following a response appear to be different phenotypes. How these two responses occur clearly requires further study.

Estrogenic metabolites of tamoxifen are produced in humans and rodents, but it is unclear that these are present in amounts where the intratumor effects of tamoxifen are often driven primarily by estrogenic compared with antiestrogenic metabolites — at least with the standard tamoxifen therapy of 20 mg/day for 5 years. Thus, other than with tumor flare, the switch to a primarily agonist response may reflect other molecular changes within the tumor cells, rather than altered tamoxifen pharmacokinetics/pharmacodynamics. Prevalence of a tamoxifen-stimulated progression is difficult to assess but would be expected to result in a beneficial response upon cessation of therapy. Withdrawal responses in patients on a standard tamoxifen regimen have been documented. For example, complete plus partial responses are seen in ≤ 10% of patients.The most common response is disease stabilization (~10%); when this is included in the estimate of overall response, withdrawal of tamoxifen has benefit in approximately 20% of patients. While the duration of these responses is generally relatively short, a small minority of patients can receive benefit lasting for up to one-year or longer (reviewed in[53]).

Measuring total intratumor concentrations of drug and metabolites does not provide a direct measure of their availability to interact with receptors. Intracellular binding proteins have been described and the sequestration of active forms of the drug would reduce its potency. The most widely studied antiestrogen binding protein is the microsomal epoxide hydrolase.[66,67] Whether this protein is a major determinant of tamoxifen activity remains unclear but studies with intracellular binding proteins establish the principle that drug sequestration can occur.

Role of ER-Action in Antihormone Resistance

The majority of breast cancers that initially respond but subsequently become resistant to primary endocrine therapy retain detectable expression of ERα, although often the levels of expression are lower than those in the primary tumor. Furthermore, responses to second and third line endocrine therapies targeting ERα are well documented, although the overall response rates and duration of responses are lower than that seen with first line endocrine therapy. These observations are consistent with a functional role for the continued ERα expression. Indeed, cells that are E2-independent, fulvestrant resistant, and tamoxifen crossresistant, are growth inhibited when ERα expression is inhibited by ERα-targeting siRNA.[68] As is the case with E2-dependent breast cancer cells, it is likely that functional ERα acts to regulate cell fate and/or cell proliferation. A role for ERβ remains unclear, in part because its expression is somewhat restricted in ERα+ breast cancers. Nonetheless, ERβ may play an important role in some breast cancers.[69] Particularly intriguing in this regard is evidence of its expression in "triple-negative" breast tumors that are otherwise considered negative for ERα expression (triple-negative tumors have no detectable expression of ERα or PgR and do not exhibit cERB-B2 amplification/overexpression) and in breast tumors arising in BRCA1 mutation carriers.[70]

Altered ERα Action as a Mechanism of Resistance

ER action, as affected by being occupied by an antiestrogen, can be modified in several ways. The presence of excess estradiol, or other estrogenic compounds may compete for binding and eventually replace estradiol as the activating/inhibiting ligand. Since the intratumor concentrations of estradiol are probably only rarely sufficiently high to compete fully with tamoxifen,[60] the replacement of tamoxifen (or any of its antiestrogenic metabolites) by estradiol is probably uncommon. However, other estrogens are likely present in tumors including environmental endocrine disruptors and dietary phytoestrogens.[71,72] The soy isoflavone genistein has received considerable attention in recent years. While genistein has a slightly greater affinity for ERβ when compared with ERα,[73,74] genistein is clearly an estrogen in human breast cancer cells.[75,76] Studies in animal models of human breast cancer suggest that genistein can reduce the apparent potency of both antiestrogens and aromatase inhibitors.[77,78] The latter is most likely due to direct competition with the antiestrogens for binding to ERα; the latter may simply reflect the ability of genistein to act as an estrogen and so bypass the metabolic block imposed on estrogen biosynthesis by the aromatase inhibitors. Evidence that this occurs in women remains scattered and a definitive resolution will require adequate study in randomized clinical trials. Currently, most medical oncologists recommend against the consumption of soy foods or isoflavone supplements in breast cancer patients, particularly for those women receiving an ERα-targeted endocrine therapy.

Mutation of the ESR1 gene or modifications of the ERα protein could also affect the ability to respond to an antiestrogen as an agonist or antagonist, or to elicit any biological response. There has been extensive analysis of the ESR1 gene and several mutations have been described that appear to affect its function in standard *in vitro* and *in vivo* assays.[79,80] Several appear to be found primarily in cell lines, and may be absent or rare in human breast cancers. Others have generated contradictory data on prevalence, e.g., the hypersensitive K303R mutation.[81–83] Despite significant study, no ERα mutations have yet been found that are of sufficient prevalence or biological relevance to be used in the routine direction of endocrine therapy (prediction) or in the assessment of outcome independent of therapy (prognosis).

Actions that alter function of the ERα protein could further affect endocrine responsiveness in breast cancer. For example, an E2-independent activation of ERα, perhaps following MAPK-dependent ERα phosphorylation (described below), could affect apparent endocrine responsiveness. Altered ER-coregulator functions that include loss of corepressors and/or gain of coactivators (see Chapter 7 for a more detailed discussion) may also participate. In both these examples, these signaling events may have different effects on tamoxifen compared with fulvestrant efficacy. The partial agonist effects of tamoxifen could be altered by changes in receptor function, such that a switch to an agonist response might come to dominate. A change in the preferential recruitment of receptor coactivators, rather than corepressors, could change tamoxifen responsiveness from antagonism to agonism. For fulvestrant, modifications of the receptor protein that would confer resistance may need to block receptor degradation and dimerization as well as induce constitutive activation. This joint effect may well occur in some models; siRNA targeting ERα is inhibitory in LCC9 cells that are ER-positive, estrogen independent, and fulvestrant resistant.[68]

ER-Independent Activities and their Role in Antihormone Resistance

ER-independent events are unlikely to be primary drivers of endocrine sensitivity since ER-negative breast tumors are almost exclusively unresponsive. Among the most widely studied ER-independent activities include the ability of antiestrogens to partition into plasma membranes and affect their fluidity and function,[84-86] and the direct binding of some antiestrogens to known intracellular proteins such as the mitochondrial epoxide hydrolase (see above) and some protein kinase C isoforms.[87]

The ability of triphenylethylene and steroidal compounds to partition into lipid membranes is a direct reflection of their physicochemical properties, primarily their lipophilicity. Once in plasma membranes, these compounds tend to stabilize at least their local microenvironments, sufficient to allow for detection of a reduction in general membrane fluidity.[84,85] Many key receptors are membrane bound, often as integral membrane proteins, and their function can be affected by changes in their lipid microenvironment. The precise role of antiestrogen-induces changes membrane in fluidity remain to be determined but at sufficiently high concentrations these changes are associated with the inhibition of proliferation in both ER-positive and ER-negative cells.[84] While it is difficult to separate these effects from those associated with ER action in ER-positive cells, these data are broadly consistent with changes in membrane fluidity playing some role in the antineoplastic effects of antiestrogens.

The role of inhibition of protein kinase C by tamoxifen has been controversial. The concentrations of tamoxifen required to inhibit global measures of protein kinase C activity is high, at least as described in some early studies, and may exceed those reached within many breast tumors.[60] However, more recent studies of specific protein kinase C isoforms, mostly studies with protein kinase C δ,[87] have been more strongly supportive of a possible role for this interaction. A direct inhibition of protein kinase C by tamoxifen may contribute to its action in sensitive cells.[88] Whether this is more broadly applicable to other structural classes of antiestrogens other than the triphenylethylenes, is less clear. It is also unclear if these activities are necessary and sufficient for antiestrogen action, or are modulators (rather than determinants) of overall sensitivity.

Molecular Signaling and the Regulation of Cell Fate

Breast cancer cells appear to have the complex programming available to respond to endocrine stressors by initiating either cell survival or cell death. For example, in sensitive cells antiestrogens can induce both an apoptotic and an autophagic cell death.[54,89-92] The molecular signaling associated with apoptosis has been widely studied. Mitochondrial function, as regulated by the balance between proapoptotic and antiapoptotic members of the BCL2 gene family, plays a central role.[93] It is not clear if all sensitive cells die through the same mechanism and/or if resistant cells that have been apparently re-sensitized die through the same or different cell fate mechanisms. Recent studies suggest that sensitive cells may die through both apoptosis and autophagy.[54,59,93-95] The dominant cell fate outcome in resistant cells that have been re-sensitized may also involve these mechanisms or may use an autophagy-associated necrotic cell death pathway.[93,95,96] Our data suggest that the precise mechanism of cell death, and the molecular regulatory events that determine this choice, are highly cell context specific and require further study.

The regulation of cell fate and mitogenesis decisions is not well understood and is likely to be much more complex that is accounted for in current models. Simple wiring diagrams of signaling transduction pathways are common in the literature; these models primarily identify genes and their putative directional connections.[53-55] While these diagrams can be useful for defining an initial topology of the signaling, and an initial roadmap for creating such a model has been proposed,[57] a predictive computational or mathematical model of cell fate regulation has yet to be generated and independently validated. Extracting the topology of ER network signaling from within gene expression and other data is an area of great current interest.[50,97,98] Rather than review all the possible signaling mechanisms that have been reported to affect endocrine responsiveness in breast cancers, we here address some of the signaling that has received the most attention to date. More specifically, we briefly discuss the principle of growth factor receptor crosstalk in the context of the EGFR superfamily and ERα activation.

EGFR Family and Antiestrogen Resistance

There are four members of the EGFR family, epidermal growth factor receptor (EGFR; HER-1; ERBB1), c-erbB2/HER-2/neu (ERBB2); ErbB3/HER-3 (ERBB3), and c-erbB4/HER-4 (ERBB4). Multiple ligands are known for EGFR including EGF, TGF-α, and amphiregulin[99] but none are known for HER-2. However, each receptor can homodimerize or form heterodimers with other family members, with HER-2 being the preferred heterodimer partner. In the context of endocrine resistance, most work has focused on HER-2, the latter being amplified and/or overexpressed in approximately 20% of breast cancers and strongly associated with a poor prognosis.[100]

While ER and HER-2 are positively correlated when the level of HER-2 expression is not overexpressed, most breast cancers that overexpress either EGFR and/or HER-2 are ER-, and so are poor candidates for ER-targeted endocrine therapies.[101] Notwithstanding, a small subset (~10%) of ER-positive breast cancers also overexpresses HER-2.[102] How these patients should be managed with endocrine therapies remains controversial. Patients with breast tumors that over express HER-2 generally a have poor prognosis, i.e., experience poor clinical outcome independent of therapeutic interventions. This association likely reflects, at least in part, a significantly higher baseline proliferation index and more aggressive biology as seen in ER+/HER-2+ compared with ER+/HER-2- tumors.[103] Thus, it is not always clear whether differences in responsiveness to endocrine therapies are driven primarily by mechanistic functions related to ER action, or simply by the biological properties that confer an aggressive phenotype and early recurrence (poor prognosis).

In some studies, HER-2 was not an independent predictor of overall survival in tamoxifen-treated patients with metastatic ER-positive breast cancer.[104] While individual studies have reported a worse outcome in these patients, the results from a meta analysis reported no significant interaction between tamoxifen treatment and HER-2 status with respect to clinical outcome.[105] Unfortunately, some of the studies to date have been small, or the number of informative cases has been relatively small, and/or testing the hypothesis relating to the role of HER-2 in tamoxifen responsiveness was not the primary goal of the study. Responsiveness to endocrine therapies is also associated with the level of ER expression (higher expression is generally associated with better response), and while the level of ER is generally lower in HER-2+ tumors,[103] clinical studies have inconsistently adjusted for this characteristic in their statistical modeling.

Data from other clinical studies have been suggestive of an advantage for AIs over tamoxifen in ER+/HER2+ patients but the data still fall short of being entirely convincing. For example, these studies did not adjust for any potential contribution of differences in tamoxifen metabolism by CYP2D6 in the tamoxifen arm(s), has been proposed as a reason that could underestimate the activity of tamoxifen — it would not affect estimates of AI activity — and so might favor an AI (see Chapter 11). Dowsett *et al.*, reported that when ER and HER-2 quantification was considered, a differential relative benefit for anastrazole over tamoxifen was not evident in the ER+/HER2+ cases; both tamoxifen and anastrazole were active in these women.[102] Thus, some reports of poor responses to endocrine therapies in ER+/HER-2+ patients may be partly explained by differences in the respective levels of ER (when this is not controlled for in the analysis) and/or by the generally more aggressive behavior of these tumors. For example, it would be inaccurate to invoke endocrine resistance as an explanation for the clinical outcomes in some of these studies. If the tumors in one group (HER-2+) were already proliferating more rapidly than HER-2- tumors, these would be expected to recur more quickly even if they showed a similar (or even better) initial response to an endocrine therapy.

Occasional improvements in response rates are reported, and similarly suggestive but inconclusive observations imply a survival benefit, for adding an anti-HER-2 therapy to an AI in ER+/HER2+ patients. For example, Kaufman *et al.*[106] reported a non-significant benefit with respect to survival and median overall survival benefit for adding trastuzumab to anastrazole. Johnston *et al.*[107] reported as non-significant improvement in overall survival with the addition of lapatinib to anastrazole.

Most of the studies discussed above have been done in patients with metastatic disease. Whether the benefits for overall survival will ultimately prove to be statistically significant and more consistent (or not) in the adjuvant setting remains an open question. Even if statistically significant, with large numbers the effects size might be small and the significant benefit seen in the population may have limited meaning for most individual patients.

Compared with clinical trial outcomes, data from experimental models are more consistent in implicating overexpression of HER-2 as a mechanistic driver of tamoxifen resistance.[108] Perhaps the most direct test of this hypothesis has come from engineering an overexpression of HER-2 in ER-positive human breast cancer cells. Overexpression of HER-2 confers resistance to tamoxifen in MCF-7 cells growing *in vitro*[109] or *in vivo*.[110] Whether this model adequately represents the 10% of ER-positive breast cancers that also overexpress HER2 is unclear, since the clinical data clearly show that some ER+/HER2+ breast cancers respond to endocrine therapy.[111] Nonetheless, these studies strongly suggest that some tamoxifen resistant ER+/HER2+ tumors may be driven mechanistically by signaling from activated HER-2.

AKT/PI3K/MAPK and ER-Phosphorylation

Several growth factors including EGFR, HER-2, and IGF1R can activate AKT→PI3K→MAPK signaling and are implicated in antiestrogen resistance.[112-115] Activation in this context is often associated with increased cell cycling and/or cell survival. The ability of an activated MAPK to then activate ERα by phosphorylation is well documented.[116,117] Whether activation of ERα is both necessary and sufficient for the mitogenic or cell survival signaling that is associated with growth factor signaling and MAPK activation is unclear.[118] These pathways can be activated in ER-negative cells where similar mitogenic actions are initiated. Since MAPK is often upregulated in proliferating versus resting cells, separating cause from effect can be difficult. Nonetheless, increased MAPK activity is associated with antiestrogen resistance.[119] There is good evidence that the potentially consequent phosphorylation of ERα may play a role in the antiestrogen resistant phenotype.[120] While it is tempting to attribute resistance solely to this or any other single signaling pathway, it is clear that the resistance phenotype is much more complex and diverse. Most signaling that affects critical cellular functions exhibit redundancy and the integration of several "pathways".[58]

CONCLUSIONS AND FUTURE PROSPECTS

While tamoxifen has been in clinical use since the early 1970s,[121] the factors that confer a resistant phenotype in ERα breast tumors remain largely unknown. Over-expression/activation of growth factors/cognate receptors such as ErbB2, EGFR, or the IGFs, low production of common tamoxifen metabolites, conversion to a tamoxifen-stimulated phenotype, or loss of ERα expression likely each contribute in some measure in a drug or tumor specific manner. However, these each appear to account for a relatively modest proportion of resistant tumors. Loss of responsiveness, a general drug resistance mechanism that affects many therapeutic strategies, may be the most common phenotype — although it may not be a single molecular phenotype.

The molecular signaling that confers cell survival, despite the continued expression of the primary molecular target (ERα) and the ability of antiestrogens to occupy this target, is an area of intense investigation. Whatever the topology of this signaling, it is likely to be complex, with significant components of redundant, feedback and feedforward signaling. The output of this signaling almost certainly determines the activated cell fate machinery — whether prodeath or prosurvival — and the decision to enter or exit cell cycle transition.[54,57,59] While the general execution components of cell death have been extensively studied, how they are selectively regulated by signaling from ERα, and how this is differentially affected by antiestrogens in sensitive and resistant cells, is still not well understood.[57] It is reasonable to anticipate that, as knowledge of these actions become better understood, we will be better placed to predict which tumors will be sensitive/resistant, and to determine how best to intervene to improve clinical outcomes in individual breast cancer patients.

In closing it is appropriate to document the practical applications of new molecular knowledge to provide benefit for women with breast cancer. The drug lapatinib that blocks the intracellular tyrosine kinase of both EGFR and HER-2 has shown efficacy to improve response rates with several antihormone therapies in the treatment of ER positive metastatic breast cancer.[122] Similarly the mTor inhibitor everolimus, that effectively blocks growth factor pathways downstream for the receptor, improves response rates and survival compared with aromatase inhibitors alone used in the treatment of metastatic breast cancer.[123,124] Although side effects may limit long term adjuvant applications, the prospects for further progress in molecular medicine are exciting.

REFERENCES

1. Jordan VC, Koerner S. (1975) Tamoxifen (ICI 46,474) and the human carcinoma 8S oestrogen receptor. *Eur J Cancer* **11**:205–206.
2. Jordan VC. (2008) Tamoxifen: catalyst for the change to targeted therapy. *Eur J Cancer* **44**:30–38.
3. Jensen EV, Jordan VC. (2003) The estrogen receptor: a model for molecular medicine. *Clin Cancer Res* **9**:1980–1989.
4. Brodie A. (1991) Aromatase and its inhibitors — an overview. *J Steroid Biochem Mol Biol* **40**:255–261.
5. Chumsri S, Howes T, Bao T, Sabnis G, Brodie A. (2011) Aromatase, aromatase inhibitors, and breast cancer. *J Steroid Biochem Mol Biol* **125**:13–22.
6. Dowsett M, Cuzick J, Ingle J, Coates A, Forbes J, Bliss J, Buyse M, Baum M, Buzdar A, Colleoni M, Coombes C, Snowdon C, Gnant M, Jakesz R, Kaufmann M, Boccardo F, Godwin J, Davies C, Peto R. (2010) Meta-analysis of breast cancer outcomes in adjuvant trials of aromatase inhibitors versus tamoxifen. *J Clin Oncol* **28**:509–518.
7. Early Breast Cancer Trialists' Collaborative Group. (2011) Relevance of breast cancer hormone receptors and other factors to the efficacy of adjuvant tamoxifen: patient-level meta-analysis of randomised trials. *Lancet* **378**:771–784.
8. van dV, Rea D, Seynaeve C, Putter H, Hasenburg A, Vannetzel JM, Paridaens R, Markopoulos C, Hozumi Y, Hille ET, Kieback DG, Asmar L, Smeets J, Nortier JW,

Hadji P, Bartlett JM, Jones SE. (2011) Adjuvant tamoxifen and exemestane in early breast cancer (TEAM): a randomised phase 3 trial. *Lancet* **377**:321–331.

9. Regan MM, Neven P, Giobbie-Hurder A, Goldhirsch A, Ejlertsen B, Mauriac L, Forbes JF, Smith I, Lang I, Wardley A, Rabaglio M, Price KN, Gelber RD, Coates AS, Thurlimann B. (2011) Assessment of letrozole and tamoxifen alone and in sequence for postmenopausal women with steroid hormone receptor-positive breast cancer: the BIG 1-98 randomised clinical trial at 8.1 years median follow-up. *Lancet Oncol* **12**:1101–1108.
10. Fisher B, Costantino JP, Wickerham DL, Redmond CK, Kavanah M, Cronin WM, Vogel V, Robidoux A, Dimitrov M, Atkins J, Daly M, Wieand S, Tan-Chiu E, Ford L, Wolmark N. (1998) Tamoxifen for prevention of breast cancer: report of the National Surgical Adjuvant Breast and Bowel Project P-1 study. *J Natl Cancer Inst* **90**:1371–1388.
11. Ferrero JM, Namer M. (1994) [Epidemiology of cancer of the breast] Epidemiologie du cancer du sein. *Arch Anat Cytol Pathol* **42**:198–205.
12. Jordan VC, Phelps E, Lindgren LU. (1987) Effects of antiestrogens on bone in castrated and intact female rats. *Breast Cancer Res Treat* **10**:31–35.
13. Gottardis MM, Jordan VC. (1987) Antitumor actions of keoxifene and tamoxifen in the N-nitrosomethylurea-induced rat mammary carcinoma model. *Cancer Res* **47**:4020–4024.
14. Lerner LJ, Jordan VC. (1990) Development of antiestrogens and their use in breast cancer: eighth Cain memorial award lecture. *Cancer Res* **50**:4177–4189.
15. Gottardis, MM, Robinson SP, Satyaswaroop PG, Jordan VC. (1988) Contrasting actions of tamoxifen on endometrial and breast tumor growth in the athymic nude mouse. *Cancer Res* **48**:812–815.
16. Fornander T, Rutqvist LE, Cedermark B. (1989) Adjuvant tamoxifen in early breast cancer: occurence of new primary tumors. *Lancet* **1**:117–121.
17. Cummings SR, Eckert S, Krueger KA, Grady D, Powles TJ, Cauley JA, Norton L, Nickelsen T, Bjarnson NH, Morrow M, Lippman ME, Black D, Glusman JE, Costa A, Jordan VC. (1999) The effect of raloxifene on risk of breast cancer in postmenopausal women; results from the MORE randomized trial. Multiple Outcomes of Raloxifene Evaluation. *J Am Med Assoc* **281**:2189–2197.
18. Martino S, Cauley JA, Barrett-Connor E, Powles TJ, Mershon J, Disch D, Secrest RJ, Cummings SR. (2004) Continuing outcomes relevant to Evista: breast cancer incidence in postmenopausal osteoporotic women in a randomized trial of raloxifene. *J Natl Cancer Inst* **96**:1751–1761.
19. Vogel VG, Costantino JP, Wickerham DL, Cronin WM, Cecchini RS, Atkins JN, Bevers TB, Fehrenbacher L, Pajon ER Jr., Wade JL III, Robidoux A, Margolese RG, James J, Lippman SM, Runowicz CD, Ganz PA, Reis SE, Caskill-Stevens W, Ford LG, Jordan VC, Wolmark N. (2006) Effects of tamoxifen vs raloxifene on the risk of developing invasive breast cancer and other disease outcomes: the NSABP Study of Tamoxifen and Raloxifene (STAR) P-2 trial. *J Am Med Assoc* **295**:2727–2741.

20. Vogel VG, Costantino JP, Wickerham DL, Cronin WM, Cecchini RS, Atkins JN, Bevers TB, Fehrenbacher L, Pajon ER, Wade JL III, Robidoux A, Margolese RG, James J, Runowicz CD, Ganz PA, Reis SE, Caskill-Stevens W, Ford LG, Jordan VC, Wolmark N. (2010) Update of the national surgical adjuvant breast and bowel project study of tamoxifen and raloxifene (STAR) P-2 trial: preventing breast cancer. *Cancer Prev Res (Phila)* **3**:696–706.
21. Clarke R, Brünner N, Katzenellenbogen BS, Thompson EW, Norman MJ, Koppi C, Paik S, Lippman ME, Dickson RB. (1989) Progression from hormone dependent to hormone independent growth in MCF-7 human breast cancer cells. *Proc Natl Acad Sci USA* **86**:3649–3653.
22. Brünner N, Frandsen TL, Holst-Hansen C, Bei M, Thompson EW, Wakeling AE, Lippman ME, Clarke R. (1993) MCF7/LCC2: A 4-hydroxytamoxifen resistant human breast cancer variant which retains sensitivity to the steroidal antiestrogen ICI 182,780. *Cancer Res* **53**:3229–3232.
23. Brünner N, Boysen B, Jirus S, Skaar TC, Holst-Hansen C, Lippman J, Frandsen T, Spang-Thomsen M, Fuqua SAW, Clarke R. (1997) MCF7/LCC9: an antiestrogen resistant MCF-7 variant in which acquired resistance to the steroidal antiestrogen ICI 182,780 confers an early crossresistance to the non-steroidal antiestrogen tamoxifen. *Cancer Res* **57**:3486–3493.
24. Sweeney ES, McDaniel RE, Maximov PY, Fan P. (2012) Models and mechanisms of acquired antihormone resistance in breast cancer: significant clinical progress despite limitations. *Horm Mol Biol Clin Invest* **9**:143–163.
25. Jordan VC. (2004) Selective estrogen receptor modulation: concept and consequences in cancer. *Cancer Cell* **5**:207–213.
26. Jordan VC. (2008) The 38th David A. Karnofsky lecture: the paradoxical actions of estrogen in breast cancer–survival or death? *J Clin Oncol* **26**:3073–3082.
27. Osborne CK, Hobbs K, Clark, GM. (1985) Effect of estrogens and antiestrogens on growth of human breast cancer cells in athymic nude mice. *Cancer Res* **45**:584–590.
28. Iino Y, Wolf DM, Langan-Fahey SM, Johnson DA, Ricchio M, Thompson ME, Jordan VC. (1991) Reversible control of oestradiol-stimulated growth of MCF-7 tumours by tamoxifen in the athymic mouse. *Br J Cancer* **64**:1019–1024.
29. Osborne CK, Coronado EB, Robinson JP. (1987) Human breast cancer in athymic nude mice: cytostatic effects of long-term antiestrogen therapy. *Eur J Cancer Clin Oncol* **23**:1189–1196.
30. Gottardis MM, Jordan VC. (1988) Development of tamoxifen-stimulated growth of MCF-7 tumors in athymic mice after long-term antiestrogen administration. *Cancer Res* **48**:5183–5187.
31. Lee ES, Schafer JM, Yao K, England G, O'Regan RM, De Los RA, Jordan VC. (2000) Cross-resistance of triphenylethylene-type antiestrogens but not ICI 182,780 in tamoxifen-stimulated breast tumors grown in athymic mice. *Clin Cancer Res* **6**:4893–4899.

32. Dauvois S, Danielian PS, White R, Parker MG. (1992) Antiestrogen ICI 164,384 reduces cellular estrogen receptor content by increasing its turnover. *Proc Natl Acad Sci USA* **89**:4037–4041.
33. Wijayaratne AL, McDonnell DP. (2001) The human estrogen receptor-alpha is a ubiquitinated protein whose stability is affected differentially by agonists, antagonists, and selective estrogen receptor modulators. *J Biol Chem* **276**:35684–35692.
34. Schafer JM, Lee ES, O'Regan RM, Yao K, Jordan VC. (2000) Rapid development of tamoxifen-stimulated mutant p53 breast tumors (T47D) in athymic mice. *Clin Cancer Res* **6**:4373–4380.
35. Schafer JM, Lee ES, Dardes RC, Bentrem D, O'Regan RM, De Los RA, Jordan VC. (2001) Analysis of cross-resistance of the selective estrogen receptor modulators arzoxifene (LY353381) and LY117018 in tamoxifen-stimulated breast cancer xenografts. *Clin Cancer Res* **7**:2505–2512.
36. Osborne CK, Pippen J, Jones SE, Parker LM, Ellis M, Come S, Gertler SZ, May JT, Burton G, Dimery I, Webster A, Morris C, Elledge R, Buzdar A. (2002) Double-blind, randomized trial comparing the efficacy and tolerability of fulvestrant versus anastrozole in postmenopausal women with advanced breast cancer progressing on prior endocrine therapy: results of a North American trial. *J Clin Oncol* **20**:3386–3395.
37. Howell A, Robertson JF, Quaresma AJ, Aschermannova A, Mauriac L, Kleeberg UR, Vergote I, Erikstein B, Webster A, Morris C. (2002) Fulvestrant, formerly ICI 182,780, is as effective as anastrozole in postmenopausal women with advanced breast cancer progressing after prior endocrine treatment. *J Clin Oncol* **20**:3396–3403.
38. Wolf DM, Jordan VC. (1993) A laboratory model to explain the survival advantage observed in patients taking adjuvant tamoxifen therapy. *Recent Results Cancer Res* **127:**23–33.
39. Yao K, Lee ES, Bentrem DJ, England G, Schafer JI, O'Regan RM, Jordan VC. (2000) Antitumor action of physiological estradiol on tamoxifen-stimulated breast tumors grown in athymic mice. *Clin Cancer Res* **6**:2028–2036.
40. Ellis MJ, Gao F, Dehdashti F, Jeffe DB, Marcom PK, Carey LA, Dickler MN, Silverman P, Fleming GF, Kommareddy A, Jamalabadi-Majidi S, Crowder R, Siegel BA. (2009) Lower-dose vs high-dose oral estradiol therapy of hormone receptor-positive, aromatase inhibitor-resistant advanced breast cancer: a phase 2 randomized study. *J Am Med Assoc* **302**:774–780.
41. Balaburski GM, Dardes RC, Johnson M, Haddad B, Zhu F, Ross EA, Sengupta S, Klein-Szanto A, Liu H, Lee ES, Kim H, Jordan VC. (2010) Raloxifene-stimulated experimental breast cancer with the paradoxical actions of estrogen to promote or prevent tumor growth: a unifying concept in anti-hormone resistance. *Int J Oncol* **37**:387–398.

42. Liu H, Lee ES, Gajdos C, Pearce ST, Chen B, Osipo C, Loweth J, McKian K, De Los RA, Wing L, Jordan VC. (2003) Apoptotic action of 17beta-estradiol in raloxifene-resistant MCF-7 cells *in vitro* and *in vivo*. *J Natl Cancer Inst* **95**:1586–1597.
43. Di LA, Jerusalem G, Petruzelka L, Torres R, Bondarenko IN, Khasanov R, Verhoeven D, Pedrini JL, Smirnova I, Lichinitser MR, Pendergrass K, Garnett S, Lindemann JP, Sapunar F, Martin M. (2010) Results of the CONFIRM phase III trial comparing fulvestrant 250 mg with fulvestrant 500 mg in postmenopausal women with estrogen receptor-positive advanced breast cancer. *J Clin Oncol* **28**:4594–4600.
44. Osipo C, Gajdos C, Liu H, Chen B, Jordan VC. (2003) Paradoxical action of fulvestrant in estradiol-induced regression of tamoxifen-stimulated breast cancer. *J Natl Cancer Inst* **95**:1597–1608.
45. Long BJ, Jelovac D, Thiantanawat A, Brodie AM. (2002) The effect of second-line antiestrogen therapy on breast tumor growth after first-line treatment with the aromatase inhibitor letrozole: long-term studies using the intratumoral aromatase postmenopausal breast cancer model. *Clin Cancer Res* **8**:2378–2388.
46. Brodie A, Macedo L, Sabnis G. (2010) Aromatase resistance mechanisms in model systems *in vivo*. *J Steroid Biochem Mol Biol* **118**:283–287.
47. Katzenellenbogen BS, Kendra KL, Norman MJ, Berthois Y. (1987) Proliferation, hormonal responsiveness, and estrogen receptor content of MCF-7 human breast cancer cells grown in the short-term and long-term absence of estrogens. *Cancer Res* **47**:4355–4360.
48. Song RX, Mor G, Naftolin F, McPherson RA, Song J, Zhang Z, Yue W, Wang J, Santen RJ. (2001) Effect of long-term estrogen deprivation on apoptotic responses of breast cancer cells to 17beta-estradiol. *J Natl Cancer Inst* **93**:1714–1723.
49. Lewis JS, Meeke K, Osipo C, Ross EA, Kidawi N, Li T, Bell E, Chandel NS, Jordan VC. (2005) Intrinsic mechanism of estradiol-induced apoptosis in breast cancer cells resistant to estrogen deprivation. *J Natl Cancer Inst* **97**:1746–1759.
50. Ariazi EA, Cunliffe HE, Lewis-Wambi JS, Slifker MJ, Willis AL, Ramos P, Tapia C, Kim HR, Yerrum S, Sharma CG, Nicolas E, Balagurunathan Y, Ross EA, Jordan VC. (2011) Estrogen induces apoptosis in estrogen deprivation-resistant breast cancer through stress responses as identified by global gene expression across time. *Proc Natl Acad Sci USA* **108**:18879–1888.
51. Early Breast Cancer Trialists Collaborative Group. (1992) Systemic treatment of early breast cancer by hormonal, cytotoxic, or immune therapy. *Lancet* **399**:1–15.
52. Early Breast Cancer Trialists' Collaborative Group. (1998) Tamoxifen for early breast cancer: an overview of the randomized trials. *Lancet* **351**:1451–1467.
53. Clarke R, Liu MC, Bouker KB, Gu Z, Lee RY, Zhu Y, Skaar TC, Gomez B, O'Brien K, Wang Y, Hilakivi-Clarke LA. (2003) Antiestrogen resistance in breast cancer and the role of estrogen receptor signaling. *Oncogene* **22**:7316–7339.

54. Clarke R, Shajahan AN, Riggins R, Cho Y, Crawford AC, Xuan J, Wang Y, Zwart A, Nehra N, Liu MC. (2009) Gene network signaling in hormone responsiveness modifies apoptosis and autophagy in breast cancer cells. *J Steroid Biochem Mol Biol* **114**:8–20.
55. Musgrove EA, Sutherland RL. (2009) Biological determinants of endocrine resistance in breast cancer. *Nat Rev Cancer* **9**:631–643.
56. Clarke R, Brünner N. (1995) Cross resistance and molecular mechanisms in antiestrogen resistance. *Endocr Related Cancer* **2**:59–72.
57. Tyson JJ, Baumann WT, Chen C, Verdugo A, Tavassoly I, Wang Y, Weiner LM, Clarke R. (2011) Dynamic modeling of oestrogen signalling and cell fate in breast cancer cells. *Nature Rev Cancer* **11**:523–532.
58. Clarke R, Ressom HW, Wang A, Xuan J, Liu MC, Gehan EA, Wang Y. (2008) The properties of very high dimensional data spaces: implications for exploring gene and protein expression data. *Nature Rev Cancer* **8**:37–49.
59. Clarke R, Cook KL, Hu R, Facey CO, Tavassoly I, Schwartz JL, Baumann WT, Tyson JJ, Xuan J, Wang Y, Warri A, Shajahan AN. (2012) Endoplasmic reticulum stress, the unfolded protein response, autophagy, and the integrated regulation of breast cancer cell fate. *Cancer Res* **72**:1321–1331.
60. Clarke R, Leonessa F, Welch JN, Skaar TC. (2001) Cellular and molecular pharmacology of antiestrogen action and resistance. *Pharmacol Rev* **53**:25–71.
61. Jordan VC, Brodie AM. (2007) Development and evolution of therapies targeted to the estrogen receptor for the treatment and prevention of breast cancer. *Steroids* **72**:7–25.
62. Miller WR. (1997) Uptake and synthesis of steroid hormones by the breast. *Endocr Related Cancer* **4**:307–311.
63. Plotkin D, Lechner JJ, Jung WE, Rosen PJ. (1978) Tamoxifen flare in advanced breast cancer. *J Am Med Assoc* **240**:2644–2646.
64. Clarysse A. (1985) Hormone-induced tumor flare. *Eur J Cancer Clin Oncol* **21**:545–547.
65. Pommier RF, Woltering EA, Keenan EJ, Fletcher WS. (1999) The mechanism of hormone-sensitive breast cancer progression on antiestrogen therapy. Implications for treatment and protocol planning. *Arch Surg* **122**:1311–1316.
66. Katzenellenbogen BS, Miller AM, Mullick A, Sheen YY. (1985) Antiestrogen action in breast cancer cells: modulation of proliferation and protein synthesis, and interaction with estrogen receptors and additional antiestrogen binding sites. *Breast Cancer Res Treat* **5**:231–243.
67. Fritz P, Murdter TE, Eichelbaum M, Siegle I, Weissert M, Zanger UM. (2001) Microsomal epoxide hydrolase expression as a predictor of tamoxifen response in primary breast cancer: a retrospective exploratory study with long-term follow-up. *J Clin Oncol* **19**:3–9.
68. Kuske B, Naughton C, Moore K, Macleod KG, Miller WR, Clarke R, Langdon SP, Cameron DA. (2006) Endocrine therapy resistance can be associated

with high estrogen receptor alpha (ERalpha) expression and reduced ERalpha phosphorylation in breast cancer models. *Endocr Relat Cancer* **13**:1121–1133.

69. Honma N, Horii R, Iwase T, Saji S, Younes M, Takubo K, Matsuura M, Ito Y, Akiyama F, Sakamoto G. (2008) Clinical importance of estrogen receptor-{beta} evaluation in breast cancer patients treated with adjuvant tamoxifen therapy. *J Clin Oncol* **26**:3727–3734.
70. Litwiniuk M, Roznowski K, Filas V, Godlewski D, Stawicka M, Kaleta R, Breborowicz J. (2008) Expression of estrogen receptor beta in the breast carcinoma of BRCA1 mutation carriers. *BMC Cancer* **8**:100.
71. Clarke R, Hilakivi-Clarke LA, Cho E, James MR, Leonessa F. (1996) Estrogens, phytoestrogens and breast cancer. *Adv Exp Biol Med* **401**:63–86.
72. Clarke R, Hilakivi-Clarke LA, Trock B. (2001) Breast cancer: dietary and environmental oestrogens. *Biologist* **48:**21–26.
73. Kuiper GG, Carlsson B, Grandien K, Enmark E, Haggblad J, Nilsson S, Gustafsson JA. (1997) Comparison of the ligand binding specificity and transcript tissue distribution of estrogen receptors alpha and beta. *Endocrinology* **138**:863–870.
74. Kuiper GG, Lemmen JG, Carlsson B, Corton JC, Safe SH, van der Saag PT, van der BB, Gustafsson JA. (1998) Interaction of estrogenic chemicals and phytoestrogens with estrogen receptor beta. *Endocrinology* **139**:4252–4263.
75. Hsieh C-Y, Santell RC, Haslam SZ, Helferich WG. (1998) Estrogenic effects of genistein on the growth of estrogen receptor-positive human breast cancer (MCF-7) cells *in vitro* and *in vivo*. *Cancer Res* **58**:3833–3844.
76. Allred CD, Allred KF, Ju YH, Virant SM, Helferich WG. (2001) Soy diets containing varying amounts of genistein stimulate growth of estrogen-dependent (MCF-7) tumors in a dose-dependent manner. *Cancer Res* **61**:5045–5050.
77. Ju YH, Doerge DR, Allred KF, Allred CD, Helferich WG. (2002) Dietary genistein negates the inhibitory effect of tamoxifen on growth of estrogen-dependent human breast cancer (MCF-7) cells implanted in athymic mice. *Cancer Res* **62**:2474–2477.
78. Ju YH, Doerge DR, Woodling KA, Hartman JA, Kwak J, Helferich WG. (2008) Dietary genistein negates the inhibitory effect of letrozole on the growth of aromatase-expressing estrogen-dependent human breast cancer cells (MCF-7Ca) *in vivo*. *Carcinogenesis* **29**:2162–2168.
79. Murphy LC, Dotzlaw H. (1989) Variant estrogen receptor mRNA species in human breast cancer biopsy samples. *Mol Endocrinol* **3**:687–693.
80. Fuqua SA. (2001) The role of estrogen receptors in breast cancer metastasis. *J Mammary Gland Biol Neoplasia* **6**:407–417.
81. Fuqua SA, Wiltschke C, Zhang QX, Borg A, Castles CG, Friedrichs WE, Hopp T, Hilsenbeck S, Mohsin S, O'Connell P, Allred DC. (2000) A hypersensitive estrogen receptor-alpha mutation in premalignant breast lesions. *Cancer Res* **60**:4026–4029.

82. Davies MP, O'Neill PA, Innes H, Sibson DR. (2005) Hypersensitive K303R oestrogen receptor-alpha variant not found in invasive carcinomas. *Breast Cancer Res* **7**:R113–R118.
83. Tokunaga E, Kimura Y, Maehara Y. (2004) No hypersensitive estrogen receptor-alpha mutation (K303R) in Japanese breast carcinomas. *Breast Cancer Res Treat.* **84**:289–292.
84. Clarke R, van den Berg HW, Murphy RF. (1990) Tamoxifen and 17b-estradiol reduce the membrane fluidity of human breast cancer cells. *J Natl Cancer Inst* **82**:1702–1705.
85. Custodio JBA, Almeida LM, Madeira VMC. (1993) The active metabolite hydroxytamoxifen of the anticancer drug tamoxifen induces structural changes in membranes. *Biochem Biophys Acta* **1153**:308–314.
86. Wiseman H, Paganga G, Halliwell B. (1993) Protective actions of tamoxifen and 4-hydroxytamoxifen against oxidative damage to low-density lipoproteins: a mechanism accounting for the cardioprotective action of tamoxifen? *Biochem J* **292**:635–638.
87. Jackson DN, Foster DA. (2004) The enigmatic protein kinase C{delta}: complex roles in cell proliferation and survival. *FASEB J* **18**:627–636.
88. O'Brian CA, Housey GM, Weinstein IB. (1988) Specific and direct binding of protein kinase C to an immobilized tamoxifen analogue. *Cancer Res* **48**:3626–3629.
89. Kyprianou N, English HF, Davidson NE, Isaacs JT. (1991) Programmed cell death during regression of the MCF-7 human breast cancer following estrogen ablation. *Cancer Res* **51**:162–166.
90. Bursch W, Ellinger A, Kienzl H, Torok L, Pandey S, Sikorska M, Walker R, Hermann RS. (1996) Active cell death induced by the anti-estrogens tamoxifen and ICI 164 384 in human mammary carcinoma cells (MCF-7) in culture: the role of autophagy. *Carcinogenesis* **17**:1595–1607.
91. Riggins R, Bouton AH Liu MC, Clarke R. (2005) Antiestrogens, aromatase inhibitors, and apoptosis in breast cancer. *Vitam Horm* **71**:201–237.
92. Schoenlein PV, Periyasamy-Thandavan S, Samaddar JS, Jackson WH, Barrett JT. (2009) Autophagy facilitates the progression of ERalpha-positive breast cancer cells to antiestrogen resistance. *Autophagy* **5**:400–403.
93. Crawford AC, RIggins RB, Shajahan AN, Zwart A, Clarke R. (2010) Co-inhibition of BCL-W and BCL2 restores antiestrogen sensitivity through BECN1 and promotes an autophagy-associated necrosis. *PLoS ONE* **5:**e8604. doi:10.1371.
94. Ellis PA, Saccani-Jotti G, Clarke R, Johnston SR, Anderson E, Howell A, A'Hern R, Salter J, Detre S, Nicholson R, Robertson J, Smith IE, Dowsett M. (1997) Induction of apoptosis by tamoxifen and ICI 182780 in primary breast cancer. *Int J Cancer* **72**:608–613.
95. Ning Y, RIggins RB, Mulla JE, Chung H, Zwart A, Clarke R. (2010) Interferon gamma restores breast cancer sensitivity to Fulvestrant by regulating STAT1,

IRF1, NFkB, BCL2 family members and sigaling to a caspase-dependent apoptosis. *Mol Cancer Ther* **9**:1274–1285.

96. Nehra R, Riggins RB, Shajahan AN, Zwart A, Crawford AC, Clarke R. (2010) BCL2 and CASP8 regulation by NFkB differentially affect mitochondrial function and cell fate in antiestrogen sensitive and resistant breast cancer cells. *FASEB J* **24**:2039–2054.
97. Chen L, Xuan J, Wang C, Shih L-M, Wang Y, Zhang Z, Hoffman EP, Clarke R. (2008) Knowledge guided multi-scale independent component analysis for biomarker identification. *BMC Bioinformatics* **9**:416 (16 pages as published on-line).
98. Zhang B, Li H, Riggins R, Zhan M, Xuan J, Zhang Z, Hoffman EP, Clarke R, Wang Y. (2009) Differential dependency network analysis to identify condition-specific topological changes in biological networks. *Bioinformatics* **25**:526–532.
99. Singh AB, Harris RC. (2005) Autocrine, paracrine and juxtacrine signaling by EGFR ligands. *Cell Signal* **17**:1183–1193.
100. Allred DC, Clark GM, Molina R, Tandon AK, Schnitt SJ, Gilchrist KW, Osborne CK, Tormey DC, McGuire WL. (1992) Overexpression of HER-2/neu and its relationship with other prognostic factors during the progression of invasive breast cancer. *Hum Pathol* **23**:974–979.
101. Pinhel I, Pinhel I, Hills M, Drury S, Drury S, Salter J, Salter J, Johnson L, Barrett-Lee P, Harris A, Dowsett M and on behalf of the NCRI Adjuvant Breast Cancer Trialists. (2009) ER and HER2 expression are positively correlated in HER2 Non-Over expressing breast cancer. *Cancer Res* **69**:703.
102. Dowsett M, Allred C, Knox J, Quinn E, Salter J, Wale C, Cuzick J, Houghton J, Williams N, Mallon E, Bishop H, Ellis I, Larsimont D, Sasano H, Carder P, Cussac AL, Knox F, Speirs V, Forbes J, Buzdar A. (2008) Relationship between quantitative estrogen and progesterone receptor expression and human epidermal growth factor receptor 2 (HER-2) status with recurrence in the arimidex, tamoxifen, alone or in combination trial. *J Clin Oncol* **26**:1059–1065.
103. Dowsett M, Harper-Wynne C, Boeddinghaus I, Salter J, Hills M, Dixon M, Ebbs S, Gui G, Sacks N, Smith I. (2001) HER-2 amplification impedes the antiproliferative effects of hormone therapy in estrogen receptor-positive primary breast cancer. *Cancer Res* **61**:8452–8458.
104. Arpino G, Green SJ, Allred DC, Lew D, Martino S, Osborne CK, Elledge RM. (2004) HER-2 Amplification, HER-1 expression, and tamoxifen response in estrogen receptor-positive metastatic breast cancer. *Clin Cancer Res* **10**:5670–5676.
105. Dhesy-Thind B, Pritchard KI, Messersmith H, O'Malley F, Elavathil L, Trudeau M. (2008) HER2/neu in systemic therapy for women with breast cancer: a systematic review. Breast *Cancer Res Treat* **109**:209–229.
106. Kaufman B, Mackey JR, Clemens MR, Bapsy PP, Vaid A, Wardley A, Tjulandin S, Jahn M, Lehle M, Feyereislova A, Revil C, Jones A. (2009) Trastuzumab plus anastrozole versus anastrozole alone for the treatment of postmenopausal women with human epidermal growth factor receptor 2-positive, hormone

receptor-positive metastatic breast cancer: results from the randomized phase III TAnDEM study. *J Clin Oncol* **27**:5529–5537.

107. Johnston S, Pippen J, Jr., Pivot X, Lichinitser M, Sadeghi S, Dieras V, Gomez HL, Romieu G, Manikhas A, Kennedy MJ, Press MF, Maltzman J, Florance A, O'Rourke L, Oliva C, Stein S, Pegram M. (2009) Lapatinib combined with letrozole versus letrozole and placebo as first-line therapy for postmenopausal hormone receptor-positive metastatic breast cancer. *J Clin Oncol* **27**:5538–5546.
108. Johnston SR. (2010) New strategies in estrogen receptor-positive breast cancer. *Clin Cancer Res* **16**:1979–1987.
109. Yang Z, Barnes CJ, Kumar R. (2004) Human epidermal growth factor receptor 2 status modulates subcellular localization of and interaction with estrogen receptor alpha in breast cancer cells. *Clin Cancer Res* **10**:3621–3628.
110. Shou J, Massarweh S, Osborne CK, Wakeling AE, Ali S, Weiss H, Schiff R. (2004) Mechanisms of tamoxifen resistance: increased estrogen receptor-HER2/neu cross-talk in ER/HER2-positive breast cancer. *J Natl Cancer Inst* **96**:926–935.
111. Cortes J, Baselga J. (2009) How to treat hormone receptor-positive, human epidermal growth factor receptor 2-amplified breast cancer. *J Clin Oncol* **27**:5492–5494.
112. Dickson RB, Lippman ME. (1995) Growth factors in breast cancer. *Endocrine Rev* **16**:559–589.
113. Wiseman LR, Johnson MD, Wakeling AE, Lykkesfeldt AE, May FEB, Westley BR. (1993) Type I IGF receptor and acquired tamoxifen resistance in oestrogen-responsive human breast cancer cells. *Eur J Cancer* **29A**:2256–2264.
114. Ellis MJ, Coop A, Singh B, Mauriac L, Llombert-Cussac A, Janicke F, Miller WR Evans DB, Dugan M, Brady C, Quebe-Fehling E, Borgs M. (2001) Letrozole is more effective neoadjuvant endocrine therapy than tamoxifen for ErbB-1- and/or ErbB-2-positive, estrogen receptor-positive primary breast cancer: evidence from a phase III randomized trial. *J Clin Oncol* **19**:3808–3816.
115. Liu Y, El-Ashry D, Chen D, Ding IYF, Kern FG. (1995) MCF-7 breast cancer cells overexpressing transfected *c-erb*B-2 have an *in vitro* growth advantage in estrogen-depleted conditions and reduced estrogen-dependence and tamoxifen-sensitivity *in vivo*. *Breast Cancer Res Treat* **34**:97–117.
116. Murphy LC, Niu Y, Snell L, Watson P. (2004) Phospho-Serine-118 estrogen receptor-{alpha} expression is associated with better disease outcome in women treated with tamoxifen. *Clin Cancer Res* **10**:5902–5906.
117. Kok M, Holm-Wigerup C, Hauptmann M, Michalides R, Stal O, Linn S, Landberg G. (2009) Estrogen receptor-alpha phosphorylation at serine-118 and tamoxifen response in breast cancer. *J Natl Cancer Inst* **101**:1725–1729.
118. Martin LA, Farmer I, Johnston SRD, Ali S, Marshall C, Dowsett M. (2003) Enhanced Estrogen receptor (ER), ERBB2, and MAPK signal transduction pathways operate

during the Adaptation of MCF-7 cells to long term estrogen deprivation. *J Biol Chem* **278**:30458–30468.

119. Donovan JCH, Milic A, Slingerland JM. (2001) Constitutive MEK/MAPK Activation leads to p27Kip1 deregulation and antiestrogen resistance in human breast cancer cells. *J Biol Chem* **276**:40888–40895.
120. Generali D, Buffa FM, Berruti A, Brizzi MP, Campo L, Bonardi S, Bersiga A, Allevi G, Milani M, Aguggini S, Papotti M, Dogliotti L, Bottini A, Harris AL, Fox SB. (2009) Phosphorylated ERalpha, HIF-1alpha, and MAPK signaling as predictors of primary endocrine treatment response and resistance in patients with breast cancer. *J Clin Oncol* **27**:227–234.
121. Cole MP, Jones CTA, Todd IDH. (1971) A new antioestrogenic agent in late breast cancer. An early clinical appraisal of ICI 46474. *Br J Cancer* **25**:270–275.
122. Riemsma R, Formes C, Amonkar M, Lykopoulos K, Diaz J, Kleisjen J, Rea D. (2012) Systematic review of lapatinib in combination with letrozole compared with other first-line treatments for hormone receptor positive (HR+) and HER2+ advanced or metastatic breast cancer (MBC). *Curr Med Res Opin* **28**: 1263–1279.
123. Baselga J, Campone M, Piccart M, Burris H, *et al.* (2012) Everolimus in postmenopausal hormone-receptor positive advanced breast cancer. *New Engl J Med* **366**:520–52ss9.
124. Bachelot T, Bourgier C, Cropet C, Ray-Coquard I, *et al.* (2012) Randomized phase II trial of everolimus in combination with tamoxifen in patients with hormone receptor-positive, human epiderman growth factor receptor 2-negative metastatic breast cancer with prior exposure to aromatase inhibitors: a GINECO study. *J Clin Oncol* **30**(22):2718–2724.

13

NOVEL SELECTIVE ESTROGEN RECEPTOR MODULATORS

Surojeet Sengupta and V. Craig Jordan

Abstract

Estrogen plays a central role in women's health and diseases. It mediates its diverse biological effects in the estrogen target tissues through estrogen receptors (ERs) alpha (α) and beta (β). Besides the several beneficial effects of estrogen in female health, it also fuels the proliferation of ERα-positive breast cancers and uterine cancers in women. As a result, targeting and selectively modulating the activity of the ERs remains the approach of choice to block the harmful effects of estrogen action in tissues involved in diseases while retaining the beneficial effects in other tissues. Selective estrogen receptor modulators (SERMs) are molecules used to achieve selective modulation of the ERs such that it functions as an estrogen antagonist at breast level by blocking the proliferation of the cancer cells while retaining the antiosteoporotic effects of estrogen and thus functioning as an estrogen agonist at the bone level. Tamoxifen, the prototypical SERM, is the first molecule to be therapeutically used for treatment as well as prevention of breast cancers. This advance in therapeutics was followed by a second SERM, raloxifene (formerly known as keoxifene), which is currently used to treat and prevent postmenopausal osteoporosis and can also prevent breast cancers as efficiently as tamoxifen. The advantage of raloxifene is that, unlike tamoxifen treatment, there is no increase in endometrial cancer in postmenopausal patients. However, tamoxifen and raloxifene treatment is also associated with some undesirable side effects. The clinical success of these two SERMs has encouraged directing the efforts toward developing new SERM molecules with better efficacy and fewer side effects. In addition, the discovery of ERβ has naturally enhanced a quest for subtype selective receptor modulators. Academic institutions as well as the pharmaceutical industry have directed their efforts toward designing novel ERβ selective modulators which can be used therapeutically to treat various diseases. This chapter summarizes the preclinical findings on novel SERMs and the current status of research and development in this field.

Keywords: SERM Mechanism; Ospemifene; Bazedoxifene; ER Beta Ligands; Lasofoxifene; Arzoxifene.

INTRODUCTION

Selective Estrogen Receptor Modulators (SERMs) are molecules which can bind to the estrogen receptor (ER) alpha and/or ER beta and function as an estrogen agonist or antagonist in diverse estrogen target cells or tissues[1] (Fig. 1). The historical perspective and current

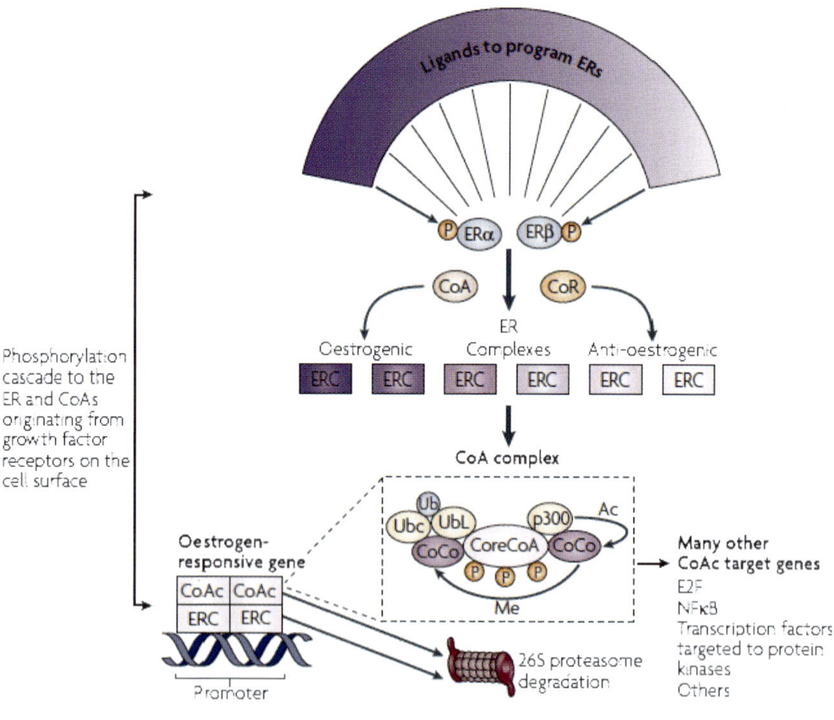

Figure 1. Molecular networks potentially influence the expression of SERM action in a target tissue. The shape of the ligands that bind to the estrogen receptors. (ERs) α

clinical applications and results are the stories in this book.The discovery of the nonsteroidal antiestrogens in the late 1950s presaged the eventual discovery of SERMs.[2,3] The first nonsteroidal antiestrogen, MER-25 (Fig. 2), showed antiestrogenic activity including promising activity as a postcoital contraceptive in rats.[4] However, MER-25 is not a SERM; it is an antiestrogen in all species tested.[4,5] The discovery of MER-25 led to the investigation of several other drugs of the same class, including the mixed isomers of clomiphene[6] and ICI 46,474 (now known as tamoxifen), both of which successfully prevented implantation in the rats.[7,8] However, the initial enthusiasm for developing these compounds as a contraceptive for humans was lost when they were completely ineffective in humans in preventing

Figure 1. (*Continued*) and β programs the complex to become an estrogenic or antiestrogenic signal.The context of the ER complex (ERC) can influence the expression of the response through the numbers of corepressors (CoRs) or coactivators (CoAs). In simple terms, a site with few CoAs or high levels of CoRs might be a dominant antiestrogenic site. However, the expression of estrogenic action is not simply the binding of the receptor complex to the promoter of the estrogen-responsive gene, but a dynamic process of CoA complex assembly and destruction.A core CoA, such as steroid receptor coactivator protein3 (SRC3), and the ERC are influenced by phosphorylation cascades that phosphorylate target sites on both complexes. The core CoA then assembles an activated multiprotein complex containing specific co-coactivators (CoCos) that might include p300, each of which has a specific enzymatic activity to be activated later. The CoA complex (CoAc) binds to the ERC at the estrogen-responsive gene promoter to switch on transcription. The CoCo proteins then perform methylation (Me) or acetylation (Ac) to activate dissociation of the complex. Simultaneously, ubiquitinylation by the bound ubiquitin-conjugating enzyme (Ubc) targets ubiquitin ligase (UbL) destruction of protein members of the complex through the 26S proteasome.The ERs are also ubiquitylated and destroyed in the 26S proteasome. Therefore, a regimented cycle of assembly, activation and destruction occurs on the basis of the preprogrammed ER complex. However, the coactivator, specifically SRC3, has ubiquitous action and can further modulate or amplify the ligand-activated trigger through many modulating genes that can consolidate and increase the stimulatory response of the ERC in a tissue. Therefore, the target tissue is programmed to express a spectrum of responses between full estrogen action and antiestrogen action on the basis of the shape of the ligand and the sophistication of the tissue-modulating network. NFκB: nuclear factor κB. (Reproduced with permission from Nature Publishing Group, Ref. #21.)

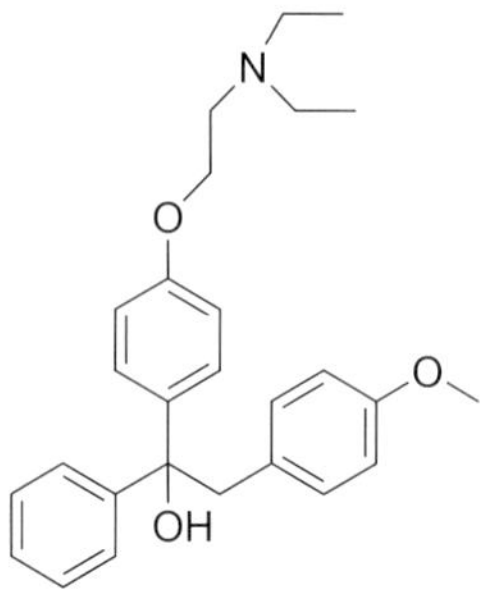

MER-25

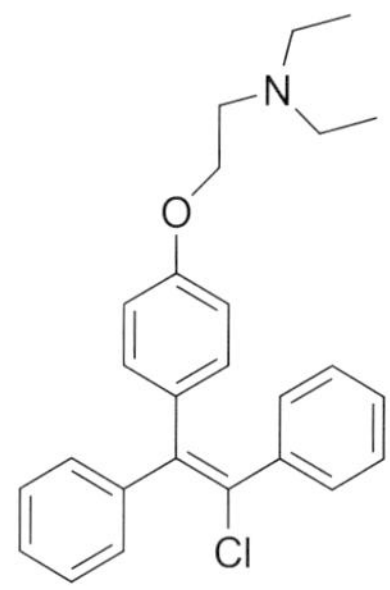

Clomiphene

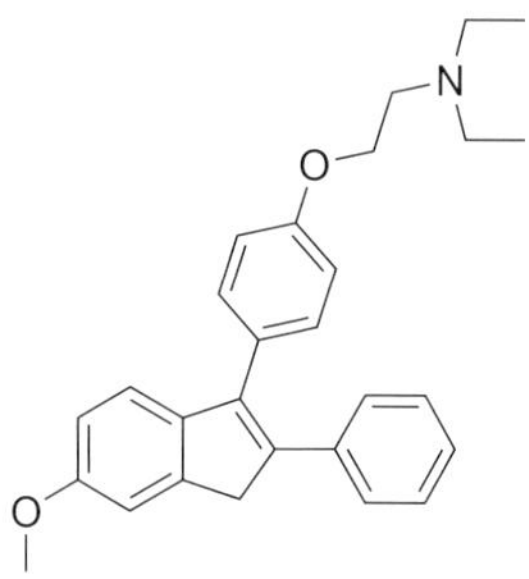

U-11,555A

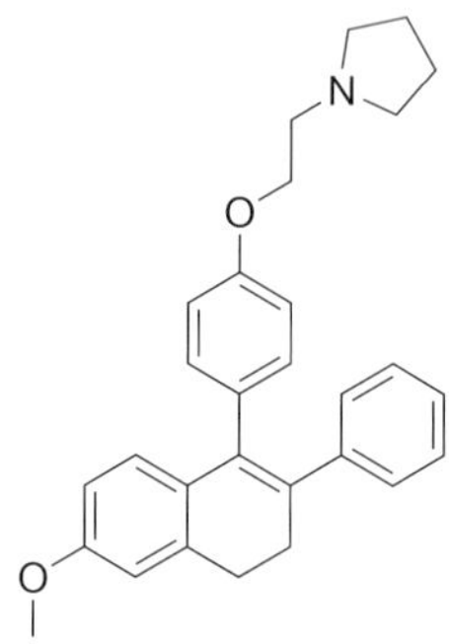

Nafoxidine
U-11,100A

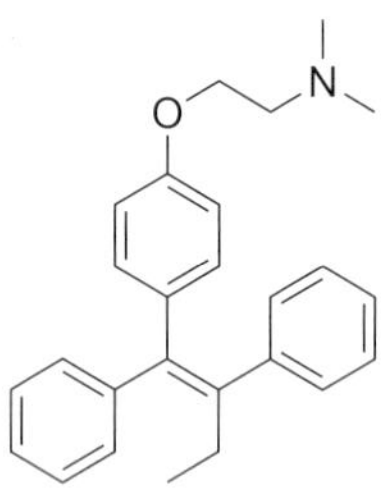

Tamoxifen
(ICI 46,474)

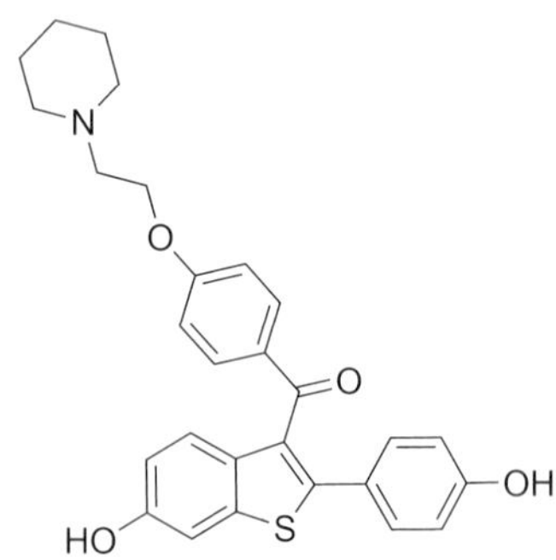

Raloxifene

Figure 2. Chemical structures of MER-25, clomiphene, U-11,555A, nafoxidine, tamoxifen and raloxifene. Note: The structure shown for clomiphene is the antiestrogenic trans isomer enclomiphene. The drug Clomid is a mixture of antiestrogenic (enclomiphene) and estrogenic (zuclomiphene) isomers.

implantation of the embryo. In fact, on the contrary, tamoxifen and clomiphene (Fig. 2) both induced ovulation in women and improved the prospects of pregnancy[9] (Chapter 5). The separated isomers of tamoxifen were reported with major differences in biological properties.[10,11] Indeed, tamoxifen was found to be estrogenic in the mouse vagina but an antiestrogen in rats.[11] Earlier, fixed ring versions of triphenylethylenes had been reported by scientists in the Upjohn Company: U-11,100A (nafoxidine) and U-11,555A (Fig. 2). Both were antiestrogenic, with antifertility properties in several species.[12,13] It is obvious from a superficial comparison of the structures in Fig. 2 that there is not much divergence of either the central core of the non-steroidal antiestrogens or SERMs. This similarity in a wide variety of "reincarnations" is the theme of this chapter.

Although the link between ovarian function and development of breast cancer was recognized in 1896,[14] the idea of an ER (now known as ER alpha) by Jensen in 1962[15] and its subsequent isolation from the rat uterus[16] established the mediator of estrogen action in the target tissues including breast cancers. This knowledge also enabled the prediction of breast tumor response to the SERM, tamoxifen, in the animal model[17] as well as breast cancer patients.[18] Thereafter, the evidence that the ER can be targeted using the SERM tamoxifen was established through a series of laboratory studies using the DMBA-induced mammary tumor model in rats. Studies showed that tamoxifen was able to inhibit the binding of estrogen to the ER in breast tumors *in vitro* and *in vivo*.[17,19,20] Tamoxifen was approved for breast cancer treatment in the United Kingdom in 1973, followed by the United States in 1977 for the treatment of advanced breast cancer in postmenopausal women.[21]

Subsequent studies demonstrated for the first time the chemopreventive properties of tamoxifen for breast cancers using the DMBA-induced rat mammary carcinoma model.[22] These laboratory findings were mimicked in clinical trials where a 50% smaller breast cancer incidence was observed in tamoxifen-treated high-risk women compared to placebo-treated controls[23] and tamoxifen was approved by The Food and Drug Administration of the United States as a chemopreventive drug for breast cancer in high-risk women in 1998.[24,25]

The clinical success of tamoxifen led to the development of another SERM, raloxifene (Fig. 2) (originally known as keoxifene), which is a benzothiophene derivative and is currently used in clinical practice as a multifunctional agent for treatment and prevention of postmenopausal osteoporosis and reduces the risk of invasive breast cancer in women.[26–28] Although raloxifene was initially investigated for breast cancer treatment its limited effectiveness and cross-resistance with tamoxifen led to the failure of its development as an anti-breast cancer agent.[29] However, the discovery of the bone-protective effects of raloxifene[30] and prevention of breast cancer by raloxifene[31] opened the door to all subsequent studies and development.[32] These paradoxical observations established the idea of "selective" actions of the nonsteroidal antiestrogens on estrogen-target tissues. The concepts generated in the laboratories have been successfully translated into the clinical setting as raloxifene was very effective in preventing postmenopausal bone loss and vertebral fractures[26] as well as around 75% risk reduction in invasive breast cancer relative to placebo treatment over a three-year period in the MORE trial.[27] Another clinical trial, known as Study of Tamoxifen And Raloxifene (STAR), which compared tamoxifen with raloxifene for the reduction of incidences of breast cancer in high-risk women, found no significant differences in the number of breast cancers between the two SERMs.[33] This established that tamoxifen and raloxifene were equally effective in reducing the breast cancer incidences in the high-risk group of women. However, in a recent update of the STAR trial[34] which followed the patients after they stopped the treatments, it was stated that women treated with raloxifene had a greater number of invasive breast cancer incidences than tamoxifen-treated women. This suggests that raloxifene may be effective in reducing the incidences of invasive breast cancers as long as women are treated with it.

NEW SERMs

Despite the therapeutic and chemopreventive effects of tamoxifen and raloxifene, they have also some undesirable side effects. Nevertheless, encouraged by the therapeutic success of tamoxifen and raloxifene, efforts are now directed toward developing new SERMs with fewer side effects and better efficacy. In addition, since the identification of ERβ in 1996,[35] ER-subtype-selective SERMs have been developed which could potentially be used as selective agents to modulate the effects of ERβ in different tissues.[36] Chemically, the SERMs represent a diverse set of structures which includes triphenylethylene, benzothiophene, chromene (benzopyran), napthalene, indole and steroids. This chapter provides a summary of the laboratory pharmacology of promising new SERMs molecules which are in various stages of clinical development, namely bazedoxifene, lasofox ifene, arzoxifene and ospemifene (Fig. 3), and then the structure–function relationships of SERMs of the future.

Bazedoxifene (TSE-424, WAY-140424)

Bazedoxifene (Fig. 3) is an indole SERM which is being actively developed to treat osteoporosis, with the potential to prevent breast cancer. In preclinical studies it did not stimulate the proliferation of ER-positive breast cancer cells MCF-7 and also inhibited the E2-induced proliferation of MCF-7 cells in a dose-dependent manner with an IC_{50} of 0.19 nM.[37] Bazedoxifene binds to ERα and ERβ with an affinity lower than for raloxifene and is also less selective for ERα than raloxifene.[37] It was able to increase the bone mass as measured by quantitative computed tomography in ovariectomized (OVX) rats as compared to control.[37] This was also confirmed in an 18-month study conducted on cynomolgus monkeys where OVX-induced bone loss was prevented with bazedoxifene treatment without adverse effects on bone structures, the lipid profile or the reproductive tract.[38,39] Bazedoxifene treatment at a dose of 5.0 mg/kg/day did not alter the uterine weight of OVX rats as compared to the control treatment group.[37] However, low doses of bazedoxifene (0.5 mg/kg/day) treatment led to a 35% increase in the uterine wet weight of OVX rats in the same study. Interestingly, bazedoxifene was also able to inhibit the raloxifene-induced stimulation of endometrial luminal epithelial cell and myometrial hypertrophy.[37] At the transcription level it did not show any agonistic effect as revealed by *in vitro* transcriptional assay using

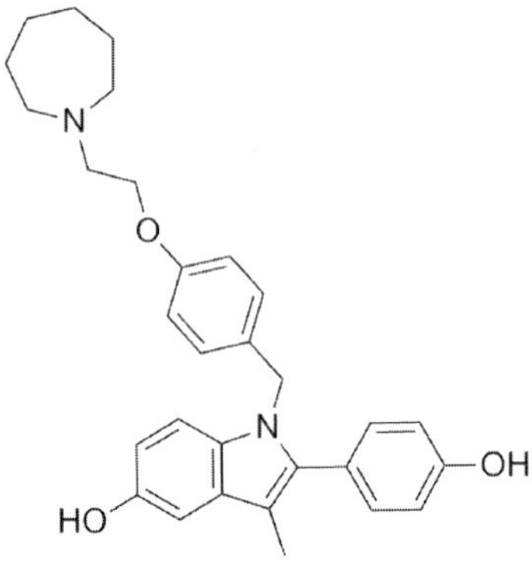

Bazedoxifene

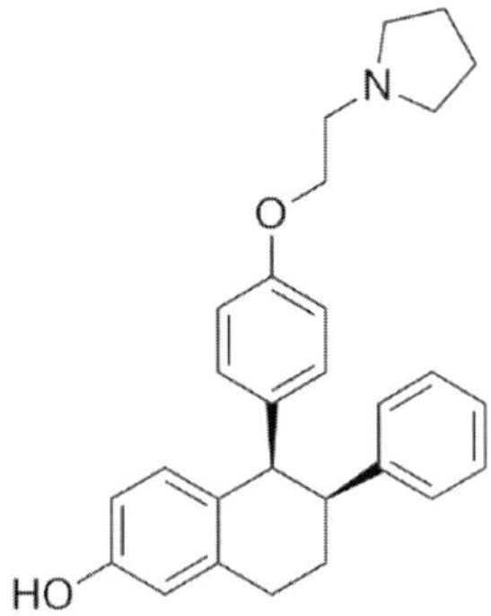

Lasofoxifene

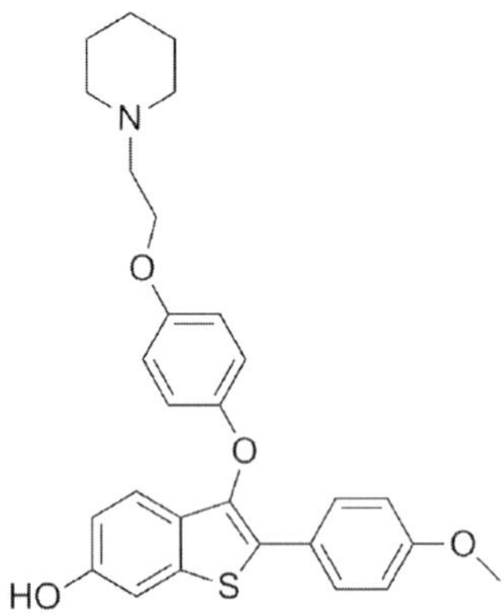

Arzoxifene

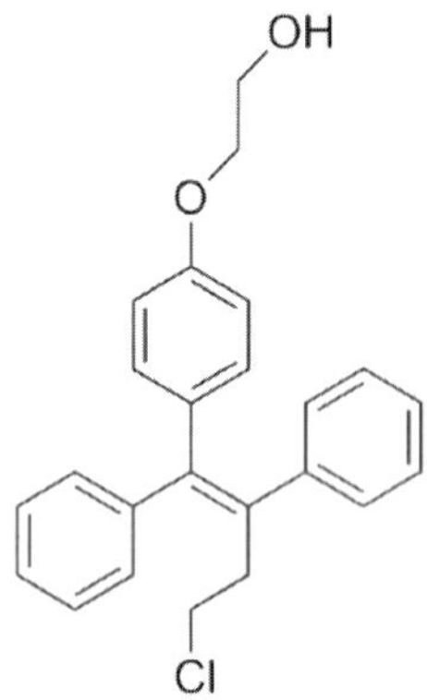

Ospemifene

Figure 3. Chemical structures of new SERMs in preclinical and clinical development.

complement component 3 promoter.[40] Clinical trials evaluating the effects on the uterus of postmenopausal women are discussed in Chapter 15, which suggests that bazedoxifene functions as an estrogen antagonist at the uterine level, particularly at higher doses.

A recent report[41] suggested that bazedoxifene is effective in inhibiting the hormone-independent growth of estrogen-deprived resistant breast cancer cells by inducing a "G-1" cell cycle blockade. This inhibition of growth by bazedoxifene was mediated through the down-regulation of cyclin D1 and ERα in the resistant cells.

An interesting pharmacology has been reported by combining bazedoxifene and conjugated estrogens. Using rodent models, it has been shown that combination of bazedoxifene and conjugated estrogens exerts positive vasomotor, lipid and skeletal responses with minimal uterine stimulation.[42] This supports the initial observations that raloxifene plus estrogen enhances the bone-preserving properties but blocks the rat uterus.[30] Bazedoxifene administration at a 3.0 mg/kg dose was able to block the conjugated-estrogen-induced increase in uterine weight without reversing the positive vasomotor action of conjugated estrogen in ovariectomized rats.[42] This combination of dose was also able to reduce the total cholesterol level while significantly increasing the total bone density of the ovariectomized animals as compared to control treatments.[42] This suggested that pairing the correct dose of bazedoxifene and conjugated estrogen might be effective in the treatment of menopausal symptoms as well as the prevention of osteoporosis. Presently it is not clear if other SERMs will also have similar effects as unique antagonistic effects of bazedoxifene were observed using gene profiling in MCF-7 breast cancer cells as compared to raloxifene and lasofoxifene.[43] This raised the possibility that different SERMs combined with conjugated estrogens may exhibit differential pharmacological effects, which merits further investigation.The results of recent clinical studies using the combination treatment of bazedoxifene and conjugated estrogens are discussed in Chapter 15. It appears that the pharmacology of the bazedoxifene and conjugated estrogen combination may prove to be safer and more efficacious in preventing osteoporosis and other undesirable postmenopausal symptoms. Further studies are warranted to examine the effectiveness of bazedoxifene alone or in combination with conjugated estrogens in breast cancer prevention.

Lasofoxifene (CP-336156, Fablyn)

Lasofoxifene (Fig. 3) is a naphthalene derivative SERM and its structure is distinctly different from those of other SERMs, with a high binding affinity for ERα and excellent bioavailability.[44] In a competitive binding assay using human ERα protein, it was found to be comparable to 17β estradiol in terms of binding *in vitro*.[45] Under *in vivo* conditions, when lasofoxifene was administered orally to immature rats no evidence of uterine stimulation was observed even at higher doses, as uterine wet weight as well as dry weight was unchanged compared to control animals. Also, no uterine hypertrophy was detected when lasofoxifene was orally administered to aged female rats.[45] The crystal structure of lasofoxifene bound to the ERα ligand-binding domain (LBD) reveals that lasofoxifene fits into the LBD pocket of ERα as 17β estradiol and other ligands and causes conformational changes similar to tamoxifen-or raloxifene-bound ERα LBDs, in which helix 12 is positioned in an "antagonist" confirmation which prevents the recruitment of coactivators to the ligand–ERα complex.[46] This structure of the lasofoxifene ER LBD explains the antagonistic behavior of lasofoxifene at the uterine level. In preclinical studies lasofoxifene was reported to be very effective in preserving bone mineral density and lowering of serum cholesterol in the OVX rat model.[44,45,47] In addition, it was also reported to prevent bone loss in aging male rats as well as orchidectomy-induced bone loss, and also had positive effects on total serum cholesterol in aging male rats.[48,49] It was not only unable to stimulate proliferation of MCF-7 (ERα-positive breast cancer cells) cells *in vitro*[44] but also prevented mammary tumors in the NMU-induced rat mammary tumor model, and the results were similar to those of tamoxifen treatment as a chemotherapeutic and chemopreventive agent.[50] The oral bioavailability of lasofoxifene is remarkably higher than that of raloxifene, as revealed by a comparative study in rats and monkeys.[44] A recent study noted[51] that lasofoxifene is extensively metabolized in rats and monkeys, with most tissues achieving maximal concentrations within 1 hr of oral administration of a single dose of carbon-14-labeled lasofoxifene. This study also reported that more than 95% of the drug and its metabolites are excreted in feces through the biliary route in both rats and monkeys, with a very small amount of the glucuronidated conjugate metabolite.[51] Indeed, the increased oral bioavailability of lasofoxifene was attributed to its being a poor substrate for glucuronidation.[44] This

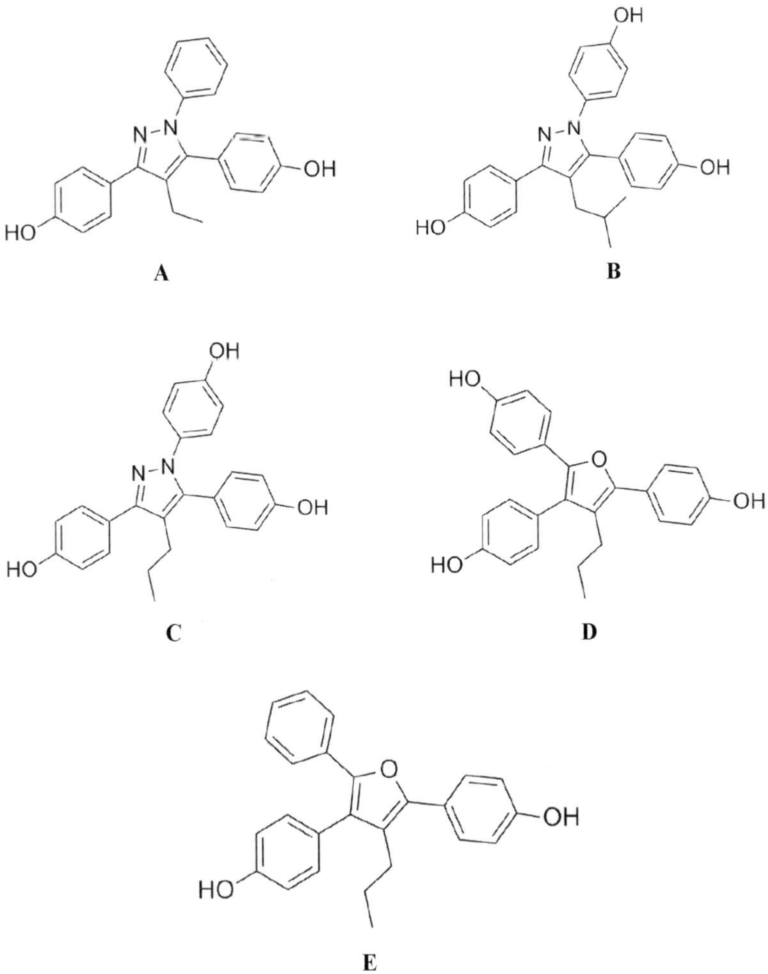

Figure 4. Different ERα-specific ligands synthesized by the Katzenellenbogen group at the University of Illinois at Urbana-Champaign.

greatest ERα subtype selectivity.[80,81] In a related study, substituted furans were investigated.[79] A triphenolic derivative (Fig. 4D) had excellent selectivity for ERα with potent agonist activity, whereas no activity was observed (agonist or antagonist) with ERβ. Interestingly, removal of the hydroxyl at the 5-phenolic substitution (Fig. 4E) resulted in estrogen-like action at ERβ. It is possible that helix 12 of ERβ can now effectively seal[75] the compound into the LBD to activate AF-2. Unfortunately, there are no x-ray crystallographic data on a full estrogen agonist liganded with the whole human ERβ, and so the concept cannot be confirmed.

Interesting leads to ERβ specificity are also being found with other simple molecules.[82] The compound DPN (2,3-bis(4-hydroxyphenyl)propionitrile (Fig. 5) has a 70-fold higher binding affinity and 170-fold higher potency in activating ERβ compared to ERα. Replacement of the CN group with acetylene or a polar function, which mimics the linear geometry and polarity of the −C≡N group, demonstrates that it is essential for ERβ selectivity. Furthermore, addition of a nitrate substitution β to the first in DPN or an *o*-methyl on the aromatic rings increases ERβ selectivity.[82]

Currently, a search for completely ERα- and ERβ-selective agents is a priority, so that new medicines can be applied to disease treatment without the concern of multiple receptor interactions. Clearly, if a drug is an antagonist at ERβ but an agonist at ERα, then this might

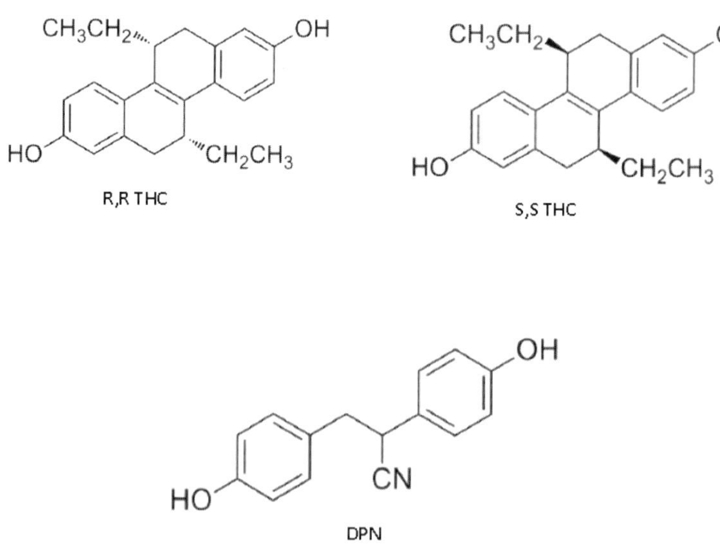

Figure 5. Potent ERβ-specific ligands synthesized by the Katzenellenbogen group at the University of Illinois at Urbana-Champaign.

prove to be unsuitable if the specific blocking of one receptor is required. In a heterogeneous distribution, ERα agonist complexes would overwhelm the antagonist ERβ complexes. However, if the drug discovery assay is an artificial signal transduction pathway based on an estrogen response element, then complex physiology could be missed. It has been suggested that tamoxifen-stimulated tumors occur through an overexpression of ERβ at the expense of ERα.[83-85] The theory proposes that tamoxifen ERβ complexes activate AP-1-responsive genes through a protein–protein interaction. The idea might open up a small but significant escape route for cell cycle blockade. However, the current compounds appear to have been tested only on an ERE reporter system. It is therefore unclear whether *R*,*R*-THC (Fig. 5) will also block the AP-1 response. Nevertheless, the developing knowledge about the crystal structure of the ERβ binding domain[75,77] will provide valuable clues for further structure–activity relationship studies. The path forward from academic laboratories to industry has now applied an integrated system of evaluation in reporter systems, cancer cell lines and animals. X-ray crystallography is becoming routine, as is ligand docking *in silico*.

The pharmaceutical industry has been particularly active in elucidating the structure–function relationships of ERα- or ERβ-specific ligands. Glaxo employed a discovery program using a peptide interacting profile (phage display) to identify a novel quinoline-based ERα ligand that had minimal activity stimulating human endometrial Ishikawa cells. The most active quinoline is illustrated in Fig. 6. The substitution of the acrylic side chain for a dimethylaminoethoxy side chain results in the stimulation of Ishikawa cells.[86] Merck Research Laboratories identified a benzoxanthin with potent selective ERα activity. The compound (Fig. 6) was 50-fold selective in competitive binding assays and 100-fold selective for ERα vs. ERβ in a transactivation assay in HEK-293 cells. The ERα selectivity may lie in the interaction of the sulphur atom of the benzoxanthin ring with the two discriminating amino acids, Leu 384 for ERα and Met 354 for ERβ, in the respective binding pockets.[87] Novartis has identified SERMs with conformationally restricted side chains that are ERα-specific based on tetrahydroisoquinoline (Fig. 6). The compound identified had little

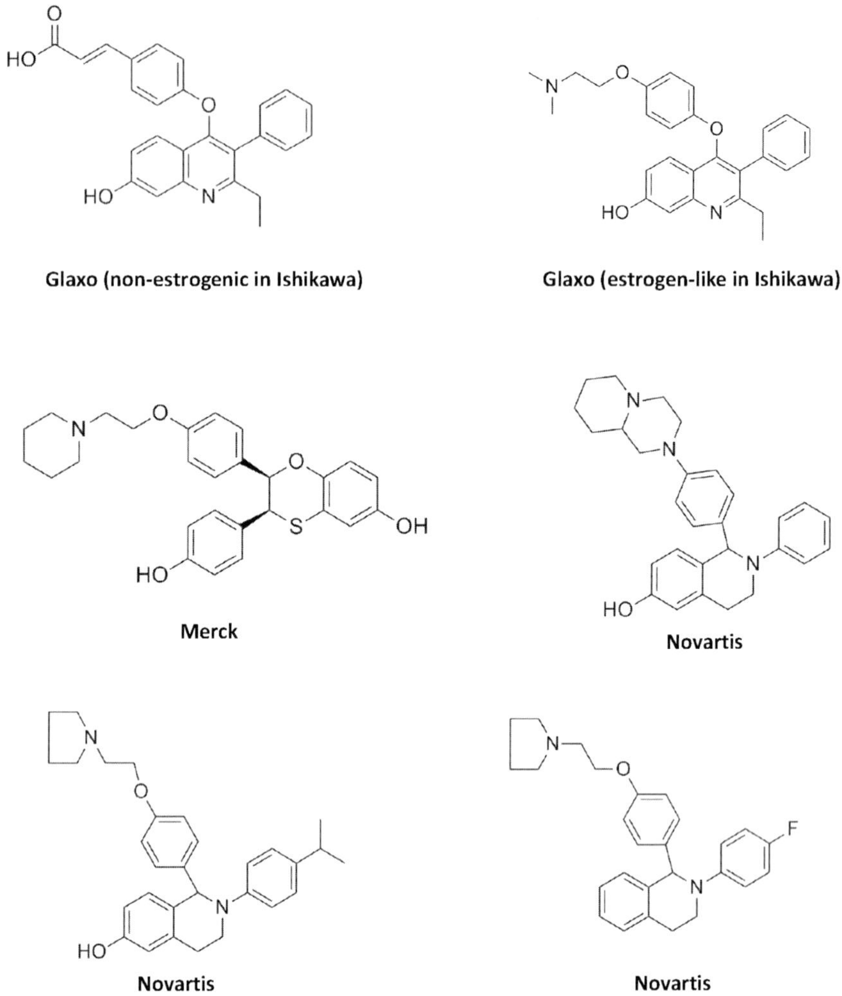

Figure 6. ERα-specific ligands developed by various pharmaceutical companies.

intrinsic activity in MCF-7 cell assays. Bioavailability was 49%.[88] Further structure–activity relationships with testing *in vitro* demonstrating that the pyrrolidine and piperidine derivatives of the p-isopropyl substitutions of the tetrahydroisoquinoline SERMs had a 50-fold selectivity for ERα vs. ERβ (Fig. 6); however, the excellent pharmacokinetic data for the *p*-fluoro derivatives made them worthy of further investigations.[89]

The design and discovery of compounds to target ERβ has received more attention than ERα. This is in part because there are no specific ERβ-modulating drugs in medical practice, and so this is an exciting area for discovery. One never knows what one will find serendipitously. Both the pharmaceutical industry and academia are involved.

Glaxo has reported a large series of 1,3,5-triazine-based ER modulators with the goal of developing a greater understanding of the pharmacology associated with subtype-selective modulation of ERβ (Fig. 7). The compounds discovered, from a lead compound with low potency and selectivity, initially had improved potency in the 10 nM range *in vitro* and a 30-fold selectivity for ERβ as antagonists but were partial agonists at ERα.[90]

Eli Lilly reported a benzopyran-selective ERβ agonist 1 (SERBA-1) (Fig. 7). The ERα and ERβ crystal structures with SERBA-1 were reported to demonstrate the reason for ERβ selectivity. In complementary reports, the Katzenellenbogen group,[91,92] Wyeth[93,94] and Schering[95] demonstrated the utility of melding x-ray crystallography, molecular modeling and structure-based drug design (SBDD). The Eli Lilly group demonstrated the biological specificity of SERBA-1 in causing involution of the mouse prostate with no effects on testicular or ventral prostate weight. Testosterone levels were not affected, which contrasts with other SERMs.[96]

Wyeth reported diphenolic azoles as potent ERβ-selective agonists, and ERB-041 (Fig. 7) as being the most selective ligand for ERβ.[97] ERB-041 shows more than 200-fold selectivity for ERβ over ERα, and it was completely devoid of any classic estrogenic action like uterotrophic effect in the immature mouse or prevention of bone mineral density loss in OVX rats.[97,98] However, ERB-041 has dramatic anti-inflammatory effect in the preclinical animal model of adjuvant-induced arthritis.[98,99] Recent reports also suggest that activating ERβ using ERB-041 can provide survival benefits in animal models of bacterial sepsis by reducing tissue damage and inhibiting transcription of proinflammatory proteins.[100]

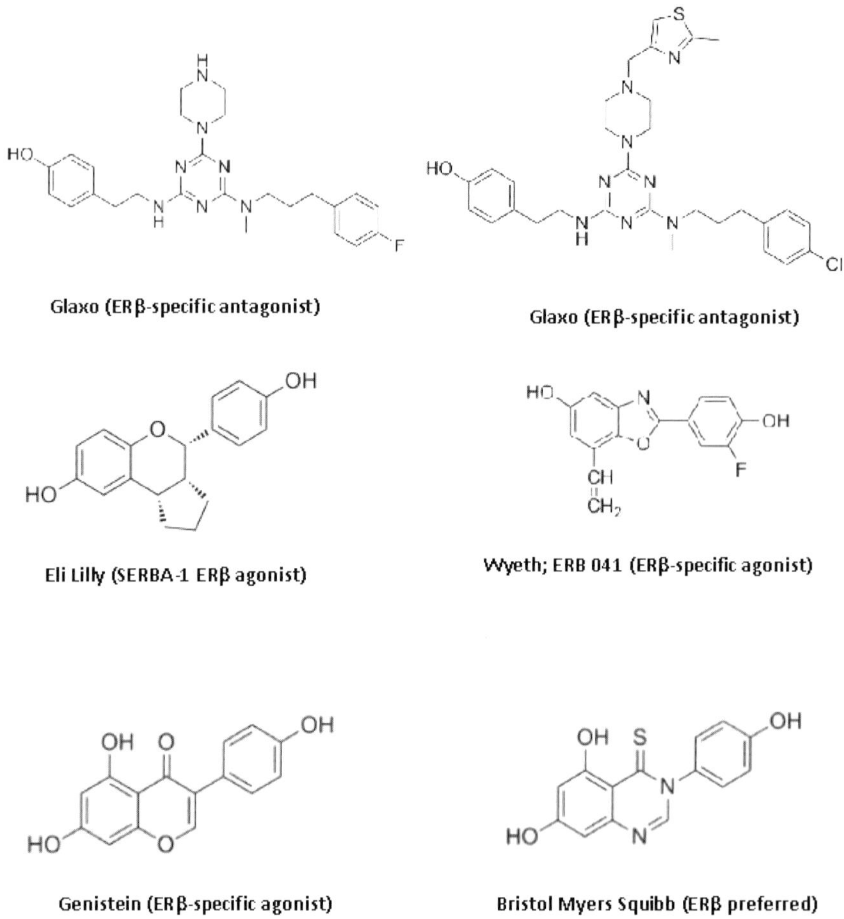

Figure 7. ERβ-specific ligands developed by various pharmaceutical companies.

Bristol Myers Squibb has returned to the knowledge that the phytoestrogen genistein (Fig. 7) has a higher affinity for ERβ than ERα[73] to discover a 3-arylquinazolinethione derivative as a selective ERβ modulator. The selectivity for ERβ is around 200-fold.[101] The compound is an agonist (Fig. 7).

Finally, a Chinese group has identified a series of ERβ ligands by virtual screening and molecular screening and molecular dynamics and bioassay from readily purchased compounds (102).

DISCOVERY OF NEW PROPERTIES OF SERMs

The original discovery of SERMs (see Chapter 15) provided both opportunities and challenges. The challenge is the control of hot flashes, as this is a quality-of-life issue. No matter how good a medicine is at preventing a few spontaneous breast cancers in a large population, or controlling the increase in fractures noted with osteoporosis, if a woman does not accept the side effects of persistent hot flashes and stops taking the medicine, there is no advance. Numerous pharmaceutical companies are addressing the issue of hot flash control to create a more acceptable SERM.

Dainippon Pharmaceuticals has employed spiro[indene-1,1'-indane]-5,5'-diol as a SERM core to design new SERMs with all the traditional SERM effects on preserving bone, reducing circulating cholesterol with minimum effects on the breast and endometrium but preventing hot flashes. This is done assessing ovariectomized rat tail skin temperature.[103] They reported a useful lead compound (Fig. 8) that merits further investigation.

Eli Lilly has studied the structure–activity relationships of oxachrysene derivative (Fig. 8) to find a compound that binds to ERα and ERβ, blocks the growth of MCF-7 breast cancer cells and Ishikawa endometrial cancer cells, and is a powerful antagonist of estrogen-stimulated increases in rat uterine weight. In ovariectomized rats, bone density is maintained and there is a decrease in circulating cholesterol levels. Most importantly, the compound was able to control temperature in a morphine-dependent rat model.[104]

Johnson and Johnson has a long-standing program to address the issue of the control of hot flashes in novel SERMs. They originally reported that a chromene-derived bisbenzopyran (Fig. 8) had all the appropriate characteristics of a SERM plus the alleviation of hot flashes (changes in the tail skin temperature during morphine withdrawal in the rat) and the alleviation of vaginal dryness.[105] Subsequently, the same group reported improved biology for a chromene-derived SERM (Fig. 8)[106] as an improved "backup" to their original phase II development of the chromene-based molecule.

Finally, two additional innovations deserve mention. Eli Lilly has described a SERM targeted to premenopausal women for the treatment

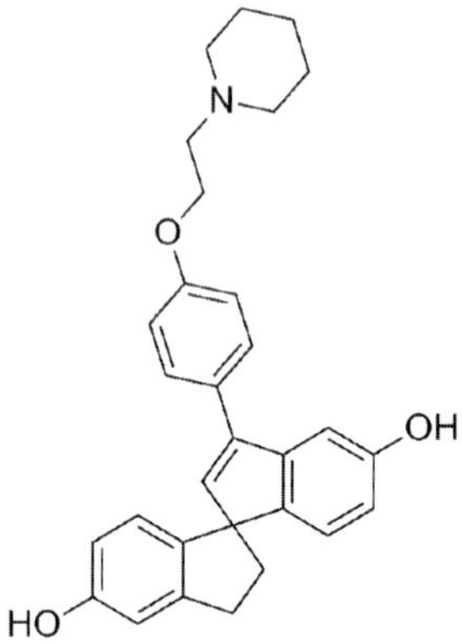

Dainippon (+ enantiomer)

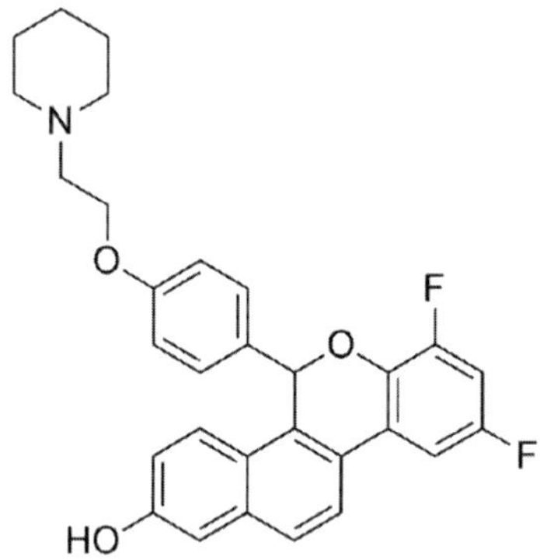

Eli Lilly ((R) –(+) enantiomer)

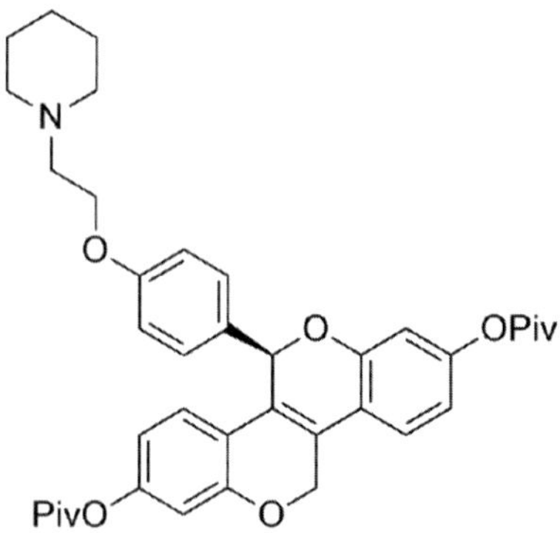

Johnson and Johnson
(R enantiomer; original clinical candidate)

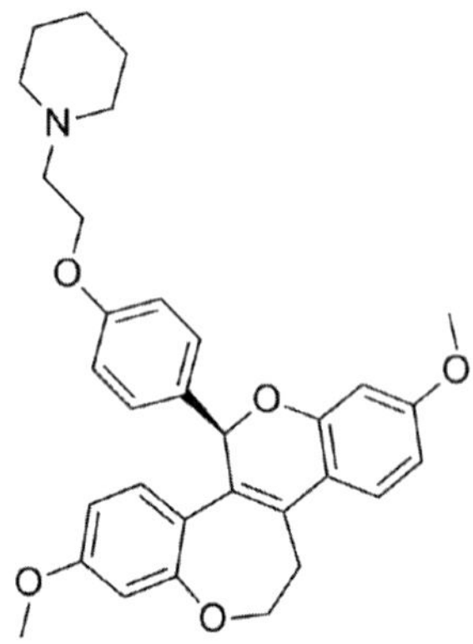

Johnson and Johnson
(R enantiomer; backup)

Figure 8. Ligands developed by various pharmaceutical companies for the treatment of "hot flashes" in postmenopausal women.

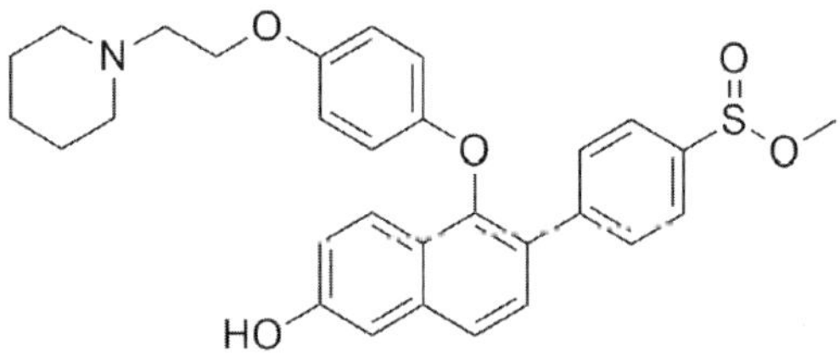

Eli Lilly (LY 2066948)

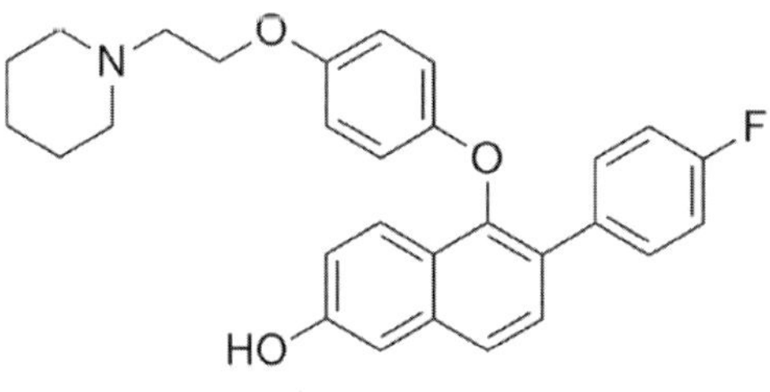

Eli Lilly (crosses blood brain barrier)

Figure 9. Chemical structure of SERM (LY 2066948) which could be used to treat leiomyomas in premenopausal women. The molecule does not cross the blood-brain barrier because of the methyl sulfone group. Exchange for fluorine allows passage of the blood-brain barrier.

of leiomyomas.[107] Their compounds are naphthalene derivatives; however, their compound of interest LY 2066948 (Fig. 9) contains a methyl sulfone moiety that permits potent antiuterotrophic activity but ensures that the compound does not cross the blood-brain barrier. They demonstrated that substitution of the methyl sulfone for fluorine (Fig. 9) that passage of the blood-brain barrier occurs, resulting in 25-fold higher circulating estrogen. Thus, LY 2066948 could be a useful agent for stopping the growth of benign leiomyomas in premenopausal women.

Alzheimer's disease is a devastating neurodegenerative condition affecting 4.5 million people in the United States. Estrogen is known to stop the degeneration of neurons but current preparations have additional toxicological issues (Chapter 4). Interestingly, there was a chance finding that the pure antiestrogen fulvestrant can mimic estrogen to stop neuronal degeneration[108] (Fig. 10).

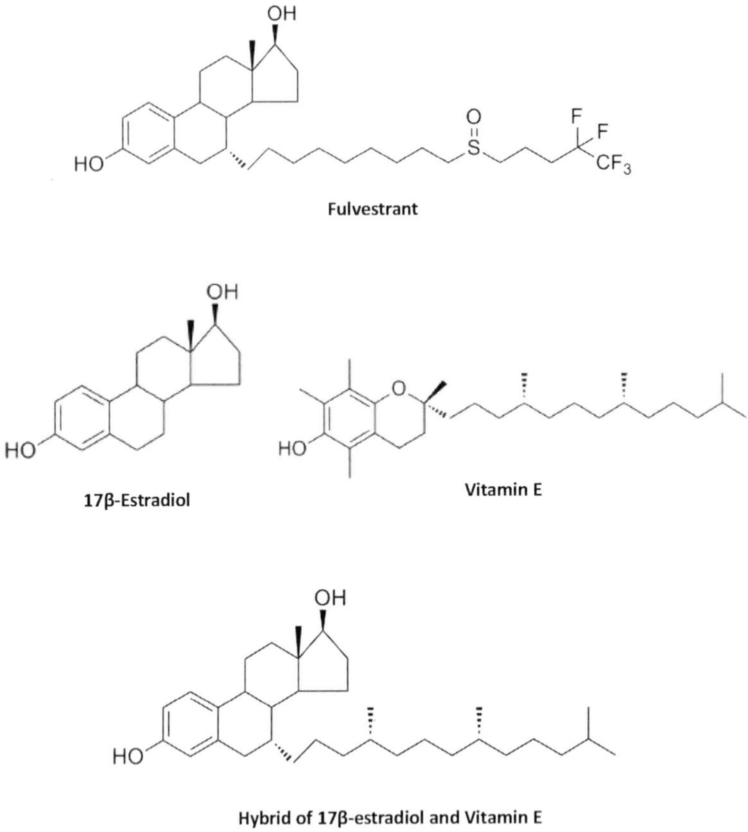

Figure 10. Chemical structures of 17β-estradiol, fulvestrant, vitamin E and the hybrid molecule of 17β-estradioland vitamin E, which is capable of crossing the blood-brain barrier and can function as a brain-specific estrogen.

Unfortunately, fulvestrant does not cross the blood-brain barrier. A switch of the similar side chain of vitamin E to the 7α position of fulvestrant demonstrates that it may be possible to create a "brain-specific" estrogen[109] (Fig. 10).

CONCLUSION

The reinvention of the prototype SERM, tamoxifen, from a failed contraceptive to a successful drug for targeted breast cancer therapy and prevention[110] has paved an encouraging path for the development of new molecules of this class with improved efficacies and a better safety profile. An estimated reduction in the mortality rate exclusively due to adjuvant tamoxifen therapy is assessed to be 19% (median).[111] Raloxifene, another SERM, is presently used as an antiosteoporotic agent for postmenopausal women and also has been effective in preventing breast cancers without causing undesirable estrogen-like actions at the uterine level. Several new SERMs are in advanced phases of development, with better efficacies not only for breast cancer but also for osteoporosis and coronary heart disease.[112,113] SERMs are now clearly recognized as multifunctional drugs which can interact with the estrogen receptor alpha and/or beta, and these specific interactions can affect various ER target tissues for different clinical indications.[114,115] These pioneering studies with SERMs have now opened up new horizons for the development of novel molecules for other members of the nuclear receptor family, namely progesterone receptor, androgen receptor, glucocorticoid receptor, etc., which can selectively target numerous disorders and diseases (Chapters 16,17).

ACKNOWLEDGEMENTS

This work (V. C. J) was supported by the Department of Defense Breast Program under award No. W81XWH-06-1-0590 Center of Excellence; subcontract under SU2C (AACR) grant No. SU2C-AACR-DT0409; the Susan G. Komen for the Cure Foundation under award No. SAC100009; GHUCCTS CTSA (Grant #UL1RR031975) and the Lombardi Comprehensive Cancer Center Support Grant (CCSG) Core Grant NIH P30 CA051008.The views and opinions of the authors do not reflect those of the U.S.Army or the Departmentof Defense.

REFERENCES

1. Sengupta S, Jordan VC. (2008) Selective estrogen modulators as an anticancer tool: mechanisms of efficiency and resistance. *Adv Exp Med Biol* **630**:206–219.
2. Jordan VC. (1984) Biochemical pharmacology of antiestrogen action. *Pharmacol Rev* **36**:245–276.
3. Lerner LJ, Jordan VC. (1990) Development of antiestrogens and their use in breast cancer: Eighth Cain Memorial Award Lecture. *Cancer Res* **50**:4177–4189.

4. Lerner LJ, Holthaus Jr, FJ, Thompson CR. (1958) A non-steroidal estrogen antiagonist 1-(p-2-diethylaminoethoxyphenyl)-1-phenyl-2-p-methoxyphenyl ethanol. *Endocrinology* **63**:295–318.
5. Lerner LJ. (1981) The first non-steroidal antioestrogen — MER25. In: eds. Sutherland RL. Jordan VC. *Non-Steroidal Antioestrogens: Molecular Pharmacology and Antitumour Activity*, pp. 1–16. Australia: Academic.
6. Holtkamp DE, Greslin JG, Root CA, Lerner LJ. (1960) Gonadotrophin inhibiting and anti-fecundity effects of chloramiphene. *Proc Soc Exp Biol Med* **105**: 197–201.
7. Harper MJ, Walpole AL. (1967) Mode of action of I.C.I. 46,474 in preventing implantation in rats. *J Endocrinol* **37**:83–92.
8. Harper MJ, Walpole AL. (1967) A new derivative of triphenylethylene: effect on implantation and mode of action in rats. *J Reprod Fertil* **13**:101–119.
9. Lunan CB, Klopper A. (1975) Antioestrogens. a review. *Clin Endocrinol (Oxf)* **4**:551–572.
10. Kilbourn BT, Mais RHB, Owston PG. (1968) Identification of isomers of a substituted triarylethylene: the crystal structure of 1-p-(2-dimethylaminoethoxyphenyl)-1,2-cis-diphenylbut-1-ene hydrobromide. *Chem Commun* 291–292.
11. Harper MJ, Walpole AL. (1966) Contrasting endocrine activities of cis and trans isomers in a series of substituted triphenylethylenes. *Nature* **212**:87.
12. Duncan GW, Lyster SC, Clark JJ, Lednicer D. (1963) Antifertility activities of two diphenyl-dihydronapthalene derivatives. *Proc Soc Exp Biol Med* **112**:439–442.
13. Duncan GW, Stucki JC, Lyster SC, Lednicer D. (1962) An orally effective mammalian antifertility agent. *Proc Soc Exp Biol Med* **109**:163–166.
14. Beatson GT. (1896) On the treatment of inoperable cases of carcinoma of the mamma: suggestions for a new method of treatment with illustrative cases. *Lancet* **2**:104–107.
15. Jensen EV, Jacobson HI. (1962) Basic guides to the mechanism of estrogen action. *Recent Prog Horm Res* **18**:387–414.
16. Toft D, Gorski J. (1966) A receptor molecule for estrogens: isolation from the rat uterus and preliminary characterization. *Proc Natl Acad Sci USA* **55**:1574–1581.
17. Jordan VC, Jaspan T. (1976) Tamoxifen as an anti-tumour agent: oestrogen binding as a predictive test for tumour response. *J Endocrinol* **68:**453–460.
18. Kiang DT, Kennedy BJ. (1977) Tamoxifen (antiestrogen) therapy in advanced breast cancer. *Ann Intern Med* **87**:687–690.
19. Jordan VC, Koerner S. (1975) Tamoxifen (ICI 46,474) and the human carcinoma 8S oestrogen receptor. *Eur J Cancer* **11**:205–206.
20. Nicholson RI, Golder MP. (1975) The effect of synthetic anti-oestrogens on the growth and biochemistry of rat mammary tumours. *Eur J Cancer* **11**:571–579.
21. Jordan VC. (2003) Tamoxifen: a most unlikely pioneering medicine. *Nat Rev Drug Discov* **2**:205–213.

22. Jordan VC. (1976) Effect of tamoxifen (ICI 46,474) on initiation and growth of DMBA-induced rat mammary carcinomata. *Eur J Cancer* **12**:419–424.
23. Fisher B, Costantino JP, Wickerham DL, Redmond CK, Kavanah M, Cronin WM, Vogel V, Robidoux A, Dimitrov N, Atkins J, Daly M, Wieand S, Tan-Chiu E, Ford L, Wolmark N. (1998) Tamoxifen for prevention of breast cancer: report of the National Surgical Adjuvant Breast and Bowel Project P-1 Study. *J Natl Cancer Inst* **90**:1371–1388.
24. Jordan VC. (2006) Tamoxifen (ICI46,474) as a targeted therapy to treat and prevent breast cancer. *Br J Pharmacol* **147** (Suppl 1): S269–S276.
25. Jordan VC. (2007) Chemoprevention of breast cancer with selective oestrogen-receptor modulators. *Nat Rev Cancer* **7**:46–53.
26. Ettinger B, Black DM, Mitlak BH, Knickerbocker RK, Nickelsen T, Genant HK, Christiansen C, Delmas PD, Zanchetta JR, Stakkestad J, Gluer CC, Krueger K, Cohen FJ, Eckert S, Ensrud KE, Avioli LV, Lips P, Cummings SR. (1999) Reduction of vertebral fracture risk in postmenopausal women with osteoporosis treated with raloxifene: results from a 3-year randomized clinical trial. Multiple Outcomes of Raloxifene Evaluation (MORE) Investigators. *JAMA* **282**:637–645.
27. Cummings SR, Eckert S, Krueger KA, Grady D, Powles TJ, Cauley JA, Norton L, Nickelsen T, Bjarnason NH, Morrow M, Lippman ME, Black D, Glusman JE, Costa A, Jordan VC. (1999) The effect of raloxifene on risk of breast cancer in postmenopausal women: results from the MORE randomized trial. *JAMA* **281**:2189–2197.
28. Jordan VC, Gapstur S, Morrow M. (2001) Selective estrogen receptor modulation and reduction in risk of breast cancer, osteoporosis, and coronary heart disease. *J Natl Cancer Inst* **93**:1449–1457.
29. Buzdar AU, Marcus C, Holmes F, Hug V, Hortobagyi G, (1988) Phase II evaluation of Ly156758 in metastatic breast cancer. *Oncology* **45**:344–345.
30. Jordan VC, Phelps E, Lindgren JU. (1987) Effects of anti-estrogens on bone in castrated and intact female rats. *Breast Cancer Res Treat* **10**:31–35.
31. Gottardis MM, Jordan VC. (1987) Antitumor actions of keoxifene and tamoxifen in the N-nitrosomethylurea-induced rat mammary carcinoma model. *Cancer Res* **47**:4020–4024.
32. Miller CP. (2002) SERMs: evolutionary chemistry, revolutionary biology. *Curr Pharm Des* **8**:2089–2111.
33. Vogel VG, Costantino JP, Wickerham DL, Cronin WM, Cecchini RS, Atkins JN, Bevers TB, Fehrenbacher L, Pajon Jr, ER, Wade 3rd, JL, Robidoux A, Margolese RG, James J, Lippman SM, Runowicz CD, Ganz PA, Reis SE, McCaskill-Stevens W, Ford LG, Jordan VC, Wolmark N, (2006) Effects of tamoxifen vs raloxifene on the risk of developing invasive breast cancer and other disease outcomes: the NSABP Study of Tamoxifen and Raloxifene (STAR) P-2 trial. *JAMA* **295**:2727–2741.
34. Vogel VG, Costantino JP, Wickerham DL, Cronin WM, Cecchini RS, Atkins JN, Bevers TB, Fehrenbacher L, Pajon ER, Wade 3rd, JL, Robidoux A, Margolese RG,

57. Sato M, Turner CH, Wang T, Adrian MD, Rowley E, Bryant HU. (1998) LY353381. HCl: a novel raloxifene analog with improved SERM potency and efficacy *in vivo*. *J Pharmacol Exp Ther* **287**:1-7.
58. Palkowitz AD, Glasebrook AL, Thrasher KJ, Hauser KL, Short LL, Phillips DL, Muehl BS, Sato M, Shetler PK, Cullinan GJ, Pell TR, Bryant HU. (1997) Discovery and synthesis of [6-hydroxy-3-[4-[2-(1-piperidinyl)ethoxy]phenoxy]-2-(4-hydroxyphenyl)]benzo[b]thiophene: a novel, highly potent, selective estrogen receptor modulator. *J Med Chem* **40**:1407-1416.
59. Suh N, Glasebrook AL, Palkowitz AD, Bryant HU, Burris LL, Starling JJ, Pearce HL, Williams C, Peer C, Wang Y, Sporn MB, (2001) Arzoxifene, a new selective estrogen receptor modulator for chemoprevention of experimental breast cancer. *Cancer Res* **61**:8412-8415.
60. Freddie CT, Larsen, SS, Bartholomaeussen M, Lykkesfeldt AE.(2004) The effect of the new SERM arzoxifene on growth and gene expression in MCF-7 breast cancer cells. *Mol Cell Endocrinol* **219**:27-36.
61. Munster PN. (2006) Arzoxifene: the development and clinical outcome of an ideal SERM. *Expert Opin Investig Drugs* **15**:317-326.
62. Schafer JM, Lee ES, Dardes RC, Bentrem D, O'Regan RM, De Los Reyes A, Jordan VC. (2001) Analysis of cross-resistance of the selective estrogen receptor modulators arzoxifene (LY353381) and LY117018 in tamoxifen-stimulated breast cancer xenografts. *Clin Cancer Res* **7**:2505-2512.
63. Ma YL, Bryant HU, Zeng Q, Palkowitz A, Jee WS, Turner CH, Sato M. (2002) Long-term dosing of arzoxifene lowers cholesterol, reduces bone turnover, and preserves bone quality in ovariectomized rats. *J Bone Miner Res* **17**:2256-2264.
64. Jordan VC, Bain RR, Brown RR, Gosden B, Santos MA. (1983) Determination and pharmacology of a new hydroxylated metabolite of tamoxifen observed in patient sera during therapy for advanced breast cancer. *Cancer Res* **43**:1446-1450.
65. Qu Q, Zheng H, Dahllund J, Laine A, Cockcroft N, Peng Z, Koskinen M, Hemminki K, Kangas L, Vaananen K, Harkonen P. (2000) Selective estrogenic effects of a novel triphenylethylene compound, FC1271a, on bone, cholesterol level, and reproductive tissues in intact and ovariectomized rats. *Endocrinology* **141**:809-820.
66. Qu Q, Harkonen PL, Vaananen HK. (1999) Comparative effects of estrogen and antiestrogens on differentiation of osteoblasts in mouse bone marrow culture. *J Cell Biochem* **73**:500-507.
67. Michael H, Harkonen PL, Kangas L, Vaananen HK, Hentunen TA. (2007) Differential effects of selective oestrogen receptor modulators (SERMs) tamoxifen, ospemifene and raloxifene on human osteoclasts *in vitro*. *Br J Pharmacol* **151**:384-395.
68. Wurz GT, Read KC, Marchisano-Karpman C, Gregg JP, Beckett LA, Yu Q, Degregorio MW. (2005) Ospemifene inhibits the growth of

dimethylbenzanthracene-induced mammary tumors in Sencar mice. *J Steroid Biochem Mol Biol* **97**:230–240.

69. Namba R, Young LJ, Maglione JE, McGoldrick ET, Liu S, Wurz GT, DeGregorio MW, Borowsky AD, MacLeod CL, Cardiff RD, Gregg JP. (2005) Selective estrogen receptor modulators inhibit growth and progression of premalignant lesions in a mouse model of ductal carcinoma *in situ*. *Breast Cancer Res* **7**:R881–R889.
70. Hellmann-Blumberg U, Taras TL, Wurz GT, DeGregorio MW. (2000) Genotoxic effects of the novel mixed antiestrogen FC-1271a in comparison to tamoxifen and toremifene. *Breast Cancer Res Treat* **60**:63–70.
71. Wurz GT, Hellmann-Blumberg U, DeGregorio MW.(2008) Pharmacologic effects of ospemifene in rhesus macaques: a pilot study. *Basic Clin Pharmacol Toxicol* **102**:552–558.
72. Mosselman S, Polman J, Dijkema R. (1996) ER beta: identification and characterization of a novel human estrogen receptor. *FEBS Lett* **392**:49–53.
73. Kuiper GG, Carlsson B, Grandien K, Enmark E, Haggblad J, Nilsson S, Gustafsson JA. (1997) Comparison of the ligand binding specificity and transcript tissue distribution of estrogen receptors alpha and beta. *Endocrinology* **138**:863–870.
74. Kuiper GG, Gustafsson JA. (1997) The novel estrogen receptor-beta subtype: potential role in the cell- and promoter-specific actions of estrogens and anti-estrogens. *FEBS Lett* **410**:87–90.
75. Pike AC, Brzozowski AM, Hubbard RE, Bonn T, Thorsell AG, Engstrom O, Ljunggren J, Gustafsson JA, Carlquist M. (1999) Structure of the ligand-binding domain of oestrogen receptor beta in the presence of a partial agonist and a full antagonist. *Embo J* **18**:4608–4618.
76. Sun J, Meyers MJ, Fink BE, Rajendran R, Katzenellenbogen JA, Katzenellenbogen BS. (1999) Novel ligands that function as selective estrogens or antiestrogens for estrogen receptor-alpha or estrogen receptor-beta. *Endocrinology* **140**:800–804.
77. Shiau AK, Barstad D, Radek JT, Meyers MJ, Nettles KW, Katzenellenbogen BS, Katzenellenbogen JA, Agard DA, Greene GL. (2002) Structural characterization of a subtype-selective ligand reveals a novel mode of estrogen receptor antagonism. *Nat Struct Biol* **9**:359–364.
78. Pike AC, Brzozowski AM, Walton J, Hubbard RE, Thorsell AG, Li YL, Gustafsson JA, Carlquist M. (2001) Structural insights into the mode of action of a pure antiestrogen. *Structure* **9**:145–153.
79. Mortensen DS, Rodriguez AL, Carlson KE, Sun J, Katzenellenbogen BS, Katzenellenbogen JA. (2001) Synthesis and biological evaluation of a novel series of furans: ligands selective for estrogen receptor alpha. *J Med Chem* **44**:3838–3848.
80. Stauffer SR, Coletta CJ, Tedesco R, Nishiguchi G, Carlson K, Sun J, Katzenellenbogen BS, Katzenellenbogen JA. (2000) Pyrazole ligands: structure–affinity/activity

relationships and estrogen receptor-alpha-selective agonists. *J Med Chem* **43**:4934–4947.

81. Fink BE, Mortensen DS, Stauffer SR, Aron ZD, Katzenellenbogen JA. (1999) Novel structural templates for estrogen-receptor ligands and prospects for combinatorial synthesis of estrogens. *Chem Biol* **6**:205–219.
82. Meyers MJ, Sun J, Carlson KE, Marriner GA, Katzenellenbogen BS, Katzenellenbogen JA. (2001) Estrogen receptor-beta potency-selective ligands: structure-activity relationship studies of diarylpropionitriles and their acetylene and polar analogues. *J Med Chem* **44**:4230–4251.
83. Speirs V, Parkes AT, Kerin MJ, Walton DS, Carleton PJ, Fox JN, Atkin SL. (1999) Coexpression of estrogen receptor alpha and beta: poor prognostic factors in human breast cancer? *Cancer Res* **59**:525–528.
84. Dotzlaw H, Leygue E, Watson PH, Murphy LC. (1999) Estrogen receptor-beta messenger RNA expression in human breast tumor biopsies: relationship to steroid receptor status and regulation by progestins. *Cancer Res* **59**: 529–532.
85. Speirs V, Malone C, Walton DS, Kerin MJ, Atkin SL. (1999) Increased expression of estrogen receptor beta mRNA in tamoxifen-resistant breast cancer patients. *Cancer Res* **59**:5421–5424.
86. Hoekstra WJ, Patel HS, Liang X, Blanc JB, Heyer DO, Willson TM, Iannone MA, Kadwell SH, Miller LA, Pearce KH, Simmons CA, Shearin J. (2005) Discovery of novel quinoline-based estrogen receptor ligands using peptide interaction profiling. *J Med Chem* **48**:2243–2247.
87. Kim S, Wu JY, Birzin ET, Frisch K, Chan W, Pai LY, Yang YT, Mosley RT, Fitzgerald PM, Sharma N, Dahllund J, Thorsell AG, DiNinno F, Rohrer SP, Schaeffer JM, Hammond ML. (2004) Estrogen receptor ligands. II. Discovery of benzoxathins as potent, selective estrogen receptor alpha modulators. *J Med Chem* **47**:2171–2175.
88. Renaud J, Bischoff SF, Buhl T, Floersheim P, Fournier B, Geiser M, Halleux C, Kallen J, Keller H, Ramage P. (2005) Selective estrogen receptor modulators with conformationally restricted side chains. Synthesis and structure–activity relationship of ERalpha-selective tetrahydroisoquinoline ligands. *J Med Chem* **48**:364–379.
89. Renaud J, Bischoff SF, Buhl T, Floersheim P, Fournier B, Halleux C, Kallen J, Keller H, Schlaeppi JM, Stark W. (2003) Estrogen receptor modulators: identification and structure–activity relationships of potent ERalpha-selective tetrahydroisoquinoline ligands. *J Med Chem* **46**:2945–2957.
90. Henke BR, Consler TG, Go N, Hale RL, Hohman DR, Jones SA, Lu AT, Moore LB, Moore JT, Orband-Miller LA, Robinett RG, Shearin J, Spearing PK, Stewart EL, Turnbull PS, Weaver SL, Williams SP, Wisely GB, Lambert MH. (2002) A new series of estrogen receptor modulators that display selectivity for estrogen receptor beta. *J Med Chem* **45**:5492–5505.

91. Sun J, Baudry J, Katzenellenbogen JA, Katzenellenbogen BS. (2003) Molecular basis for the subtype discrimination of the estrogen receptor-beta-selective ligand, diarylpropionitrile. *Mol Endocrinol* **17**:247–258.
92. Compton DR, Sheng S, Carlson KE, Rebacz NA, Lee IY, Katzenellenbogen BS, Katzenellenbogen JA. (2004) Pyrazolo[1,5-a]pyrimidines: estrogen receptor ligands possessing estrogen receptor beta antagonist activity. *J Med Chem* **47**:5872–5893.
93. Manas ES, Unwalla RJ, Xu ZB, Malamas MS, Miller CP, Harris HA, Hsiao C, Akopian T, Hum WT, Malakian K, Wolfrom S, Bapat A, Bhat RA, Stahl ML, Somers WS, Alvarez JC. (2004) Structure-based design of estrogen receptor-beta selective ligands. *J Am Chem Soc* **126**:15106–15119.
94. Mewshaw RE, Edsall Jr, RJ, Yang C, Manas ES, Xu ZB, Henderson RA, Keith Jr, JC, Harris HA. (2005) ERbeta ligands. 3. Exploiting two binding orientations of the 2-phenylnaphthalene scaffold to achieve ERbeta selectivity. *J Med Chem* **48**:3953–3979.
95. Hillisch A, Peters O, Kosemund D, Muller G, Walter A, Schneider B, Reddersen G, Elger W, Fritzemeier KH. (2004) Dissecting physiological roles of estrogen receptor alpha and beta with potent selective ligands from structure-based design. *Mol Endocrinol* **18**:1599–1609.
96. Norman BH, Dodge JA, Richardson TI, Borromeo PS, Lugar CW, Jones SA, Chen K, Wang Y, Durst GL, Barr RJ, Montrose-Rafizadeh C, Osborne HE, Amos RM, Guo S, Boodhoo A, Krishnan V. (2006) Benzopyrans are selective estrogen receptor beta agonists with novel activity in models of benign prostatic hyperplasia. *J Med Chem* **49**:6155–6157.
97. Malamas MS, Manas ES, McDevitt RE, Gunawan I, Xu ZB, Collini MD, Miller CP, Dinh T, Henderson RA, Keith Jr, JC, Harris HA. (2004) Design and synthesis of aryl diphenolic azoles as potent and selective estrogen receptor-beta ligands. *J Med Chem* **47**:5021–5040.
98. Harris HA, Albert LM, Leathurby Y, Malamas MS, Mewshaw RE, Miller CP, Kharode YP, Marzolf J, Komm BS, Winneker RC, Frail DE, Henderson RA, Zhu Y, Keith Jr, JC. (2003) Evaluation of an estrogen receptor-beta agonist in animal models of human disease. *Endocrinology* **144**:4241–4249.
99. Follettie MT, Pinard M, Keith Jr, JC. Wang L, Chelsky D, Hayward C, Kearney P, Thibault P, Paramithiotis E, Dorner AJ, Harris HA. (2006) Organ messenger ribonucleic acid and plasma proteome changes in the adjuvant-induced arthritis model: responses to disease induction and therapy with the estrogen receptor-beta selective agonist ERB-041. *Endocrinology* **147**:714–723.
100. Christaki E, Opal SM, Keith Jr, JC, Kessinian N, Palardy JE, Parejo NA, Lavallie E, Racie L, Mounts W, Malamas MS, Mewshaw RE, Harris HA, Vlasuk GP. (2010) Estrogen receptor beta agonism increases survival in experimentally induced sepsis and ameliorates the genomic sepsis signature: a pharmacogenomic study. *J Infect Dis* **201**:1250–1257.

101. Gungor T, Chen Y, Golla R, Ma Z, Corte JR, Northrop JP, Bin B, Dickson JK, Stouch T, Zhou R, Johnson SE, Seethala R, Feyen JH. (2006) Synthesis and characterization of 3-arylquinazolinone and 3-arylquinazolinethione derivatives as selective estrogen receptor beta modulators. *J Med Chem* **49:**2440–2455.
102. J. Shen, C. Tan, Y. Zhang, X. Li, W. Li, J. Huang, X. Shen Tang. Y. (2010) Discovery of potent ligands for estrogen receptor beta by structure-based virtual screening. *J Med Chem* **53:**5361–5365.
103. Watanabe N, Ikeno A, Minato H, Nakagawa H, Kohayakawa C, Tsuji J. (2003) Discovery and preclinical characterization of (+)-3-[4-(1-piperidinoethoxy)phenyl]spiro[indene-1,1'-indane]-5,5'-diol hydrochloride: a promising nonsteroidal estrogen receptor agonist for hot flush. *J Med Chem* **46**:3961–3964.
104. Wallace OB, Lauwers KS, Dodge JA, May SA, Calvin JR, Hinklin R, Bryant HU, Shetler PK, Adrian MD, Geiser AG, Sato M, Burris TP. (2006) A selective estrogen receptor modulator for the treatment of hot flushes. *J Med Chem* **49**:843–846.
105. Jain N, Kanojia RM, Xu J, Jian-Zhong G, Pacia E, Lai MT, Du F, Musto A, Allan G, Hahn D, Lundeen S, Sui Z. (2006) Novel chromene-derived selective estrogen receptor modulators useful for alleviating hot flushes and vaginal dryness. *J Med Chem* **49**:3056–3059.
106. Jain N, Xu J, Kanojia RM, Du F, Jian-Zhong G, Pacia E, Lai MT, Musto A, Allan G, Reuman M, Li X, Hahn D, Cousineau M, Peng S, Ritchie D, Russell R, Lundeen S, Sui Z. (2009) Identification and structure–activity relationships of chromene-derived selective estrogen receptor modulators for treatment of postmenopausal symptoms. *J Med Chem* **52**:7544–7569.
107. Hummel CW, Geiser AG, Bryant HU, Cohen IR, Dally RD, Fong KC, Frank SA, Hinklin R, Jones SA, Lewis G, McCann DJ, Rudmann DG, Shepherd TA, Tian H, Wallace OB, Wang M, Wang Y, Dodge JA. (2005) A selective estrogen receptor modulator designed for the treatment of uterine leiomyoma with unique tissue specificity for uterus and ovaries in rats. *J Med Chem* **48**:6772–6775.
108. Zhao L, O'Neill K, Brinton. RD. (2006) Estrogenic agonist activity of ICI 182,780 (Faslodex) in hippocampal neurons: implications for basic science understanding of estrogen signaling and development of estrogen modulators with a dual therapeutic profile. *J Pharmacol Exp Ther* **319**:1124–1132.
109. Zhao L, Jin C, Mao Z, Gopinathan MB, Rehder K, Brinton RD. (2007) Design, synthesis, and estrogenic activity of a novel estrogen receptor modulator — a hybrid structure of 17beta-estradiol and vitamin E in hippocampal neurons. *J Med Chem* **50**:4471–4481.
110. Jordan VC. (2008) Tamoxifen: catalyst for the change to targeted therapy. *Eur J Cancer* **44**:30–38.
111. Berry DA, Cronin KA, Plevritis SK, Fryback DG, Clarke L, Zelen M, Mandelblatt JS, Yakovlev AY, Habbema JD, Feuer EJ. (2005) Effect of screening and adjuvant therapy on mortality from breast cancer. *N Engl J Med* **353**:1784–1792.

112. Ariazi E, Jordan VC. (2008) Estrogen receptors as therapeutic targets in breast cancer. In: *Nuclear Receptors as Drug Targets*, eds. Ottow E, and H. Weinmann. H. Wiley-VCH, pp. 127-199.
113. Peng J, Sengupta S, Jordan VC, (2009) Potential of selective estrogen receptor modulators as treatments and preventives of breast cancer. *Anticancer Agents Med Chem* **9**:481-499.
114. Jordan VC. (2003) Antiestrogens and selective estrogen receptor modulators as multifunctional medicines. 2. Clinical considerations and new agents. *J Med Chem* **46**:1081-1111.
115. Jordan VC. (2003) Antiestrogens and selective estrogen receptor modulators as multifunctional medicines. 1. Receptor interactions. *J Med Chem* **46**:883-908.

14

ENDOCRINE PREVENTION OF BREAST CANCER

Trevor J. Powles, Russell E. McDaniel and V. Craig Jordan

Abstract

During the past four decades, the prospects of developing medicines to reduce the risk of breast cancer in healthy women have gone from a laboratory concept to a clinical reality. Tamoxifen, the pioneering selective estrogen receptor modulator (SERM), was initially shown to prevent rodent mammary carcinogenesis and the data were used to suggest prospective placebo-controlled clinical trials. Tamoxifen reduced the incidence of breast cancer by 50%; but, most importantly, this reduction versus placebo was maintained for a decade after stopping the drug. A small but significant increase in endometrial cancer in postmenopausal women taking tamoxifen, but not premenopausal women, makes tamoxifen the chemopreventive of choice for the high risk premenopausal woman. The finding that the SERM raloxifene maintains bone density in postmenopausal women but decreases breast cancer incidence without an increase in endometrial cancer makes raloxifene the chemopreventive of choice in postmenopausal women. A range of new and novel SERMs has completed clinical testing.

Keywords: SERMs; Tamoxifen; Raloxifene; Osteoporosis; Chemoprevention; Endometrial Cancer.

INTRODUCTION

The idea of the prevention of breast cancer is not new, but significant practical progress has been made, through translational research, to make the idea feasible in some women. It is now possible to reduce the incidence of breast cancer through the inhibition of estrogen action.

Professor Antoine Lacassagne[1] stated a vision for the prevention of breast cancer at the annual meeting of the American Association of Cancer Research in Boston in 1936: "If one accepts the consideration of adenocarcinoma of the breast as the consequence of a special hereditary sensibility to the proliferative actions of oestrone, one is led to imagine a therapeutic preventative for subjects predisposed by their heredity to this cancer. It would consist — perhaps in the very near future when the knowledge and use of hormones will be better understood — in the suitable use of a hormone antagonistic or excretory, to prevent the stagnation of oestrone in the ducts of the breast."

But no agent that was "antagonistic to prevent the stagnation of oestrone in the breast" was available to the clinician for clinical trial until tamoxifen.[2,3] Tamoxifen (Fig. 1) became the "antiestrogen" of

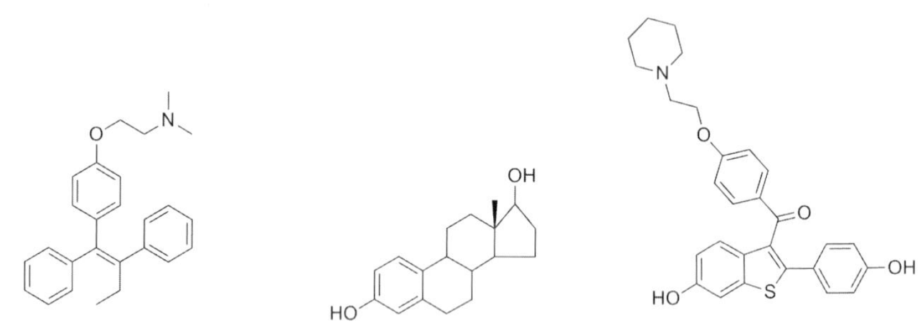

Tamoxifen
Originally ICI 46,474
a failed "morning after" pill

17 β-Estradiol

Raloxifene
originally Keoxifene
a failed breast cancer drug

Figure 1. A comparison of the structures of the potent steroidal estrogen 17β-estradiol and the nonsteroidal antiestrogens (now called SERMs) tamoxifen and raloxifene. Tamoxifen and raloxifene are both approved in the U.S. for reducing the risk of breast cancer in high risk pre- and postmenopausal women and postmenopausal women alone, respectively.

choice because (a) there was a large body of basic biological evidence that this was a valid hypothesis to test,[4-10] (b) tamoxifen was noted to reduce the incidence of contralateral breast cancer when used as an adjuvant therapy to treat micrometastases from the original primary tumor[11] and, most importantly, (c) there was a huge and expanding clinical experience with tamoxifen as a long term treatment for node-positive and node-negative breast cancer. The latter point was important, as the majority of patients with estrogen receptor (ER)-positive node-negative breast cancers are cured by surgery (plus radiation) alone and so five years of adjuvant tamoxifen was essentially being used in the majority of "well women."[12,13]

The treatment trials database and translational research were essential for addressing the hypothesis that tamoxifen, a nonsteroidal antiestrogen, could effectively block the genesis and growth of ER-positive breast cancer but would be ineffective against the growth of ER-negative disease. Nevertheless, in the 1980s, estrogen was also considered to be an essential component of women's health by maintaining bone density and preventing coronary heart disease. Thus, if tamoxifen, an antiestrogen, prevented the development and growth of ER-positive breast cancer in half a dozen high risk women per year per thousand,[14] hundreds of other women in the selected population might subsequently develop osteoporosis and coronary heart disease. The intervention with tamoxifen would be detrimental to public health. The good news was that tamoxifen was not an antiestrogen everywhere; it was the lead compound of the drug group that selectively modulated ER target tissues around the body. This discovery ultimately facilitated the development of a new strategy for the utilization of new SERMs as chemopreventives in breast cancer.

SERM ACTION IN THE LABORATORY

The original work to investigate the target site pharmacology of tamoxifen in the laboratory was to provide a database with which to predict clinical outcomes and safety for future chemoprevention trials. Historically, in the 1960s, there was general interest in the chance finding that nonsteroidal antiestrogens lowered circulating cholesterol.

postmenopausal patients using long term adjuvant tamoxifen therapy.[24,25] Subsequent laboratory studies showed that less estrogenic antiestrogens such as raloxifene or ICI 164,384, a "pure" antiestrogen, were unable to cause endometrial cancer growth.[26]

SERM ACTION IN CLINICAL TRIALS

Overall, the developing interest in conducting chemoprevention studies[27-29] was complemented both by the clinical pharmacology trials at the Royal Marsden Hospital and by the Wisconsin Tamoxifen Study.

The first pilot chemoprevention study was initiated by TJP at the Royal Marsden Hospital in the early 1980s.[30] This study grew over the years of accrual and interestingly showed benefit at 20 years for those women taking tamoxifen for 8 years following recruitment.[31] All subsequent studies benefitted from the evaluation of the clinical pharmacology of tamoxifen.

Studies at the Royal Marsden Hospital methodically addressed a series of clinical issues. The concern about the potential increase in endometrial cancer with tamoxifen use in postmenopausal women prompted an evaluation of endometrial histopathology in women taking tamoxifen.[32] Concerns about tamoxifen and an increase in coronary heart disease were addressed in a study of circulating lipids and C-reactive protein, a predictor of risk for coronary heart disease.[33] Bone density was improved in postmenopausal women but slightly reduced in premenopausal women.[34]

Tamoxifen maintains bone density in ovariectomized rats[17,20,35] and this counterintuitive laboratory result for an "antiestrogen" formed the scientific basis for the Wisconsin Tamoxifen Study. The clinical study was a double-blind placebo-controlled trial to establish the actions of two years' tamoxifen on bone density in postmenopausal patients with node-negative breast cancer (at the time of recruitment, these patients were several years postdiagnosis and surgery as no adjuvant treatment was the standard of care). Tamoxifen significantly improved bone density compared to placebo treatment.[36]

During the 1980s, the Wisconsin Tamoxifen Study followed up the question of tamoxifen treatment lowering circulating cholesterol in postmenopausal patients[37,38] and noted a decrease in low density lipoprotein cholesterol but no effect on high density lipoprotein cholesterol.There was certainly some initial enthusiasm that there would be a significant decrease in coronary heart disease, but despite some encouraging reports[39–41] no consistent decrease in coronary events has been noted in the Oxford Overview Analysis for tamoxifen treatment.

Thus, tamoxifen was estrogen-like, by lowering circulating cholesterol and maintaining bone density, so that tamoxifen might provide benefit for women enrolled in a chemoprevention trial. The anticancer actions of tamoxifen were well established and supported by the inhibition of mammary carcinogenesis in rat[6,21] and mouse[42] models. But an increase in the incidence in endometrial cancer was a predictable concern, based on earlier work[23,24] before major clinical trials of chemoprevention in breast cancer started. Also, the finding that tamoxifen was a hepatocarcinogen in specific rat strains[43] was of significance toxicologically for safety reasons in any chemopreventive trials; however, it is fair to say that no evidence either at that time[44,45] or subsequently has demonstrated hepatocarcinogenesis in humans with tamoxifen.

The pivotal chemoprevention study was the Fisher P-1 study (Fig. 2), conducted by the National Surgical Adjuvant Breast and Bowel

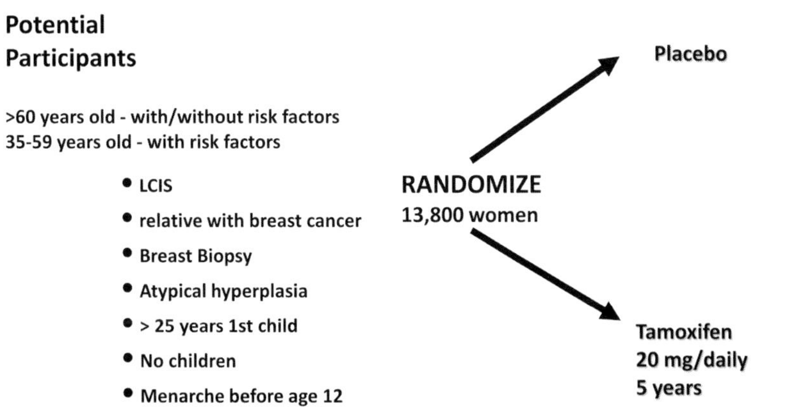

Figure 2. The risk requirements for recruitment to the National Surgical Breast and Bowel Project (NSABP)/National Cancer Institute (NCI) study P-1 to determine the worth of tamoxifen for preventing breast cancer in high risk pre- and postmenopausal women.[28]

Project (NSABP).[28] This landmark study was an adequately powered prospective, placebo-controlled trial primarily used by the Food and Drug Administration as evidence to approve tamoxifen for the reduction of risk of breast cancer in pre- and post-menopausal women at high risk for the disease.

There are significant additional benefits for women at risk for breast cancer nested within the results of the P-1 prevention trial during treatment with tamoxifen.There were fewer fractures, but this was not significant overall. Tamoxifen reduces ER–positive invasive breast cancer incidence by 50%, and the same is true for ductal carcinoma *in situ* (DCIS).[28] Benefits of breast chemoprevention last for years following the cessation of treatment[46] and this has been confirmed by other studies.[47] This is clearly a consistent long term "antitumor action" of tamoxifen imprinted following therapy, as noted by the sustained antitumor effect of tamoxifen following adjuvant treatment.[48,49] We will comment further on the new concept of "imprinting" in the "SERM Summary."

Despite extensive testing, tamoxifen is seen as presenting the well woman with significant risks such as endometrial cancer and blood clots (it must be stressed only in postmenopausal women)[28] (Fig. 3). There is also the nagging concern about rat hepatocarcinoma. Tamoxifen has a human carcinogen black box designation in the U.S. With all these uncertainties, clearly another strategy for chemoprevention was necessary for an appropriate science-based advance in public health. This was obvious[50] even before the NSABP trial had been launched in the early 1990s[28] but tamoxifen was the only medicine available with sufficient clinical trial experience to move forward into chemoprevention. Nevertheless, the recognition of SERMs in the laboratory[50] also catalyzed a change in the development of another nonsteroidal antiestrogen, keoxifene (Fig. 1). Keoxifene was initially investigated in the 1980s as a competitor for tamoxifen as a breast cancer drug but failed to advance in development as it failed in clinical trial.[51] Surprisingly, it also maintained bone density in rats, similar to tamoxifen, but was significantly less uterotrophic than tamoxifen:[17,52] This would later translate to a reduced risk of endometrial cancer in all subsequent clinical trials. The name was changed from keoxifene to raloxifene (Fig. 1).

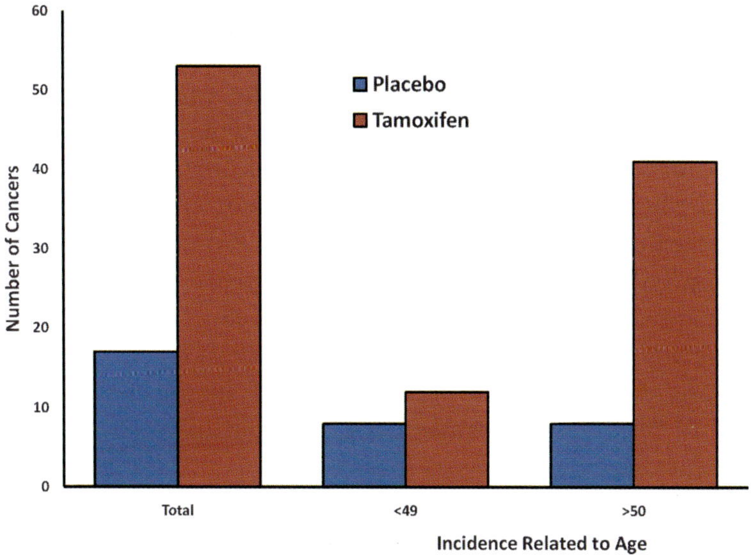

Figure 3. The total and age-related incidence of endometrial cancer in the NSABP/NCI P-1 chemoprevention trial.[46] Premenopausal women have no increased risk of developing endometrial cancer during or following five years of tamoxifen treatment.

Keoxifene prevented mammary cancer in rats but, because of poor pharmacokinetics and rapid excretion, it does not have the sustained actions of tamoxifen.[6] Continuous therapy was necessary. Thus, the scene was set for a move away from a broad therapeutic strategy with tamoxifen administered to high risk populations where a few ER-positive invasive breast cancers can be prevented but the majority of women are exposed to side effects with no benefit to balance the risks. In response, a "roadmap" was created based on laboratory science and the emerging clinical trial data that would significantly advance women's health.

A PLAN TO USE SERMS TO PREVENT MULTIPLE DISEASES IN WOMEN

A plan to prevent breast cancer as a public health initiative was initially described at the First International Chemoprevention Meeting in New York in 1987.[53] It is reasonable simply to state the proposal, published from the 1987 meeting[53] and subsequently refined and presented again at the annual meeting of the American Association for Cancer Research in San Francisco in 1989:[50] "The majority of breast cancer occurs unexpectedly and from unknown origin. Great efforts are being focused on the identification of a population of high-risk women to test 'chemopreventive' agents. But, are resources being used less than optimally? An alternative would be to seize on the developing clues provided by an extensive clinical investigation of available antiestrogens. Could analogues be developed to treat osteoporosis or even retard the development of atherosclerosis? If this proved to be true, then a majority of women in general would be treated for these conditions as soon as menopause occurred. Should the agent also retain anti-breast tumour actions, then it might be expected to act as a chemosuppressive on all developing breast cancers if these have an evolution from hormone-dependent disease to hormone-independent disease. A bold commitment to drug discovery and clinical pharmacology will potentially place us in a key position to prevent the development of breast cancer by the end of this century."[53]

The vision of the concept was refined and focused by 1990:[50] "We have obtained valuable clinical information about this group of drugs that can be applied in other disease states. Research does not travel in straight lines and observations in one field of science often become major discoveries in another. Important clues have been garnered about the effects of tamoxifen on bone and lipids, so apparently, derivatives could find targeted applications to retard osteoporosis or atherosclerosis. The ubiquitous application of novel compounds to prevent diseases associated with the progressive changes after menopause may, as a side effect, significantly retard the development of breast cancer. The target population would be postmenopausal women in general, thereby avoiding the requirement to select a high-risk group to prevent breast cancer." This concept is exactly what has been translated to clinical practice[54,55]: use a SERM (raloxifene) to treat osteoporosis and reduce the incidence of breast cancer as a beneficial side effect.[54-56]

THE SERMS SURFACE IN CLINICAL PRACTICE

Raloxifene is the pioneering SERM approved for the prevention of osteoporosis around the world. The pivotal registration trial was the Multiple Outcomes of Raloxifene Evaluation (MORE) trial. Raloxifene reduced spine fractures by 50% compared to placebo.[56] A separate analysis of breast cancer incidence demonstrated a 76% decrease in the incidence of invasive positive breast cancer (Fig. 4) over the three-year evaluation. There was no increase in endometrial cancer but DCIS remained unaffected.[54] A long-running trial — Raloxifene Use for the Heart (RUTH) — to examine whether coronary heart events could be reduced in high risk populations did not show benefit for raloxifene.[57] Looked at another way, it showed little harm, but coronary heart disease (CHD) in a high risk population was unaffected.

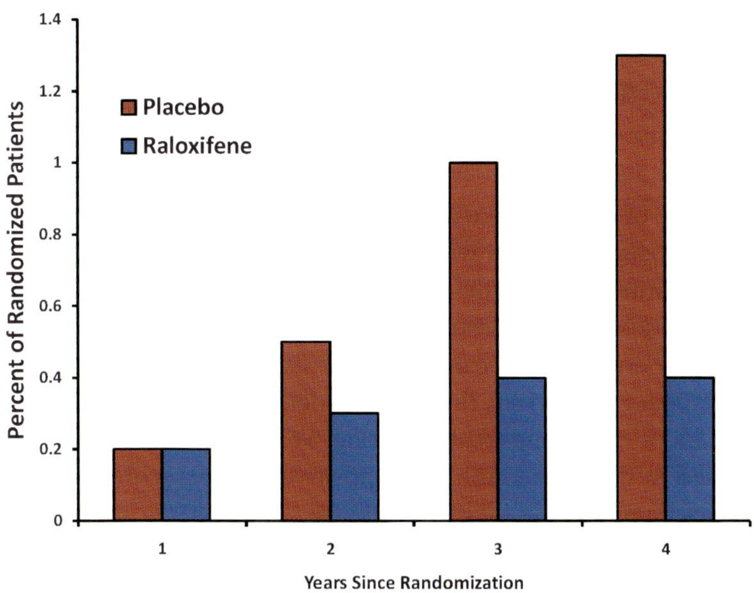

Figure 4. The annual accumulative incidence breast cancers represented as a percentage of affected randomized patients in the Multiple Outcomes of Raloxifene Evaluation (MORE), which randomized women with an increased risk of osteoporotic fractures to placebo (2576 women) or raloxifene (5129 women).[54]

ACKNOWLEDGEMENTS

This work (V.C.J.) was supported by the Department of Defense Breast Program under Award number W81XWH-06-1-0590 Center of Excellence; subcontract under the SU2C (AACR) grant No. SU2C-AACR-DT0409; the Susan G. Komen for the Cure Foundation under award No. SAC100009 and the Lombardi Comprehensive Cancer 1095 Center Support Grant (CCSG) Core Grant NIH P30 CA051008. The views and opinions of the authors do not reflect those of the U.S. Army or the Department of Defense.

REFERENCES

1. Lacassagne A. (1936) Hormonal pathogenesis of adenocarcinoma of the breast. *Am J Cancer* **27**:217–25.
2. Jordan VC. (2003) Tamoxifen: a most unlikely pioneering medicine. *Nat Rev Drug Discov* **2**:205–213.
3. Jordan VC. (2006) Tamoxifen (ICI46,474) as a targeted therapy to treat and prevent breast cancer. *Br J Pharmacol* **147** (Suppl 1):S269–S276.
4. Jordan VC, Dowse LJ. (1976) Tamoxifen as an anti-tumour agent: effect on oestrogen binding. *J Endocrinol* **68**:297–303.
5. Jordan VC, Allen KE, Dix CJ. (1980) Pharmacology of tamoxifen in laboratory animals. *Cancer Treat Rep* **64**:745–759.
6. Gottardis MM, Jordan VC. (1987) Antitumor actions of keoxifene and tamoxifen in the N-nitrosomethylurea-induced rat mammary carcinoma model. *Cancer Res* **47**: 4020–4024.
7. Jordan VC, Lababidi MK, Mirecki DM. (1990) Anti-oestrogenic and anti-tumour properties of prolonged tamoxifen therapy in C3H/OUJ mice. *Eur J Cancer* **26**:718–721.
8. Jordan VC. (1991) Are we giving tamoxifen for too long? *Eur J Cancer* **27**:1070–1072.
9. Jordan V. (1974) Antitumour activity of the antioestrogen ICI 46,474 (tamoxifen) in the dimethyl benzanthracene (DMBA)-induced rat mammary carcinoma model. *J Steroid Biochem* **5**:354.
10. Jordan VC, Fritz NF, Langan-Fahey S, Thompson M, Tormey DC. (1991) Alteration of endocrine parameters in premenopausal women with breast cancer during long-term adjuvant therapy with tamoxifen as the single agent. *J Natl Cancer Inst* **83**:1488–1491.
11. Cuzick J, Baum M. (1985) Tamoxifen and contralateral breast cancer. *Lancet* **2**:282.

12. SCTO. (1987) Adjuvant tamoxifen in the management of operable breast cancer: the Scottish Trial. Report from the Breast Cancer Trials Committee, Scottish Cancer Trials Office (MRC), Edinburgh. *Lancet* **2**:171–175.
13. Fisher B, Costantino J, Redmond C, Poisson R, Bowman D, Couture J, Dimitrov NV, Wolmark N, Wickerham DL, Fisher ER, *et al.* (1989) A randomized clinical trial evaluating tamoxifen in the treatment of patients with node-negative breast cancer who have estrogen-receptor-positive tumors. *N Engl J Med* **320**:479–484.
14. Gail MH, Brinton LA, Byar DP, Corle DK, Green SB, Schairer C, Mulvihill JJ. (1989) Projecting individualized probabilities of developing breast cancer for white females who are being examined annually. *J Natl Cancer Inst* **81**:1879–1886.
15. Harper MJ, Walpole AL. (1967) A new derivative of triphenylethylene: effect on implantation and mode of action in rats. *J Reprod Fertil* **13**:101–119.
16. Consensus conference. (1985) Adjuvant chemotherapy for breast cancer. *JAMA* **254**:3461–3463.
17. Jordan VC, Phelps E, Lindgren JU. (1987) Effects of anti-estrogens on bone in castrated and intact female rats. *Breast Cancer Res Treat* **10**:31–35.
18. Beall PT, Misra LK, Young RL, Spjut HJ, Evans HJ, LeBlanc A. (1984) Clomiphene protects against osteoporosis in the mature ovariectomized rat. *Calcif Tissue Int* **36**:123–125.
19. Turner RT, Vandersteenhoven JJ, Bell NH. (1987) The effects of ovariectomy and 17 beta-estradiol on cortical bone histomorphometry in growing rats. *J Bone Miner Res* **2**:115–122.
20. Turner RT, Wakley GK, Hannon KS, Bell NH. (1987) Tamoxifen prevents the skeletal effects of ovarian hormone deficiency in rats. *J Bone Miner Res* **2**:449–456.
21. Jordan VC. (1976) Effect of tamoxifen (ICI 46,474) on initiation and growth of DMBA-induced rat mammary carcinomata. *Eur J Cancer* **12**:419–424.
22. Jordan VC, Robinson SP. (1987) Species-specific pharmacology of antiestrogens: role of metabolism. *Fed Proc* **46**:1870–1874.
23. Gottardis MM, Robinson SP, Satyaswaroop PG, Jordan VC. (1988) Contrasting actions of tamoxifen on endometrial and breast tumor growth in the athymic mouse. *Cancer Res* **48**:812–815.
24. Fornander T, Rutqvist LE, Cedermark B, Glas U, Mattsson A, Silfversward C, Skoog L, Somell A, Theve T, Wilking N, *et al.* (1989) Adjuvant tamoxifen in early breast cancer: occurrence of new primary cancers. *Lancet* **1**:117–120.
25. Fisher B, Costantino JP, Redmond CK, Fisher ER, Wickerham DL, Cronin WM. (1994) Endometrial cancer in tamoxifen-treated breast cancer patients: findings from the National Surgical Adjuvant Breast and Bowel Project (NSABP) B-14. *J Natl Cancer Inst* **86**:527–537.
26. Gottardis MM, Ricchio ME, Satyaswaroop PG, Jordan VC. (1990) Effect of steroidal and nonsteroidal antiestrogens on the growth of a tamoxifen-stimulated

human endometrial carcinoma (EnCa101) in athymic mice. *Cancer Res* **50**:3189–3192.

27. Powles TJ, Eeles R, Ashley S, Easton D, Chang J, Dowsett M, Tidy A, Viggers J, Davey J. (1998) Interim analysis of the incidence of breast cancer in the Royal Marsden Hospital tamoxifen randomised chemoprevention trial. *Lancet* **352**:98–101.
28. Fisher B, Costantino JP, Wickerham DL, Redmond CK, Kavanah M, Cronin WM, Vogel V, Robidoux A, Dimitrov N, Atkins J, Daly M, Wieand S, Tan-Chiu E, Ford L, Wolmark N. (1998) Tamoxifen for prevention of breast cancer: report of the National Surgical Adjuvant Breast and Bowel Project P-1 Study. *J Natl Cancer Inst* **90**:1371–1388.
29. Veronesi U, Maisonneuve P, Costa A, Sacchini V, Maltoni C, Robertson C, Rotmensz N, Boyle P. (1998) Prevention of breast cancer with tamoxifen: preliminary findings from the Italian randomised trial among hysterectomised women. Italian Tamoxifen Prevention Study. *Lancet* **352**:93–97.
30. Powles TJ, Hardy JR, Ashley SE, Farrington GM, Cosgrove D, Davey JB, Dowsett M, McKinna JA, Nash AG, Sinnett HD, *et al.* (1989) A pilot trial to evaluate the acute toxicity and feasibility of tamoxifen for prevention of breast cancer. *Br J Cancer* **60**:126–131.
31. Powles TJ, Ashley S, Tidy A, Smith IE, Dowsett M. (2007) Twenty-year follow-up of the Royal Marsden randomized, double-blinded tamoxifen breast cancer prevention trial. *J Natl Cancer Inst* **99**:283–290.
32. Kedar RP, Bourne TH, Powles TJ, Collins WP, Ashley SE, Cosgrove DO, Campbell S. (1994) Effects of tamoxifen on uterus and ovaries of postmenopausal women in a randomised breast cancer prevention trial. *Lancet* **343**:1318–321.
33. Powles TJ, Jones AL, Ashley SE, O'Brien ME, Tidy VA, Treleavan J, Cosgrove D, Nash AG, Sacks N, Baum M, *et al.* (1994) The Royal Marsden Hospital pilot tamoxifen chemoprevention trial. *Breast Cancer Res Treat* **31**:73–82.
34. Powles TJ, Hickish T, Kanis JA, Tidy A, Ashley S. (1996) Effect of tamoxifen on bone mineral density measured by dual-energy x-ray absorptiometry in healthy premenopausal and postmenopausal women. *J Clin Oncol* **14**:78–84.
35. Turner RT, Wakley GK, Hannon KS, Bell NH. (1988) Tamoxifen inhibits osteoclast-mediated resorption of trabecular bone in ovarian hormone-deficient rats. *Endocrinology* **122**:1146–1150.
36. Love RR, Mazess RB, Barden HS, Epstein S, Newcomb PA, Jordan VC, Carbone PP, DeMets DL. (1992) Effects of tamoxifen on bone mineral density in postmenopausal women with breast cancer. *N Engl J Med* **326**:852–856.
37. Love RR, Newcomb PA, Wiebe DA, Surawicz TS, Jordan VC, Carbone PP, DeMets DL. (1990) Effects of tamoxifen therapy on lipid and lipoprotein levels in postmenopausal patients with node-negative breast cancer. *J Natl Cancer Inst* **82**:1327–1332.
38. Love RR, Wiebe DA, Newcomb PA, Cameron L, Leventhal H, Jordan VC, Feyzi J, DeMets DL. (1991) Effects of tamoxifen on cardiovascular risk factors in postmenopausal women. *Ann Intern Med* **115**:860–864.

39. Nordenskjöld B, Rosell J, Rutqvist LE, Malmström PO, Bergh J, Bengtsson NO, Hatschek T, Wallgren A, Carstensen J. (2005) Coronary heart disease mortality after 5 years of adjuvant tamoxifen therapy: results from a randomzed trial. *J Natl Cancer Inst* **97**:1609–1610.
40. Hackshaw A, Roughton M, Forsyth S, Monson K, Reczko K, Sainsbury R, Baum M. (2011) Long-term benefits of 5 years of tamoxifen: 10-year follow-up of a large randomized trial in women at least 50 years of age with early breast cancer. *J Clin Oncol* **29**:1657–1663.
41. McDonald CC, Stewart HJ. (1991) Fatal myocardial infarction in the Scottish adjuvant tamoxifen trial. The Scottish Breast Cancer Committee. *BMJ* **303**:435–437.
42. Jordan VC, Lababidi MK, Langan-Fahey S. (1991) Suppression of mouse mammary tumorigenesis by long-term tamoxifen therapy. *J Natl Cancer Inst* **83**:492–496.
43. Greaves P, Goonetilleke R, Nunn G, Topham J, Orton T. (1993) Two-year carcinogenicity study of tamoxifen in Alderley Park Wistar-derived rats. *Cancer Res* **53**:3919–3924.
44. Jordan VC. (1995) What if tamoxifen (ICI 46,474) had been found to produce rat liver tumors in 1973? A personal perspective. *Ann Oncol* **6**:29–34.
45. Jordan VC, Morrow M. (1994) Should clinicians be concerned about the carcinogenic potential of tamoxifen? *Eur J Cancer* **30A**:1714–1721.
46. Fisher B, Costantino JP, Wickerham DL, Cecchini RS, Cronin WM, Robidoux A, Bevers TB, Kavanah MT, Atkins JN, Margolese RG, Runowicz CD, James JM, Ford LG, Wolmark N. (2005) Tamoxifen for the prevention of breast cancer: current status of the National Surgical Adjuvant Breast and Bowel Project P-1 study. *J Natl Cancer Inst* **97**:1652–1662.
47. Cuzick J, Forbes JF, Sestak I, Cawthorn S, Hamed H, Holli K, Howell A. (2007) Long-term results of tamoxifen prophylaxis for breast cancer-—96-month follow-up of the randomized IBIS-I trial. *J Natl Cancer Inst* **99**:272–282.
48. EBCTG. (2011) Relevance of breast cancer hormone receptors and other factors to the efficacy of adjuvant tamoxifen: patient-level meta-analysis of randomised trials. *Lancet* **378**:771–784.
49. EBCTCG. (2005) Effects of chemotherapy and hormonal therapy for early breast cancer on recurrence and 15-year survival: an overview of the randomised trials. *Lancet* **365**:1687–1717.
50. Lerner LJ, Jordan VC. (1990) Development of antiestrogens and their use in breast cancer: Eighth Cain Memorial Award Lecture. *Cancer Res* **50**:4177–4189.
51. Lewis JS, Jordan VC. (2006) Case histories: raloxifene. In: *Comprehensive Medicinal Chemistry II*, Vol. 8, eds. Taylor J, Triggle D. Oxford: Elsevier, pp. 103–121.
52. Black LJ, Jones CD, Falcone JF. (1983) Antagonism of estrogen action with a new benzothiophene-derived antiestrogen. *Life Sci* **32**:1031–1036.
53. Jordan VC. (1988) Chemosuppression of breast cancer with tamoxifen: laboratory evidence and future clinical investigations. *Cancer Invest* **6**:589–595.

54. Cummings SR, Eckert S, Krueger KA, Grady D, Powles TJ, Cauley JA, Norton L, Nickelsen T, Bjarnason NH, Morrow M, Lippman ME, Black D, Glusman JE, Costa A, Jordan VC. (1999) The effect of raloxifene on risk of breast cancer in postmenopausal women: results from the MORE randomized trial. Multiple Outcomes of Raloxifene Evaluation. *JAMA* **281**:2189–2197.
55. Martino S, Cauley JA, Barrett-Connor E, Powles TJ, Mershon J, Disch D, Secrest RJ, Cummings SR. (2004) Continuing outcomes relevant to Evista: breast cancer incidence in postmenopausal osteoporotic women in a randomized trial of raloxifene. *J Natl Cancer Inst* **96**:1751–1761.
56. Ettinger B, Black DM, Mitlak BH, Knickerbocker RK, Nickelsen T, Genant HK, Christiansen C, Delmas PD, Zanchetta JR, Stakkestad J, Gluer CC, Krueger K, Cohen FJ, Eckert S, Ensrud KE, Avioli LV, Lips P, Cummings SR. (1999) Reduction of vertebral fracture risk in postmenopausal women with osteoporosis treated with raloxifene: results from a 3-year randomized clinical trial. Multiple Outcomes of Raloxifene Evaluation (MORE) Investigators. *JAMA* **282**:637–645.
57. Barrett-Connor E, Mosca L, Collins P, Geiger MJ, Grady D, Kornitzer M, McNabb MA, Wenger NK. (2006) Effects of raloxifene on cardiovascular events and breast cancer in postmenopausal women. *N Engl J Med* **355**:125–137.
58. Vogel VG, Costantino JP, Wickerham DL, Cronin WM, Cecchini RS, Atkins JN, Bevers TB, Fehrenbacher L, Pajon ER Jr, Wade JL 3rd, Robidoux A, Margolese RG, James J, Lippman SM, Runowicz CD, Ganz PA, Reis SE, McCaskill-Stevens W, Ford LG, Jordan VC, Wolmark N. (2006) Effects of tamoxifen vs. raloxifene on the risk of developing invasive breast cancer and other disease outcomes: the NSABP Study of Tamoxifen and Raloxifene (STAR) P-2 trial. *Jama* **295**:2727–2741.
59. Vogel VG, Costantino JP, Wickerham DL, *et al.* (2010) Update of the National Surgical Adjuvant Breast and Bowel Project Study of Tamoxifen and Raloxifene (STAR) P-2 Trial: preventing breast cancer. *Cancer Prev Res (Phila)* **3**: 696–706.
60. Baumler E. (1984) *Paul Ehrlich, Scientist for Life*. New York: Holmes & Meier, p. 288.
61. Jordan VC. (2003) Antiestrogens and selective estrogen receptor modulators as multifunctional medicines. 2. Clinical considerations and new agents. *J Med Chem* **46**:1081–1111.
62. Jordan VC. (2003) Antiestrogens and selective estrogen receptor modulators as multifunctional medicines. 1. Receptor interactions. *J Med Chem* **46**:883–908.
63. Qu Q, Harkonen PL, Vaananen HK. (1999) Comparative effects of estrogen and antiestrogens on differentiation of osteoblasts in mouse bone marrow culture. *J Cell Biochem* **73**:500–507.
64. Qu Q, Zheng H, Dahllund J, *et al.* (2000) Selective estrogenic effects of a novel triphenylethylene compound, FC1271a, on bone, cholesterol level, and reproductive tissues in intact and ovariectomized rats. *Endocrinology* **141**:809–820.

65. Gennari L, Merlotti D, Valleggi F, Nuti R. (2009) Ospemifene use in postmenopausal women. *Expert Opin Investig Drugs* **18**:839–849.
66. Rutanen EM, Heikkinen J, Halonen K, *et al.* (2003) Effects of ospemifene, a novel SERM, on hormones, genital tract, climacteric symptoms, and quality of life in postmenopausal women: a double-blind, randomized trial. *Menopause* **10**:433–439.
67. Bachmann GA, Komi JO, Group OS. (2010) Ospemifene effectively treats vulvovaginal atrophy in postmenopausal women: results from a pivotal phase 3 study. *Menopause* **17**:480–486.
68. Komi J, Lankinen KS, Harkonen P, *et al.* (2005) Effects of ospemifene and raloxifene on hormonal status, lipids, genital tract, and tolerability in postmenopausal women. *Menopause* **12**:202–209.
69. Palkowitz AD, Glasebrook AL, Thrasher KJ, *et al.* (1997) Discovery and synthesis of [6-hydroxy-3-[4-[2-(1-piperidinyl)ethoxy]phenoxy]-2-(4-hydroxyphenyl)]benzo[b]thiophene: a novel, highly potent, selective estrogen receptor modulator. *J Med Chem* **40**:1407–1416.
70. Kendler DL, Palacios S, Cox DA, *et al.* (2012) Arzoxifene versus raloxifene: effect on bone and safety parameters in postmenopausal women with osteoporosis. *Osteoporos Int* **23**:1091–1101.
71. Deshmane V, Krishnamurthy S, Melemed AS, Peterson P, Buzdar AU. (2007) Phase III double-blind trial of arzoxifene compared with tamoxifen for locally advanced or metastatic breast cancer. *J Clin Oncol* **25**:4967–4973.
72. Gruber C, Gruber D. (2004) Bazedoxifene (Wyeth). *Curr Opin Investig Drugs* **5**:1086–1093.
73. Miller CP, Collini MD, Tran BD, *et al.* (2001) Design, synthesis, and preclinical characterization of novel, highly selective indole estrogens. *J Med Chem* **44**: 1654–1657.
74. Komm BS, Kharode YP, Bodine PV, *et al.* (2005) Bazedoxifene acetate: a selective estrogen receptor modulator with improved selectivity. *Endocrinology* **146**:3999–4008.
75. Black LJ, Sato M, Rowley ER, *et al.* (1994) Raloxifene (LY139481 HCI) prevents bone loss and reduces serum cholesterol without causing uterine hypertrophy in ovariectomized rats. *J Clin Invest* **93**:63–69.
76. Ke HZ, Paralkar VM, Grasser WA, *et al.* (1998) Effects of CP-336,156, a new, nonsteroidal estrogen agonist/antagonist, on bone, serum cholesterol, uterus and body composition in rat models. *Endocrinology* **139**:2068–2076.
77. Miller PD, Chines AA, Christiansen C, *et al.* (2008) Effects of bazedoxifene on BMD and bone turnover in postmenopausal women: 2-yr results of a randomized, double-blind, placebo-, and active-controlled study. *J Bone Miner Res* **23**:525–535.
78. Pinkerton JV, Archer DF, Utian WH, *et al.* (2009) Bazedoxifene effects on the reproductive tract in postmenopausal women at risk for osteoporosis. *Menopause* **16**:1102–1108.

79. Silverman SL, Christiansen C, Genant HK, *et al.* (2008) Efficacy of bazedoxifene in reducing new vertebral fracture risk in postmenopausal women with osteoporosis: results from a 3-year, randomized, placebo-, and active-controlled clinical trial. *J Bone Miner Res* **23**:1923–1934.
80. de Villiers TJ, Chines AA, Palacios S, *et al.* (2011) Safety and tolerability of bazedoxifene in postmenopausal women with osteoporosis: results of a 5-year, randomized, placebo-controlled phase 3 trial. *Osteoporos Int* **22**:567–576.
81. Grady D, Ettinger B, Moscarelli E, *et al.* (2004) Safety and adverse effects associated with raloxifene: multiple outcomes of raloxifene evaluation. *Obstet Gynecol* **104**:837–844.
82. Cummings SR, Ensrud K, Delmas PD, *et al.* (2010) Lasofoxifene in postmenopausal women with osteoporosis. *N Engl J Med* **362**:686–696.
83 http://www.ukmi.nhs.uk/applications/ndo/record_view_open.asp?newDrugID=3184 Af. (Cited 2011, 29 Sep.) UK Medicines Information New Drugs Online. New Drugs Online Report for lasofoxifene.
84. Maeda T, Ke HZ, Simmons H, Thompson D. (2004) Lasofoxifene, a next generation estrogen receptor modulator: preclinical studies. *Clin Calcium* **14**:85–93.
85. Moffett A, Ettinger M, Bolognese M, *et al.* (2004) Lasofoxifene, a next generation SERM, is effective in preventing loss of BMD and reducing LDL-C in postmenopausal women. *J Bone Min Res* **19**:S96.
86. Lee A, Radecki D, Wolter K. (2005) Lasofoxifene Phase II and Phase III clinical trial design and strategy. *J Bone Min Res* **20**:M384.
87. Eastell R, Reid DM, Vukicevic S, *et al.* (2008) The effects of lasofoxifene on bone turnover markers: the PEARL trial. *J Bone Min Res* **23(Suppl 1)**: S81:
88. Cummings SR, Ensrud K, Delmas PD, LaCroix AZ, Vukicevic S, Reid DM, Goldstein S, Sriram U, Lee A, Thompson J, Armstrong RA, Thompson DD, Powles T, Zanchetta J, Kendler D, Neven P, Eastell R. (2010) Lasofoxifene in postmenopausal women with osteoporosis. *N Engl J Med* **362**:686–696.
89. LaCroix AZ, Powles TJ, Osborne CK, Wolter K, Thompson JR, Thompson DD, Allred DC, Armstrong R, Cummings SR, Eastell R, Ensrud KE, Goss P, Lee A, Neven P, Reid DM, Curto M, Vukicevic S. (2010) Breast cancer incidence in the randomized PEARL trial of lasofoxifene in postmenopausal osteoporotic women. *J Natl Cancer Inst* **102**:1706–1715.
90. McClung M SE, Cummings S, *et al.* (2005) Lasofoxifene increased BMD of the spine and hip and decreased bone turnover markers in postmenopausal women with low or normal BMD. *J Bone Min Res* **20**:S97.
91. Davidson M, Moffett A, Welty F, *et al.* (2005) Extraskeletal effects of lasofoxifene on postmenopausal women. *J Bone Min Res* **20**:S173.
92. McClung MR, Siris E, Cummings S, *et. al.* (2006) Prevention of bone loss in postmenopausal women treated with lasofoxifene compared with raloxifene. *Menopause* **13**:377–386.

93. Goss PE, Ingle JN, Ales-Martinez JE, Cheung AM, Chlebowski RT, Wactawski-Wende J, McTiernan A, Robbins J, Johnson KC, Martin LW, Winquist E, Sarto GE, Garber JE, Fabian CJ, Pujol P, Maunsell E, Farmer P, Gelmon KA, Tu D, Richardson H. (2011) Exemestane for breast-cancer prevention in postmenopausal women. *N Engl J Med* **364**:2381–2391.
94. Cuppone F, Bria E, Verma S, Pritchard KI, Gandhi S, Carlini P, Milella M, Nistico C, Terzoli E, Cognetti F, Giannarelli D. (2008) Do adjuvant aromatase inhibitors increase the cardiovascular risk in postmenopausal women with early breast cancer? Meta-analysis of randomized trials. *Cancer* **112**:260–267.
95. Chlebowski RT, Kuller LH, Prentice RL, Stefanick ML, Manson JE, Gass M, Aragaki AK, Ockene JK, Lane DS, Sarto GE, Rajkovic A, Schenken R, Hendrix SL, Ravdin PM, Rohan TE, Yasmeen S, Anderson G. (2009) Breast cancer after use of estrogen plus progestin in postmenopausal women. *N Engl J Med* **360**:573–587.
96. Anderson GL, Limacher M, Assaf AR, Bassford T, Beresford SA, Black H, Bonds D, Brunner R, Brzyski R, Caan B, Chlebowski R, Curb D, Gass M, Hays J, Heiss G, Hendrix S, Howard BV, Hsia J, Hubbell A, Jackson R, Johnson KC, Judd H, Kotchen JM, Kuller L, LaCroix AZ, Lane D, Langer RD, Lasser N, Lewis CE, Manson J, Margolis K, Ockene J, O'Sullivan MJ, Phillips L, Prentice RL, Ritenbaugh C, Robbins J, Rossouw JE, Sarto G, Stefanick ML, Van Horn L, Wactawski-Wende J, Wallace R, Wassertheil-Smoller S. (2004) Effects of conjugated equine estrogen in postmenopausal women with hysterectomy: the Women's Health Initiative randomized controlled trial. *JAMA* **291**:1701–1712.
97. LaCroix AZ, Chlebowski RT, Manson JE, Aragaki AK, Johnson KC, Martin L, Margolis KL, Stefanick ML, Brzyski R, Curb JD, Howard BV, Lewis CE, Wactawski-Wende J. (2011) Health outcomes after stopping conjugated equine estrogens among postmenopausal women with prior hysterectomy: a randomized controlled trial. *JAMA* **305**:1305–1314.
98. Anderson GL, Chlebowski RT, Aragaki AK, Kuller LH, Manson JE, Gass M, Bluhm E, Hubbell FA, Lane D, Martin L, Ockene J, Rohan T, Schenken R, Wactawski-Wende J. (2012) Conjugated equine oestrogen and breast cancer incidence and mortality in postmenopausal women with hysterectomy: extended follow-up of the Women's Health Initiative Randomised Placebo-Controlled Trial. *Lancet Oncology* **13**:476–486.
99. Jordan VC, Ford LG. (2011) Paradoxical clinical effect of estrogen on breast cancer risk: a "new" biology of estrogen-induced apoptosis. *Cancer Prev Res (Phila)* **4**:633–637.
100. Jordan VC. (2004) Selective estrogen receptor modulation: concept and consequences in cancer. *Cancer Cell* **5**:207–213.
101. Ellis MJ, Gao F, Dehdashti F, Jeffe DB, Marcom PK, Carey LA, Dickler MN, Silverman P, Fleming GF, Kommareddy A, Jamalabadi-Majidi S, Crowder R, Siegel BA. (2009) Lower-dose vs. high-dose oral estradiol therapy of hormone

receptor-positive, aromatase inhibitor-resistant advanced breast cancer: a phase 2 randomized study. *JAMA* **302**:774-780.

102. Ariazi EA, Cunliffe HE, Lewis-Wambi JS, Slifker MJ, Willis AL, Ramos P, Tapia C, Kim HR, Yerrum S, Sharma CG, Nicolas E, Balagurunathan Y, Ross EA, Jordan VC. (2011) Estrogen induces apoptosis in estrogen deprivation-resistant breast cancer through stress responses as identified by global gene expression across time. *Proc Natl Acad Sci USA* **108**:18879-18886.

15

SELECTIVE ANDROGEN RECEPTOR MODULATORS AS IMPROVED ANDROGEN THERAPY FOR WOMEN

Christopher C. Coss and James T. Dalton

Abstract

Selective androgen receptor modulators (SARMs) are a new class of anabolic agents that can increase muscle mass and improve physical function without unwanted effects on the prostate, skin, or hair that are commonly associated with steroidal androgens. These agents may have therapeutic utility in both women and men with muscle wasting.

Keywords: Androgen Receptor (AR); Anabolic; Androgenic; Menopause; Climacteric; Libido; Testosterone; Chronic Obstructive Pulmonary Disease (COPD); Polycystic Ovarian Syndrome (PCOS); Body Composition; Selective Androgen Receptor Modulator (SARM).

INTRODUCTION

Androgens play diverse roles in female physiology, ranging from sexual organ development and libido to musculoskeletal homeostasis. Throughout most of a woman's lifetime, androgens and androgenic prohormones circulate in appreciable amounts, operating either directly on the androgen receptor or as estrogenic precursors. A number of both iatrogenic and pathogenic states of androgen dysfunction are well characterized, offering clear opportunities for pharmacologic intervention. Androgen administration in women has an established history of success but its therapeutic potential has been limited by unwanted virilizing side effects. The recent discovery and clinical development of nonsteroidal, tissue-selective androgen receptor modulators with reduced potential for virilization has sparked renewed interest in androgen therapy for women.

ANDROGENS IN NORMAL FEMALE PHYSIOLOGY

Understanding of female androgen physiology lags far behind what is known in men. However, near-ubiquitous expression of the androgen receptor in female tissues,[1] appreciable circulating androgens and preandrogenic hormones (Table 1),[2] strong correlations between serum testosterone and several pathologies,[3,4] and the success of exogenous androgen administration in treating hypoandrogenic disease states[5,6] combine to make a strong case for the importance of androgens in women. Some have estimated, based on summation of circulating androgen metabolites, that total androgen production in women can approach nearly two-thirds that in men.[7]

Androgen Production

In reproductively competent women, the most prevalent circulating hormones generally classified as androgens include dehydroepiandrosterone sulfate (DHEAS), dehydroepiandrosterone (DHEA), androstenedione (A), testosterone (T), and dihydotestosterone (DHT) (Fig. 1). DHEAS, DHEA, and A all have negligible affinity for the androgen receptor, requiring further enzymatic conversion to T and DHT to elicit their androgenic effects. DHEAS and DHEA are primarily secreted by the adrenal glands, whereas A and T are secreted directly by both the adrenals and the ovaries. Nearly half of circulating A is derived from bioconversion of DHEA-S via DHEA. Likewise, the greater majority of circulating T is the result of A bioconversion. DHT, the most potent endogenous androgen, is made almost exclusively via peripheral conversion of T. Estrogens require androgenic intermediates but circulate in two- to fivefold excess on average in premenopausal women when compared to men (Table 1).

Table 1. Approximate mean total circulating levels of hormones in men and women. (Combined data from Ref.10, 113 and 114.)

	Women (nmol/L)		Men (nmol/L)
	Premenopausal	Postmenopausal	
DHEA-S	5500	2500	2000
DHEA	17	15	10
Androstenedione (A)			
Follicular	3.0	2.50	4
Luteal	3.8		
Testosterone (T)			
Follicular	0.70	0.60	12
Luteal	0.90		
Estrone (E1)			
Follicular	0.14	0.10	0.13
Luteal	0.29		
Estradiol (E2)			
Follicular	0.16	0.04	0.10
Luteal	0.44		

Like other steroidogenesis, androgen production is tightly controlled both spatially and temporally by numerous well-established endocrine, paracrine, and intracrine signals.[2,7] A family of steroidogenic enzymes, expressed to varying degrees in many tissues, drive steroid hormone conversion, affording remarkable flexibility in local androgenic milieus.[7] The importance of these enzymes is best highlighted by their association with disease in cases of both congenital absence and abundance.[8,9]

Several temporal phenomena govern rates of androgen production throughout a women's lifetime. In premenopausal women, ovarian cyclicity and its accompanying luteinizing hormone (LH) patterns result in a midcycle peak of both circulating A and T (Table 1). Upon follicular depletion at the onset of menopause, E2 production plummets, resulting in a release of negative feedback at the pituitary. A consequent LH increase drives increased ovarian A and T production into

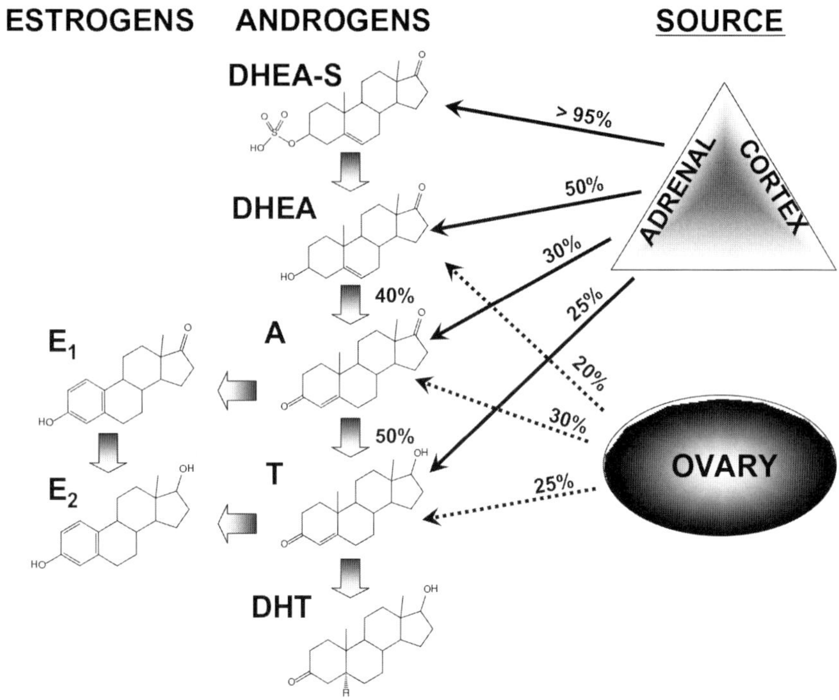

Figure 1. Percentage contributions of circulating hormones in premenopausal women. (Adapted from Ref.10.)

later life. Finally, adrenal androgen production peaks in a woman's mid-20s, declining steadily thereafter. As DHEA-S and DHEA contribute substantially to circulating A and T levels, the net result is a steady postmenopausal decline in circulating androgens.[10]

Androgen Function

Great insight into the myriad physiological roles androgens play has come from androgen receptor knockout animals and, by corollary, humans with androgen receptor deficiencies. As a sex-linked gene residing on the X chromosome, the androgen receptor is present as a single copy in males. Functional mutations in the androgen receptor are phenotypically manifest in numerous ways, with the complete absence of function in men resulting in testicular feminization and infertility.[11] Thus, deleterious AR mutations are seldom passed on, women are usually asymptomatic heterozygotes, and homozygous AR-null females are a genetic impossibility. Modern Cre-loxP recombination has overcome this natural obstacle, permitting the generation of ARKO female mice and new insight into female androgen physiology.[12-14]

The Reproductive System

The requirement for full AR function in the proper development of male sexual organs is well established. Female ARKO mice have a much subtler phenotype, with outwardly normal reproductive anatomies. However, reproduction itself is compromised in ARKO females which produce reduced litter sizes when compared to AR-wt (wild-type) littermate controls and display delayed and lengthened estrous cycles. By 40 weeks of age, ARKO female mice were infertile as the result of severely compromised folliculogenesis.[13] To examine potential extraovarian influences which the androgen receptor might have on this reproductive abnormality, Walters *et al.* surgically transplanted ovaries from ARKO females into AR-wt mice and vice versa.[12] The AR-wt females with ARKO ovaries displayed fertility and menses that were nearly indistinguishable from those of normal mice. However, the AR-wt ovaries, when placed in an ARKO female mouse host, showed delayed, lengthened estrous cycles and compromised folliculogeneis. Also, progesterone levels in ARKO females are less than 50% of their normal counterparts, implicating a "luteal phase defect" which can result from a suboptimal preovulatory gonadotrophin surge and/or reduced luteotrophic support.[14] These findings strongly support a critical extraovarian role for androgens, likely of neuroendocrine function, within the hypothalamic-pituitary-gonadal axis.

Much like ovarian function, AR signaling appears requisite in normal uterine physiology. In the same experimental paradigm, AR-wt hosts with ARKO ovaries showed reduced total uterine, endometrial, and myometrial areas when compared to normal controls.[12] Even further reductions in all measures were reported in ARKO hosts with

AR-wt ovaries. Interestingly, though total uterine size was reduced by compromised AR function, no differences in the success of embryonic implantation or in the health of implanted pups were seen.[12] Taken together these results argue that androgens play an important role in uterine development but not uterine function.

To date no reports of sexual behavior dysfunction in ARKO female animal models exist. Testosterone administration has a well-documented positive impact on libido in climacteric (surgical or natural) women.[15] The degree to which indirect effects of T, namely aromatization to E2, drive increased female libido is unknown. The absence of sexual dysfunction in female ARKO mice would argue that direct activation of the androgen receptor is irrelevant in normal female sexual behavior. However, measurable levels of the androgen receptor are expressed in portions of the female brain governing libido.[16] Also, nonaromatizable androgens, specifically nonsteroidal selective androgen receptor modulators (SARMs), show positive effects on libido in ovariectomized rat models.[17] The role of androgens in female libido remains controversial.

Androgens and Bone

Bone development and homeostasis are remarkably complex processes requiring intricate coordination between numerous cell types. The study of individual cellular components is compromised by removal from the skeletal microenvironment and researchers' limited ability to reproduce elaborate paracrine signaling in culture. Compounding these inherent investigational challenges are the highly localized expression of numerous nuclear hormone receptors and steroidogenic enzymes even within the same cell type.[18] It is generally accepted that androgens and estrogens provide critical endocrine signals to skeletal cells during both development and adult life. Reduced circulating androgens or estrogens, in both men and women, have been linked to net bone loss though the specifics of this deficit are disputed, and sexually dimorphic.[19] Elucidating androgens' role in normal bone physiology is further confounded by muscular anabolism and its potential contribution to mechanical loading requisite in normal bone growth.[18] Current mechanistic insight is drawn largely from experiments using cultured skeletal cells of various lineages.

The bone remodeling cycle affords consistent bone strength by balancing resorption and synthesis in response to a plethora of mechanical and endocrine/paracine signals. Androgens have been implicated, both directly and indirectly, in the proper function of three critical skeletal cell types within the bone compartment; osteoblasts, osteoclasts and osteocytes (Fig. 2). Androgen receptor expression has been reported in all three cell types in a temporally and spatially dependent manner, irrespective of gender.[18,20] Of the three, receptor expression is greatest in osteoblasts at sites of new bone formation. Direct androgenic effects in cultured osteoblast progenitor cells, as measured by nonaromatizable androgen treatment (DHT), include increased cellular proliferation and differentiation. Likewise, androgen treatment has been shown to counter apoptosis in both osteoblasts and osteocytes.[18] The presence of a direct androgenic

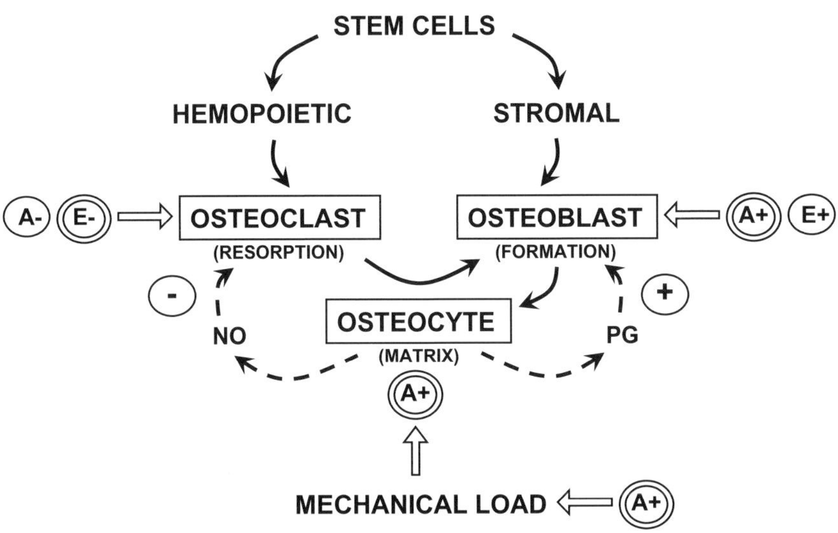

Figure 2. Bone remodeling, sex steroid and mechanical loading. E, — estrogen; A, — androgen; NO, — nitric oxide; PG, — prostaglandin; +, stimulatory; -, inhibitory; double circle, — primary activity; single circle, — secondary activity. [Reproduced from *J Bone Miner Res* (2000) **15**:301-307, with permission from the American Society for Bone and Mineral Research.[115]]

effect in osteoclasts is more controversial, though treatment of cultured osteoclasts with DHT resulted in reduced resorptive capacity in several species.[21] The equally important indirect effects of androgens on these cells are thought to occur in multiple ways, including T aromatizaton to E2, modulation of secreted cytokines and growth factors expressed locally in bone, and through stimulation of hematopoetic stem cells whose potential fate includes osteoclasts.[18] The relative contributions of direct versus indirect androgen signaling are likely a function of species, age, hormonal status and gender.

The composite effects of hormone depletion on bone are readily apparent in ovariectomized (OVX) female rats. Surgical ovariectomy drastically reduces circulating T and E2, resulting in net bone loss due primarily to increased cancellous bone turnover.[22] Interestingly, both T and DHT treatment have marked sparing effects on cancellous bone in OVX females.[23,24] However, in this model, DHT's efficacy is reduced compared to T, arguing the importance aromatization to E2 for full bone maintenance in females.

When one is considering bone phenotypes in both ARKO and double estrogen receptor knockout (DERKO) female mice, an apparent secondary role for androgens in female bone homeostasis emerges. Intact female DERKO mice show reduced cancellous bone when compared to wt controls.[25] Both T and DHT treatment in OVX DERKO female mice increase cancellous bone mass, presumably in the absence of any estrogen signaling.[18,26] It bears mentioning that in these studies the cancellous bone mass in androgen-treated castrates relative to wt intact animals was not reported. Interestingly, ARKO female mice have no evident bone phenotype.[13] This sharply contrasts with the severe osteopenia present in age-matched ARKO males. When taken together, direct androgenic action has a clear ability to spare bone, though estrogenic signaling appears both necessary and sufficient for full bone maintenance in female rodents.

Androgens and Body Composition

The nitrogen-retaining, i.e. anabolic, properties of androgen preparations have been known since the 1930s.[10] Incredibly, definitive studies showing dose-dependent increases in muscle size, muscle strength, and fat-free mass (FFM) resulting from testosterone supplementation lagged 50 years behind this initial discovery. Numerous confounding factors, including poorly controlled diet and exercise regimens, nonrandomized or blinded study designs, and relatively low doses fueled controversy around exogenous androgens' anabolic ability for nearly a half century. It is now generally accepted that androgens produce direct anabolic effects on skeletal muscle, though the cellular mechanisms underlying these effects are still under intense investigation.[27,28]

Astonishingly, AR expression is scant in skeletal muscle cells and their precursors. Relative to other androgen target tissues, namely prostate, androgen binding in rodent muscle is low, implying minimal receptor expression.[29] In fact, human muscle AR expression was not routinely detectable via immunostaining until signal amplification techniques were employed.[30] Muscle AR expression has since been shown to depend on multiple factors, including gender, muscle group, location of myonuclei within the muscle fiber, and exercise regimen, to name but a few.[31] Low level AR expression has also been reported in cultured satellite cells, a primary muscle precursor cell, which was shown to increase in response to testosterone treatment.[32]

Androgen-induced increases in muscle size are associated with protein accretion, hypertrophy of both type I and II muscle fibers, and an increase in the number of myonuclei and satellite cells.[28] Studies where subjects were fed radio-labeled amino acids demonstrated that testosterone administration both ameliorates protein loss during fasting and stimulates protein anabolism, with feeding resulting in a net protein gain.[33] Satellite cell proliferation is required for increases in both myonuclear number and muscle fiber hypertrophy.[34] Thus, androgens' proliferative effects on satellite cells are likely paramount in their anabolic activity.[28] Additionally, both T and DHT have been shown to promote myogenic and inhibit adipogenic differentiation of cultured pluripotent mesenchymal stem cells.[35] Testosterone has also been reported to inhibit lipid uptake and lipoprotein lipase activity in adipocytes while concomitantly promoting lipolysis.[36] These results amount to a model of androgen action on body composition promoting both muscle growth and preventing fat deposition (Fig. 3).

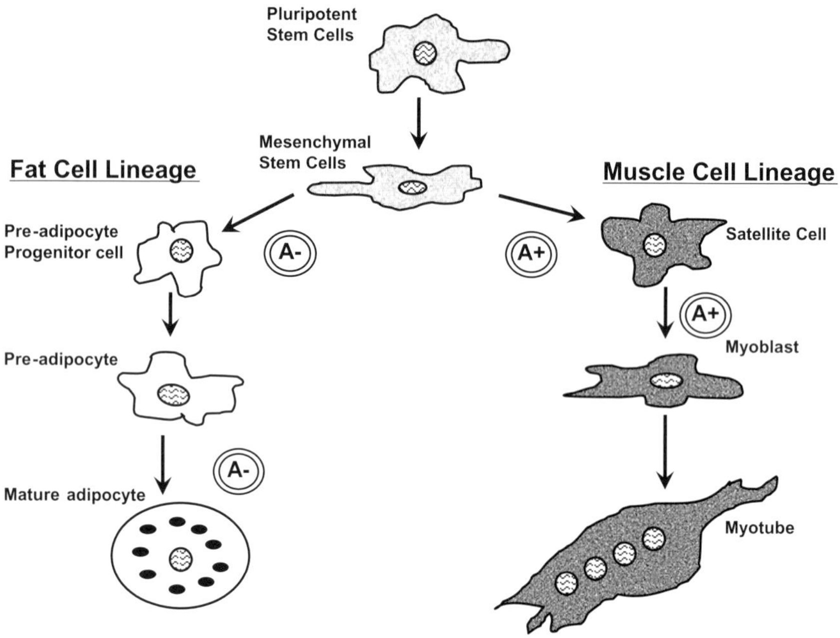

Figure 3. Model of androgen action on body composition. A, — androgen. (Adapted with permission from Bhasin *et al.*[28])

Nearly all studies of androgens' effect on body composition have been performed on males. No reports of altered lean mass or fat mass in female ARKO mice exist. ARKO males, however, present with profound late onset obesity.[13] Hypoandrogenic women show significant increases in FFM in response to androgen therapy.[37] Yet, pathologically elevated androgens in women often include the comorbidities of obesity, metabolic disorder and diabetes.[38] Significant sexual dimorphism is suggested by the limited work done on ARKO mice, but further study precludes establishing a clear relationship between endogenous androgens and female body composition.

Androgens and Hair/Skin

Androgens affect several functions of human skin, including sebaceous gland growth and differentiation, hair growth and wound healing. AR expression has been reported in epidermal and follicular keratinocytes, sebocytes, sweat gland cells, dermal papilla cells, dermal fibroblasts, endothelial cells and genital melatinocytes.[39] To date no reports of skin phenotypes in ARKO rodent models exist. Furthermore, the exact nature of androgen function within these cell types is still an area of active investigation.[39] Current understanding of androgen function in the skin compartment is derived largely from clinical associations.

As in other organs, local concentrations of androgens in skin are tightly regulated by the expression of synthetic and metabolic enzymes. Cutaneous expression of 5α-reductase in both men and women governs the conversion of T to the most potent endogenous androgen, 5α-dihydrotestosterone (DHT). The importance of pronounced 5α-reductase expression in hair follicles is demonstrated by the absence of facial hair growth and alopecia in naturally deficient men.[40] Likewise, in hyperandrogenic women, increased cutaneous DHT concentrations are associated with hirsutism and androgenic alopecia.[41]

Androgens also promote sebocyte proliferation and sebum production within the sebaceous gland. When present in excess, they drive abnormal follicular epithelial cell desquamation, contributing to comedone development and subsequent bacterial colonization.[41] The resulting acne vulgaris is associated with increased circulating as well as local androgen production. DHT production is also implicated in sebaceous gland dysfunction by the correspondence of increased facial acne with increased 5α-reducatse activity in facial sebaceous glands relative to other areas of skin.[42]

Androgens and Erythropoiesis

The stimulation of hemoglobin production and bone marrow activity following androgen administration to an intact male rat was first reported in 1941. In the 30 years that followed, androgens' effects on the hematopoietic system were an area of active research.[43] Since that time androgen therapy to this end has fallen out of favor, replaced by recombinant erythropoient EPO, though stimulation of erythropoiesis is still useful as an ancillary androgen activity in several disease states.

No reports of hematopoietic dysfunction in ARKO animal models exist but the extent to which blood phenotypes have been considered is not clear. Androgens have been shown to promote erythroid differentiation and thus prevent leucocyte differentiation via their action on bone marrow stem cells.[44] The expression of AR has been detected in bone marrow erythroid cells. Also, testosterone administration to cultured erythrocytes stimulates colony formation dose-dependently and enhances their differentiation to EPO-responsive cells, in an effect reversible by antiandrogen treatment.[45] This suggest that androgen cotherapy could potentially improve a patient's response to recombinant EPO administration.

ANDROGEN DYSFUNCTION IN FEMALE DISEASE

A large body of research suggests that androgen function in women extends beyond reproduction and estrogenic precursors. Accordingly, diseases states associated with androgen dysfunction in women encompass both reproductive and nonreproductive sequele. The importance of proper androgen homeostasis in the healthy human female is unmistakable when one is considering the consequences of androgen dysfunction.

Androgen Excess (Hyperandrogenism)

Polycystic Ovarian Syndrome

Polycystic ovarian syndrome (PCOS) is the most common female endocrine disorder, with a prevalence of nearly 7% among reproductively competent women.[46] Current diagnostic criteria include biochemical or clinical hyperandrogenism as well ovulatory dysfunction and the polycystic ovarian morphology for which the syndrome is named. The etiology of PCOS is poorly understood but its heritable nature implies a strong genetic component or even fetal programming, as some have hypothesized.[47] Several factors have been implicated in PCOS androgen dysfunction, including aberrant microsomal steroidogenesis, exaggerated ovarian responsiveness to

LH/hcGH, and increased adrenal androgen production secondary to hyperinsulinemia.[48]

PCOS presents with incredible clinical heterogeneity though ovarian or adrenal hyperandrogenism is found in 60%–80% of patients.[46] The symptoms of hyperandrogenism associated with PCOS include unwanted hair growth, hirsutism, seborrhea and mild androgenic alopecia. Oddly, other classic virilizing effects identified with androgen excess in women, including masculanization of musculature, severe male pattern baldness and clitoromegaly, are rarely associated with PCOS.[49]

A diagnosis of PCOS predisposes women for increased risks of infertility, amenorrhea, endometrial carcinoma, obesity, type II diabetes mellitus, dyslipidemia, hypertension, and possibly cardiovascular disease.[49] It is unknown if hyperandrogenism is causative in PCOS, secondary to its many comorbidities, or some amalgamation of the two. Prenatal androgen exposure in mammals can result in ovulatory dysfunction, closely mimicking many symptoms of PCOS. Yet, treatment of PCOS patients with insulin-sensitizing agents ameliorates many of their symptoms, including reducing circulating androgens.[47] Further study is required to clarify the precise role of androgens in the etiology of PCOS.

Non-PCOS Hyperandrogenism

Idiopathic hyperandrogenism, the second-most-common disorder of androgen excess, is essentially a "catch-all" for hyperandrogenic women who do not fulfill the PCOS diagnostic criteria and lack elevated 17-hydroxyprogesterone (17OHP) levels.[50] Elevated 17OHP production is a hallmark of nonclassical adrenal hyperplasia (NCAH) or 21-hydroxylase deficiency, constituting the third-most-common disorder of androgen excess. The enzymatic blockade in NCAH results in elevated adrenal androgen production in an effort to maintain normal cortisol levels.[3] While greater than 70% of clinically hyperandrogenic women are diagnosed with PCOS, only 16% are considered to be idiopathically hyperandrogenic and less than 5% 21-hydroxylase-deficient. Idiopathic hyperandrogenic women have normal ovarian function and present with reduced BMI and incidence of obesity when compared to their PCOS counterparts, even though their level of androgen excess is quite similar. NCAH patients, however, present with far greater excess in circulating total T, free T and DHEAS than PCOS women, but curiously have reduced ovarian dysfunction.[50]

Androgen Deficiency (Hypoandrogenism)

While female androgen excess is clinically defined, androgen deficiency in women remains nebulous. As previously mentioned, a decline in ovarian androgen production precedes menopause and continues throughout postmenopausal life. This drop occurs such that serum T levels in a 60-year-old woman are half that of a 40-year-old, who in turn has reduced circulating androgens when compared to healthy women in their second decade. Thus, postmenopausal women could readily be considered androgen-deficient, as some advocate.[51,52] The significance of ovarian androgen contribution, both pre- and postmenopause, is even further highlighted in surgically menopausal (oophrectomized) women. These patients have significantly lower androgen levels than age-matched naturally menopausal women and report severe menopausal symptoms at nearly double the rate of their naturally climacteric counterparts.[4] Hypoandrogenism of adrenal origin is quite rare. Hypopituitarism and related adrenal insufficiencies do, however, result in reduced circulating hormones, including an androgen deficit.[6,53] Identifying the contributions of hypoandrogenism to the larger pathology in these patients is challenging, in light of their level of endocrine dysfunction.

Iatrogenic hypoandrogenism in women is common. Both hormone replacement therapy (HRT) in postmenopausal women and oral contraceptives in reproductively competent women have been shown to lower serum T.[10,54] In both cases, LH suppression and SHBG induction in response to exogenous estrogen administration synergize to reduce bioavailable T.[10] Despite the many potential etiologies for androgen deficiency in women, little agreement exists on what constitutes "deficient" circulating levels of androgens and the symptoms therein.

A group of women's health experts have proposed a female androgen insufficiency syndrome (FAIS). FAIS is defined as low serum testosterone combined with reduced libido, diminished well-being, lowered mood and persistent unexplained fatigue.[52] As many of these symptoms are shared with estrogen deficiency in women, FAIS diagnoses are limited to properly estrogenized women. Numerous studies support the efficacy of T in ameliorating the aforementioned symptoms in potentially FAIS women.[51,52] However, strong dissent exists among clinicians who argue that longer studies are required in determining the safety of exogenous T administration, especially in older women.[55]

FEMALE ANDROGEN THERAPY PAST AND PRESENT

Androgens have been used clinically in women for over 50 years, demonstrating efficacy in both reproductive and extrareproductive disease states. Traditionally, administering androgens to women has always come at the risk of virilization. "Androgen replacement" regimens have gained favor of late as lower physiological doses of androgen produce fewer unwanted side effects, apparently at the cost of efficacy in some studies. Nevertheless, even a brief survey of androgens' clinical history in women offers insight into their ample therapeutic potential in the female population.

Androgen Therapy in Female Disease

Menopause

As previously discussed, a steady decline in circulating androgens accompanies drastic reductions in available estrogens during the menopausal transition. Overlaid on this natural decline are iatrogenic deficiencies accompanying common female HRT. A growing body of evidence has indicated that androgen supplementation (primarily T) in addition to estrogenic HRT shows increased efficacy in ameliorating multiple menopausal symptoms when compared to estrogens alone.[15]

Perhaps the most-investigated indication for androgen therapy in climacteric women is libido. Initial reports concerning the benefits of methyltestosterone on low female libido date back to the 1950s.[56] Since that time, literally dozens of clinical trials have shown that the addition of androgen to estrogen treatment improved numerous facets of female sexual function, including libido, sexual desire, arousal, frequency and satisfaction.[15] Clinicians have further correlated improvements in female sexual function with both reduced circulating SHBG and increased serum T in these patients. Recently the FDA effectively ended the development of a testosterone patch specifically formulated for low libido in menopausal women, citing insufficient long term safety data, though similar products are widely used in Europe.[57,58]

Androgens have been reported to prevent bone loss in established osteoporosis.[59] This effect is seen with androgen monotherapy as well as in combination with estrogens in postmenopausal women.[7] Some studies have suggested that anabolic steroids can promote bone formation, thus potentially alleviating an existing bone deficit.[60] This finding is significant, in that existing first line osteoporosis therapies are antiresorptive and thus unable to reverse pretreatment bone loss.[61] Much like libido, several studies show that when given with estrogens, androgens demonstrate increased efficacy over estrogens alone in improving bone mineral density in climacteric patients.[7]

Hot flashes and other vasomotor symptoms are reported by 80% of climacteric women and are the most common reason for these patients seeking medical care.[62] Androgen cotherapy has been found to relieve hot flashes in surgically menopausal women who did not respond to estrogen alone.[63] Currently the sole FDA-approved androgen therapy for women is a product composed of methyltestosterone combined with estrogen esters for the treatment of vasomotor symptoms unresponsive to estrogen monotherapy (Estratest, Solvay Pharmaceuticals Inc.).[15]

Breast Cancer

Several androgen modalities have been employed as adjuvant therapy in breast cancer.[7] Clinical successes have included higher response rates and a longer time to disease progression when androgens were coadministered with an antiestrogen as compared to antiestrogen alone.[64] There is a dearth of preclinical evidence to support an antiproliferative role for androgens in female breast tissue.[7] Also, AR expression was detected in nearly half of primary breast cancers.[65] Unfortunately, the high doses of androgens required for efficacy in breast malignancy came at the cost of virilization. The successful development of highly efficacious aromatase inhibitors, combined with androgens' unwelcome side effects, has prevented the further use of androgens as an endocrine adjuvant for breast cancer.[66]

Androgen Therapy in Muscle Wasting

Muscle wasting is a catabolic state secondary to a number of serious chronic pathologies often copresenting with hypogonadism. Frailty resulting from substantial declines in muscle mass can result in profound loss in quality of life and reduced survival.[67] Due to androgens' beneficial effects on body composition, androgen therapy has been evaluated in a number of patient populations experiencing severe muscle loss.

HIV/AIDS

Profound weight loss is a common comorbidity of HIV infection.[68] To date studies on female HIV patients have been confined to doses of testosterone resulting in near-physiologic serum T levels.[68,69] These studies' findings were mixed, ranging from no effect to strength gains in the absence of significant increases in body weight. However, increases in lower-body muscle strength failed to translate into significant functional benefit as measured by a 6 min walk test, one potential measure of functional performance.[70] The doses administered were appreciably lower than those given in similar successful studies on males and did not result in significant side effects, perhaps at the cost of efficacy.[67]

Chronic Obstructive Pulmonary Disease

Weight loss and muscle wasting are correlated with disease severity, impaired mobility, lower quality of life and higher mortality in chronic obstructive pulmonary disease (COPD) patients.[71] Elevated resting energy expenditure, decreased dietary intake and glucocorticoid therapy all contribute to muscle loss in COPD.[67,71] Decline in lean mass is exacerbated by a reduced capacity for exercise and deconditioning, establishing a morbid synergy. Although few studies have evaluated androgen intervention in female COPD, likely due to concerns about side effects at anabolic doses, a variety of studies indicate that anabolic therapy promotes gains in FFM and is capable of improving some measures of pulmonary function in patients with mild-to-moderate COPD. Schols *et al.*[72] showed that androgen administration in combination with dietary intervention demonstrated greater weight gains than nutritional supplementation alone in both men and women. The weight gains reported were due to increases in FFM and accompanied improvements in maximal inspiratory pressure. This investigation included an exercise regimen common in COPD rehabilitation.[72] However, a recent trial (including both men and women with severe COPD), designed to evaluate androgen-mediated improvement in walking distance outside of structured rehabilitation, was ended early due to lack of efficacy.[71] Although the lack of an exercise requirement in the more recent study may argue for the necessity of simultaneous pulmonary rehabilitation during anabolic therapy in COPD patients, reliance on an endpoint of muscle endurance (i.e. a 6 min walking distance) may have obscured benefits in muscle strength and maximal inspiratory pressure that were observed in these patients, as equivalent doses were administered biweekly in both studies.

End Stage Renal Disease

Patients who are on hemodialysis also commonly experience muscle wasting.[73] Androgens' ability to stimulate erythropoiesis has been evaluated in the context of treating anemia associated with end stage renal disease (ESRD). However, potential improvements in body composition or functional benefits have been a secondary consideration. Trials have shown that androgen therapy in male hemodialysis patients provided both significant increases in lean body mass and a physical performance benefit.[74] Also, a small study showed that exercise alone in both male and female dialysis patients provided physical improvements coupled with a functional benefit.[75] A more recent examination of potential synergism between androgen administration and exercise in both men and women concluded that androgen therapy increased lean mass in excess of what was seen in exercise alone. However, strength gains and significant functional benefit were found to be the result of resistance training.[73] In this study women were treated with the same androgen as in the COPD studies, but at four times the dose. Specific effects in women were not reported.

Androgen Therapy in Hematopoietic Disease

Androgens' ability to stimulate erythropoiesis has long been exploited in the treatment of both primary and secondary anemias.[10] Some of the first patients to receive an androgen for this reason were anemic women with breast cancer.[76] As was the case in the treatment of breast cancer, the doses required for efficacy resulted in significant virilization.[77] Interestingly, when compared head-to-head with recombinant EPO therapy, androgens provided similar improvements in anemia, with an anabolic benefit in male patients.[78] However, due to unacceptable side effect profiles, androgens have been replaced by EPO in treating anemic women.

Therapeutic Limitations of Steroidal Androgen Modalities

Despite the vast therapeutic potential of androgens in women, only one androgen preparation is approved by the FDA for use in women. The recent highly publicized failure of Proctor and Gamble's testosterone patch for low female libido, Intrinsa, reflects the medical community's tepid relationship with steroidal hormone therapy in women.[58] Some have suggested that the non-life-threatening nature of low libido created a higher hurdle to approval. While not outwardly mentioned in the FDA's decision, it is likely this played a role in the regulatory agency's request for more long term safety data in otherwise healthy women. Therapeutic need aside, Intrinsa's development and subsequent failure are illustrative of many current limitations of female androgen therapy.

Formulation

Testosterone has very low oral bioavailability and is rapidly cleared from the body.[10] Numerous chemical modifications to T have been shown to retain pharmacologic activity while improving pharmacokinetics (Table 2). A limited number of orally available T analogs have been developed; most of them are now clinically obsolete, due to highly variable interindividual exposures and hepatotoxicity.[10] During the development of a T formulation for low libido, both oral 17α-methyltestosterone and a number of parenteral formulations demonstrated efficacy.[58] However, 17α-methyltestosterone administration resulted in undesirable changes in lipids, while both intramuscular injections and surgically implanted pellets of T esters resulted in superphysiological and unpredictable serum T.[57] Increased reports of virilization from these modalities were substantiated by a recent safety assessment of exogenous T therapy in European women. A survey of over 2000 woman using various parenteral (exclusively non-transdermal) T products showed increased virilization when compared to age-matched controls.[79] Ultimately, a transdermal patch was developed and employed in phase II and III trials. This modality provided a consistent, extended release of unmodified T, resulting in serum levels in the normal range sufficient for efficacy in low libido, with few reports of androgenic side effects.[57]

Aromatization

In the wake of the Women's Health Initiative (WHI), FDA experts voiced serious concerns about administering T to women already receiving estradiol.[58] WHI investigators reported increases in both breast cancer risk and adverse cardiovascular events in women receiving estrogen-progesterone HRT when compared to placebo, causing great public unrest.[80] One potential worry with Intrinsa was the addition of unknown amounts of estradiol via aromatization of exogenous

Table 2. Clinically evaluated androgens in women.

Testosterone Analogs

Name	Compound structure	Routes	Indications
Testosterone	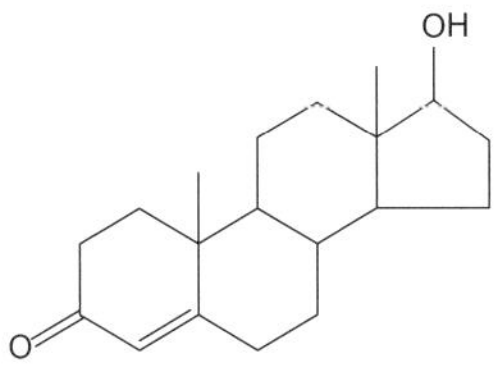	Implant Patch Cream	Libido Muscle wasting (HIV) Osteoporosis
Testosterone propionate	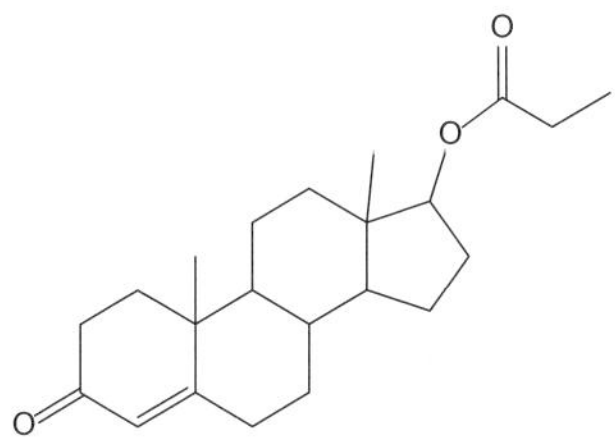	IM depot	Breast cancer
Testosterone enanthanate	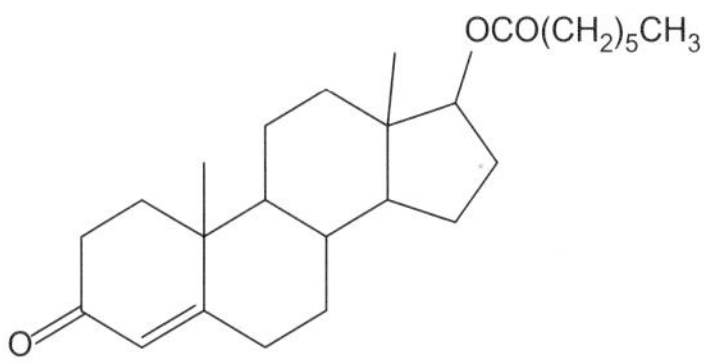	IM depot	Anemia
Testosterone undeconate	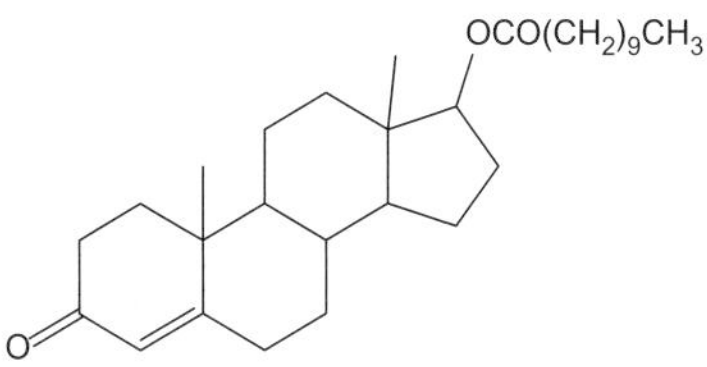	Oral IM depot	Libido
17α-methyl-testosterone	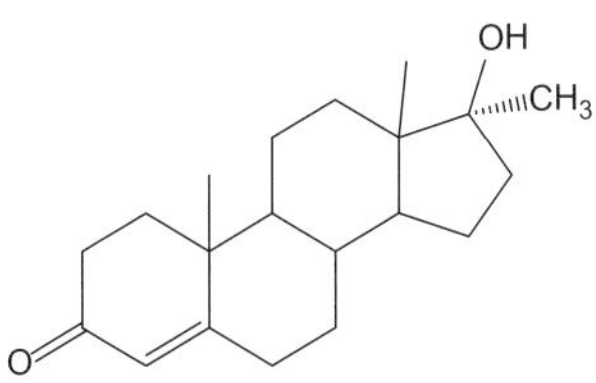	Oral	Libido Hot flashes (combo w/ CEE)
Fluoxymesterone	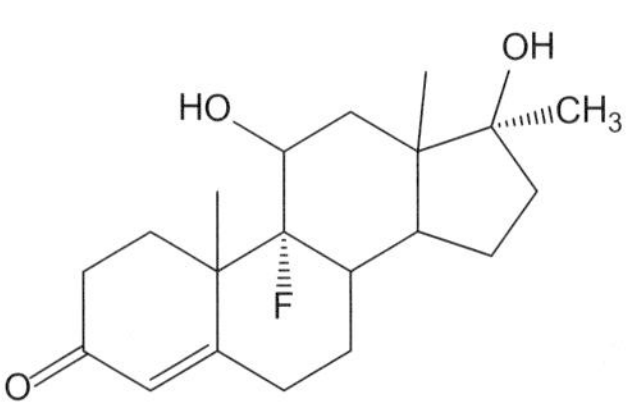	Oral	Breast cancer

(Continued)

Table 2. (*Continued*)

	Name	Compound structure	Routes	Indications
Anabolic Steroids	Nandrolone deconate	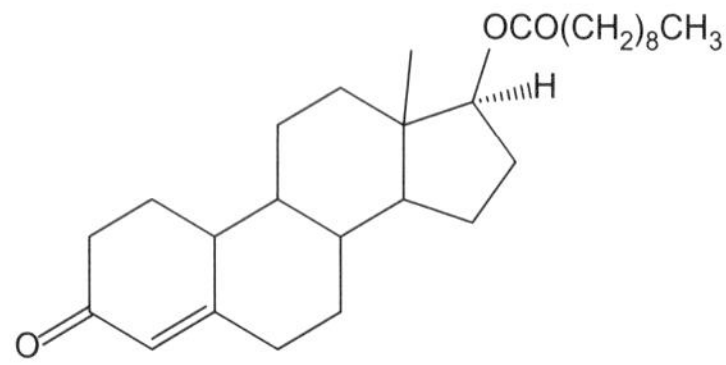	IM depot	Libido Muscle wasting (HIV) Muscle wasting (COPD) Muscle wasting (ESRD) Anemia Osteoporosis
	Oxymetholone	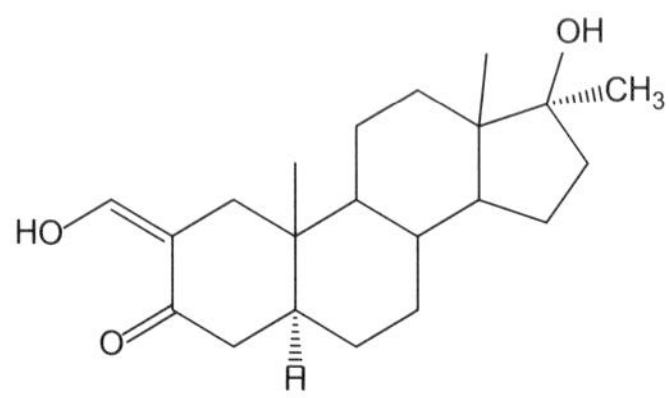	Oral	Muscle wasting (HIV) Anemia

T.[81] Though it is unlikely that consequential amounts of E2 are produced at the serum T levels reported during clinical evaluation of Intrinsa,[81–83] adding yet another "sex steroid" to the mix in climacteric women was deemed too risky given the relatively short six-month trials. Though there is no shortage of criticism concerning the design and findings of WHI,[84,85] and a growing argument for the safety of androgen therapy in women,[79] the regulatory culture appears nevertheless to be entrenched.

Improved Androgen Therapy for Women

Initially, larger doses of androgens were shown to be efficacious in treating low libido but were accompanied by side effects. Intrinsa seemingly stuck a balance by delivering physiological doses of T while retaining efficacy and essentially eliminating androgenic side effects. Unfortunately, many clinical indications for androgens in women require superphysiological doses for efficacy. A reduction in virilization potential would enable effective androgen treatment in a much larger group of indicated women. Likewise, conversion to estradiol was cause for consternation, warranted or not, in long term administration of T to climacteric women. In this case and others, a nonaromatizable androgen could assuage the concerns of many clinicians. Oral androgens, though more convenient, have a history of poor pharmacokinetics and liver toxicity. A well-tolerated orally available androgen could eliminate the need for patches, injections or surgically implanted pellets. The recent discovery and development of selective androgen receptor modulators (SARMs) has made significant inroads by combining many of these desirable properties.

SARMs — EXPANDING FEMALE ANDROGEN THERAPY

Efforts to modulate steroidal androgen pharmacology by augmenting the 19-carbon chemophore date back to the 1940s. Early researchers experienced some success in promoting desired anabolic efficacy while limiting unwanted, off-target effects. It was not until discovery efforts moved beyond the steroidal template that the full potential of anabolic therapy was apparent.

Discovery and Characterization of Nonsteroidal Androgen Receptor Agonists

The first nonsteroidal AR agonists, the aryl propionamides and quinolinones, were discovered independently in the late 1990s by scientists at the University of Tennessee and Ligand Pharmaceuticals, respectively (Table 3).[86,87] In each case, novel agonistic activity was discovered while the researchers were in pursuit of androgen receptor antagonists. Both pharmacophores showed low nanomolar affinity for the androgen receptor and, importantly, full agonism in cellular activity assays.[86,88] Though the molecular pharmacology of these ligands was quite similar to that of DHT, their nonsteroidal backbone provided several key benefits *in vivo*. First, they were not substrates for 5α reduction to DHT — the enzymatic amplification of androgen signaling which is thought to drive virilization in women. Second,

Table 3. Nonsteroidal AR ligands and activities. RBA, relative binding affinity (DHT-100%).

Name	Compound Structure	RBA (%)	Activity	Refs.
R-bicalutamide	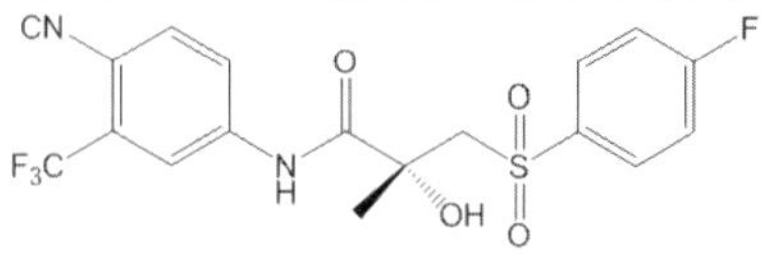	0.4	Antagonist	116
Aryl proprionamide SARM (GTx)	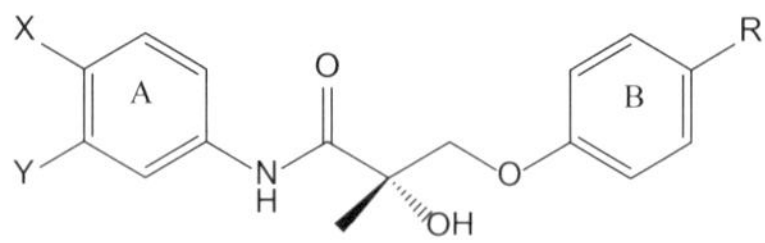	1–11	Partial to full agonists	91, 117, 118
Tricylic quinoline (ligand)	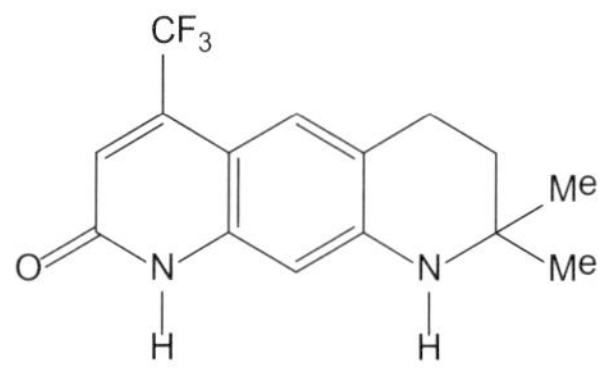	14	Antagonist	119
Tricylic quinoline SARM (ligand)	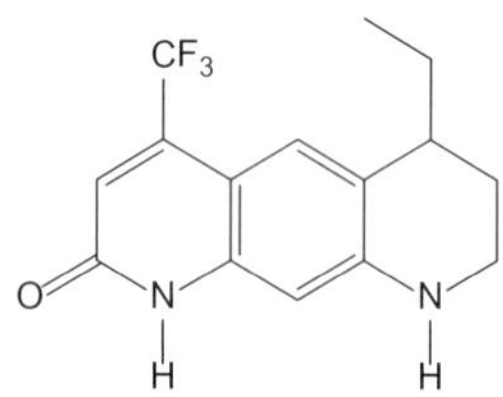	18	Full agonist	88

they were not substrates for aromatase conversion to E2 — the enzymatic culprit driving feminization in men. Third, they possessed structural flexibility affording improved pharmacokinetics with high bioavailability and extended half-lives.[89]

The potential for these new molecules to revolutionize androgen therapy was recognized immediately. Negro-Vilar of Ligand Pharmaceuticals spoke on this point in 1999, suggesting that "androgen therapy was about to experience a fundamental change, both in extent of use and in the range of applications".[90] In outlining the ideal SARM, he went on to suggest that tissue selectivity could be achieved with AR ligands, drawing parallels to the more established field of SERMs.

Tissue Selectivity

The initial report of SARMs' unique *in vivo* pharmacology came from Yin *et al.*[91] who first characterized the pharmacokinetics and pharmacodynamics of a novel series of aryl propionamides in 2003. Employing an immature castrated rat, or Hershberger, model, they studied the ability of testosterone or SARMs, to restore levator ani muscle or ventral prostate mass to intact, sham-operated control levels (Fig. 4). This choice of animal model was based on Hershberger *et al's.*[92] observation that both the levator ani muscle and the ventral prostate rapidly atrophy in response to castration in the young male rat. Likewise, exogenous androgen administration to the hormone-depleted animal results in dose-dependent, robust hypertrophy of both tissues as a measure of the anabolic (levator ani response) and androgenic (prostate, seminal vesicle response) potential of the ligand in question. In the example presented here, when delivered by osmotic pump over 14 days, S-4, an aryl propionamide SARM, maintained levator ani muscle mass at the two highest dose rates tested while very minimally stimulating prostate and seminal vesicles (Fig. 4b). Testosterone propionate, by comparison, robustly stimulated both androgenic and anabolic tissues in the same experimental paradigm, with near-identical ED_{50} values in the levator ani muscle and ventral prostate (Fig. 4a). It was clear from these early experiments that anabolic and androgenic activities could be dissociated.

Since the first reports of tissue selectivity, several SARM scaffolds have shown disparate anabolic and androgenic activities.[89] The exact means by which this tissue selectivity is achieved is currently debated, though numerous plausible mechanisms have been suggested. Some evidence exists that, much like SERMs, AR:SARM transcriptional complexes are ligand- and tissue-specific.[93] This model suggests that a tissue-selective SARM bound to the androgen receptor preferentially interacts with a different group of transcription factors in an

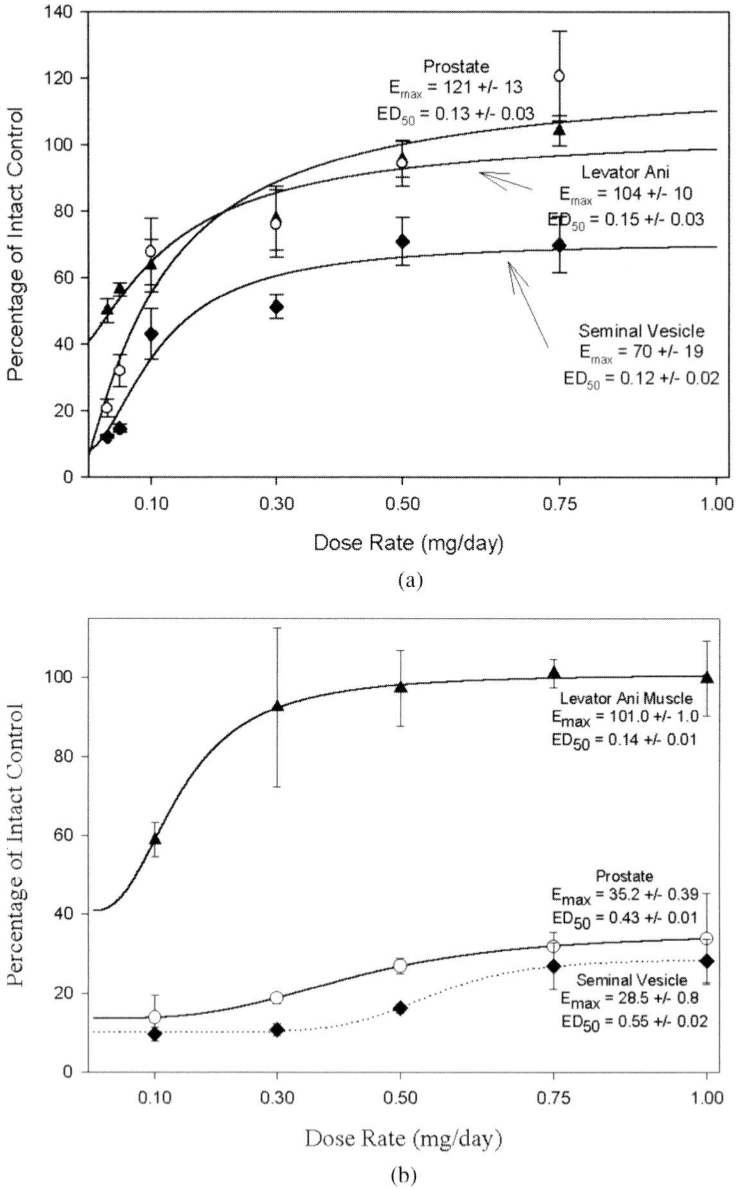

Figure 4. Tissue selectivity of aryl propionamide SARMs. Testosterone propionate (a) vs. S-4 (b). (Reprinted with permission from Yin *et al.*[91]).

androgenic tissue when compared to an anabolic tissue, resulting in tissue- specific transcriptional outcomes. Other evidence suggests that SARMs and steroids differentially affect nongenomic signaling cascades in cellular models of anabolic and androgenic tissues.[94] Yet another hypothesis argues that the lack of amplification via 5α reduction in androgenic tissues is paramount in SARM pharmacology.[95] In this model, SARMs act like testosterone, not DHT, in 5α-reductase-expressing tissues. As androgenic tissues express 5α-reductase, SARMs fall short of full activity in these compartments, thus demonstrating tissue-selective action.

Insufficient data exist for any of these models to stand alone. It is likely that all three of these molecular mechanisms contribute to the tissue selectivity of SARMs. However, elucidating just how SARMs work is more than just an intriguing scientific exercise. Understanding the driving force behind SARMs' pharmacology will greatly inform rational design of future SARMs for specific disease states or patient sub populations.

Preclinical Promise

Beyond tissue selectivity, SARMs have shown efficacy as androgens in numerous rodent models of disease.[96] Of particular interest to women's health are several studies showing SARM effects on bone in OVX female rats. Using a tetrahydroquinoline (THQ) SARM, Hanada *et al.*[97] from Kaken Pharmaceuticals showed similar efficacy to DHT in sparing both cancellous and cortical bone mineral density when compared to vehicle, though different doses were used. In the same study, strong evidence for the anabolic potential of SARM on bone was provided by marked effects on the periosteal mineral apposition rate and cortical bone mechanical strength. In both cases SARM outperformed E2. In a much longer study using an orally available aryl propionamide SARM, Kearbey *et al.*[98] demonstrated dose-dependent sparing effects of lumbar vertebra BMD reaching complete maintenance at the highest dose tested. This effect was reversible by cotreatment with a potent AR antagonist and was accompanied by resistance to fracture similar to intact controls as measured by femoral breaking strength. In a recent work, Vajda *et al.*[99] of Ligand Pharmaceuticals used a SARM of undisclosed structure alongside a bisphosphonate (Alendronate) to cotreat OVX rats. In all measurements, the combination was either equiefficacious as a single agent or provided additive benefits. SARM treatment was shown to increase total body and gastrocnemius weight in these animals while simultaneously decreasing fat pad mass. These results are of particular interest as recent clinical efforts combining another osteoanabolic agent, parathyroid hormone, and a bisphosphonate failed to show synergism.[100]

SARMs have also shown efficacy in rodent models of female libido. Using a pyrazoline derivative SARM, Allan *et al.*[17] from Johnson & Johnson demonstrated increased female sexual activity following a week of SARM administration. In this partner preference study, treated OVX females were exposed to castrated or sexually experienced intact males. SARM-treated females preferred intact males, in an effect that trended with dose and was comparable to testosterone-treated controls.

SARMs have also demonstrated measurable efficacy in hormone-replete disease models, taking advantage of their tissue selectivity to antagonize unwanted endogenous androgen action in androgenic tissues. In intact male rats, SARMs have been shown to decrease prostate size, with little to no effect on muscle.[101] Likewise, a SARM has been shown to reduce prostate tumor xenograft size in a castrated mouse in the presence of exogenous testosterone administered to stimulate tumor growth.[102] One could envision exploiting similar pharmacology in diseases of female androgen excess. In this case a SARM could potentially ameliorate the hirsutism etc. associated with hyperandrogenism without depriving the patient of androgens' important anabolic effects on muscle and bone.

Theoretically, SARMs could be used in any situation calling for an AR ligand not requiring estrogenic contributions. In just a decade of development, a continuum of SARM activity has emerged. Grades of peripheral tissue selectivity and anabolic potential are apparent, creating the opportunity to consider the merits of employing a particular SARM in one disease state, or a patient population over another.[103] Long term clinical goals are geared toward extended treatment in otherwise healthy individuals, a treatment paradigm previously unconsidered due to androgenic side effects, especially in women.

Clinical Investigation: The Next Generation of Androgen Therapy

The need for an androgen therapy with reduced virilization potential was addressed before the discovery of nonsteroidal AR agonists. The anabolic steroid nandrolone is subject to 5α reduction, but the resulting metabolite is a less, not more, potent androgen (Table 2).[104] Nandrolone's reduced androgenic profile caused it to gain favor clinically for female patients, but its widespread use is limited by moderate virilization, liver toxicity, parenteral dosing and significant prostagenic activity.[105-107] SARMs demonstrate greatly reduced cross-reactivity against other nuclear hormone receptors and to date no significant liver toxicity has been associated with nonsteroidal AR agonist administration, further adding to the benefits of abandoning the steroidal pharmacophore.[96] As hoped, SARMs are also thus far devoid of androgenic effects in the clinic.

Current Clinical Indications

Sacropenia, or an age-related decline in lean body mass, is a growing clinical concern in an aging population.[108] Currently, definitions of this disease state are varied and its etiology is poorly defined, but loss of muscle mass, more so than increased age, is associated with institutionalization, disability and, ultimately, mortality.[103] An anabolic agent with reduced side effects, providing a functional benefit, is the ideal therapy to address this largely unmet need. Scientists from Merck and Co., Inc. reported that a SARM, MK-0773, and testosterone increased lean body mass in healthy older men made hypogonadal with a gonadotrophin-releasing hormone agonist.[109] Although MK-0773 reduced prostate volume, while testosterone increased it, both MK-0773 and T led to significant increases in serum-prostate-specific antigen (PSA). In a recently completed phase IIa clinical trial, Ostarine™ (GTx Inc., Merck

Inc.) demonstrated the ability to significantly increase lean body mass and decrease fat mass in healthy elderly men and postmenopausal women.[110] These positive changes in body composition were accompanied by a performance benefit as measured by stair climbing following less than three months of oral once-daily dosing. The absence of serum PSA elevation in male patients or hirsutism in women demonstrated a selective anabolic benefit. Though this initial clinical effort was small, the positive results suggest SARMs' utility in age-related functional decline and muscle loss in both men and women. Other similar efforts by Pharmacopeia (age-related functional decline) and Ligand Pharmaceuticals (frailty) have completed phase I trials.

Ligand Pharmaceuticals, in a collaboration with Abbott, has also started early clinical evaluation of their lead SARM in osteoporosis of both sexes.[103] Aside from direct anabolic effects on bone, increasing lean mass in osteoporotic women could be of particular benefit as climacteric women lose nearly 4% of their muscle mass during the first three years of menopause.[20] With muscle contributing greater than 70% of the mechanical strain known to be requisite in bone maintenance, the combined effects of a SARM on body composition would likely be synergistic in osteoporosis. Increased muscle mass and strength could also lead to reduction in falls, a major morbidity in osteoporosis.[111] When taken together, a combined osteo- and myoanabolic agent such as a SARM stands to make great strides in reducing the impact of osteoporosis in aging men and women.

Ostarine™ was also clinically investigated for the treatment of cancer cachexia. In this follow-up phase IIb study, 68 men and 46 women with non-small-cell lung cancer, colorectal cancer, breast cancer, non-Hodgkin's lymphoma or chronic lymphocytic leukemia were randomized to receive placebo or 3 mg of Ostarine.[112] Increases in lean body mass, decreases in fat mass and improvements in stair climbing power were observed for both genders. Transient and small increases in liver transaminases were observed, but Ostarine was generally well tolerated. In the treatment of cancer cachexia, or disease-associated muscle loss of any etiology, a SARM offers the same anabolic and functional benefits as existing androgen regimens with a reduced side effect profile.

Long Term Safety

Due to the lack of long term exogenous androgen administration in women, there is lack of safety data with which to gauge risk.[15] While so far some side effects have been addressed by the development of nonsteroidal AR agonists, a number of open questions remain concerning the long term effects of androgens on liver function and cardiovascular parameters. In the limited clinical evaluations to date, reductions in serum lipids were reported though LDL/HDL ratios remained in the low risk category. Also, some changes in liver function tests, specifically ALT inductions, were reported following SARM treatment. It is likely that the first prospective assessment of long term androgen administration in women on these parameters will involve a SARM.

SARMs — LOOKING FORWARD

Androgens' great therapeutic promise in women is suggested by a growing understanding of androgen biology coupled with increasing clinical successes. A constant fear of side effects greatly curtailed expanded use of androgens in women, and still contributes to dogmatic regulatory opinion. With the development of improved androgens, with reduced side effects, clinicians will be afforded new tools to assess the relative risks and merits of androgen administration. The parallels in the molecular mechanism and ability to achieve tissue-selective pharmacologic effects between SERMs and SARMs suggest that SARMs stand poised to usher in a new era of androgen therapy in men and women.

REFERENCES

1. Wilson CM, McPhaul MJ. (1996) A and B forms of the androgen receptor are expressed in a variety of human tissues. *Mol Cell Endocrinol* **120**:51–57.
2. Burger HG. (2002) Androgen production in women. *Fertil Steril* **77** (Suppl 4):S3–S5.
3. Carmina E. (2006) Ovarian and adrenal hyperandrogenism. *Ann NY Acad Sci* **1092**:130–137.

4. Sarrel PM. (2002) Androgen deficiency: menopause and estrogen-related factors. *Fertil Steril* 77 (Suppl 4) :S63–67.
5. Arlt W. (2006) Androgen therapy in women. *Eur J Endocrinol* **154**:1–11.
6. Zang H, Davis SR. (2008) Androgen replacement therapy in androgen-deficient women with hypopituitarism. *Drugs* **68**:2085–2093.
7. Labrie F, Luu-The V, Labrie C, Belanger A, Simard J, Lin SX, Pelletier G. (2003) Endocrine and intracrine sources of androgens in women: inhibition of breast cancer and other roles of androgens and their precursor dehydroepiandrosterone. *Endocr Rev* **24**:152–182.
8. Miller WL. (2005) Disorders of androgen synthesis — from cholesterol to dehydroepiandrosterone. *Med Princ Pract* **14** (Suppl 1):58–68.
9. Ghayee HK, Auchus RJ. (2007) Basic concepts and recent developments in human steroid hormone biosynthesis. *Rev Endocr Metab Disord* **8**:289–300.
10. Neischlag E, Behre HM, Nieschlag S. (2004) *Testosterone: Action, Deficiency, Substitution*. Cambridge University Press.
11. Galani A, Kitsiou-Tzeli S, Sofokleous C, Kanavakis E, Kalpini-Mavrou A. (2008) Androgen insensitivity syndrome: clinical features and molecular defects. Hormones (Athens) **7**:217–229.
12. Walters KA, McTavish KJ, Seneviratne MG, Jimenez M, McMahon AC, Allan CM, Salamonsen LA, Handelsman DJ. (2009) Subfertile female androgen receptor knockout mice exhibit defects in neuroendocrine signaling, intraovarian function, and uterine development but not uterine function. *Endocrinology* **150**:3274–3282.
13. Matsumoto T, Shiina H, Kawano H, Sato T, Kato S. (2008) Androgen receptor functions in male and female physiology. *J Steroid Biochem Mol Biol* **109**:236–241.
14. Hu YC, Wang PH, Yeh S, Wang RS, Xie C, Xu Q, Zhou X, Chao HT, Tsai MY, Chang C. (2004) Subfertility and defective folliculogenesis in female mice lacking androgen receptor. *Proc Natl Acad Sci USA* **101**:11209–11214.
15. Bolour S, Braunstein G. (2005) Testosterone therapy in women: a review. *Int J Impot Res* **17**:399–408.
16. Bixo M, Backstrom T, Winblad B, Andersson A. (1995) Estradiol and testosterone in specific regions of the human female brain in different endocrine states. *J Steroid Biochem Mol Biol* **55**:297–303.
17. Allan GF, Tannenbaum P, Sbriscia T, Linton O, Lai MT, Haynes-Johnson D, Bhattacharjee S, Zhang X, Sui Z, Lundeen SG. (2007) A selective androgen receptor modulator with minimal prostate hypertrophic activity enhances lean body mass in male rats and stimulates sexual behavior in female rats. *Endocrine* **32**:41–51.
18. Vanderschueren D, Vandenput L, Boonen S, Lindberg MK, Bouillon R, Ohlsson C. (2004) Androgens and bone. *Endocr Rev* **25**:389–425.

19. Kawano H, Sato T, Yamada T, Matsumoto T, Sekine K, Watanabe T, Nakamura T, Fukuda T, Yoshimura K, Yoshizawa T, Aihara K, Yamamoto Y, Nakamichi Y, Metzger D, Chambon P, Nakamura K, Kawaguchi H, Kato S. (2003) Suppressive function of androgen receptor in bone resorption. *Proc Natl Acad Sci USA* **100**:9416–9421.
20. Notelovitz M. (2002) Androgen effects on bone and muscle. *Fertil Steril* **77** (Suppl 4):S34–S41.
21. Pederson L, Kremer M, Judd J, Pascoe D, Spelsberg TC, Riggs BL, Oursler MJ. (1999) Androgens regulate bone resorption activity of isolated osteoclasts *in vitro*. *Proc Natl Acad Sci USA* **96**:505–510.
22. Wronski TJ, Walsh CC, Ignaszewski LA. (1986) Histologic evidence for osteopenia and increased bone turnover in ovariectomized rats. *Bone* **7**:119–123.
23. Tobias JH, Gallagher A, Chambers TJ. (1994) 5 alpha-dihydrotestosterone partially restores cancellous bone volume in osteopenic ovariectomized rats. *Am J Physiol* **267**:E853–859.
24. Ederveen AG, Kloosterboer HJ. (2001) Tibolone exerts its protective effect on trabecular bone loss through the estrogen receptor. *J Bone Miner Res* **16**:1651–1657.
25. Sims NA, Dupont S, Krust A, Clement-Lacroix P, Minet D, Resche-Rigon M, Gaillard-Kelly M, Baron R. (2002) Deletion of estrogen receptors reveals a regulatory role for estrogen receptors-beta in bone remodeling in females but not in males. *Bone* **30**:18–25.
26. Sims NA, Clement-Lacroix P, Minet D, Fraslon-Vanhulle C, Gaillard-Kelly M, Resche-Rigon M, Baron R. (2003) A functional androgen receptor is not sufficient to allow estradiol to protect bone after gonadectomy in estradiol receptor–deficient mice. *J Clin Invest* **111**:1319–1327.
27. Herbst KL, Bhasin S. (2004) Testosterone action on skeletal muscle. *Curr Opin Clin Nutr Metab Care* **7**:271–277.
28. Bhasin S, Taylor WE, Singh R, Artaza J, Sinha-Hikim I, Jasuja R, Choi H, Gonzalez-Cadavid NF. (2003) The mechanisms of androgen effects on body composition: mesenchymal pluripotent cell as the target of androgen action. *J Gerontol A Biol Sci Med Sci* **58**:M1103–1110.
29. Krieg M. (1976) Characterization of the androgen receptor in the skeletal muscle of the rat. *Steroids* **28**:261–274.
30. Kadi F, Bonnerud P, Eriksson A, Thornell LE. (2000) The expression of androgen receptors in human neck and limb muscles: effects of training and elf-administration of androgenic-anabolic steroids. *Histochem Cell Biol* **113**:25–29.
31. Kadi F. (2008) Cellular and molecular mechanisms responsible for the action of testosterone on human skeletal muscle. A basis for illegal performance enhancement. *Br J Pharmacol* **154**:522–528.
32. Sinha-Hikim I, Taylor WE, Gonzalez-Cadavid NF, Zheng W, Bhasin S. (2004) Androgen receptor in human skeletal muscle and cultured muscle satellite cells: up-regulation by androgen treatment. *J Clin Endocrinol Metab* **89**:5245–5255.

33. Ferrando AA, Sheffield-Moore M, Paddon-Jones D, Wolfe RR, Urban RJ. (2003) Differential anabolic effects of testosterone and amino acid feeding in older men. *J Clin Endocrinol Metab* **88**:358–362.
34. Hawke TJ, Garry DJ. (2001) Myogenic satellite cells: physiology to molecular biology. *J Appl Physiol* **91**:534–551.
35. Singh R, Artaza JN, Taylor WE, Gonzalez-Cadavid NF, Bhasin S. (2003) Androgens stimulate myogenic differentiation and inhibit adipogenesis in C3H 10T1/2 pluripotent cells through an androgen receptor-mediated pathway. *Endocrinology* **144**:5081-5088.
36. Tenover JS. (1992) Effects of testosterone supplementation in the aging male. *J Clin Endocrinol Metab* **75**:1092–1098.
37. Miller KK, Biller BM, Beauregard C, Lipman JG, Jones J, Schoenfeld D, Sherman JC, Swearingen B, Loeffler J, Klibanski A. (2006) Effects of testosterone replacement in androgen-deficient women with hypopituitarism: a randomized, double-blind, placebo-controlled study. *J Clin Endocrinol Metab* **91**:1683–1690.
38. Liepa GU, Sengupta A, Karsies D. (2008) Polycystic ovary syndrome (PCOS) and other androgen excess-related conditions: can changes in dietary intake make a difference? *Nutr Clin Pract* **23**:63–71.
39. Zouboulis CC, Chen WC, Thornton MJ, Qin K, Rosenfield R. (2007) Sexual hormones in human skin. *Horm Metab Res* **39**:85-95.
40. Randall VA, Thornton MJ, Hamada K, Messenger AG. (1992) Mechanism of androgen action in cultured dermal papilla cells derived from human hair follicles with varying responses to androgens *in vivo*. *J Invest Dermatol* **98**:86S-91S.
41. Essah PA, Wickham EP, 3rd, Nunley JR, Nestler JE. (2006) Dermatology of androgen-related disorders. *Clin Dermatol* **24**:289–298.
42. Thiboutot D. (2004) Acne: hormonal concepts and therapy. *Clin Dermatol* **22**:419–428.
43. Shahidi NT. (1973) Androgens and erythropoiesis. *N Engl J Med* **289**:72–80.
44. Kozlov VA, Tsyrlova IG, Zhuravkin IN. (1979) Different effect of testosterone on polypotential stem hematopoietic stem cells and immunocompetent B-lymphocytes. *Zh Mikrobiol Epidemiol Immunobiol* **9**:72–76.
45. Malgor LA, Valsecia M, Verges E, De Markowsky EE. (1998) Blockade of the *in vitro* effects of testosterone and erythropoietin on Cfu-E and Bfu-E proliferation by pretreatment of the donor rats with cyproterone and flutamide. *Acta Physiol Pharmacol Ther Latinoam* **48**:99–105.
46. Diamanti-Kandarakis E. (2008) Polycystic ovarian syndrome: pathophysiology, molecular aspects and clinical implications. *Expert Rev Mol Med* **10**:e3.
47. Abbott DH, Barnett DK, Bruns CM, Dumesic DA. (2005) Androgen excess fetal programming of female reproduction: a developmental aetiology for polycystic ovary syndrome? *Hum Reprod Update* **11**:357–374.
48. Nisenblat V, Norman RJ. (2009) Androgens and polycystic ovary syndrome. *Curr Opin Endocrinol Diabetes Obes* **16**:224–231.

49. Azziz R, Carmina E, Dewailly D, Diamanti-Kandarakis E, Escobar-Morreale HF, Futterweit W, Janssen OE, Legro RS, Norman RJ, Taylor AE, Witchel SF. (2009) The Androgen Excess and PCOS Society criteria for the polycystic ovary syndrome: the complete task force report. *Fertil Steril* **91**:456–488.
50. Carmina E, Rosato F, Janni A, Rizzo M, Longo RA. (2006) Extensive clinical experience: relative prevalence of different androgen excess disorders in 950 women referred because of clinical hyperandrogenism. *J Clin Endocrinol Metab* **91**:2–6.
51. Papalia MA, Davis SR. (2003) What is the rationale for androgen therapy for women? *Treat Endocrinol* **2**:77–84.
52. Rivera-Woll LM, Papalia M, Davis SR, Burger HG. (2004) Androgen insufficiency in women: diagnostic and therapeutic implications. *Hum Reprod Update* **10**:421–432.
53. Lovas K, Husebye ES. (2008) Replacement therapy for Addison's disease: recent developments. *Expert Opin Investig Drugs* **17**:497–509.
54. Thorneycroft IH, Stanczyk FZ, Bradshaw KD, Ballagh SA, Nichols M, Weber ME. (1999) Effect of low-dose oral contraceptives on androgenic markers and acne. *Contraception* **60**:255–262.
55. Padero MC, Bhasin S, Friedman TC. (2002) Androgen supplementation in older women: too much hype, not enough data. *J Am Geriatr Soc* **50**:1131–1140.
56. Greenblatt RB, Barfield WE, Garner JF, Calk GL, Harrod JP. Jr. (1950) Evaluation of an estrogen, androgen, estrogen–androgen combination, and a placebo in the treatment of the menopause. *J Clin Endocrinol Metab* **10**:1547–1558.
57. Basaria S, Dobs AS. (2006) Clinical review: controversies regarding transdermal androgen therapy in postmenopausal women. *J Clin Endocrinol Metab* **91**:4743–4752.
58. Spark RF. (2005) Intrinsa fails to impress FDA advisory panel. *Int J Impot Res* **17**:283–284.
59. Henneman PH, Wallach S. (1957) A review of the prolonged use of estrogens and androgens in postmenopausal and senile osteoporosis. *AMA Arch Intern Med* **100**:715–723.
60. Need AG, Horowitz M, Bridges A, Morris HA, Nordin BE. (1989) Effects of nandrolone decanoate and antiresorptive therapy on vertebral density in osteoporotic postmenopausal women. *Arch Intern Med* **149**:57–60.
61. Cole Z, Dennison E, Cooper C. (2008) Update on the treatment of post-menopausal osteoporosis. *Br Med Bull* **86**:129–143.
62. Shen W, Stearns V. (2009) Treatment strategies for hot flushes. *Expert Opin Pharmacother* **10**:1133–1144.
63. Sherwin BB, Gelfand MM. (1984) Effects of parenteral administration of estrogen and androgen on plasma hormone levels and hot flushes in the surgical menopause. *Am J Obstet Gynecol* **148**:552–557.
64. Ingle JN, Twito DI, Schaid DJ, Cullinan SA, Krook JE, Mailliard JA, Tschetter LK, Long HJ, Gerstner JG, Windschitl HE, *et al.* (1991) Combination

hormonal therapy with tamoxifen plus fluoxymesterone versus tamoxifen alone in postmenopausal women with metastatic breast cancer. an updated analysis. *Cancer* **67**:886–891.

65. Bryan RM, Mercer RJ, Bennett RC, Rennie GC, Lie TH, Morgan FJ. (1984) Androgen receptors in breast cancer. *Cancer* **54**:2436–2440.
66. Hennessy BT, Pusztai L. (2005) Adjuvant therapy for breast cancer. *Minerva Ginecol* **57**:305–326.
67. Bhasin S, Calof OM, Storer TW, Lee ML, Mazer NA, Jasuja R, Montori VM, Gao W, Dalton JT. (2006) Drug insight: testosterone and selective androgen receptor modulators as anabolic therapies for chronic illness and aging. *Nat Clin Pract Endocrinol Metab* **2**:146–159.
68. Johns K, Beddall MJ, Corrin RC. (2005) Anabolic steroids for the treatment of weight loss in HIV-infected individuals. *Cochrane Database Syst Rev* CD005483.
69. Miller KK. (2009) Androgen deficiency: effects on body composition. *Pituitary* **12**:116–124.
70. Dolan S, Wilkie S, Aliabadi N, Sullivan MP, Basgoz N, Davis B, Grinspoon S. (2004) Effects of testosterone administration in human immunodeficiency virus-infected women with low weight: a randomized placebo-controlled study. *Arch Intern Med* **164**:897–904.
71. Sharma S, Arneja A, McLean L, Duerksen D, Leslie W, Sciberras D, Lertzman M. (2008) Anabolic steroids in COPD: a review and preliminary results of a randomized trial. *Chron Respir Dis* **5**:169–176.
72. Schols AM, Soeters PB, Mostert R, Pluymers RJ, Wouters EF. (1995) Physiologic effects of nutritional support and anabolic steroids in patients with chronic obstructive pulmonary disease: a placebo-controlled randomized trial. *Am J Respir Crit Care Med* **152**:1268–1274.
73. Johansen KL, Painter PL, Sakkas GK, Gordon P, Doyle J, Shubert T. (2006) Effects of resistance exercise training and nandrolone decanoate on body composition and muscle function among patients who receive hemodialysis: a randomized, controlled trial. *J Am Soc Nephrol* **17**:2307–2314.
74. Johansen KL, Mulligan K, Schambelan M. (1999) Anabolic effects of nandrolone decanoate in patients receiving dialysis: a randomized controlled trial. *JAMA* **281**:1275–1281.
75. Headley S, Germain M, Mailloux P, Mulhern J, Ashworth B, Burris J, Brewer B, Nindl BC, Coughlin M, Welles R, Jones M. (2002) Resistance training improves strength and functional measures in patients with end-stage renal disease. *Am J Kidney Dis* **40**:355–364.
76. Kennedy BJ, Gilbertsen AS. (1957) Increased erythropoiesis induced by androgenic-hormone therapy. *N Engl J Med* **256**:719–726.
77. Watson AJ. (1989) Adverse effects of therapy for the correction of anemia in hemodialysis patients. *Semin Nephrol* **9**:30–34.

78. Teruel JL, Marcen R, Navarro-Antolin J, Aguilera A, Fernandez-Juarez G, Ortuno J. (1996) Androgen versus erythropoietin for the treatment of anemia in hemodialyzed patients: a prospective study. *J Am Soc Nephrol* **7**:140–144.
79. Van Staa TP, Sprafka JM. (2009) Study of adverse outcomes in women using testosterone therapy. *Maturitas* **62**:76–80.
80. Rossouw JE, Anderson GL, Prentice RL, LaCroix AZ, Kooperberg C, Stefanick ML, Jackson RD, Beresford SA, Howard BV, Johnson KC, Kotchen JM, Ockene J. (2002) Risks and benefits of estrogen plus progestin in healthy postmenopausal women: principal results from the Women's Health Initiative randomized controlled trial. *JAMA* **288**:321–333.
81. Guay A. (2005) Commentary on androgen deficiency in women and the FDA advisory board's recent decision to request more safety data. *Int J Impot Res* **17**:375–376.
82. Somboonporn W, Davis SR. (2004) Testosterone effects on the breast: implications for testosterone therapy for women. *Endocr Rev* **25**:374–388.
83. Shufelt CL, Braunstein GD. (2008) Testosterone and the breast. *Menopause Int* **14**:117–122.
84. Kingsberg SA. (2005) The testosterone patch for women. *Int J Impot Res* **17**:465–466.
85. Bluming AZ, Tavris C. (2009) Hormone replacement therapy: real concerns and false alarms. *Cancer J* **15**:93–104.
86. Dalton JT, Mukherjee A, Zhu Z, Kirkovsky L, Miller DD. (1998) Discovery of nonsteroidal androgens. *Biochem Biophys Res Commun* **244**:1–4.
87. Edwards JP, West SJ, Pooley CL, Marschke KB, Farmer LJ, Jones TK. (1998) New nonsteroidal androgen receptor modulators based on 4-(trifluoromethyl)-2(1H)-pyrrolidino[3,2-g] quinolinone. *Bioorg Med Chem Lett* **8**:745–750.
88. Hamann LG, Mani NS, Davis RL, Wang XN, Marschke KB, Jones TK. (1999) Discovery of a potent, orally active, nonsteroidal androgen receptor agonist: 4-ethyl-1,2,3,4-tetrahydro-6-(trifluoromethyl)-8-pyridono[5,6-g]-quinoline (LG121071). *J Med Chem* **42**:210–212.
89. Gao W, Kim J, Dalton JT. (2006) Pharmacokinetics and pharmacodynamics of nonsteroidal androgen receptor ligands. *Pharm Res* **23**:1641–1658.
90. Negro-Vilar A. (1999) Selective androgen receptor modulators (SARMs): a novel approach to androgen therapy for the new millennium. *J Clin Endocrinol Metab* **84**:3459–3462.
91. Yin D, Gao W, Kearbey JD, Xu H, Chung K, He Y, Marhefka CA, Veverka KA, Miller DD, Dalton JT. (2003) Pharmacodynamics of selective androgen receptor modulators. *J Pharmacol Exp Ther* **304**:1334–1340.
92. Hershberger LG, Shipley EG, Meyer RK. (1953) Myotrophic activity of 19-nortestosterone and other steroids determined by modified levator ani muscle method. *Proc Soc Exp Biol Med* **83**:175–180.

ionamides is a major structural determinant of *in vivo* disposition and activity of selective androgen receptor modulators. *J Pharmacol Exp Ther* **315**:230–239.

118. Chen J, Hwang DJ, Bohl CE, Miller DD, Dalton JT. (2005) A selective androgen receptor modulator for hormonal male contraception. *J Pharmacol Exp Ther* **312**:546–553.
119. Hamann LG, Higuchi RI, Zhi L, Edwards JP, Wang XN, Marschke KB, Kong JW, Farmer LJ, Jones TK. (1998) Synthesis and biological activity of a novel series of nonsteroidal, peripherally selective androgen receptor antagonists derived from 1,2-dihydropyridono[5,6-g]quinolines. *J Med Chem* **41**:623–639.

16

AN EMERGING PRINCIPLE: SELECTIVE NUCLEAR RECEPTOR MODULATORS

Ping Fan and V. Craig Jordan

Abstract

Ligands for the nuclear receptor superfamily control many aspects of biology, including development, reproduction, and homeostasis, through regulation of the transcriptional activity of their cognate nuclear receptors (NRs). These receptors now represent one of the most important targets for therapeutic drug development. Selective nuclear receptor modulators (SNRMs) are receptor ligands that exhibit agonistic or antagonistic characterization in a cell and tissue context-dependent manner. The prototypical SNRM is tamoxifen, which, as a selective estrogen receptor modulator (SERM), can activate or inhibit estrogen receptor action with tissue selectivity. The success of SERMs led to the emergence of the concept to develop tissue-selective drugs for all members of the NR superfamily. We will review the current progress in developing NR modulators from designing synthetic compounds that mimic the full function of cognate ligands to developing compounds with tissue selectivity that modulate the functional activity of an NR in a manner that is distinct from that of cognate ligands. The goal is to reduce undesirable side effects.

Keywords: Nuclear Receptors (NRs); Selective Estrogen Receptor Modulators (SERMs); Selective Androgen Receptor Modulators (SARMs); Selective Progesterone Receptor Modulators (SPRMs); Selective Glucocorticoid Receptor Modulators (SGRMs); Selective Mineralocorticoid Receptor Modulators (SMRMs); Selective Thyroid Receptor Modulators (STRMs); Selective Peroxisome-Proliferator-Activated Receptor Modulators (SPPARMs).

THE SITUATION BEFORE 1985: EMERGENCE OF THE FIRST TISSUE SELECTIVE NR MODULATOR — TAMOXIFEN

The link between ovarian function and the development of breast cancer has been known for more than a century.[1] The initial clinical observations were subsequently complemented by the laboratory finding that early ovariectomy can reduce the incidence of mammary cancer in high-incidence strains of mice (A. E. Lathrop and L. Loeb, 1916),which led to speculation that an antagonist of estrogen accumulation could prevent the breast cancer (A. Lacassagne, 1936).In 1958, Leonard Lerner reported the antiestrogenic properties of the first nonsteroidal antiestrogen, ethamoxytriphetol, and subsequently clomiphene. Remarkably, both compounds were effective postcoital contraceptives in laboratory animals, and so one possible application was in fertility control.[2] Ethamoxytriphetol proved too toxic for clinical use, and in women clomiphene had exactly the opposite antifertility action noted in rats and induced ovulation and enhanced fertility in subfertile women. Although the scientists envisaged many clinical applications for antiestrogens, the application of ethamoxytriphetol, clomiphene and nafoxidine as experimental treatments for advanced breast cancer were all discontinued because of extensive side effects.[2]

Tamoxifen (ICI 46,474) was initially identified in the early 1960s in a program designed to develop a contraceptive. These antiestrogens had failed to fulfill their promise as postcoital contraceptives and were unlikely to be developed further by the pharmaceutical industry.[3] In 1972, that perspective started to change and ICI 46,474 was subsequently reinvented as the first targeted therapy for breast cancer.[5] However, the clinical evaluation of tamoxifen in advanced breast cancer was not as popular as cytotoxic chemotherapy throughout the 1970s.[3] About 30 years later, enthusiasm for the nonspecific approach of chemotherapy was replaced by the new generation target therapies.[3] Tamoxifen was found to block estrogen binding to the human tumor estrogen receptor (ER) and could almost completely prevent rat mammary carcinogenesis. Most importantly, it had less clinical side effects compared with other endocrine therapies. The idea that tamoxifen should be applied as a long-term adjuvant therapy for patients with ER-positive primary breast cancer first appeared through *the* systematic animal experiments.[2,3] Tamoxifen has been the gold standard for the endocrine treatment of all stages of ER-positive breast cancer and the World Health Organization lists it as an essential drug for the treatment of breast cancer.[2,3] Hundreds of thousands of women now have enhanced survivorship after their diagnosis of ER-positive breast cancer. Tamoxifen also became the first cancer chemopreventive to be approved by the Food and Drug Administration (FDA) for the reduction of breast cancer incidence in both pre- and postmenopausal women at high risk. However, patient care was affected by the recognition in the laboratory that tamoxifen would potentially increase the incidence of endometrial cancer during long-term adjuvant therapy. At that time, a failed breast cancer drug, keoxifene (now renamed raloxifene), was found to maintain bone density of rats (estrogenic action) while simultaneously preventing mammary carcinogenesis (antiestrogenic action).[4] The second-generation SERM, raloxifene, is effective against osteoporosis in postmenopausal women and prevents breast cancer as effectively as tamoxifen, but without an increase in endometrial cancer. Although SERMs have been available for clinical use for 50 years, major efforts are still focused on making new modulators with a better therapeutic profile and fewer side effects (Table 1).

CHANGE IN THINKING POST-SERMS

The SERM concept has spread to develop tissue-selective drugs for several members of the NR.[5-7] Of the 48 NRs, 24–27 bind ligands with a characterized ligand-binding domain (LBD) although orphan NRs do not need a small ligand for gene regulation.[8] The members of this family are divided into three classes, with class I containing receptors for estrogen, progesterone, mineralocorticoids, glucocorticoids and androgens. Receptors for vitamin D, retinoids and thyroids are placed in class II. Class III receptors are those for which ligands have not yet been identified, and are hence classified as "orphans." However, in recent years, natural and synthetic ligands for many of these orphan receptors have been uncovered.[8] The method of targeting NRs in general is too large a topic to cover adequately in this chapter, so the

Table 1. Progress in the development of SERMs with tissue selectivity.

	Tamoxifen	Raloxifene	Lasofoxifene
Structure	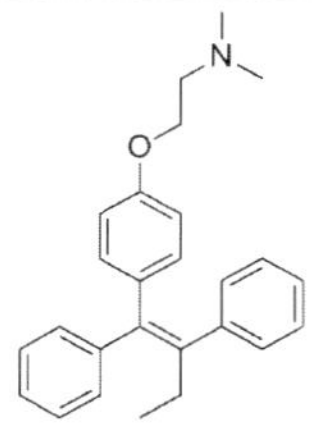	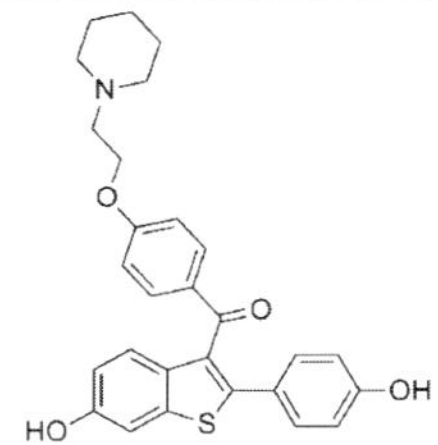	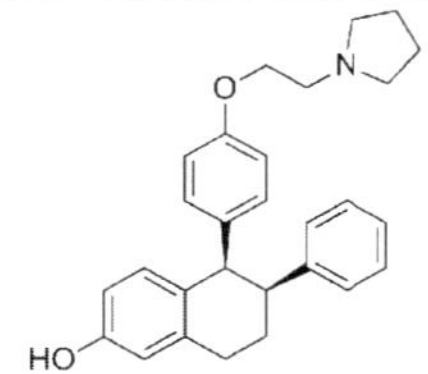
ER antagonists for breast cancer	Treatment and prevention of ER positive breast cancer	Prevention of ER positive breast cancer	Prevention of ER positive breast cancer
ER agonist for bone	Maintain the bone density of postmenopausal women	Prevention of vertebral fractures	Higher potency than first two. Prevention of vertebral and nonvertebral fractures
ER agonist to reduce cardiovascular risk	No	No	Yes
ER agonist to cause endometrial cancer	Yes	No	No

interested reader is referred to the clinical and translational articles for further information. We have chosen instead to focus our discussion on compounds that selectively target androgen receptors (SARMs), progesterone receptors (SPRMs), glucocorticoid receptors (SGRMs), mineralocorticoid receptors (SMRMs), thyroid receptors (STRMs) and peroxisome proliferator-activated receptors (SPPARMs).

SARMs (Selective Androgen Receptor Modulators)

The androgen receptor (AR) plays a critical role in the function of several organs, including primary and accessory sexual organs, skeletal muscle and bone, making it a desirable therapeutic target. Testosterone was the first anabolic androgen to be used clinically which exerts its nonselective anabolic and androgenic effects via the AR.[9] Though administration of steroidal androgens improves muscle mass and bone mineral density, they also have significant undesired effects leading to increased prostate size, acne, and effects on serum lipids. In the latter half of the 20th century, the chemical scaffold of testosterone was modified extensively, producing many steroidal androgens.[9] Although some enhancements of the anabolic effects were obtained relative to testosterone, it was never possible to adequately dissociate the anabolic and androgenic activities of these compounds, and they came to be referred to as anabolic-androgenic steroids (AASs). The AAS and testosterone are substrates for 5α reductase and aromatase, causing a variety of untoward side effects, which limits their use for therapeutic purposes. Many of the AASs have been withdrawn as licensed products.[9] This encourages the pursuit of tissue-selective androgen receptor modulators (SARMs). Interest in the development and therapeutic potential of SARMs has increased dramatically over the past decade[10] (see Chapter 17).

Tissue-selective antiandrogen SARMs could be used to prevent or treat benign prostate hypertrophy or prostate cancer without blocking anabolic androgen effects on bone, muscle or the CNS. Oral nonsteroidal antiandrogens were developed in the 1970s, and include bicalutamide, flutamide and nilutamide (Fig. 1). These antiandrogens have high specificity for the AR, but lack tissue selectivity,

hippocampus and adipose tissue. It is most closely related to the GR, with which it shares a common ancestor.[43] The MR is capable of binding multiple classes of steroids with high affinity, including mineralocorticoids, aldosterone and deoxycorticosterone, the glucocorticoids, cortisol (in humans) or corticosterone (in rodents), and progesterone.[44] While aldosterone is considered the primary physiological MR ligand in humans, in some tissues cortisol may be the primary ligand for the MR, whereas the PR behaves as a predominant antagonist.[45]

MR activation plays an important role in the pathophysiology of hypertension and cardiac fibrosis.[44] Although there is clinical utility in blocking MR action, the usefulness of available MR antagonists is limited because of cross-reactivity with the AR and PR (spironolactone) or possibly by low affinity for MR (eplerenone).[46] These observations indicate that a receptor-specific higher-affinity antagonist with appropriate bioavailability would give a greater response. Two large clinical trials have demonstrated that MR blockade can significantly reduce morbidity and mortality in patients with heart failure.[44] However, the use of MR antagonists is also limited by the adverse renal effect of hyperkalemia.[47] A tissue-selective MR antagonist that targets only heart failure while avoiding unwanted renal effects would thus be highly desirable. Specific coregulator–MR interactions may also be a suitable target in the rational design of tissue-specific MR modulators, as has been described for other steroid receptors.[44] However, the search for MR-interacting coregulators is less advanced, partly due to the difficulty of purifying a stable full-length MR.[48] The number of coregulators identified to date for the MR is very limited compared with other nuclear receptors.[44] All these data outlined above point to clinically important roles for a variety of selective MR modulators: antagonists with selectivity for the MR versus other steroid receptors, antagonists with improved affinity for the MR and improved potency, and antagonists to the MR with tissue selectivity.[49]

STRMs (Selective Thyroid Receptor Modulators)

Thyroid hormone receptors (TRs) exert profound effects on development, metabolism and multiple specific organ functions. Most of the actions of thyroid hormones (T_4) are mediated by TRs. T_4 is carried in the circulation as a complex with thyroid-binding globulin (TBG). It dissociates from TBG and is then transported across the cell membrane and converted to the active form of the hormone, T_3[50] (Fig. 6). The heart is an important target for thyroid hormone action. T_3 increases the heart rate, speed, and force of systolic contraction.[51] It can increase the metabolic rate and lowers plasma lipid levels.[52] Thyroid hormones have some actions that might be useful therapeutically, but others that are deleterious. Potential therapeutically useful actions include those to induce weight loss and lower plasma cholesterol levels. Potential deleterious actions are particularly on the heart to induce tachycardia and arrhythmia. There are two isoforms of TRs, TRα and TRβ, and these two TRs exhibit marked differences in the

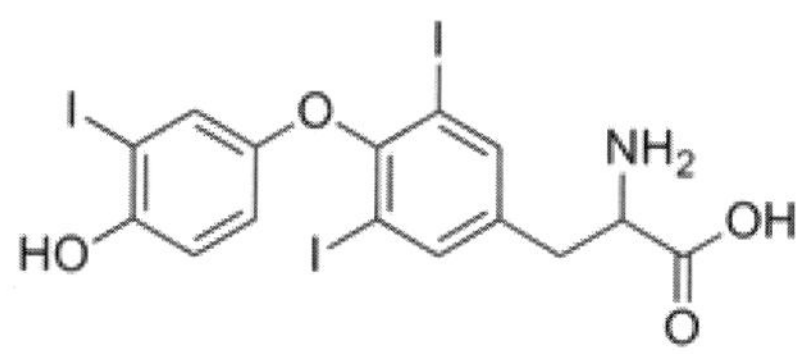

T_3

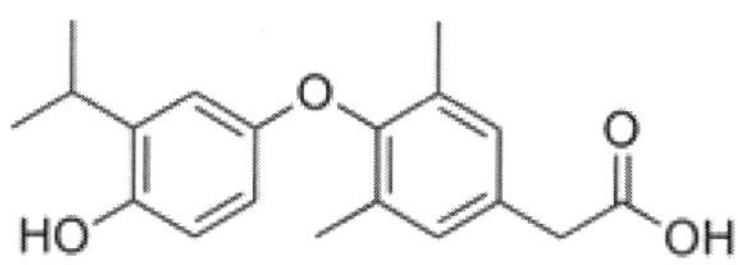

GC-1

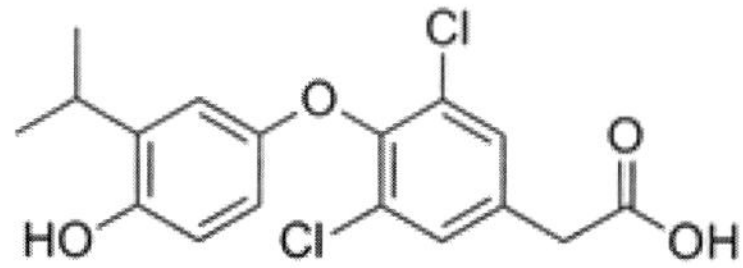

KB-141

Figure 6. Structures of thyroid hormone T_3 and thyroid receptor beta (TRβ) selective modulators GC-1 and KB-141.

expression pattern; for example, TRα transcripts are abundant in the brain and heart, whereas high levels of TRβ are detected in the liver.[53] It has been known that the heart rate is regulated by TRα, whereas many other effects are regulated by TRβ. Effects of thyroid hormones on the metabolic rate are mediated through both the α and β isoforms, and on cholesterol almost exclusively through the β isoform. [49] Most thyroid hormones do not discriminate between the two TRs, and bind them with similar affinities. Thus, there is reason to pursue selective TR modulator (STRM) function, and several agents have been shown to have some β-selective, hepatic selective and/or cardiac-sparing activities.[54] First-generation TRβ-selective agonists, GC-1 (also known as sobetirome, produced by Quattrx) and KB 141 (produced by KaroBio), have been designed and synthesized that have high affinity for the TRs and are selective in both the binding and activation functions for TRβ over TRα (Fig. 6).[55] These TRβ-selective compounds show promise as therapeutics for inducing weight loss and lowering plasma lipid levels while avoiding the unwanted effects, such as that on the heart rate.[55] They also have potential applications in atherosclerosis, obesity and type II diabetes.[56] GC-1 is now in human trials. More recently, TR-selective analogs that capitalize on the selectivity of the TRβ isoform and liver uptake have been reported.[56] Two of them are in clinical trials: KB 2115 (developed by KaroBio) and MB 07811 (produced by Metabasis). Both have been described as TRβ-selective ligands that are preferentially taken up by the liver.[56] The ligands behave in the predicted manner: serum cholesterol is reduced and the heart rate is unaffected at therapeutic doses.

Selective antagonists to TRs would be useful for medical therapy and research. Excess thyroid hormone production by the thyroid gland or administration of exogenous thyroid hormones results in hyperthyroidism or thyrotoxicosis, marked by an increased heart rate with possible atrial arrhythmias and heart failure, muscle wasting, osteoporosis in postmenopausal women, and other symptoms.[57] The current standard treatments of endogenous hyperthyroidism can have serious manifestations in the elderly, and have remained essentially unchanged for more than 30 years.[58] Directly antagonizing the effect of the thyroid hormone at the receptor level may be a significant improvement for the treatment of hyperthyroid patients. No TRα-selective ligands are available right now, which potentially would further ameliorate the problem with cardiac disease connected with hyperthyroidism.[58] Design of TR antagonist ligands has been based on the attachment of a large extension group to a TR agonist whose structure would perturb formation of the TR-coactivator-binding surface.[59] Despite its significant potential use, no TR antagonist has reached clinical application.

SPPARMs (Selective Peroxisome-Proliferator-Activated Receptor Modulators)

Peroxisome-proliferator-activated receptors (PPARs) are ligand-activated transcription factors belonging to the nuclear hormone receptor superfamily. Three PPAR isotypes have been identified: PPAR-γ, PPAR-α and PPAR-δ (also known as PPAR-β), which are involved in the regulation of glucose and lipid homeostasis.[60] The antidyslipidemic activity of these compounds results chiefly from the induction of hepatic lipid uptake and catabolism. Research studies demonstrate that PPAR-α and PPAR-γ ligands might also have important anti-inflammatory, vasoprotective actions.[61] Understanding of PPAR-δ function has been augmented through a series of preclinical studies in which PPAR-δ activation diminished metabolic derangements and obesity, apparently by increasing lipid combustion in skeletal muscle.[62] PPARs were originally named for their ability to induce hepatic peroxisome proliferation in mice in response to xenobiotic stimuli. PPARα is mainly found in the liver and plays a pivotal role in the regulation of cellular uptake, activation and β oxidation of fatty acids and in lipoprotein metabolism.[63] PPARγ is expressed primarily in adipose tissue and acts as a transcription factor regulating adipocyte differentiation and glucose homeostasis.[64] There are currently no important drugs in clinical use which target PPAR-δ, whereas PPAR-α and PPAR-γ agonistic drugs are the most prominent targets for the treatment of metabolic diseases. The thiazolidinedione (TZD) antidiabetic agents rosiglitazone and pioglitazone are PPAR-γ agonists whose insulin-sensitizing actions are mediated largely by pleiotropic effects

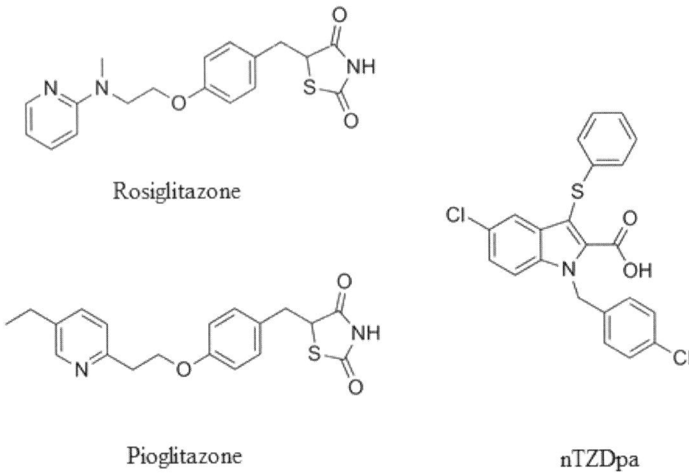

Figure 7. Structures of PPAR-γ agonists TZDs derivatives (rosiglitazone and piogliazone) and selective PPAR-γ modulator nTZDpa.

in adipose tissue (Fig. 7). The fibrate antiatherosclerotic, hypolipidemic agents such as fenofibrate and gemfibrozil are PPAR-α agonists.[65] Insulin resistance, characterized by reduced responsiveness to normal circulating concentrations of insulin, is a common feature of almost all patients with type II diabetes (T2DM). This disorder is also strongly associated with other metabolic syndromes, including hypertension, dyslipidemias and polycystic ovary syndrome (PCOS). TZDs are a new class of orally active drugs that are designed to enhance the actions of insulin through a combination of metabolic actions, including the partitioning of lipid stores and the regulation of metabolic and inflammatory mediators called adipokines.[66] To date, PPAR agonists have been therapeutically used in patients with T2DM and in patients with reproductive disorders such as PCOS.[67] TZDs exert their antidiabetic effects by selectively activating PPAR-γ, which controls glucose and fatty acid metabolism. Although the TZDs rosiglitazone and pioglitazone, which are full agonists of PPAR-γ, have significant clinical antidiabetic efficacy, their use is associated with adverse events, including plasma volume expansion, hemodilution, edema, increased adiposity and weight gain.[68] In addition, TZDs induce cardiac enlargement in some preclinical species.[69] These undesirable effects, and the potential to cause congestive heart failure in the considerable subset of diabetic patients with underlying cardiopathies, limit the clinical application of TZDs. Therefore, laboratory research evidence points toward a possible separation of unwanted effects of PPAR-γ agonists from its beneficial antidiabetic effects by using selective PPAR-γ modulators (SPPARγMs).[70] A significant step forward occurred with the description of nTZDpa, a potent, selective, non-TZD SPPARγM.[70] nTZDpa (Fig. 7) is a partial agonist at PPAR-γ in cell-based transactivation assays, and has decreased adipogenic activity and an attenuated gene signature in adipocytes in culture. These novel activities result from the unique physical interaction of nTZDpa with PPAR-γ, which provides diminished conformational stability of the receptor compared with full agonists. A new non-TZD SPPARγM, FK 614, is currently in phase II clinical trials; it has insulin-sensitizing activity which is similar to that of PPAR-γ full agonists in diabetic mice but which induces less hemodilution and cardiomegaly in rats.[71] Halofenate is known as a hypolipidemic and hypouricemic agent which acts as SPPARγM *in vitro* and *in vivo* preclinical experiments.[72] One research group recently demonstrated that modifications in the structure of pirinixic acid led to a subtype-selective PPAR-γ agonist and enhanced the development of a selective PPAR-γ modulator initially guided by molecular docking studies.[73] All of these findings supported the belief that SPPARγMs might provide more tolerable therapy for T2DM than the currently available TZDs in the near future.

CHALLENGES FOR THE FUTURE

It is just over 20 years since the first steroid receptor cDNAs were cloned, a development that led to the birth of a superfamily of ligand-activated transcription factors: the nuclear receptors (NRs). These molecules act as sensors of the extracellular and intracellular environment and play crucial roles in controlling development, differentiation, metabolic homeostasis, and reproduction.[74] The process of deciphering the molecular mechanisms of other ligands' action in the target tissues has tended to lag behind the pathfinder estrogen.[1] Nevertheless, the basic model for the regulation of nuclear hormone receptor action is consistent but the details of these hormones are distinctly different from estrogen action, which in turn created novel therapeutic opportunities to stop the biological actions of each active steroidal agent. These receptors now represent one of the most important targets for therapeutic drug development.[75] Current efforts regarding NR drug development continue to focus on improving the function and tissue selectivity of drug candidates to reduce undesirable side effects. The crosstalk among NRs complicates the discovery of selective modulators to different NRs. Recent developments suggest that NRs have many more faces with the potential to express a range of different functionalities, depending on factors that include the tissue type, ligand type, receptor variants, cofactor surroundings and target gene promoters. This behavior of NRs has made the development of safer ligands, which trigger only the expression of a desirable subset of genes, a real challenge.[76] Thus, it must be stressed that more knowledge-based fundamental research is needed to ensure the design and development of selective NR modulators capable of reaching the clinic.

ACKNOWLEDGEMENTS

This work (V. C. J.) was supported by the Department of Defense Breast Program under award No. W81XWH-06-1-0590 Center of Excellence; subcontract under SU2C (AACR) grant No. SU2C-AACR-DT0409; the Susan G. Komen for the Cure Foundation under award No. SAC100009; GHUCCTS CTSA (Grant #UL1RR031975) and the Lombardi Comprehensive Cancer Center Support Grant (CCSG) Core Grant NIH P30 CA051008. The views and opinions of the authors do not reflect those of the U.S. Army or the Department of Defense.

REFERENCES

1. Jordan VC. (2009) A century of deciphering the control mechanisms of sex steroid action in breast and prostate cancer: the origins of targeted therapy and chemoprevention. *Cancer Res* **69**:1243–1254.

2. Jordan VC. (2008) The 38th David A. Karnofsky Lecture: the paradoxical actions of estrogen in breast cancer — survival or death. *J Clin Oncol* **26**:3073–3082.
3. Jordan VC. (2003) Tamoxifen: a most unlikely pioneering medicine. *Nat Rev Drug Discov* **2**:205–213.
4. Jordan VC. (2008) The rise of raloxifene and the fall of invasive breast cancer. *J Natl Cancer Inst* **100**:831–833.
5. Jordan VC. (2008) Tamoxifen: catalyst for the change to targeted therapy. *Eur J Cancer* **44**:30–38.
6. Jordan VC. (2001) Selective estrogen receptor modulation: a personal perspective. *Cancer Res* **61**:5683–5687.
7. Jordan VC, O'Malley BW. (2007) Selective estrogen-receptor modulators and antihormonal resistance in breast cancer. *J Clin Oncol* **25**:5815–5824.
8. Tsai MJ, O'Malley BW. (1994) Molecular mechanisms of action of steroid/thyroid receptor superfamily members. *Annu Rev Biochem* **63**:451–486.
9. Mohler ML, Bohl CE, Jones A, Coss CC, Narayanan R, He Y, Hwang DJ, Dalton JT, Miller DD. (2009) Nonsteroidal selective androgen receptor modulators (SARMs): dissociating the anabolic and androgenic activities of the androgen receptor for therapeutic benefit. *J Med Chem* **52**:3597–3617.
10. Segal S, Narayanan R, Dalton JT. (2006) Therapeutic potential of the SARMs: revisiting the androgen receptor for drug discovery. *Expert Opin Invest Drugs* **15**:377–387.
11. Clarke BL, Khosla S. (2009) New selective estrogen and androgen receptor modulators. *Curr Opin Rheumatol* **21**:374–379.
12. Allan G, Lai MT, Sbriscia T, Linton O, Haynes-Johnson D, Bhattacharjee S, Dodds R, Fiordeliso J, Lanter J, Sui Z, Lundeen S. (2007) A selective androgen receptor modulator that reduces prostate tumor size and prevents orchidectomy-induced bone loss in rats. *J Steroid Biochem Mol Biol* **103**:76–83.
13. Dalton JT, Mukherjee A, Zhu Z, Kirkovsky L, Miller DD. (1998) Discovery of nonsteroidal androgens. *Biochem Biophys Res Commun* **244**:1–4.
14. Edwards JP, West SJ, Pooley CL, Marschke KB, Farmer LJ, Jones TK. (1998) New nonsteroidal androgen receptor modulators based on 4-(trifluoromethyl)-2(1H)-pyrrolidino[3,2-g] quinolinone. *Bioorg Med Chem Lett* **8**:745–750.
15. Narayanan R, Mohler ML, Bohl CE, Miller DD, Dalton JT. (2008) Selective androgen receptor modulators in preclinical and clinical development. *Nucl Recept Signal* **6**:1–36.
16. Marhefka CA, Gao W, Chung K, Kim J, He Y, Yin D, Bohl C, Dalton JT, Miller DD. (2004) Design, synthesis, and biological characterization of metabolically stable selective androgen receptor modulators. *J Med Chem* **47**:993–998.
17. Gao W, Dalton JT. (2007) Expanding the therapeutic use of androgens via selective androgen receptor modulators (SARMs). *Drug Discov Today* **12**:241–248.

18. Hamann LG, Higuchi RI, Zhi L, Edwards JP, Wang XN, Marschke KB, Kong JW, Farmer LJ, Jones TK. (1998) Synthesis and biological activity of a novel series of nonsteroidal, peripherally selective androgen receptor antagonists derived from 1,2-dihydropyridono [5,6-*g*]quinolines. *J Med Chem* **41**:623–639.
19. Schindler AE, Campagnoli C, Druckmann R, Huber J, Pasqualini JR, Schweppe KW, Thijssen JHH. (2003) Classification and pharmacology of progestins. *Maturitas* **46**:S7–S16.
20. Chwalisz K, Garg R, Brenner R, Slayden O, Winkel C, Elger W. (2006) Role of nonhuman primate models in the discovery and clinical development of selective progesterone receptor modulators (SPRMs). *Reprod Biol Endocrinol* 49(Suppl 1):S8.
21. Murphy AA, Kettel LM, Morales AJ, Roberts VJ, Yen SS. (1993) Regression of uterine leiomyomata in response to the antiprogesterone RU 486. *J Clin Endocrinol Metab* **76**:513–517.
22. Chabbert-Buffet N, Meduri G, Bouchard P, Spitz IM. (2005) Selective progesterone receptor modulators and progesterone antagonists: mechanisms of action and clinical applications. *Hum Reprod Update* **11**:293–307.
23. DeManno D, Elger W, Garg R, Lee R, Schneider B, Hess-Stumpp H, Schubert G, Chwalisz K. (2003) Asoprisnil (J867): a selective progesterone receptor modulator for gynecological therapy. *Steroids* **68:**1019–1032.
24. Chwalisz K, Perez MC, Demanno D, Winkel C, Schubert G, Elger W. (2005) Selective progesterone receptor modulator development and use in the treatment of leiomyomata and endometriosis. *Endocr Rev* **26**(3): 423–438.
25. Chwalisz K, Brenner RM, Fuhrmann UU, Hess-Stumpp H, Elger W. (2000) Antiproliferative effects of progesterone antagonists and progesterone receptor modulators on the endometrium. Steroids 65:741–751.
26. Chwalisz K, Elger W, Stickler T, Mattia-Goldberg C, Larsen L. (2005) The effects of 1-month administration of asoprisnil (J867), a selective progesterone receptor modulator, in healthy premenopausal women. *Hum Reprod* **20:**1090–1099.
27. Chwalisz K, Larsen L, Mattia-Goldberg C, Edmonds A, Elger W, Winkel CA. (2007) A randomized, controlled trial of asoprisnil, a novel selective progesterone receptor modulator, in women with uterine leiomyomata. *Fertil Steril* **87**:1399–1412.
28. Tabata Y, Iizuka Y, Shinei R, Kurihara K, Okonogi T, Hoshiko S, Kurata Y. (2003) CP8668, a novel orally active nonsteroidal progesterone receptor modulator with tetrahydrobenzindolone skeleton. *Eur J Pharmacol* **461**:73–78.
29. Sapolsky RM, Romero LM, Munck AU. (2000) How do glucocorticoids influence stress responses? integrating permissive, suppressive, stimulatory, and preparative actions. *Endocr Rev* **21**:55–89.
30. De Bosscher K, Haegeman G, Elewaut D. (2010) Targeting inflammation using selective glucocorticoid receptor modulators. *Curr Opin Pharmacol* **10**:497–504.

31. Spitz IM, Bardin CW. (1993) Mifepristone (RU 486) — a modulator of progestin and glucocorticoid action. *N Engl J Med* **329**:404–412.
32. Friedman JE, Sun Y, Ishizuka T, Farrell CJ, McCormack SE, Herron LM, Hakimi P, Lechner P, Yun JS. (1997) Phosphoenolpyruvate carboxykinase (GTP) gene transcription and hyperglycemia are regulated by glucocorticoids in genetically obese *db/db* transgenic mice. *J Biol Chem* **272**:31475–31481.
33. Schäcke H, Berger M, Hansson TG, McKerrecher D, Rehwinkel H. (2008) Dissociated non-steroidal glucocorticoid receptor modulators: an update on new compounds. *Expert Opin Ther Patents* **18**:339–352.
34. Hudson AR, Roach SL, Higuchi RI. (2008) Recent developments in the discovery of selective glucocorticoid receptor modulators (SGRMs). *Curr Topics Med Chem* **8**:750–765.
35. De Bosscher K, Van Craenenbroeck K, Meijer OC, Haegeman G. (2008) Selective transrepression versus transactivation mechanisms by glucocorticoid receptor modulators in stress and immune systems. *Eur J Pharmacol* **583**:290–302.
36. Coghlan MJ, Jacobson PB, Lane B, Nakane M, Lin CW, Elmore SW, Kym PR, Luly JR, Carter GW, Turner *R* (2003). A novel antiinflammatory maintains glucocorticoid efficacy with reduced side effects. *Mol Endocrinol* **17**:860–869.
37. Doggrell S. (2003) Is AL-438 likely to have fewer side effects than the glucocorticoids? *Expert Opin Invest Drugs* **12**:1227–1229.
38. López FJ, Ardecky RJ, Bebo B, Benbatoul K, De Grandpre L, Liu S, Leibowitz MD, Marschke K, Rosen J, Rungta D. (2008). LGD-5552, an antiinflammatory glucocorticoid receptor ligand with reduced side effects, *in vivo*. *Endocrinology* **149**:2080–2089.
39. Takahashi H, Razavi H, Thomson D. (2008) Recent progress in the discovery of novel glucocorticoid receptor modulators. *Curr Top Med Chem* **8**:521–530.
40. Opherk C, Tronche F, Kellendonk C, Kohlmüller D, Schulze A, Schmid W, Schütz G. (2004) Inactivation of the glucocorticoid receptor in hepatocytes leads to fasting hypoglycemia and ameliorates hyperglycemia in streptozotocin-induced diabetes mellitus. *Mol Endocrinol* **16**:1346–1353.
41. Mohler ML, He Y, Wu Z, Hong SS, Miller DD. (2007) Non-steroidal glucocorticoid receptor antagonists: the race to replace RU-486 for anti-glucocorticoid therapy. *Expert Opin Ther Pat* **17**:59–81.
42. Bookout AL, Jeong Y, Downes M, Yu RT, Evans RM, Mangelsdorf DJ. (2006) Anatomical profiling of nuclear receptor expression reveals a hierarchical transcriptional network. *Cell* **126**:789–799.
43. Ortlund EA, Bridgham JT, Redinbo MR, Thornton JW. (2007) Crystal structure of an ancient protein: evolution by conformational epistasis. *Science* **317**:1544–1548.
44. Yang J, Young MJ. (2009) The mineralocorticoid receptor and its coregulators. *J Mol Endocrinol* **43**:53–64.

45. Fagart J, Wurtz JM, Souque A, Hellal-Levy C, Moras D, Rafestin-Oblin ME. (1998) Antagonism in the human mineralocorticoid receptor. *EMBO J* **17**: 3317–3325.
46. McMahon EG. (2001) Recent studies with eplerenone, a novel selective aldosterone receptor antagonist. *Curr Opin Pharmaco* **1**:190–196.
47. Juurlink DN, Mamdani MM, Lee DS, Kopp A, Austin PC, Laupacis A, Redelmeier DA. (2004) Rates of hyperkalemia after publication of the randomized aldactone evaluation study. *N Engl J Med* **351**:543–551.
48. Clyne CD, Chang C-Y, Safi R, Fuller PJ, McDonnell DP, Young MJ. (2009) Purification and characterization of recombinant human mineralocorticoid receptor. *Mol Cell Endocrinol* **302**:81–85.
49. Baxter JD, Funder JW, Apriletti JW, Webb P. (2004) Towards selectively modulating mineralocorticoid receptor function: lessons from other systems. *Mol Cell Endocrinol* **217**:151–165.
50. Yen PM. (2001) Physiological and molecular basis of thyroid hormone action. *Physiol Rev* **81**:1097–1042.
51. Graettinger JS, Muenster JJ, Selvertone LA, Campbell JA. (1959) A correlation of clinical and hemodynamic studies in patients with hyperthyroidism with and without congestive heart failure. *J Clin Invest* **38**:1316–1327.
52. Underwood AH, Emmett JC, Ellis D, Flynn SB, Leeson PD, Benson GM, Novelli R, Pearce NJ, Shah VP. (1986) A thyromimetic that decreases plasma cholesterol levels without increasing cardiac activity. *Nature* **324**:425–429.
53. Bookout AL, Jeong Y, Downes M, Yu RT, Evans RM, Mangelsdorf DJ. (2006) Anatomical profiling of nuclear receptor expression reveals a hierarchical transcriptional network. *Cell* **126**:789–799.
54. Grover GJ, Mellstrom K, Ye L, Malm J, Li YL, Bladh LG, Sleph PG, Smith MA, George R, Vennstrom B, Mookhtiar K, Horvath R., Speelman J, Egan D, Baxter JD. (2003) Selective thyroid hormone receptor-beta activation: a strategy for reduction of weight, cholesterol, and lipoprotein (a) with reduced cardiovascular liability. *Proc Natl Acad Sci USA* **100**:10067–10072.
55. Yoshihara HA, Scanlan TS. (2003) Selective thyroid hormone receptor modulators. *Curr Top Med Chem* **3**:1601–1616.
56. Baxter JD, Webb P. (2009) Thyroid hormone mimetics: potential applications in atherosclerosis, obesity and type 2 diabetes. *Nat Rev Drug Discov* **8**:308–320.
57. Webb P. (2004) Selective activators of thyroid hormone receptors. *Expert Opin Investig Drugs* **13**:489–500.
58. Malm J, Färnegârdh M, Grover GJ, Ladenson PW. (2009) Thyroid hormone antagonists: potential medical applications and structure–activity relationships. *Curr Med Chem* **16**:3258–3266.
59. Schapira M, Raaka BM, Das S, Fan L, Totrov M, Zhou Z, Wilson SR, Abagyan R, Samuels HH. (2003) Discovery of diverse thyroid hormone receptor antagonists by high-throughput docking. *Proc Natl Acad Sci USA* **100**:7354–7359.

60. Kliewer SA, Forman BM, Blumberg B, Ong ES, Borgmeyer U, Mangelsdorf DJ, Umesono K, Evans RM. (1994) Differential expression and activation of a family of murine peroxisome proliferator-activated receptors. *Proc Natl Acad Sci USA* **91**:7355–7359.
61. Ricote M, Li AC, Willson TM, Kelly CJ, Glass CK. (1998) The peroxisome proliferator-activated receptor-γ is a negative regulator of macrophage activation. *Nature* **391**:79–82.
62. Evans RM, Barish GD, Wang Y-X. (2004) PPARs and the complex journey to obesity. *Nat Med* **10**:355–361.
63. Auboeuf D, Rieusset J, Fajas L, Vallier P, Frering V, Riou JP, Staels B, Auwerx J, Laville M, Vidal H. (1997) Tissue distribution and quantification of the expression of mRNAs of peroxisome proliferator-activated receptors and liver X receptor-alpha in humans: no alteration in adipose tissue of obese and NIDDM patients. *Diabetes* **46**:1319–1327.
64. Tontonoz P, Hu E, Spiegelman BM. (1994) Stimulation of adipogenesis in fibroblasts by PPAR gamma 2, a lipid-activated transcription factor. *Cell* **79**:1147–1156.
65. Kliewer SA, Sundseth SS, Jones SA, Brown PJ, Wisely GB, Koble CS, Devchand P, Wahli W, Willson TM, Lenhard JM, Lehmann JM. (1997) Fatty acids and eicosanoids regulate gene expression through direct interactions with peroxisome proliferator-activated receptors alpha and gamma. *Proc Natl Acad Sci USA* **94**:4318–4323.
66. Olefsky JM, Saltiel AR. (2000) PPAR gamma and the treatment of insulin resistance. *Trends Endocrinol Metab* **11**:362–368.
67. Nolan JJ, Ludvik B, Beerdsen P, Joyce M, Olefsky J. (1994) Improvement in glucose tolerance and insulin resistance in obese subjects treated with troglitazone. *N Engl J Med* **331**:1188–1193.
68. Yki-Jarvinen H. (2004) Thiazolidinediones. *N Engl J Med* **351**:1106–1118.
69. Berger JP, Petro AE, Macnaul KL, Kelly LJ, Zhang BB, Richards K, Elbrecht A, Johnson BA, Zhou G, Doebber TW, Biswas C, Parikh M, Sharma N, Tanen MR, Thompson GM, Ventre J, Adams AD, Mosley R, Surwit RS, Moller DE. (2003) Distinct properties and advantages of a novel peroxisome proliferator-activated receptor [gamma] selective modulator. *Mol Endocrino* **17**:662–676.
70. Berger JP, Akiyama TE, Meinke PT. (2005) PPARs: therapeutic targets for metabolic disease. *Trends Pharmacol Sci* **26**:244–251.
71. Minoura H, Takeshita S, Kimura C, Hirosumi J, Takakura S, Kawamura I, Seki J, Manda T, Mutoh S. (2007) Mechanism by which a novel non-thiazolidinedione peroxisome proliferator activated receptor gamma agonist, FK614, ameliorates insulin resistance in Zucker fatty rats. *Diabetes Obes Metab* **9**:369–378.
72. Allen T, Zhang F, Moodie SA, Clemens LE, Smith A, Gregoire F, Bell A, Muscat GE, Gustafson TA. (2006) Halofenate is a selective peroxisome proliferator-activated receptor γ modulator with antidiabetic activity. *Diabetes* **55**:2523–2533.

73. Thieme TM, Steri R, Proschak E, Paulke A, Schneider G, Schubert-Zsilavecz M. (2010) Rational design of a pirinixic acid derivative that acts as subtype-selective PPAR gamma modulator. *Bioorg Med Chem Lett* **20**(8):2469–2473.
74. McEwan IJ. (2009) Nuclear receptors: one big family. *Methods Mol Biol* **505**:3–18.
75. Chen T. (2009) Nuclear receptor drug discovery. *Curr Opin Chem Biol* **12**: 418–426.
76. Lonard DM, O'Malley BW. (2007) Nuclear receptor coregulators: judges, juries, and executioners of cellular regulation. *Mol Cell* **27**:691–700.

Appendices

"Should we ask anybody who is undertaking a major project in science, in the heat of the fight, what drives and pushes him so relentlessly, he will never think of an external goal; it is the passion of the hunter and soldier…the stimulus of the fight with its setbacks."

Friedrich Miescher
Discoverer of DNA
1844-1895

APPENDIX 1

SCIENTIFIC SURVIVAL SUGGESTIONS

V. Craig Jordan

This book represents an ongoing evolution of interconnected and complementary research areas that I have experienced over the past four decades, during my scientific training and as the leader of my "Tamoxifen Teams" throughout Europe (Leeds University, U.K., and Bern, Switzerland) and the United States (the Worcester Foundation for Experimental Biology, University of Wisconsin Madison; Northwestern University Chicago; the Fox Chase Cancer Center Philadelphia; and currently at the Lombardi Comprehensive Cancer Center Georgetown University, Washington, D.C.) In the 1970s, I used to say, "Have brain, will travel." (With apologies to Paladin and *Have Gun, Will Travel,* the United States television series in the late 1950s.)

This has been an exciting and sometimes challenging career with excellent staff and students, and the 40 years have provided me with a number of useful personal lessons that some have said deserve to be passed on to the next generation:

(1) *Carefully decide on your mentor or supervisor for your doctorate.* Choosing a Ph.D. supervisor is critical. This individual will be your mentor and guide maybe for life. Several qualifications for the mentor and supervisor are required for success and you should ask yourself important questions: Has this individual done this before or will their efforts with me end in failure? Has this mentor already earned tenure? If your mentor is cast out, then you are also gone, with nothing. If the faculty member has

not yet achieved tenure, ask yourself whether they have the adequate funds, experience and vision to run a laboratory for five years (or at least demonstrate they have that potential). Are there skilled individuals in the laboratory who can teach you new laboratory skills rapidly? Trial and error is a poor path to progress, especially today, with few resources and accelerated deadlines. Can you ever see your mentor? Once a month is fine but the phantom supervisor can turn into the phantom Ph.D! Be cautious about the faculty member who will tell you their plan for you to finish in a couple of years. No project is what it seems on paper. That is just talk. Experiments are the conversation with Nature. The answers are reality. Nature does not lie. The Ph.D path I chose in the Department of Pharmacology was to crystallize the ER. Well, that did not work—but it was great on paper (see "The Foundation of a Career," at the beginning of this book)! The good news was that my Ph.D. supervisor, Dr. Edward Clark, a senior tenured faculty member, was a wonderful teacher of pharmaceutical chemistry, and had encyclopedic knowledge and a long row of bound Ph.D. theses in his office. I had a scholarship from the Medical Research Council, so all was well (except no one cared initially what I had learned). For my own students — each is a special individual who excelled at what they chose to do. What we achieved together is next in Appendix 2 and as individuals in Appendix 3.

(2) *Train yourself to ascend in science.* Your personality, dedication, achievements and past experience will determine whether you have trained yourself to be ready and are talent-spotted. In my own case, career risks were essential for profiting from opportunities. This was, in fact, training to ascend. Change is good for scientific development, which in turn provides academic and personal security. New environments and challenges provide invaluable experience to apply in the next stage of your career development. As a young scientist, I wanted the University to provide help and security. It was obvious to me at least that I should be talent-spotted. My chairman replied, "You keep talking about the University — there is no such thing. It is a group of

men with an agenda and you are not in it." I have carried this philosophy with me ever since. At the Robert H. Lurie Comprehensive Cancer Center later in my career, my director, Steve Rosen, saw the opportunity for a world-class breast program. He had "an agenda and I was in it"! He was right. We had $50 million invested in our Breast Cancer Research and Treatment Program alone in a decade.

(3) *Get documentation that says what you can do.* At my grammar school in Cheshire, where I first learnt the harsh realities of competitive academic success or failure, a teacher once told our class, "Get as many pieces of paper as possible that say what you did, what you can do and how good you are." Invaluable advice to be talent-spotted, but it does not pay the bills. It tells the story of what you did and not what you will do in the future. It does, however, provide greater security and promise for further investment in your laboratory.

(4) *Experiments are a conversation with Nature.* You define a question and, based on the selection of an appropriate model to answer your question in humans, you isolate the parameter of interest as a single variable with appropriate controls. The controls are critical for proving you can do what others before you have done. If the answer to your question does not give you what you anticipate, then your premise may be wrong, but only if all the controls are perfect. Nature does not lie. Several of my students, over the years, presented me with data that did not comply with my "mental model" of what should happen. By way of example, Clive Dix (1976) first showed me the high-affinity binding of 4-hydroxytamoxifen to the ER. I told him to go away and do it again, as no antiestrogen has the same binding affinity to the ER as estradiol. He was right! It was a discovery. Doug Wolf (1991) first showed me that physiologic estradiol induces tumor regression in tamoxifen-resistant breast cancer. He tried everything to explain why he could not reproduce a previous student's (Marco Gottardis) work. The tumor had changed with years of retransplantation and it was a discovery. The new biology of estrogen-induced apoptosis was born in 1992 but, like a

child, took 10 years to mature. Change in science takes time and must be sustained with determination.

(5) *Follow your instincts about what you want to achieve: have a goal.* I have never followed the fashion in medical science. I only wanted to participate in developing medicines to help people. That is pharmacology. Pharmacology and chemistry were always my passion, and my goal was to advocate change in medical practice. Through molecules, models and mechanism, drugs were reinvented—tamoxifen a failed contraceptive, and raloxifene (formerly keoxifene), a failed breast cancer drug — and became pioneering medicines. These drugs, their success, and the recognition and discovery of SERMs are the central theme of this book. Eat, sleep and work at your passion. It is a privilege to participate, not a right.

(6) *Know an opportunity when you are presented with change.* Opportunities occur rarely, but when they occur, they must be seized with alacrity. It is the only way forward. The opportunities to learn at the Worcester Foundation in America (1972–74) and the Ludwig Institute for Cancer Research in Switzerland (1979–80) were both career-changing experiences second to none. But both were gambles and not easy decisions to embrace with a young family, overseas for a year or two. However, I did not truly understand the process of identifying opportunity until Wisconsin. It was decided that I would inherit the space, staff and directorship of the Steroid Receptor Laboratory of a departing senior faculty member. At the time, the prospect was overwhelming as my laboratory space would be half a floor of the six-floor Comprehensive Cancer Center. I would double my staff overnight and the Clinical Steroid Receptor Laboratory required absolute quality control and focus on tracking all clinical samples and reports. I phoned my mentor Bill McGuire in San Antonio to explain to him my circumstances and to say that I did not feel equal to the task. He told me, "You are looking at the task incorrectly — this is an opportunity…." Now I understood. Wisconsin was where we discovered SERMs and estrogen-induced apoptosis. I was "talent-spotted," and I advanced

from Assistant Professor through Associate Professor (selected as an H. I. Romnes Faculty Fellow) to become Professor of Human Oncology and Pharmacology in five years. I was 38 years old. I became Director of the Breast Cancer Research and Treatment Program for the WCCC. This was their flagship program in the Cancer Center. Very good decision to go to Wisconsin.

(7) *Give chances to allow your team members to excel.* I had the good fortune to have dedicated teachers whose job description they stated as "helping pupils (or later, at university, students), to attain their full potential" (www.175heroes.org.uk/ronnie_kaye). Theirs was a system of a "gut feeling" and giving people a chance to excel. Give opportunities without self-interest. This means you must ensure a team member leaves better than when they joined. This means not only your graduate students and postdoctoral fellows, but also your technical staff. Their inclusion in publications is essential for enhancing their career prospects. For them, remember lesson #3.

(8) *Understand it is your responsibility to obtain support for your science.* My desire as a teenager to develop medicines to treat cancer naturally drew me to the astonishing contributions of Professor Paul Ehrlich and his creation of synthetic chemical therapies (chemotherapy) to treat disease successfully. Ehrlich gave us one more lesson of science as true today as it was in 1890: the 4 Gs — *Glück*, *Geduld*, *Geschick* and *Geld*. Medical science requires luck (*Glück*), but patience (*Geduld*) is critical. Recognition of the discovery through luck must also be consolidated during the skilled (*Geschick*) conversation with Nature. But the whole of this process requires money (*Geld*). Ehrlich was perhaps the first to create a mosaic of funding sources from the government, private philanthropy (the Speyer Stiftung) and the pharmaceutical industry (Hoechst). In the same way, each young scientist must become expert at "selling" the project by writing grants or advocating opportunities to philanthropic organizations or the pharmaceutical industry.

(9) *Science can be distracting, but publish or perish.* Experiments are fun and can be exciting as your conversation with Nature develops. Unfortunately, some scientists only see experiments as the goal and their attention keeps wandering from experiment to experiment. They become distressed once they realize that they have not cured cancer and keep trying more experiments. However, the profession of science is about communication so that you can participate and contribute to the web of knowledge (see xxxii–xxxiii). Unless you have the discipline to write every day, create "a story so far" and publish your results in an appropriate, refereed journal, it is as if you never did your experiment. No one knows! Some young scientists delude themselves that abstracts at meetings are science. This is true only if a refereed paper is hot on the heels of the abstract after the meeting! Only refereed papers are relevant for jobs, promotions and grants.

(10) *When you publish, your colleagues and competitors will often ignore your landmark work!* Publications are the record of scientific ideas and progress. Nevertheless, you should realize that others need to create their careers and will not feel it necessary to refer to your laboratory research or ideas. It is a truism that you will experience: "Not practical; won't work," followed by "Even if it does work, not relevant," followed by "OK it works, but not important" and, finally, "Fantastic! You've cured cancer. I thought of it first!" If you choose to be a translational scientist, you should understand at the outset that the clinical publication is the "discovery" and will be heralded as a breakthrough, and your contribution may not be included as the justification for the transition from the laboratory to the clinical trial. However, your contribution, over time, with discipline and enthusiasm, will be documented in the literature and provide retrospective evidence to propel your career forward. This and your talent and flexibility to adapt to new environments and opportunities will ensure your recognition but only if you (or someone else ascribes it to you) have written it down.

APPENDIX 2

DECADES OF DISCOVERY: THE SERM STORY — THE ST. GALLEN PRIZE

V. Craig Jordan

"I have but one lamp by which my feet are guided, and that is the lamp of experience. I know no way of judging of the future but by the past."

— *Patrick Henry, the first elected governor of Virginia, 1775*

Selective estrogen receptor modulators (SERMs) are a well-established drug group in medicine. They are also unique, as their mechanisms depend on differentially switching on and switching off target sites around a woman's body — selectively[1]! Tamoxifen is the pioneering SERM[2] and the first medicine to be tested and approved for reducing the risk of any cancer.[3] The problem to be solved was how to identify and treat the appropriate high risk women to reduce or eliminate their risk of developing breast cancer. The population-based models[4] could focus down to a few thousand high risk women, but breast carcinogenesis would only be subverted in a few dozen. These lucky few did not know who they were, amongst the thousands who were treated and who would never get breast cancer. Unfortunately, the strategy to apply the pioneering SERM, tamoxifen, for population-based chemoprevention was flawed at the outset, as laboratory and clinical evidence predicted that there was an elevated risk of an increase in endometrial cancer for postmenopausal women.[5,6] This was a slight but significant risk. Women worried. A range of other side effects (e.g. blood clots, cataracts, menopausal symptoms) would also be experienced by the many to benefit the few. The situation changed dramatically with the discovery that the two "lead" SERMs, tamoxifen and raloxifene, maintained bone density in laboratory animals, but also prevented mammary carcinogenesis.[7,8] In addition, raloxifene was less uterotrophic than tamoxifen. Would there be no endometrial risk? With the recognition of SERMs in the 1980s, a unique public health strategy was possible.

The new strategy was stated (twice) in the literature, which provided a simple roadmap for the pharmaceutical industry to follow (eventually!). *Are we looking in the wrong place? The majority of breast cancers occur unexpectedly and from unknown origin. Great efforts are being focused upon the identification of a population of high risk women to test "chemopreventive" agents. But are resources being used less than optimally? The problem is much greater than the current horizon. Indeed, even if we had the best chemopreventive for a minority of selected women, the overall impact on the disease might be negligible. An alternative would be to seize upon the developing clues provided by an extensive clinical investigation of available antiestrogens. Could analogs be developed to treat osteoporosis or even retard the development of atherosclerosis? If this proved to be true, then a majority of women in general could be treated for these conditions as soon as menopause occurred. Should the agent also retain antibreast tumor actions, then it might be expected to act as a chemosuppressive on all developing breast cancers if these have an evolution from hormone-dependent to hormone-independent disease. A bold commitment to drug discovery and clinical pharmacology will potentially place us in a key position to prevent the development of breast cancer by the end of this century.*[9]

And subsequently: *Is this the end of the possible applications for antiestrogens? Certainly not! We have obtained valuable clinical information about this group of drugs that can be applied in other disease states. Research does not travel in straight lines and observations in one field of science often become major discoveries in another. Important clues have been garnered about the effects of tamoxifen on bone and lipids, so it is possible that derivatives could find targeted applications to retard osteoporosis or atherosclerosis. The ubiquitous application of novel compounds to prevent diseases associated with the progressive changes after menopause may, as a side effect, significantly retard the development of breast cancer. The target population would be postmenopausal women in general, thereby avoiding the requirement to select a high risk group to prevent breast cancer.* [10]

Today, new SERMs hold the promise of fulfilling the stated prediction from two decades ago. Lasofoxifene,[11] for example, is approved in the E.U. for the prevention of osteoporosis in osteopenic women, but at the same time, it reduces the incidence of breast cancer, coronary heart events, strokes and endometrial cancer. Raloxifene, the pioneering SERM to prevent both breast cancer and osteoporosis,[12,13] is not as robust in its SERM pharmacology (there is no effect on coronary events or strokes) as lasofoxifene. Remarkably, lasofoxifene is 100 times more potent as a SERM; raloxifene is recommended at 60 mg daily but lasofoxifene is effective at 0.5 mg daily!

With this background of the current success of SERMs, my goal is to guide the reader through an evolution of ideas. History is often written as the achievement of dynasties. But, as with dynasties, the dogma of the preceding dynasty in medicine must be overcome, not by sudden force, but by unrelenting pressure and the reason of evidence. Only tenacity can change medicine through ideas as the standard of care is maintained and jealously guarded by the dynasty.

My early catalytic role in the evolution of our story is well-documented in the refereed literature.[2,14] Suffice it to say that, as a pharmacologist, I had a passion to develop drugs to treat cancer. But where to start? By a series of accidents, I met the right people at the right time, but the career choice to study the pharmacology of non-steroidal antiestrogens for my Ph.D. was then seen as a dead end. They were failed contraceptives and of only academic interest. But this was the point of a Ph.D. in Britain — training in the research method with a medical research scholarship. Thus, we enter the first of our four decades.

THE 1970s: THE REINVENTION OF TAMOXIFEN AS THE GOLD STANDARD FOR THE TREATMENT AND PREVENTION OF BREAST CANCER

The dynasty to be defeated in the opening years of the 1970s was combination cytotoxic chemotherapy. Chemotherapy was king, fresh from the victory over childhood leukemia and poised to "MOPP" up Hodgkin's disease. It was reasoned by the dynasty: If only the right combination of agents could be found in the lexicon of options, a cure would be assured. No one was advocating antihormone (or, as it was termed, "hormone therapy"!) research and treatment.

I saw an opportunity to develop a failed contraceptive, ICI 46,474, further than was originally believed it could go. In 1972, ICI 46,474 was abandoned by the pharmaceutical industry for continuing clinical testing because there was no profit to be made. Nevertheless, the meeting between me and the Head of the Fertility Control Program, Arthur Walpole (or "Walop," as he was affectionately called), proved to be critical to our story. He examined my Ph.D. at Leeds University, but ensured I had the resources at the Worcester Foundation in Shrewsbury, MA, USA and Leeds University to create a clinical strategy for this orphan drug. He ensured it was put on the market, — now tell us how to use it! The strategy I conceived and implemented is shown in Fig. 1.

The strategy was based on three principles:

(1) Target the tumor ER with tamoxifen;
(2) Give long term adjuvant tamoxifen therapy;
(3) Plan for chemoprevention.

All these principles were unpopular at the outset, but persistence and hundreds of evidence-based lectures around the world to my clinical colleagues slowly defeated the dynasty of combination cytotoxic chemotherapy to cure breast cancer. "Antihormone therapy" became the treatment of choice, with long term adjuvant tamoxifen therapy targeted to the tumor ER (the first targeted therapy).

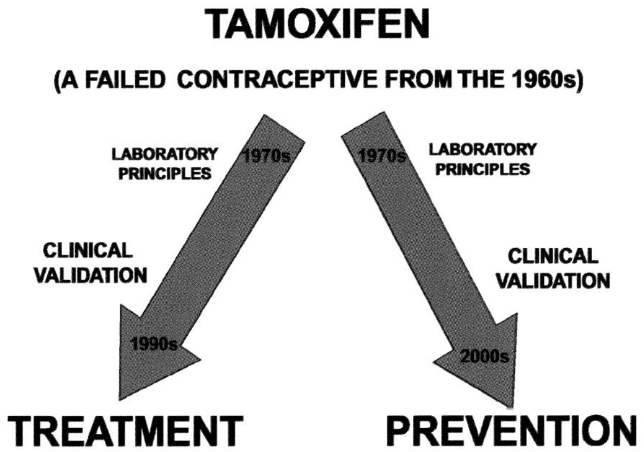

Figure 1. The translational development of tamoxifen for breast cancer. Tamoxifen was originally a failed oral contraceptive, ICI 46,474, that was abandoned by the pharmaceutical industry in 1972. The successful development of tamoxifen required three key components: (1) targeting the tumor ER with tamoxifen; (2) giving long term adjuvant tamoxifen therapy; (3) planning for chemoprevention. As a result of targeting the tumor ER with tamoxifen, antihormone therapy became the treatment of choice for long term adjuvant tamoxifen therapy. This has been successfully validated through clinical trials[15] and has saved the lives of millions of women around the world.

How bad was the first decade of discovery? If I may be so bold at this point to tell a story of my friend and colleague Steven E. Jones, M.D. When I started my international journey to advocate my principles for adjuvant antihormone therapy, Steven was in Arizona, co-director of the Adjuvant Therapy of Cancer Meeting in the 1970s. I was setting up the Ludwig Institute in Bern, Switzerland and was invited to present my new ideas about the use of tamoxifen at their 1979 meeting. There I was, sandwiched between the greats of cytotoxic chemotherapy, Vince DeVita and Bernie Fisher. I, in contrast, was advocating a stealth attack on breast cancer with tamoxifen that by comparison had no side effects. Little hope, one would think, but the plan succeeded. Two decades later, Steven Jones rose at a meeting in Washington and started his talk by declaring, "Craig Jordan was correct." Through the clinical trial mechanism, it has now been proven that long term (five years) adjuvant tamoxifen treatment targeted to the tumor ER has enhanced the survival of millions of women worldwide.[15] An orphan drug that is cheap and easy to administer has saved, and continues to save, hundreds of thousands of lives annually.

Now our story turns to the second decade with the "new" fashion in oncology — chemoprevention.

THE 1980s: SERMS SURFACE

The idea of preventing cancer became a popular fashion in the 1980s. This is a noble goal and one of the primary goals of cancer research, but the goal has proven hard to address. The idea as applied to breast cancer has its origin with the French scientist Professor Antoine Lacassagne, who stated, in his lecture at the Annual Meeting of the American Association for Cancer Research in Boston (1936), "If one accepts the consideration of adenocarcinoma of the breast as a consequence of a special hereditary sensibility to the proliferative action of oestrone, one is led to imagine a therapeutic preventive for subjects predisposed by their heredity to this cancer....".[16] However, Lacassagne's evidence was based on oophorectomy of mice from strains that develop a high incidence of mammary cancer and there were no mechanisms or compounds to advance and address the question. This would have to wait another quarter century, for the serendipitous discovery of the nonsteroidal antiestrogens.[10]

Tamoxifen was advanced for testing as a potential chemopreventive for breast cancer in the early 1980s based on three facts:

(1) There was laboratory evidence that tamoxifen would prevent rat mammary carcinogenesis.[17-19]
(2) Tamoxifen was becoming widely used in medicine to treat breast cancer, so it was argued that side effects were known and anticipated. This was not really true, as it took translational research[5] to draw the attention of the clinical community to the small risk of endometrial cancer.[6]
(3) Tamoxifen, when used as an adjuvant therapy, reduced the incidence of contralateral breast cancer.[20]

Nevertheless, there was a major toxicological (and ethical) issue over treating well women with a drug classified as a "nonsteroidal antiestrogen."[21] If, as was believed at the time, estrogen was good for building bone and reducing the risk of coronary heart disease, what would be the value of the chemoprevention strategy that prevents breast cancer but condemns women to an elevated risk of crushing osteoporosis or fatal coronary heart disease? To address the concern, laboratory studies were initiated to evaluate the pharmacology of tamoxifen on estrogen target tissues.

Studies on rats demonstrated that both tamoxifen and the failed and discontinued breast cancer drug, raloxifene (then known as keoxifene),[22] both maintained bone density in ovariectomized rats[8] and prevented rat mammary carcinogenesis.[7] However, raloxifene was not as effective as tamoxifen, probably because of poor pharmacodynamics, i.e. raloxifene does not accumulate, is rapidly excreted and there is only 2% bioavailability by the oral route of administration.[23] This pharmacological fact was to surface clinically following clinical trials 20 years later (see next section).

A pattern was emerging in the mid-1980s concerning the pharmacology of the nonsteroidal antiestrogens clomiphene, tamoxifen and raloxifene. The facts that led to the SERM concept being described in my laboratory can now be summarized:

(1) Clomiphene, a mixture of estrogenic *cis* and antiestrogenic *trans* geometric isomers, has bone-preserving properties in the ovariectomized rat.[24] Clomiphene had been tested as a breast cancer drug in patients,[25] but the manufacturer declined to advance development based on potential problems with cholesterol metabolism and a concern about cataracts. The drug remained the gold standard for the induction of ovulation where only five-day courses were given.[26]

(2) The fact that clomiphene was an impure mixture of estrogenic and antiestrogenic isomers made the bone-preserving effects uncertain. The estrogenic isomer might have been the favored pharmacologic agent at bone. In contrast, tamoxifen is the pure antiestrogenic trans isomer that preserves bone[8] and raloxifene

is a fixed ring structure that is exclusively antiestrogenic (very weakly estrogenic) in the uterus but estrogenic in bone.[8]

(3) Both tamoxifen and raloxifene are antitumor agents in rat mammary carcinogenesis.[7]

(4) Tamoxifen stimulates endometrial cancer growth (and mouse uterine growth) but blocks estradiol-stimulated growth of breast cancer transplanted in the same immune-deficient animal.[5,27] This experiment demonstrates target site specificity.

(5) Tamoxifen lowers circulating cholesterol in the rat[28] and this property was included in the initial patent application, which read, "*The alkene derivatives of the invention are useful for the modification of the endocrine status in man and animals and they may be useful for the control of hormone-dependent tumours or for the management of the sexual cycle and aberrations thereof. They also have useful hypocholesterolaemic activity.*"[2]

The claims as a breast cancer drug were denied and required to be omitted in the United States patent application until eventually a patent was awarded in 1985 by the Court of Appeals. In other words, tamoxifen was tested and marketed in America initially without patent protection for a dozen years. But nobody cared, as there was little possibility of success, either as a therapy or commercially (or so everybody thought!).

Thus, based on all these data, primarily from my laboratory, the SERM concept surfaced and the roadmap for clinical development started, as noted previously.[9,10] These data were the scientific basis of the Wisconsin Tamoxifen Study, initiated in the late 1980s to evaluate the pharmacology of tamoxifen on bone density and circulating cholesterol. It was the proven clinical translation of the tamoxifen (SERM) concept to preserve bone density[29] and lower circulating cholesterol[30,31] that awakened the sleeping pharmaceutical industry to develop raloxifene to prevent and treat osteoporosis in postmenopausal women. This started with the "magical" patenting in 1992 of raloxifene for this indication[21] and the publication of laboratory studies confirming my work on the SERM actions of raloxifene in rats.[32]

THE 1990s: RALOXIFENE'S PROMISE BECOMES A REALITY

During the 1990s, I transitioned from my focus on laboratory investigations with SERMs to a role of "scientific resource" for major clinical trials. I was invited by Eli Lilly to chair their Oncology Advisory Committee, which had responsibilities to adjudicate breast cancer detection in their initial osteoporosis trial, Multiple Outcomes with Raloxifene Evaluation (MORE). Subsequently, Dr. Norman Wolmark would invite me to be the scientific chair of the largest breast chemoprevention study — the Study of Tamoxifen and Raloxifene (STAR).

The MORE trial recruited 7705 postmenopausal women with osteoporosis to be randomized to placebo, 60 or 120 mg raloxifene daily. Raloxifene reduced fractures of the spine by 40% over the initial three-year evaluation period.[12] In our parallel evaluation of the incidence of breast cancer, there was a significant decrease in the incidence of ER-positive breast cancer by 70% with no increase in endometrial cancer.[12] The laboratory concept of SERMs[10] translated to the clinic. Women being treated for osteoporosis would develop less breast cancer if they took raloxifene. But here was an important pharmacological point — it was proven that they must keep taking raloxifene to obtain benefit. This laboratory principle noted with rapidly excreted SERMs in the 1980s[7,18] was to emerge as a clinical fact from the STAR trial after treatment stopped (see later).

What happened to tamoxifen in chemoprevention? Professor Trevor Powles initiated the first pilot toxicity study of tamoxifen in high risk women in the early 1980s,[33,] but it was Dr. Bernard Fisher who successfully conducted the first major randomized placebo-controlled clinical chemoprevention trial of tamoxifen in women at high risk for breast cancer. All preclinical predictors were confirmed — tamoxifen reduced the incidence of breast cancer, increased the incidence of endometrial cancer in postmenopausal women, and there was a decrease, though not significant, in the fracture rate.[3,34] Unanticipated information (though prior clinical studies suggested an effect) was an increase in operations for cataracts. The other fact consistent with the overview analysis of clinical trials for adjuvant therapeutic tamoxifen[15] was that tamoxifen alone produced a long term beneficial effect to suppress the development of breast cancer more than a decade after tamoxifen therapy stopped.[35] We will return to the science behind this observation later. As we noted, the fact that tamoxifen increased endometrial cancer in postmenopausal women now caused a turn to raloxifene, which had no increased endometrial cancer in the MORE trial.[12]

The STAR trial pitted tamoxifen 20 mg daily against raloxifene 60 mg daily for five years to compare and contrast efficacy and side effects for the reduction of breast cancer incidence in high risk postmenopausal women. As an aside, I was often asked how I would feel if raloxifene was found to be superior to tamoxifen. Happy — as the science of both drugs came from my laboratory and both drugs had to be reinvented as useful medicine after being essentially discarded by industry: tamoxifen, a failed contraceptive, and raloxifene, a failed breast cancer drug. The first analysis of the STAR trial showed equivalent efficacy to reduce the incidence of breast cancer by 50%.[13] However, side effects were decreased with raloxifene. In particular, there was less endometrial hyperplasia with raloxifene and fewer hysterectomies. Operations for cataracts were fewer for raloxifene. This analysis was conducted during raloxifene therapy,[13] but the subsequent analysis conducted after therapy had stopped[36] demonstrated tamoxifen had a sustained antitumor action whereas there was a reduced efficacy (75%) for raloxifene. The drugs were different with their pharmacology and raloxifene must be given indefinitely.

We conclude that the fact that raloxifene is a drug with low bioavailability and therefore the pharmacodynamics to concentrate at the target site — the effect on the breast tissue is reduced. If sustained, local concentrations of tamoxifen and raloxifene are different and the elevated concentrations of tamoxifen drive the cell population to evolve differently than those exposed to low levels of raloxifene; thus, there will be consequences for tumorigenesis and the evolution of drug resistance. We hypothesize that the low levels of raloxifene remain therapeutically "antiestrogenic" for the duration of therapy but the endogenous estrogen from the woman's own body causes nascent tumor regrowth.

In contrast, the sustained high concentrations of tamoxifen locally in the breast cause a change in the evolution of the breast cancer cell population that in some way leaves an "antitumor memory" for years after therapy stops — but how? This leads us to the final decade of discovery: estrogen-induced apoptosis.

THE 2000s: ESTROGEN-INDUCED APOPTOSIS?

The first chemical therapy to treat any cancer successfully was the use of high dose estrogen therapy to treat metastatic breast cancer in postmenopausal patients.[37] High dose estrogen therapy became the standard of care until the introduction of tamoxifen in the 1970s.[10,38] At the end of his career, Sir Alexander Haddow, FRS, presented the inaugural Karnofsky Lecture, where he expressed his disappointment about the lack of progress in understanding mechanisms: "...the extraordinary extent of tumour regression observed in perhaps 1% of postmenopausal cases (with estrogen) has always been regarded as of major theoretical importance, and it is a matter for some disappointment that so much of the underlying mechanisms continues to elude us...."[39] What was known was that the high dose estrogen therapy was the more effective as a breast cancer treatment the further away the patient was from menopause—but why?

The advance in our understanding was to await an examination of model systems in the laboratory to decipher the mechanisms of antihormone drug resistance (Fig. 2). The whole topic has recently been summarized,[40] but the facts must be stated to illustrate how transparency in Nature can occur through unanticipated results in another area of research.

The first transplantable model of resistance to long term tamoxifen therapy demonstrated unique qualities. Acquired resistance is evidenced by tamoxifen-stimulated (actually SERM-stimulated, as it turns out) growth. Tumors grow because of tamoxifen, not in spite of tamoxifen, as occurs with all other anticancer agents. What was even more surprising was the fact that when tamoxifen treatment is stopped, estrogen can again stimulate growth. This model replicates tamoxifen resistance during the treatment of ER-positive metastatic

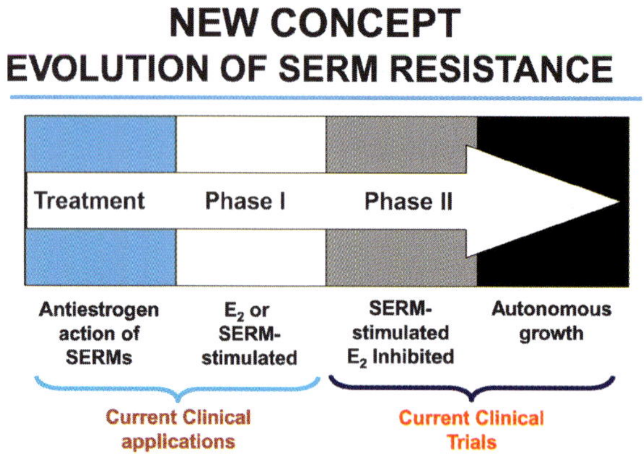

Figure 2. The evolution of drug resistance to SERMs. Acquired resistance occurs during long-term treatment with a SERM and is evidenced by SERM-stimulated breast tumor growth. Tumors also continue to exploit estrogen for growth when the SERM is stopped, so a dual signal transduction process develops. The pure antiestrogen, fulvestrant, destroys the ER and prevents tumor growth in SERM-resistant disease. This phase of drug resistance is referred to as phase I resistance. Continued exposure to a SERM results in continued SERM-stimulated growth (phase II), but eventually autonomous growth occurs that is unresponsive to fulvestrant or aromatase inhibitors. The event that distinguishes phase I from phase II acquired resistance is a remarkable switching mechanism that now causes apoptosis, rather than growth, with physiologic levels of estrogen. These distinct phases of laboratory drug resistance have their clinical parallels and this new knowledge is being integrated into the treatment plan.

breast cancer: resistance occurs within a couple of years, estrogen or tamoxifen is required for continued growth and estrogen withdrawal or fulvestrant (the pure antiestrogen that causes destruction of the ER) is an appropriate second line therapy. What the model of acquired resistance did not do was explain how it was possible to use five years of adjuvant tamoxifen therapy to treat patients selectively. If the laboratory model was correct, and had been available at the time long term adjuvant therapy was planned as a treatment strategy, then no one would consider treatment longer than a year for adjuvant antihormone therapy. It would obviously be dangerous for patients. The same argument was used in the 1970s by the clinical community. Tamoxifen cannot control metastatic breast cancer for, on average, more than two years, so one cannot give long term (greater than five years) adjuvant tamoxifen. We were missing something fundamental about the biology of micrometastatic breast cancer exposed to long term tamoxifen therapy.

The breakthrough in understanding came through serendipity and, as always, with outstanding graduate students possessing exceptional laboratory skills. The model of acquired resistance to tamoxifen could only be maintained by serial transplantation in successive generations of tamoxifen-treated athymic mice. We were unable to transfer the tumors to cell culture for study, so we had to pay the expense of preserving the only naturally developed model of resistance to tamoxifen had to be born. That, as it turned out, was the good, new and an opportunity for future discovery.

The acquired drug resistance to tamoxifen evolves in an environment of tamoxifen (Fig. 2). Retransplantation of tumors into further tamoxifen-treated mice caused adapted cell populations that depended on tamoxifen for growth (phase I), but as the survival networks became reconfigured, a vulnerability emerged (phase II). After five years of exposure to tamoxifen, the resulting tumor no longer saw estrogen as a survival signal, but as an apoptotic trigger (phase II).

In 1992, these data were presented for the first time at the St. Gallen Breast Cancer Conference.[41] The hypothesis advanced was that the termination of tamoxifen, at the correct time, was important for the woman's own estrogen to destroy the microfoci of appropriately prepared target cells. This new biology of physiologic estrogen causing apoptosis was the reason for the enhanced survivorship of patients treated with a full five years of tamoxifen. As a result, a shift in thinking occurred and the clinical trial community subsequently exploited the concept, now published in the refereed literature,[42] that therapeutic estrogen or indeed "physiologic estrogen" in the form of low dose estrogen replacement therapy (ERT) could cause the correctly configured tumors with acquired antihormone resistance to regress.[43,44] The estrogen dynasty originally deposed, struck back. Nature answered as well. Today, there is much interest in the paradoxical actions of physiologic estrogen in breast cancer.[45,46] Recent results from the Women's Health Initiative demonstrate a reduction in the

Table 1. Decades of discovery.

- The 1970's:The re-invention of tamoxifen as the "gold standard" for the treatment and prevention of breast cancer
- The 1980's: SERMs surface
- The 1990's: Raloxifene's Promise is a Reality
- The 2000's: Estrogen-induced apoptosis?

incidence of breast cancer for hysterectomized, postmenopausal women who take long term estrogen replacement therapy.[47] Like our tamoxifen story, the effects persist for years following the cessation of ERT.[48] Practical advances, not only in the therapy of cancer but pre-emptively with "natural" chemoprevention, may result from these findings in the future.

In summary, I have mapped out the decades of discovery that emerged from a single quest some 40 years ago — to develop a drug useful for the treatment and prevention of breast cancer. At the time, I could count on the fingers of one hand the people who were interested in the quest. Virtually no one cared, and it was not going to happen. But science is not like that; as in politics, ideas have their time but it is really about people and a passion to keep the flame of truth alive. I am immensely grateful to Professor Hans-Jöerg Senn and his committee for selecting me to receive the St. Gallen Prize for Advances in Breast Cancer Research (Fig. 3). Thanks also go to my friends and colleagues Aron Goldhirsch and Richard Gelber. We all started our personal journeys together in Bern, Switzerland in the late 1970s and we have remained friends and colleagues to this day. Most importantly, I am grateful for the 40 years of the "Tamoxifen Teams" who worked and trained with me in my laboratories at Leeds University (U.K.), WFEB (U.S.), the Ludwig Institute for Cancer Research (Bern; Switzerland), the University of Wisconsin (Madison), Northwestern University (Chicago), the Fox Chase Cancer Center (Philadelphia) and the Lombardi Comprehensive Cancer Center (Washington, D.C.). I had the privilege to guide their lives and they turned the ideas we conceived into lives saved around the world.

The quotation at the top this essay was the one I used to open my prize lecture in St. Gallen, on 16 March 2011: "I have but one lamp by

Figure 3. The St. Gallen Prize for Clinical Breast Cancer Research Lecture at the 12th St. Gallen International Breast Cancer Conference, St. Gallen, Switzerland, March 16, 2011. The St. Gallen Breast Cancer Award is given every two years to a scientist who has made exceptional contributions to breast cancer research, and is considered to be the most prestigious award in breast cancer. Professor Hans-Joerg Senn, the founder of the St. Gallen Oncology Conferences, presented Dr. Jordan with this award, and then the latter delivered his prize lecture: "Evolution of Long-Term Adjuvant Anti-hormone Therapy: Consequences and Opportunities."

which my feet are guided, and that is the lamp of experience. I know no way of judging of the future but by the past." My lamp was tamoxifen. However, this journey, as I hope I have illustrated, is so much more than the successful development of tamoxifen for the adjuvant treatment of breast cancer. It is about a way of constructing a conversation with Nature, with the goal of defeating a powerful enemy within us — cancer.

ACKNOWLEDGEMENTS

This work (V.C.J.) was supported by the Department of Defense Breast Program under award No. W81XWH-06-1-0590 Center of Excellence (this interdisciplinary research grant supports research into estrogen-induced apoptosis in breast cancer); subcontract under SU2C (AACR) grant No.

SU2C-AACR-DT0409; the Susan G. Komen for the Cure Foundation under award No. SAC100009, GHUCCTS CTSA (Grant #UL1RR031975) and the Lombardi Comprehensive Cancer Center Support Grant (CCSG) Core Grant NIH P30 CA051008 from the National Cancer Institute. The content is solely the responsibility of the author and does not necessarily represent the official views of the National Cancer Institute or the National Institutes of Health. The views and opinions of the author do not reflect those of the U.S. Army or the Department of Defense.

REFERENCES

1. Jordan VC. (2001) Selective estrogen receptor modulation: a personal perspective. *Cancer Res* **61**:5683–5687.
2. Jordan VC. (2003) Tamoxifen: a most unlikely pioneering medicine. *Nat Rev Drug Discov* **2**:205–213.
3. Fisher B, Costantino JP, Wickerham DL, Redmond CK, *et al.* (1998) Tamoxifen for prevention of breast cancer: report of the National Surgical Adjuvant Breast and Bowel Project P-1 Study. *J Natl Cancer Inst* **90**:1371–1388.
4. Gail MH, Brinton LA, Byar DP, Corle DK, *et al.* (1989) Projecting individualized probabilities of developing breast cancer for white females who are being examined annually. *J Natl Cancer Inst* **81**:1879–1886.
5. Gottardis MM, Robinson SP, Satyaswaroop PG, Jordan VC. (1988) Contrasting actions of tamoxifen on endometrial and breast tumor growth in the athymic mouse. *Cancer Res* **48**:812–815.
6. Fornander T, Rutqvist LE, Cedermark B, Glas U, *et al.* (1989) Adjuvant tamoxifen in early breast cancer: occurrence of new primary cancers. *Lancet* **1**:117–120.
7. Gottardis MM, Jordan VC. (1987) Antitumor actions of keoxifene and tamoxifen in the N-nitrosomethylurea-induced rat mammary carcinoma model. *Cancer Res* **47**:4020–4024.
8. Jordan VC, Phelps E, Lindgren JU. (1987) Effects of anti-estrogens on bone in castrated and intact female rats. *Breast Cancer Res Treat* **10**:31–35.
9. Jordan VC. (1988) Chemosuppression of breast cancer with tamoxifen: laboratory evidence and future clinical investigations. *Cancer Invest* **6**:589–595.
10. Lerner LJ, Jordan VC. (1990) Development of antiestrogens and their use in breast cancer: Eighth Cain Memorial Award Lecture. *Cancer Res* **50**:4177–4189.
11. Cummings SR, Ensrud K, Delmas PD, LaCroix AZ, *et al.* (2010) Lasofoxifene in postmenopausal women with osteoporosis. *N Engl J Med* **362**:686–696.
12. Cummings SR, Eckert S, Krueger KA, Grady D, *et al.* (1999) The effect of raloxifene on risk of breast cancer in postmenopausal women: results from the MORE randomized trial. Multiple Outcomes of Raloxifene Evaluation. *JAMA* **281**: 2189–2197.

13. Vogel VG, Costantino JP, Wickerham DL, Cronin WM, *et al.* (2006) Effects of tamoxifen vs. raloxifene on the risk of developing invasive breast cancer and other disease outcomes: the NSABP Study of Tamoxifen and Raloxifene (STAR) P-2 trial. *JAMA* **295**:2727–2741.
14. Jordan VC. (2006) Tamoxifen (ICI 46,474) as a targeted therapy to treat and prevent breast cancer. *Br J Pharmacol* **147**:S269–S276.
15. EBCTCG. (2005). Effects of chemotherapy and hormonal therapy for early breast cancer on recurrence and 15-year survival: an overview of the randomised trials. *Lancet* **365**:1687–1717.
16. Lacassagne A. (1936) Hormonal pathogenesis of adenocarcinoma of the breast. *Am J Cancer* **27**:217–225.
17. Jordan VC. (1976) Effect of tamoxifen (ICI 46,474) on initiation and growth of DMBA-induced rat mammary carcinomata. *Eur J Cancer* **12**:419–424.
18. Jordan VC, Allen KE. (1980). Evaluation of the antitumour activity of the non-steroidal antioestrogen monohydroxytamoxifen in the DMBA-induced rat mammary carcinoma model. *Eur J Cancer* **16**:239–251.
19. Jordan VC, Allen KE, Dix CJ. (1980) Pharmacology of tamoxifen in laboratory animals. *Cancer Treat Rep* **64**:745–759.
20. Cuzick J, Baum M. (1985) Tamoxifen and contralateral breast cancer. *Lancet* **2**:282.
21. Jordan VC. (1984) Biochemical pharmacology of antiestrogen action. *Pharmacol Rev* **36**:245–276.
22. Lewis JS, Jordan VC. (2006) Case histories: raloxifene. *Comprehensive Medicinal Chemistry II* **8**:103–121.
23. Snyder KR, Sparano N, Malinowski JM. (2000) Raloxifene hydrochloride. *Am J Health Syst Pharm* **57**:1669–1675; quiz 1676–1668.
24. Beall PT, Misra LK, Young RL, Spjut HJ, *et al.* (1984) Clomiphene protects against osteoporosis in the mature ovariectomized rat. *Calcif Tissue Int* **36**:123–125.
25. Herbst AL, Griffiths CT, Kistner RW. (1964) Clomiphene citrate (Nsc-35770) in disseminated mammary carcinoma. *Cancer Chemother Rep* **43**:39–41.
26. ASRM PC. (2006) Use of clomiphene citrate in women. *Fertil Steril* **86**:S187–S193.
27. Jordan VC, Robinson SP. (1987) Species-specific pharmacology of antiestrogens: role of metabolism. *Fed Proc* **46**:1870–1874.
28. Harper MJ, Walpole AL. (1967) A new derivative of triphenylethylene: effect on implantation and mode of action in rats. *J Reprod Fertil* **13**:101–119.
29. Love RR, Mazess RB, Barden HS, Epstein S. (1992) Effects of tamoxifen on bone mineral density in postmenopausal women with breast cancer. *N Engl J Med* **326**:852–856.
30. Love RR, Newcomb PA, Wiebe DA, Surawicz TS, *et al.* (1990) Effects of tamoxifen therapy on lipid and lipoprotein levels in postmenopausal patients with node-negative breast cancer. *J Natl Cancer Inst* **82**:1327–1332.
31. Love RR, Wiebe DA, Newcomb PA, Cameron L, *et al.* (1991) Effects of tamoxifen on cardiovascular risk factors in postmenopausal women. *Ann Intern Med* **115**:860–864.

32. Black LJ, M Sato, *et al.* (1994). Raloxifene (LY139481 HCI) prevents bone loss and reduces serum cholesterol without causing uterine hypertrophy in ovariectomized rats. *J Clin Invest* **93**(1):63–69.
33. Powles TJ, Hardy JR, Ashley SE, Farrington GM, *et al.* (1989) A pilot trial to evaluate the acute toxicity and feasibility of tamoxifen for prevention of breast cancer. *Br J Cancer* **60**:126–131.
34. Fisher B, Costantino JP, Wickerham DL, Cecchini RS, *et al.* (2005) Tamoxifen for the prevention of breast cancer: current status of the National Surgical Adjuvant Breast and Bowel Project P-1 study. *J Natl Cancer Inst* **97**:1652–1662.
35. Powles TJ, Ashley S, Tidy A, Smith IE, *et al.* (2007) Twenty-year follow-up of the Royal Marsden randomized, double-blinded tamoxifen breast cancer prevention trial. *J Natl Cancer Inst* **99**:283–290.
36. Vogel VG, Costantino JP, Wickerham DL, Cronin WM, *et al.* (2010) Update of the National Surgical Adjuvant Breast and Bowel Project Study of Tamoxifen and Raloxifene (STAR) P-2 Trial: Preventing Breast Cancer. *Cancer Prev Res (Phila)* **3**:696–706.
37. Haddow A, Watkinson JM, Paterson E, Koller PC, *et al.* (1944) Influence of synthetic oestrogens on advanced malignant disease. *Br Med J* **2**:393–398.
38. Jordan VC. (2009) A century of deciphering the control mechanisms of sex steroid action in breast and prostate cancer: the origins of targeted therapy and chemoprevention. *Cancer Res* **69**:1243–1254.
39. Haddow A. (1970) Thoughts on chemical therapy. *Cancer* **26**:737–754.
40. Jordan VC, Ford LG. (2011) Paradoxical clinical effect of estrogen on breast cancer risk: a "new" biology of estrogen-induced apoptosis. *Cancer Prev Res (Phila)* **4**:633–637.
41. Wolf DM, Jordan VC. (1993) A laboratory model to explain the survival advantage observed in patients taking adjuvant tamoxifen therapy. *Recent Results Cancer Res* **127**:23–33.
42. Yao K, Lee ES, Bentrem DJ, England G, *et al.* (2000) Antitumor action of physiological estradiol on tamoxifen-stimulated breast tumors grown in athymic mice. *Clin Cancer Res* **6**:2028–2036.
43. Lonning PE, Taylor PD, Anker G, Iddon J, *et al.* (2001) High-dose estrogen treatment in postmenopausal breast cancer patients heavily exposed to endocrine therapy. *Breast Cancer Res Treat* **67**:111–116.
44. Ellis MJ, Gao F, Dehdashti F, Jeffe DB, *et al.* (2009) Lower-dose vs. high-dose oral estradiol therapy of hormone receptor–positive, aromatase inhibitor–resistant advanced breast cancer: a phase 2 randomized study. *JAMA* **302**:774–780.
45. Lewis-Wambi JS, Jordan VC. (2009) Estrogen regulation of apoptosis: how can one hormone stimulate and inhibit? *Breast Cancer Res* **11**:206.
46. Maximov PY, Lewis-Wambi JS, Jordan VC. (2009) The paradox of oestradiol-induced breast cancer cell growth and apoptosis. *Curr Signal Transduct Ther* **4**:88–102.

47. Stefanick ML, Anderson GL, Margolis KL, Hendrix SL, *et al.* (2006) Effects of conjugated equine estrogens on breast cancer and mammography screening in postmenopausal women with hysterectomy. *JAMA* **295**:1647–1657.
48. LaCroix AZ, Chlebowski RT, Manson JE, Aragaki AK, *et al.* (2011) Health outcomes after stopping conjugated equine estrogens among postmenopausal women with prior hysterectomy: a randomized controlled trial. *JAMA* **305**:1305–1314.

POST SCRIPT

On May 11, 2011, Ambassador Sager and his wife, Christine, hosted a celebration at the Swiss Ambassador's residence in Washington, DC, to recognize the internationally significant achievements of Dr. V. Craig Jordan, Research Director of the Lombardi Comprehensive Cancer Center, Washington, DC, and Professor Hans-Joerg Senn, the Director of the St. Gallen Conferences for the treatment and prevention of breast cancer in Switzerland. Both honorees received awards and citations from Ambassador Sager during the reception and dinner.

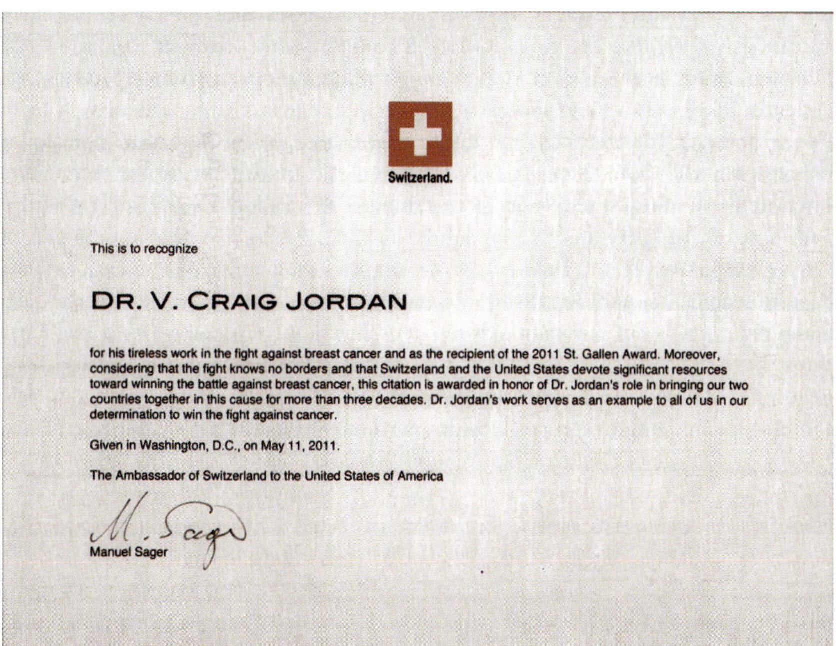
Switzerland.

This is to recognize

DR. V. CRAIG JORDAN

for his tireless work in the fight against breast cancer and as the recipient of the 2011 St. Gallen Award. Moreover, considering that the fight knows no borders and that Switzerland and the United States devote significant resources toward winning the battle against breast cancer, this citation is awarded in honor of Dr. Jordan's role in bringing our two countries together in this cause for more than three decades. Dr. Jordan's work serves as an example to all of us in our determination to win the fight against cancer.

Given in Washington, D.C., on May 11, 2011.

The Ambassador of Switzerland to the United States of America

Manuel Sager

The celebration at Ambassador Sager's Residence on the evening of May 11, 2011. The Event drew Members of the Diplomatic Community, Members of Congress, Senior Representatives and Presidents from the American Association for Cancer Research (Dr. Judy Garber), the American Society for Clinical Oncology (Dr. Sandra Swain), Representatives from Susan G. Komen For The Cure, American Airlines, Novartis, the Director (Dr. Louis Weiner) and Faculty from the Lombardi Comprehensive Cancer Center and Lois Trench-Hines (and her husband, George), the first drug monitor for tamoxifen in the United States, who worked with Dr. Jordan prior to the registration of tamoxifen in the United States in December 1977. Pictured above are Dr. Jordan and Professor Senn **(center)** and members of the Lombardi Comprehensive Cancer Center "Tamoxifen Team": **Front Row**: Dr. Ifeyinwa Obiorah (PhD Student), Dr. Ping Fan (Research Assistant Professor), Helen Kim (Laboratory Manager), Dr. Surojeet Sengupta (Research Assistant Professor), Elizabeth Sweeney (PhD Student). **Back row**: Julia Tijerina (Executive Assistant), Dr. Philipp Maximov (Postdoctoral Fellow), Russell McDaniel (Research Assistant II), Dr. SeungSang Ko (Visiting Associate Professor from the Cheil General Hospital & Women's Health Care Center, KwanDong University College of Medicine, Seoul, South Korea).

APPENDIX 3

AN ACCOUNT OF STUDENTS OBTAINING A PH.D. DEGREE (OR AN M.D. FOR PHYSICIANS IN THE BRITISH SYSTEM) WHILE IN THE TAMOXIFEN TEAM OVER THE LAST 30 YEARS

V. Craig Jordan

Each student came to read for a Ph.D. with only a general outline of the area of research to be addressed. I have never subscribed to the predictable, logical, methodical, stepwise approach to science demanded by granting agencies. "Year three we will -----!" I subscribe to the words of Miessner at the front of the Appendices. I believe the process is a conversation with Nature and a "battle" to be won, with all its setbacks. Battles can never be predicted after the first clash; it is the response you make to the replies of Nature that predicts success or failure in science. The Duke of Wellington was asked by his second-in-command, the Earl of Uxbridge, before the Battle of Waterloo: What was the plan to be continued should he be killed? "To defeat the French...." was the simple reply from the Duke.

As a mentor and supervisor, you have essential responsibilities to prepare the prospective scientist for the conversation (battle) ahead with Nature. Here is my approach. Each student must learn the practical side of science from the experts in the laboratory. These are usually skilled technicians who perform experiments in cell culture or maintain the cell culture facility all day, every day. Senior postdoctoral fellows or junior faculty in the team must teach the more complex techniques.

There is no conversation in science if the student does not know the language, so it is critical at the outset that they complete a published review or two (no more) in either a book or a journal. This is the best way to learn about the scope of the literature and its translation to practical benefit for society. Cathy Murphy[1] and Jennifer MacGregor-Schafer[2] each completed "heroic" reviews in the refereed literature. However, their outstanding efforts have been cited 379 and 454 times, respectively. I usually set two or three practical tasks for each Ph.D. project, with the anticipation that at least one will be fruitful. Remarkably, almost always the student produces data that are worthy of publication from each task. The student develops unique findings on new mechanisms or provides valuable new tools for future generations of students to exploit.

To illustrate these points, I have chosen to describe the accomplishments of my students over the decades with what was found, what was the publication, what were their total citation numbers (as of the date of this book) and how many publications the individual produced overall.The integrated data are summarized in a table at the end of their individual entries (Table 1).

CLIVE J. DIX, DEPARTMENT OF PHARMACOLOGY, UNIVERSITY OF LEEDS, 1976–1979; ICI PHARMACEUTICALS DIVISION RESEARCH SCHOLAR

- Jordan VC, Dix CJ, Allen KE. (1979) The effectiveness of long term tamoxifen treatment in a laboratory model for adjuvant hormone therapy of breast cancer. eds. Salmon SE, Jones SE, In: *Adjuvant Therapy of Cancer II*, Philadelphia: Grune & Stratton, pp. 19-26.
- Dix CJ, Jordan VC. (1980) Subcellular effects of mono-hydroxytamoxifen in the rat uterus: steroid receptors and mitosis. *J Endocrinol* **85**:393-404.
- Dix CJ, Jordan VC. (1980) Modulation of rat uterine steroid hormone receptors by estrogen and antiestrogen. *Endocrinol* **107**:2011-2020.

Table 1. Graduate students in the Tamoxifen Team.

Name	No. of publications	No. of citations	Most prominent job
Clive J. Dix	17	635	Co-Founder and Chief Executive Officer, PowderMed, Ltd. (U.K.)
Anna T. Riegel	12	156	Cecilia Fisher Rudman-endowed Professor of Oncology, Associate Director for Cancer Research Education, Georgetown University (Washington, D.C.)
Stewart D. Lyman	7	113	Director, Extramural Research, Immunex Corporation (Seattle, WA)
Ethel M. Cormier	7	104	Associate Director, Beauty Technology Division, Proctor & Gamble (Cincinnati, OH)
Marco M. Gottardis	24	1485	Executive Director, Oncology Discovery Disease Sciences and Biologics, Bristol Myers Squibb (Princeton, NJ)
Catherine S. Murphy	12	762	Technical Director, Gene Therapy Clinical Trial, Wallace-Kettering Neurological Institute (Kettering, OH)
Meei-Huey Jeng	10	547	Professor of Medicine, Indiana University, Purdue University Indianapolis (Indianapolis, IN)
Shun-Yuan Jiang	11	691	Professor, Department of Research, Buddhist Tzu Chi General Hospital Taipei Branch (New Taipei City, Taiwan) Professor, Cancer Research Laboratory, Tri-Service General Hospital (Taipei, Taiwan)
Doug M. Wolf	18	515	Senior Director/Team Leader, Oncology Medical Affairs, Pfizer (New York, NY)

(continued)

Table 1. *(continued)*

Name	No. of publications	No. of citations	Most prominent job
John J. Pink	9	403	Assistant Professor, General Medical Sciences (Oncology), Director of the Translational Research Core Facility, Case Comprehensive Cancer Center, Case Western Reserve University (Cleveland, OH) Assistant Professor, Department of Radiation Oncology, Case Western Reserve University (Cleveland, OH)
William H. Catherino	14	346	Associate Professor, Department of Obstetrics and Gynecology Uniformed Services of the Health Sciences (Bethesda, MD) Associate Program Director, Division of Reproductive Endocrinology and Infertility, Program in Adult and Reproductive Endocrinology, National Institute for Child Health and Human Development, National Institutes of Health (Bethesda, MD)
Jennifer I. MacGregor-Schafer	30	935	Managing Director, Cancer InSight, LLC
Ruth M. O'Regan	25	435	Professor and Vice-Chair for Educational Affairs, Department of Hematology and Medical Oncology, Emory University (Atlanta, GA)
Rita C. Dardes	17	212	Associate Professor, Department of Gynecology, Federal University of São Paolo (Brazil)
Philipp Y. Maximov*	3	4	Susan G. Komen for the Cure International Breast Cancer Postdoctoral Fellow, Georgetown University (Washington, D.C.)

*Special circumstances.

Defined the pharmacology of the potent tamoxifen metabolite 4-hydroxytamoxifen, which became the "standard" antiestrogen and laboratory reagent for the next 30 years and was used by others to crystallize the antiestrogenic complex with the ligand-binding domain of the ER.[3] Defined the benefits of long term tamoxifen therapy as a potential adjuvant therapy to treat breast cancer. A very important paper at a meeting that defined the future direction of adjuvant tamoxifen therapy as a long term treatment.

ANNA T. RIEGEL (*NÉE* TATE), McARDLE LABORATORY FOR CANCER RESEARCH, UNIVERSITY OF WISCONSIN, 1980–1983; FULBRIGHT HAYS SCHOLAR

- Tate AC, Greene GL, DeSombre ER, Jensen EV, Jordan VC. (1984) Differences between estrogen- and antiestrogen-estrogen receptor complexes identified with an antibody raised against the estrogen receptor. *Cancer Res* **44**:1012-1018.
- Tate AC, Jordan VC. (1984) Nuclear [3H]4-hydroxytamoxifen (4-OHTAM) and [3H]estradiol (E2)-estrogen receptor complexes in the MCF-7 breast cancer and GH3 pituitary tumor cell lines. *Mol Cell Endocrinol* **36**:211-219.
- Jordan VC, Tate AC, Lyman SD, Gosden B, Wolf M, Bain RR, Welshons WV. (1985) Rat uterine growth and induction of progesterone receptor without estrogen receptor translocation. *Endocrinol* **116**:1845-1857.

Used both polyclonal and monoclonal antibodies generated by Geoffrey Greene at the Ben May Laboratory, University of Chicago in a collaborative study with Dr. Elwood V. Jensen to compare and contrast the properties of the estradiol and antiestrogen ER complex. These data supported our proposed model for estrogen and antiestrogen action using ER structure-function relationships at a target gene.[4] The model seemed to be essentially correct (see Chapter 6). The pharmacology of tamoxifen analogs (with very low affinity for ER) that could not be metabolically activated was used to investigate the growth and progesterone receptor synthesis in the immature rat uterus *in vivo*. The standard cell disruption techniques showed that the ER was not present in the nuclear compartment; only an unoccupied ER was present in the cytosol. This was evidence that the technology to demonstrate ER localization was always faulty or suspect. Work by others (e.g. Greene) using the first monoclonal antibodies established that the ER was a nuclear protein that "fell out of the nucleus" once cells were disrupted. The tamoxifen analogs that weakly bind to the nuclear ER would turn on growth and progesterone receptor synthesis, but when the tissue was disrupted the ligand fell off the ER in the nucleus and the unoccupied receptor fell out of the nucleus into the cytosol. This paper is a wonderful example of the Heisenberg 'uncertainty' principle, which declares that the technique used to determine location in and of itself alters the structure and location of the target being observed in Nature.

STEWART D. LYMAN, McARDLE LABORATORY, UNIVERSITY OF WISCONSIN, 1982–1984

- Lyman SD, Jordan VC. (1985) Metabolism of tamoxifen and its uterotrophic activity. *Biochem Pharmacol* **34**:2787-2794.
- Lyman SD, Jordan VC. (1985) Possible mechanisms for the agonist actions of tamoxifen and the antagonist actions of MER-25 (ethamoxytriphetol) in the mouse uterus. *Biochem Pharmacol* **34**:2795-2806.

Conducted a wonderful systematic study of the actions of non-steroidal antiestrogens through the ER in mouse uterine models. This really created an extensive database for the subsequent understanding of the selective estrogenic and antiestrogenic actions of non-steroidal antiestrogens that became selective ER modulation!

ETHEL M. CORMIER, MOLECULAR AND CELLULAR BIOLOGY GRADUATE PROGRAM, UNIVERSITY OF WISCONSIN, 1982–1988

- Cormier EM, Jordan VC. (1989) Contrasting ability of antiestrogens to inhibit MCF-7 growth stimulated by estradiol or epidermal growth factor. *Eur J Cancer Clin Oncol* **25**:57-63.

- Cormier EM, Wolf MF, Jordan VC. (1989) Decrease in estradiol-stimulated progesterone receptor production in MCF-7 cells by epidermal growth factor and possible clinical implication for paracrine-regulated breast cancer growth. *Cancer Res* **49**:576–580.

In the early 1980s, growth factor signaling became a 'hot' topic. Compared and contrasted the efficacy of antiestrogens to inhibit estrogen or epidermal growth factor (EGF)-stimulated growth. Antiestrogens did not block EGF-stimulated growth and EGF reduced estrogen-stimulated induction of the progesterone recep tor. Highlighted resistance mechanisms[5] for antiestrogens that were subsequently confirmed.[6]

MARCO M. GOTTARDIS, DEPARTMENT OF HUMAN ONCOLOGY, UNIVERSITY OF WISCONSIN, 1983–1989

- Gottardis MM, Jordan VC. (1987) The antitumor actions of keoxifene (raloxifene) and tamoxifen in the N-nitrosomethylurea-induced rat mammary carcinoma model. *Cancer Res* **47**: 4020–4024.
- Gottardis MM, Robinson SP, Satyaswaroop PG, Jordan VC. (1988) Contrasting actions of tamoxifen on endometrial and breast tumor growth in the athymic mouse. *Cancer Res* **48**:812–815.
- Gottardis MM, Jordan VC. (1988) Development of tamoxifen-stimulated growth of MCF-7 tumors in athymic mice after long-term antiestrogen administration. *Cancer Res* **48**:5183–5187.
- Gottardis MM, Jiang SY, Jeng MH, Jordan VC. (1989) Inhibition of tamoxifen-stimulated growth of an MCF-7 tumor variant in athymic mice by novel steroidal antiestrogens. *Cancer Res* **49**:4090–4093.

Contributed four major findings for translation into medical practice. Was the first to show that raloxifene can prevent rat mammary carcinogenesis, a drug that is now FDA-approved for breast cancer risk reduction in postmenopausal women. Was the first to illustrate that patients using adjuvant tamoxifen may have an increased incidence of endometrial cancer; subsequently demonstrated in clinical trial.[7] Was the first to develop a transplantable model *in vivo* of acquired resistance to tamoxifen and showed that tumors grew in response to either tamoxifen or estrogen (a unique finding supporting the use of aromatase inhibitors as a treatment after tamoxifen failure), and was the first to use a pure antiestrogen to illustrate the potential advantage for the treatment of tamoxifen resistance (subsequently proven in clinical trial).

CATHERINE S. MURPHY, DEPARTMENT OF HUMAN ONCOLOGY, UNIVERSITY OF WISCONSIN, 1984–1990

- Murphy CS, Meisner LF, Wu SQ, Jordan VC. (1989) Short-and long-term estrogen deprivation of T47D human breast cancer cells in culture. *Eur J Cancer Clin Oncol* **25**:1777-1788.
- Murphy CS, Langan-Fahey SM, McCague R, Jordan VC. (1990) Structure-function relationships of hydroxylated metabolites of tamoxifen that control the proliferation of estrogen-responsive T47D breast cancer cells *in vitro*. *Mol Pharmacol* **38**:737-743.
- Murphy CS, Parker CJ, McCague R, Jordan VC. (1991) Structure/activity relationships of nonisomerizable derivatives of tamoxifen: importance of hydroxyl group and side chain positioning for biological activity. *Mol Pharmacol* **39**:421-428.

Was the first to demonstrate that ER-positive breast cancer cells had the potential to lose the ER during estrogen deprivation (see John J. Pink). Used structure-function relationships and medicinal chemistry to prove that both geometric isomers of hydroxylated tamoxifen derivative 4-hydroxytamoxifen were antiestrogens. Before this definitive study, it was assumed that the cis isomer of 4-OH tamoxifen was fully estrogenic and this hypothesis was used to create a putative mechanism of resistance.[8] Mapped out the structure-function relationships of antiestrogens to regulate breast cancer cell proliferation prior to x-ray crystallography, subsequently refining and defining the conclusions.

MEEI-HUEY JENG, DEPARTMENT OF HUMAN ONCOLOGY, UNIVERSITY OF WISCONSIN, 1987–1992

- Jeng MH, Jordan VC. (1991) Growth stimulation and differential regulation of TGFβ1, TGFβ2 and TGFβ3 mRNA levels by norethindrone in MCF-7 human breast cancer cells. *Mol Endocrinol* **5**:1120–1128.
- Jeng MH, Parker CJ, Jordan VC. (1992) Estrogenic potential of progestins in oral contraceptives to stimulate human breast cancer cell proliferation. *Cancer Res* **52**:6539–6546.
- Jeng MH, Langan-Fahey SM, Jordan VC. (1993) Estrogenic actions of RU486 in hormone-responsive MCF-7 human breast cancer cells. *Endocrinol* **132**:2622–2630.
- Jeng MH, Jiang SY, Jordan VC. (1994) Paradoxical regulation of estrogen-dependent growth factor gene expression in estrogen receptor (ER)-negative human breast cancer cells stably expressing ER. *Cancer Lett* **82**:123–128.

Growth factors again! This time the regulation of the TGFβ family. Insisted (constantly!) on conducting a study of progestins, which eventually broke my resolve (and, I thought, my better judgment), so I gave her a synthetic panel of progestins to test in her cells. Discovered the estrogen-like qualities of the progestins that are 19 nor-testosterone derivatives and the estrogen-like properties of the antiprogestin RU486. Discovered the "paradoxical" growth factor gene control by estrogens in ER cells, stably transfected with the ER. The growth inhibitor TGFβ went down and the growth stimulator TGFα went up — growth stopped! No idea why this stops transfectants from growing. However, we have used estradiol-stimulated TGFα induction as a gene marker ever since.

SHUN-YUAN JIANG, DEPARTMENT OF HUMAN ONCOLOGY, UNIVERSITY OF WISCONSIN, 1987–1992; SCHOLARSHIP FROM THE TAIWANESE MINISTRY OF DEFENSE

- Jiang SY, Jordan VC. (1992) Growth regulation of estrogen receptor-negative breast cancer cells transfected with cDNA's for estrogen receptor. *J Natl Cancer Inst* **84**:580–591.

- Jiang SY, Wolf DM, Yingling JM, Chang C, Jordan VC. (1992) An estrogen receptor-positive MCF-7 clone that is resistant to antiestrogens and estradiol. *Mol Cell Endocrinol* **90**:77–80.
- Pink JJ, Jiang SY, Fritsch M, Jordan VC. (1995) An estrogen-independent MCF-7 breast cancer cell line which contains a novel 80-kilodalton estrogen receptor-related protein. *Cancer Res* **55**:2583–2590.

Was the first to stably transfect the ER into ER-negative breast cancer cells. Other molecular biology groups had failed. I presented the first results at a symposium organized by the late Bill McGuire at the Endocrine Society meeting in Washington, D.C., June 19–22, 1991. A prominent member of the cancer endocrinology scientific community congratulated me with the following words: "You are the least likely person anyone would predict to do this!" Created two cloned MCF-7 derivative cell lines, MCF-7:5C and MCF-7:2A, in response to estrogen withdrawal. This advance was to subsequently position us to study acquired resistance to aromatase inhibitors 10 years later. But, at the time, nothing too exciting was found. We put them in the liquid nitrogen tank until Joan Lewis-Wambi, as a postdoctoral fellow, changed all our previous understanding of the biology of the cells by changing the serum in which the cells grow. Gave us tools to understand E_2-induced apoptosis (Chapter 2).

DOUG M. WOLF, DEPARTMENT OF HUMAN ONCOLOGY, UNIVERSITY OF WISCONSIN, 1988–1993; SUSAN G. KOMEN GRADUATE STUDENT

- Wolf DM, Langan-Fahey SM, Parker CJ, McCague R, Jordan VC. (1993) Investigation of the mechanism of tamoxifen-stimulated breast tumor growth with non-isomerizable analogs of tamoxifen and metabolites. *J Natl Cancer Inst* **85**:806–812.
- Wolf DM, Jordan VC. (1994) The estrogen receptor from a tamoxifen-stimulated MCF-7 tumor variant contains a point mutation in the ligand binding domain. *Breast Cancer Res Treat* **31**:129–138.

- Wolf DM, Jordan VC. (1993) A laboratory model to explain the survival advantage observed in patients taking adjuvent tamoxifen therapy. *Recent Results Cancer Res* **127**:23–33.

Used medicinal chemistry to demonstrate that tamoxifen was not converted to estrogenic metabolites by isomerization that caused acquired resistance, as had been suggested previously.[8] Discovered the first natural mutation of the ER (D351Y) in an experimental tamoxifen-stimulated breast tumor, which was subsequently developed in the laboratory to be the first mutation to change the pharmacology of the antiestrogen raloxifene to become estrogen-like.[9,10] Based on subsequent x-ray crystallography, D351 was discovered to be a critical surface amino acid for modulating the antiestrogenic side chain of both tamoxifen and raloxifene as estrogens/antiestrogens (Chapter 6). Was the first to discover that long-term tamoxifen treatment creates a vulnerability to physiologic estrogen in breast cancer that causes apoptosis and tumor regression. The "Tamoxifen Team" kept the project going with philanthropy, but it took me 15 years to get competitive funding. Now medicine has changed again, with new potential for treatment and prevention (Chapters 2 and 4). Thank goodness for Susan G. Komen for the Cure Grant SAC100009 (2010 to the present) and the Department of Defense Center of Excellence Grant W81XWH-06-1-0590 (2006–2014).

JOHN J. PINK, DEPARTMENT OF HUMAN ONCOLOGY, UNIVERSITY OF WISCONSIN, 1990–1995

- Murphy C, Pink JJ, Jordan VC. (1990) Characterization of a receptor-negative, hormone-nonresponsive clone derived from T47D human breast cancer cell line kept under estrogen-free conditions. *Cancer Res* **50**:7285–7292.
- Pink JJ, Jiang SY, Fritsch M, Jordan VC. (1995) An estrogen-independent MCF-7 breast cancer cell line which contains a novel 80-kilodalton estrogen receptor-related protein. *Cancer Res* **55**:2583–2590.

- Pink JJ, Wu SQ, Wolf DM, Bilimoria MM, Jordan VC. (1996) A novel 80-kilodalton human estrogen receptor containing a duplication of exons 6 and 7. *Nucleic Acids Res* **24**:962–969.
- Pink JJ, Jordan VC. (1996) Models of estrogen receptor regulation by estrogens and antiestrogens in breast cancer cell lines. *Cancer Res* **56**:2321–2330.
- Pink JJ, Bilimoria MM, Assikis VJ, Jordan VC. (1996) Irreversible loss of the estrogen receptor in T47D breast cancer cells following prolonged estrogen deprivation. *Br J Cancer* **74**:1227–1236.
- Pink JJ, Fritsch M, Bilimoria MM, Assikis VJ, Jordan VC. (1997) Cloning and characterization of a 77-kilodalton estrogen receptor isolated from a human breast cancer cell line. *Br J Cancer* **75**:17–27.

Recruited to my laboratory at the Wisconsin Comprehensive Cancer Center to introduce new techniques in molecular biology into the lab and enhance the productivity of C. Murphy, S.Y. Jiang and M. H. Jeng. Did just that, but cloned and confirmed the first ER-negative breast cancer cells from ER-positive cells by estrogen, deprivation, described the biology of the MCF-7:2A cells, described the different mechanisms for ER regulation, and cloned and characterized the first high-molecular-weight ER protein in breast cancer.

WILLIAM H. CATHERINO, DEPARTMENT OF HUMAN ONCOLOGY, UNIVERSITY OF WISCONSIN, M.D./PH.D. PROGRAM, 1991–1995

- Catherino WH, Jeng MH, Jordan VC. (1993) Norgestrel and gestodene stimulate breast cancer cell growth through an estrogen receptor-mediated mechanism. *Br J Cancer* **67**:945–952.
- Catherino WH, Jordan VC. (1995) Increasing the number of tandem estrogen response elements increases the estrogenic activity of a tamoxifen analogue. *Cancer Lett* **92**:39–47.
- Catherino WH, Wolf DM, Jordan VC. (1995) A naturally occurring estrogen receptor mutation results in increased estrogenicity of a tamoxifen analog. *Mol Endocrinol* **9**:1053–1063.

With great self-confidence, created the BC-2 (Bill Catherino, of course) cell line by stably transfecting the asp351tyr ER into ER-negative MDA-MB-231 breast cancer cells. This created another essential cell model for all subsequent work on the predictable modulation of the estrogenic and antiestrogenic actions of the SERM-ER complex. Continued work on progestins and breast cancer growth with the new progestin, gestodene, a third generation oral contraceptive. Most importantly, created an ERE luciferase construct widely distributed and used in the research community for a decade.

JENNIFER I. MACGREGOR–SCHAFER, NORTHWESTERN UNIVERSITY, DEPARTMENT OF DEFENSE GRADUATE STUDENT TRAINING PROGRAM, CHICAGO, IL, 1995–2001

- MacGregor JI, Liu H, Tonetti DA, Jordan VC. (1999) The interaction of raloxifene and the active metabolite of the antiestrogen EM-800 with the human estrogen receptor (ER). *Cancer Res* **59**:4308–4313.
- MacGregor-Schafer JI, Liu H, Bentrem D, Zapf J, Jordan VC. (2000) Allosteric silencing of activating function 1 in the 4-hydroxytamoxifen-estrogen receptor complex by substituting glycine for aspartate at amino acid 351. *Cancer Res* **60**:5097–5105.
- MacGregor-Schafer JI, Lee E-S, O'Regan RM, Yao K, Jordan VC. (2000) Rapid development of tamoxifen-stimulated mutant p53 breast tumors (T47D) in athymic mice. *Clin Cancer Res* **6**:4373–4380.

Advanced the understanding of SERM-ER complex relationships, proving that a new nonsteroidal "pure antiestrogen," EM-652, was actually a raloxifene lookalike binding in the ER. The hypothesis for Jennifer to test was the result of Dr. Hiro Takei (an associate professor of surgery visiting from Japan) making me a slide of the published orientation of raloxifene in the ligand-binding domain of the ER in 1998. During the presentation of my lecture, I turned to look at the slide on the screen and realized that EM-652, which had previously always been drawn like a "pure antiestrogen," was actually raloxifene-like when redrawn upside-down and backward. Later, EM-800 was confirmed to be a SERM that maintained bone density in rats. No surprises in the conversation with Nature here! Created the JM-6 (Jennifer MacGregor, naturally) stably transfected cell line with Asp351Gly ER in MDA-MB-231 cells. Proved that the surface of the ligand-binding domain can communicate with AF-1 (previously this had not been thought to happen) and provided a tool for enabling subclassification of planar estrogens (class I) and bulky nonplanar estrogens (class II).[11] In our lab-speak of the original crocodile model,[12] jaws (helix 12) closed and jaws opened, respectively. Simultaneously created the first tamoxifen-resistant cell line with ER-positive T47D cells in athymic mice. Three projects all successful!

RUTH M. O'REGAN, UNIVERSITY COLLEGE, DUBLIN, IRELAND, 1996–2000 (FACULTY IN MEDICAL ONCOLOGY, NORTHWESTERN UNIVERSITY, CHICAGO, IL)

- O'Regan R, Cisneros A, England GM, MacGregor JI, Muenzer HD, Assikis VJ, Bilimoria MM, Piette M, Dragan YP, Pitot HC, Chatterton R, Jordan VC. (1998) Effects of the antiestrogens, tamoxifen, toremifene and ICI 182,780, on endometrial cancer growth *in vivo*. *J Natl Cancer Inst* **90**:1552-1558.
- O'Regan RM, Gajdos C, Dardes RC, De Los Reyes A, Park W, Rademaker AW, Jordan VC. (2002) Effects of raloxifene after tamoxifen on breast and endometrial tumor growth in athymic mice. *J Natl Cancer Inst* **94**:274-283.
- O'Regan RM, Osipo C, Ariazi E, Lee ES, Meeke K, Morris C, Bertucci A, Sarker MAB, Grigg R, Jordan VC. (2006) Development and therapeutic options for the treatment of raloxifene-stimulated breast cancer in athymic mice. *Clin Cancer Res* **12**:2255-2263.

A medical oncologist who chose to complete a Doctor of Medicine degree at University College in Dublin through research published from work completed in my laboratory. Three clinically relevant publications; each had practical applications in therapeutics and healthcare for cancer patients. The use of raloxifene after adjuvant tamoxifen was controversial, but she demonstrated that it was an unwise clinical intervention.

RITA C. DARDES, DEPARTMENT OF OBSTETRICS AND GYNECOLOGY, UNIVERSITY OF SAO PAULO, BRAZIL, 1998–2001

- Dardes RC, Bentrem D, O'Regan RM, MacGregor-Schafer J, Jordan VC. (2001) Effects of the new selective estrogen receptor modulator LY353381.HCI (arzoxifene) on human endometrial cancer growth in athymic mice. *Clin Cancer Res* 7:4149–4155.
- Dardes RC, O'Regan R, Gajdos C, Robinson SP, Bentrem D, de los Reyes A, Jordan VC. (2002) Effects of a new clinically relevant antiestrogen (GW 5638) related to tamoxifen on breast and endometrial cancer growth *in vivo*. *Clin Cancer Res* **8**:1995–2001.

A gynecologist from the University of Sao Paulo, Brazil, who completed her Ph.D. as the first-ever Avon International Scholar. The day I was phoned by Dr. Joe Bertino (the Chair of the Landon Prize Committee) that I was to share the inaugural Dorothy P. Landon/AACR Prize with Elwood Jensen, I was in Florida, invited to attend a global meeting of Avon managers and have dinner with their CEO, Andrea Jung. During the dinner in the evening, I was introduced as "the father of tamoxifen" and asked to suggest to the audience how cancer research could be advanced by Avon. You have to be able to think on your feet as you walk to the podium! I stated that they should sponsor young researchers coming to America for research who would return to their countries, where they could be supported and enhance progress in cancer treatment in their countries. I spoke to the Avon audience of Rita, not knowing that Brazil was Avon's second-largest market in the world. She became the first success as an official Avon Scholar. I traveled to Sao Paulo to witness her public Ph.D. examination. Her successful translational research identified her as a talented medical scientist back in Brazil, where she became the Medical Director for Women's Health at the Avon Institute, as well as a successful clinician at the Federal

University of Sao Paulo. Others subsequently advanced the idea to the AACR to establish a major formal program of travel to America for scientific training and return to their country of origin. Always remember lesson #7, Appendix 1, as it has only good outcomes.

PHILIPP MAXIMOV, RUSSIAN STATE MEDICAL UNIVERSITY, RUSSIA, 2006–2010

- Maximov PY, Myers CB, Curpan RF, Lewis-Wambi JS, Jordan VC. (2010) Structure-function relationships of estrogenic triphenylethylenes related to endoxifen and 4-hydroxytamoxifen. *J Med Chem* **53**:3273-3283.
- Maximov P, Sengupta S, Lewis-Wambi JS, Kim HR, Curpan RF, Jordan VC. (2011) The conformation of the estrogen receptor directs estrogen-induced apoptosis in breast cancer: a hypothesis. *Horm Mol Bio Clin Investig* **5**:27-34.

The first creation of ER-negative cell lines containing the transfected wild-type ER gene (S. Y. Jiang), the discovery of the natural mutant ER Asp351Tyr (D. Wolf), the transfection of this gene into ER-negative cell lines (W. Catherino) and adaptation to be an Asp351Gly ER transfected into an ER-negative cell line (J. MacGregor-Schafer) all ultimately resulted in the creation of a new classification of natural and synthetic estrogens based on planarity.[11] These advances with laboratory models converged to discover that the shape of the estrogen-ER complex is the controlling factor in E_2-induced apoptosis. I traveled to Moscow to witness his public Ph.D. examination. He survived an onslaught from a chairman of pharmacology, who felt that it was all too advanced and outside the scope of their expertise to be examined. He responded magnificently (all of this was being conducted in Russian to a filled lecture room) and the 20-member committee voted overwhelmingly in Philipp's favor, with only one dissent. The celebration with his family was a night to remember; especially meeting Philipp's uncle, Denis V. Manturov, Minister of Trade and Industry in President Putin's government. He is continuing his research at the Lombardi Comprehensive Cancer Center as a Susan G.

The 2011 Tamoxifen Team (left to right): Ifeyinwa Obiorah, Michael Biarnes, Dr. Ping Fan, Dr. Surojeet Sengupta, Elizabeth Sweeney, Puspanjali Bhatta, Helen Kim, Dr. V. Craig Jordan, Fadeke Agboke, Alisse Hannaford, Dr. Philipp Maximov, Russell McDaniel, Eric Jernigan, Julia Tijenira.

Komen for the Cure Foundation International Postdoctoral Fellow. His research theme is ongoing in the laboratory, being advanced by two current graduate students, Ifeyinwa Obiorah (Chapters 4 and 5) and Elizabeth Sweeney, both supported by scholarships from the Susan G. Komen for the Cure Foundation.

REFERENCES

1. Jordan VC, Murphy CS. (1990) Endocrine pharmacology of antiestrogens as antitumor agents. *Endocr Rev* **11**:578–610.
2. MacGregor-Schafer JI, Jordan VC. (1998) Basic guides to the mechanisms of antiestrogen action. *Pharmacol Rev* **50**:151–196.
3. Shiau AK, Barstad D, Loria PM, Cheng L, *et al.* (1998) The structural basis of estrogen receptor/coactivator recognition and the antagonism of this interaction by tamoxifen. *Cell* **95**:927–937.
4. Lieberman ME, Gorski J, Jordan VC. (1983) An estrogen receptor model to describe the regulation of prolactin synthesis by antiestrogens *in vitro*. *J Biol Chem* **258**:4741–4745.

5. Robinson SP, Jordan VC. (1989) The paracrine stimulation of MCF-7 cells by MDA-MB-231 cells: possible role in antiestrogen failure. *Eur J Cancer Clin Oncol* **25**:493–497.
6. Osborne CK, Bardou V, Hopp TA, Chamness GC, *et al.* (2003) Role of the estrogen receptor co-activator AIB1 (SRC-3) and HER-2/neu in tamoxifen resistance in breast cancer. *J Natl Cancer Inst* **95**:353–361.
7. Fornander T, Rutqvist LE, Cedermark B, Glas U, *et al.* (1989) Adjuvant tamoxifen in early breast cancer: occurrence of new primary cancers. *Lancet* **1**:117–120.
8. Osborne CK, Coronado E, Alfred DC, Wiebe V, *et al.* (1991) Acquired tamoxifen resistance: correlation with reduced breast tumor levels of tamoxifen and isomerization of trans-4-hydroxytamoxifen. *J Natl Cancer Inst* **83**:1477–1482.
9. Levenson AS, Catherino WH, Jordan VC. (1997) Estrogenic activity is increased for an antiestrogen by a natural mutation of the estrogen receptor. *J Steroid Biochem Mol Bio* **60**:261–268.
10. Levenson AS, Jordan VC. (1998) The key to the antiestrogenic mechanism of raloxifene is amino acid 351 (aspartate) in the estrogen receptor. *Cancer Res* **58**:1872–1875.
11. Jordan VC, Schafer JM, Levenson AS, Liu H, *et al.* (2001) Molecular classification of estrogens. *Cancer Res* **61**:6619–6623.
12. Jordan VC. (1987) Laboratory models of breast cancer to aid the elucidation of antiestrogen action. *J Lab Clin Med* **109**:267–277.

INDEX